Acta Physica Austriaca
Supplementum XXVII

Proceedings of the
XXIV. Internationale Universitätswochen für Kernphysik 1985
der Karl-Franzens-Universität Graz
at Schladming (Steiermark, Austria)
February 20th–March 1st, 1985

Sponsored by
Bundesministerium für Wissenschaft und Forschung
Steiermärkische Landesregierung
International Centre for Theoretical Physics, Trieste
Sektion Industrie der Kammer der
Gewerblichen Wirtschaft für Steiermark
Steyr-Daimler-Puch A.G.

1985

Springer-Verlag
Wien New York

Nucleon-Nucleon and Nucleon-Antinucleon Interactions

Edited by
H. Mitter and W. Plessas, Graz

With 215 Figures

1985

Springer-Verlag
Wien New York

Organizing Committee

Chairman

H. Mitter
Institut für Theoretische Physik
der Universität Graz

Committee Members

W. Plessas
L. Mathelitsch
F. Widder

Secretaries

Mrs. E. Neuhold
Mrs. M. Ploberger

Softcover reprint of the hardcover 1st edition 1985

ISSN 0065-1559
ISBN-13:978-3-7091-8832-3 e-ISBN-13:978-3-7091-8830-9
DOI: 10.1007/978-3-7091-8830-9

CONTENTS

PREFACE

This volume contains the Proceedings of the "XXIV. Internationale Universitätswochen für Kernphysik" held in Schladming, Austria, in February 1985. It consists of the written versions of the lectures (3-4 hours) given at this winter school and includes also most of the seminars (30-50 minutes) presented. In choosing the topic for the 1985 meeting, our aim was to give an account of the present understanding of the nucleon-nucleon as well as nucleon-antinucleon interactions. This field, which is of definite relevance in nuclear and particle physics, has witnessed a rapid development in recent times both in theory and experiment. New evidence has emerged in the whole range from low to extremely high energies. It was an exciting experience to bring together knowledge from the very domains of nuclear and high-energy physics as well as to meet the respective researchers. Thanks to the efforts of the lecturers, who did a splendid job in presenting the lectures and in preparing their lecture notes, a comprehensive insight into the hadronic interaction between nucleons and anti-nucleons was achieved. The lecture notes were reconsidered by the authors after the meeting and are now being published in their final form. The seminars mainly dealt with specific topics currently under investigation within this rather wide field. We are grateful to all authors for their efforts, as they made it possible to speed up the publication of these proceedings.

Above all we are pleased to thank the Austrian Ministry for Science and Research and the Government of Styria for their generous sponsorship, which provided the financial basis for the school also this year. We gratefully acknowledge support from the ICTP Trieste, the Chamber of Commerce of Styria, and the Steyr-Daimler-Puch Company. Thanks are also due to many members of our institute for taking part in the organization and we are especially grateful to Mrs. E. Neuhold for her careful typing of these notes.

H. Mitter

W. Plessas

Acta Physica Austriaca, Suppl. XXVII, 3–89 (1985)

NUCLEON-NUCLEON INTERACTION BELOW 1 GeV+

by

I. SLAUS
Rudjer Bošković Institute
41001 Zagreb, Croatia, Yugoslavia

"Now the smallest particles of matter may cohere by the strongest attraction and compose bigger particles of weaker virtue, and many of these may cohere and compose bigger particles whose virtue is still weaker, and so on .."

Isaac Newton

1. INTRODUCTION

In 1935 Yukawa described [1] the nuclear force as an exchange of mesons between nucleons. Since the discovery of the pion in 1947 and its identification as a Yukawa particle, the structure of nucleons, the form of the interaction, and the electromagnetic and weak phenomena in nuclei have all been related to meson degrees-of-freedom [2]. The picture [3] of the nucleon was of an "inside" (several tenth of fm) with states of baryon-antibaryon pairs and many mesons and an "outside" composed of a cloud of virtual pions. The meson theory has been developed in analogy

+) Lectures given at the XXIV. Internationale Universitätswochen für Kernphysik, Schladming, Austria, February 20-March 1, 1985.

to quantum electrodynamics. Though the strong coupling between hadrons made impossible any calculation at short distances, the meson theory can provide a quantitative account of phenomena dominated by the process of one-pion exchange (OPE) and thus a significant step was made by formulating a strategy [4] of first understanding OPE, then the region of the exchanges of two pions (TPE) and of pionic resonances (one-boson exchange - OBE). The long range and most of the intermediate range of nucleon-nucleon (NN) force are explained in terms of OPE and TPE (and/or OBE). The lack of understanding of short-range phenomena was hidden behind the veil of the repulsive effects of ρ and ω mesons exchanges. Low-energy deuteron photodisintegration data provide evidence that this picture of the NN force is not valid for distances shorter than 1.5 fm. Fig.1 shows that standard NN force models cannot fit the data, but that additional nonnucleonic states at $r \lesssim 1.6$ fm are necessary [5].

In 1964 Gell-Mann and Zweig proposed that hadrons are made of quarks. Besides other properties quarks carry color and interact among themselves through a color-gauge field-gluons. This interaction is weaker at shorter distances and stronger at larger distances resulting in asymptotic freedom at high-momentum transfer (short distances) and in the confinement of quarks into color singlet at small momentum transfer (large distances). Therefore, at large distances only hadronic degrees of freedom need be considered.

In the quark model the nucleon-nucleon (NN) force is not a fundamental interaction of nature but a shielded remnant of a true fundamental color force between quarks. In nuclear physics we live in infrared slavery [6] and it might be reasonable to develop a sort of hybrid model approach [7] where the long and intermediate range of nuclear force are given by the meson exchange and the short

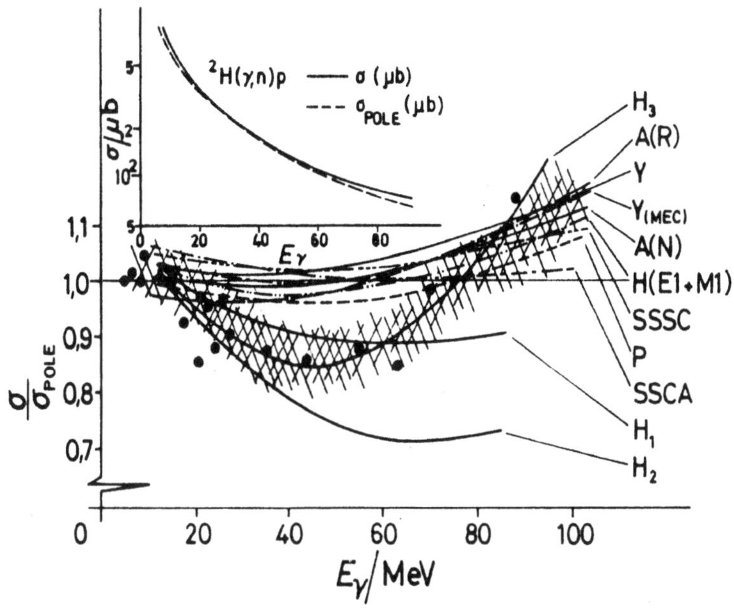

Fig. 1.: Inset: the calculation of the total cross section for the deuteron photodisintegration. The "pole cross section" represents the result of the calculation using the dispersion analysis keeping the deuteron and nucleon poles in the intermediate state in the zero-range approximation. This approximation using M1 and E1 transitions neglects the deuteron D state, employs only the asymptotic form of the S state, neglects the final-state interaction and the S and P waves and neglects meson-exchange currents, isobars and relativistic effects. The results of the calculation performed without these approximations is labeled by σ. The ratio σ/σ_{pole} is plotted in Fig. 1. Calculations have been done for various NN potentials: H(E1+M1), A(R), and Y with hard-core Yale potential, P with Hamada-Johnston, SSCA and SSSC with super-soft-core potentials, Y(MEC) including meson-exchange currents and A(R) including isobars in the deuteron. None of these calculations fits the ex-

Fig. 1.: ctd.

perimental data (the cross-hatched band represents the data with a width of about ±2 standard deviations). It seems that no familiar variation in the features of the NN force: core, D-state probability, meson-exchange currents and isobars, can account for the data. Curve H_3 is calculated by adding a non-nucleonic description of the deuteron for $r \lesssim 1.6$ fm (its contribution is about 20 %), which reasonably fits the data. Curves H_1 and H_2 also include non-nucleonic degrees [5].

range by quark and gluon exchanges. Indeed, one such hybrid model [7] gives an excellent description of phase parameters up to 300 MeV and of the deuteron ground state wave function. There could be a finite transition region where two mechanisms overlap. This overlap of mechanisms could be compared to the cluster structure in nuclear physics [8]. Also the amplitude of quark constituents of hadrons do not have to vanish abruptly. This allows the tunneling of quark waves between nucleons and it could make a significant contribution to the binding energy of nuclei [9].

Of course, quarks themselves might have structure - they could be composed of preons.

2. THE NUCLEON-NUCLEON SCATTERING

The nucleon-nucleon scattering data are the most abundant and most accurate scattering data in nuclear and particle physics. Remarkable accelerator facilities, e.g. TUNL, Erlangen, Tsukuba, Wisconsin, Karlsruhe, IUCF, TRIUMF, SIN, LAMPF, ZGS, Saclay, RCNP, KEK, Gatchina, Dubna, CERN, and Fermilab have enabled precise measurements from very low energies up to several hundred GeV. A construction of a superconducting supercollider TeV-accelerator is proposed [10]. If one would build an accelerator for 10 000 TeV, it would wrap the entire equator of the Earth. Though lunar laboratories and satelite accelerators have been contemplated (and their cost is presumably about 200 G$ or of the order of the US gross national expenditure for R&D per year) in order to achieve energies of 10^{15} GeV, an accelerator should be a hundred times bigger than the Solar system [11].

The early studies of nucleon-nucleon (NN) scattering have demonstrated that the NN forces are strongly spin dependent. At very low energy the total cross section σ for neutron-proton (np) scattering is:

$$\sigma = \frac{4\pi}{k^2} \sin^2\delta_o , \qquad (1)$$

where k is the relative wave number and δ_o is the S wave phase shift. The np system exists also as a bound system: the deuteron; and one can relate the scattering parameters to the properties of the ground state of the deuteron. The cross section for zero-energy neutrons is then given by

$$\sigma = 4\pi R^2 \left[1- \frac{r_o}{2R}\right]^{-2} , \qquad (2)$$

where $R = \frac{\hbar}{\sqrt{MB}}$ and r_o is the np effective range; M is the nucleon mass and B is the deuteron binding energy. Since the smallest value for r_o is zero and the largest is R, it follows that

$$2.33b < \sigma < 9.32b. \tag{3}$$

The experimental result is

$$\sigma_{exp} = (20.36 \pm 0.10)b. \tag{4}$$

The disagreement between the values in (3) and in (4) shows that there is a fundamental error in the assumption about nuclear forces used to derive the expression (3). The derivation is based on the assumptions that the forces have a short range and that the scattering can be described by parameters characterizing the deuteron ground state. Wigner has solved [12] the problem realizing that in the deuteron the spins of the neutron and the proton are parallel, while the scattering cross section measures uncorrelated spins:

$$\sigma = \frac{1}{4}\sigma_s + \frac{3}{4}\sigma_t , \tag{5}$$

where subscripts s and t indicate singlet and triplet np states. Using (4) for σ and (3) for σ_t in (5) it follows that

$$\sigma_s = 53 - 75b \tag{6}$$

Thus, n-p forces are strongly spin dependent.

The investigation of spin dependence of the NN force at higher energies requires polarized beams, polarized targets, and polarimeters to measure the polarization of the recoiled particles. We define the spin directions N, L, S of the particles as shown in Fig.2. The NN scattering observables

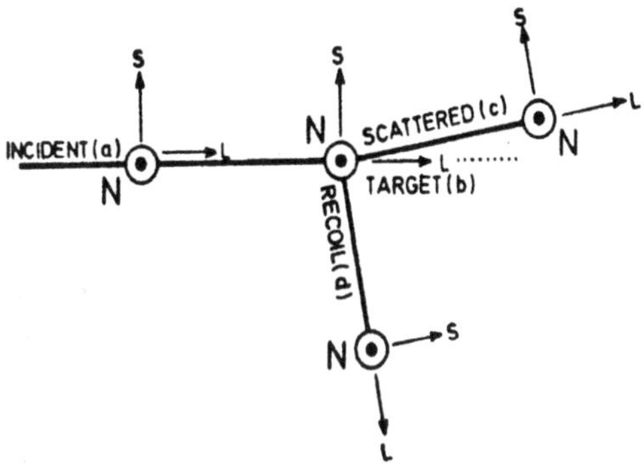

Fig. 2.: Definitions of direction N, L, and S for the incident, scattered, and recoil particles and for the target.

Table 1

(Beam, Target, Scatt., Recoil)	Symbol	Description
(0,0; 0,0)	σ	Cross Section
(0,0; i,0)	P	Polarization
(i,0; 0,0)	A	Analyzing power
(i,j; 0,0) or (0,0; i,j)	A_{ij} or C_{ij}	Correlation tensor
(i,0; 0,j)	K_{ij}	Polarization-transfer tensor
(i,0; j,0)	D_{ij}	Depolarization tensor
(i,j; 0,k)	H_{ijk}	High

i,j,k stand for N,L and S, i.e. A_{NN} = (N,N; 0,0), K_{LS} = (L,0; 0,S) and H_{LSN} = (L,S; 0,N).

are summarized in Table 1.

Differences between total cross sections for anti-parallel and parallel (longitudinal) states are:

$$\Delta\sigma_L = \sigma^{Tot}(\overset{\rightarrow}{\leftarrow}) - \sigma^{Tot}(\overset{\rightarrow}{\rightarrow}) \tag{7}$$

and (transverse)

$$\Delta\sigma_T = \sigma^{Tot}(\uparrow\downarrow) - \sigma^{Tot}(\uparrow\uparrow) \tag{8}$$

The strong spin dependence at intermediate energies is shown [13] in Fig. 3.

Now we briefly describe the polarized proton beam - polarized proton target facilities at ZGS and SIN:

ZGS[14]: The spin of polarized protons that emerged from the 12.5 GeV Argonne Zero-Gradient Synchrotron (the accelerator was shut down on Oct. 1, 1979) was in the N direction. Superconducting solenoids (with fields of e.g. 24 T-m at 12 GeV/c) were used to rotate the spin from the N to the S direction.
The longitudinally polarized beam is produced by a bending magnet with a vertical field to precess the proton spin until their polarization is parallell to the beam momentum. The sign of beam polarization is flipped on alternate pulses and this is essential to reduce the systematic error. The intensity of the polarized beam was below 10^7/pulse. The polarized proton target was 2x2x8 cm ethylene glycol doped with $K_2Cr_2O_7$ and maintaned at 0.4 K. For free protons in the target the polarization was 0.8-0.9. Two kinds of superconducting magnets are used to provide three directions: N, S, and L of target spins. The spin analysis of scattered and recoiled particles was done by a carbon polarimeter with MWPC.

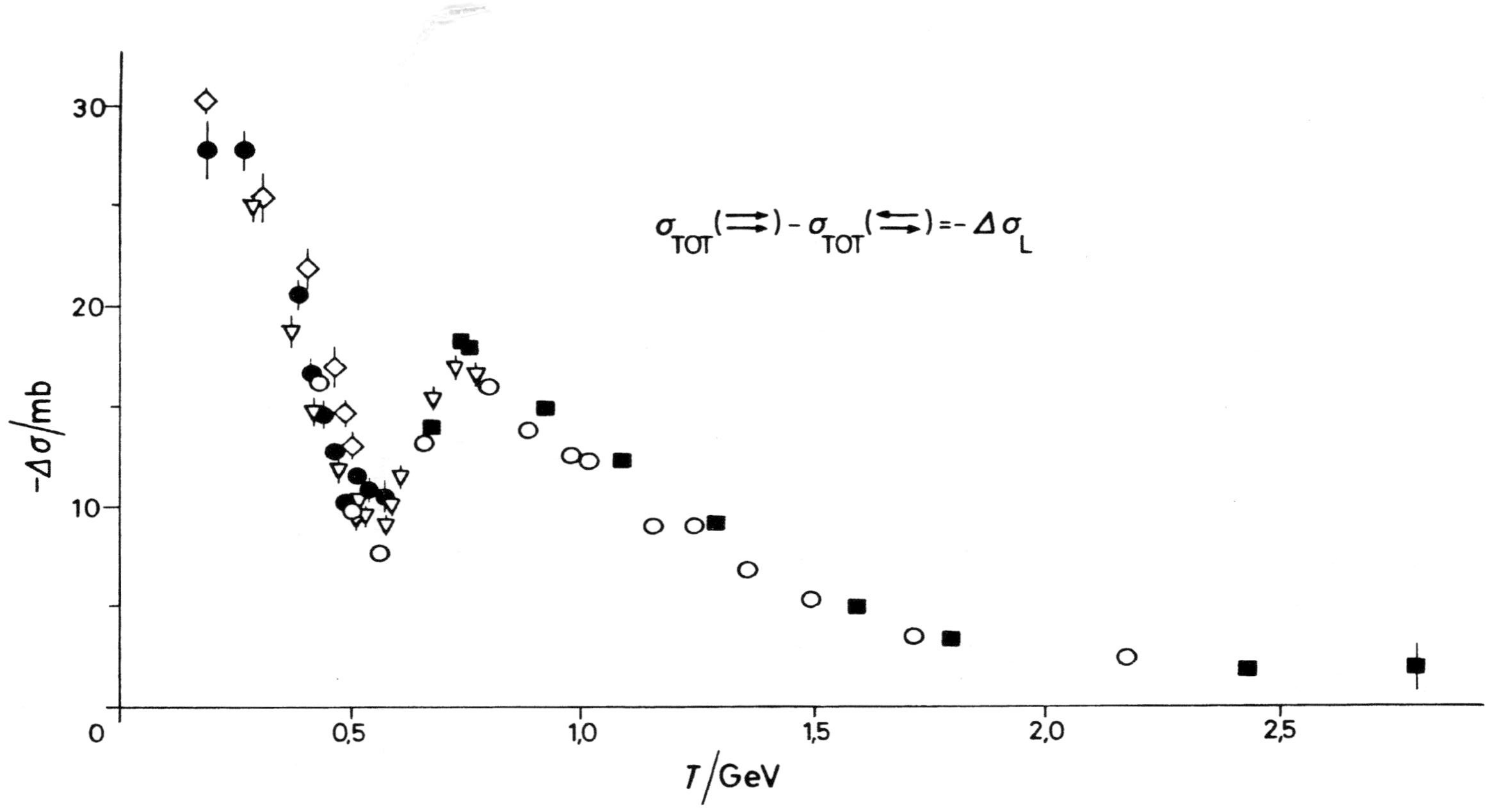

Fig. 3.: The compilation of existing measurements (ref.13). Different symbols represent different data.

SIN[15]: Polarized ion source gave 10^6 (85±2)% polarized protons per second. A 5 T superconducting solenoid in combination with deflecting magnets is used to control the orientation of beam polarization. One can obtain N, L, and S polarized protons. A conventional butanol target polarized by dynamic nuclear orientation is used. Typical polarization ∿0.6.

2.1 Partial Wave Expansion

In the absence of interaction the solution of the Schrödinger equation is the plane wave:

$$\phi(\vec{r}) = V^{-1/2} e^{i\vec{k}\vec{r}} = V^{-1/2} \sum_{\ell} i^{\ell}(2\ell+1) P_{\ell}(\cos\theta) j_{\ell}(kr) =$$

$$= V^{-1/2} \sum_{\ell} \frac{i^{\ell}}{2} (2\ell+1) P_{\ell}(\cos\theta)[h_{\ell}(kr) + h_{\ell}^{*}(kr)] ,$$

where $h_{\ell}(kr) \to \exp[i(kr - \frac{1}{2}\ell\pi)]/ikr$ as $r\to\infty$.

The part $\phi_{out} = V^{-1/2} \sum_{\ell} \frac{i^{\ell}}{2} (2\ell+1) P_{\ell}(\cos\theta) h_{\ell}(kr) = \frac{1}{k} \sum_{\ell} c_{\ell} O_{\ell o}(\vec{r})$

has the probability-current density:

$$\vec{j}_{out} = \frac{h}{2iM} (\phi^{*}_{out} \vec{\nabla} \phi_{out} - \phi_{out} \vec{\nabla} \phi^{*}_{out}) \to v\vec{r} \qquad (9)$$

as $r\to\infty$, pointing radially outward.

$$\phi_{out}(\vec{r}) \to \exp[i(kr - \frac{\ell\pi}{2})]/ikr, \text{ i.e.}$$

spherically outgoing waves.

The complete plane wave can be written in terms of ingoing and outgoing waves:

$$\phi(\vec{r}) = \frac{1}{k} \sum_{\ell} c_{\ell} [I_{\ell_o} - O_{\ell_o}] \qquad (10)$$

with

$$c_\ell = i\sqrt{\pi(2\ell+1)} \quad ; \quad I_{\ell m} = \frac{i^\ell Y_{\ell m}}{r\sqrt{v}}[ikr\, h_\ell(kr)]^*$$

and

$$O_{\ell m} = \frac{i^\ell Y_{\ell m}}{r\sqrt{v}}\, i\, kr\, h_\ell(kr).$$

The interaction changes the outgoing waves and thus the total wave function becomes:

$$\Psi(r) \to \frac{1}{k}\sum_\ell c_\ell[I_{\ell_o} - S_\ell O_{\ell_o}]\,. \tag{11}$$

The conservation principle: what goes in must come out, which is valid for real potentials, implies that

$$S_\ell^* S_\ell = 1 \text{ and thus } S_\ell = e^{2i\delta_\ell}. \tag{12}$$

The wave function $\Psi(\vec{r})$ can be written asymptotically as:

$$\Psi(\vec{r}) = \frac{1}{k}\sum_\ell c_\ell[I_{\ell o} - O_{\ell o} + (1-S_\ell)O_{\ell o}] = \phi(\vec{r}) + \frac{e^{ikr}}{r\sqrt{v}} f(\theta),$$

where

$$f(\theta) = \frac{1}{k}\sum_\ell c_\ell(1-S_\ell)Y_{\ell_o} = \sum_\ell (2\ell+1) f_\ell P_\ell(\cos\theta).$$

The differential elastic scattering cross section is:

$$(\frac{d\sigma}{d\Omega})_{el} = |f(\theta)|^2 = |\sum_{\ell} (2\ell+1) \frac{1}{k} e^{i\delta_\ell}\sin\delta_\ell P_\ell|^2 \tag{13}$$

and the total elastic cross section:

$$\sigma_{el} = \int (\frac{d\sigma}{d\Omega})_{el} \, d\Omega = \frac{4\pi}{k^2} \sum_{\ell} (2\ell+1)\sin^2\delta_\ell \; ; \tag{14}$$

since

$$\text{Im } f(0^o) = \sum_{\ell} (2\ell+1) \frac{\sin^2\delta_\ell}{k}$$

it follows the optical theorem:

$$\sigma_{el} = \frac{4\pi}{k} \text{ Im } f(0^o) \tag{15}$$

and the Wick's limit:

$$\sigma(0^o) > |\text{Im } f(0^o)|^2 = \frac{k^2}{16\pi^2} \sigma_{el}^2 \; . \tag{16}$$

Relation (15) states that the total elastic cross section can be obtained from a measurement of the transmission loss through a target in the forward direction.

Nucleons are spin-1/2 particles and the total spin of a two-nucleon system can be either 0 or 1. The foregoing discussion was for S = 0, but it can be generalized to spin-triplet partial waves. Two waves with the same total angular momentum J and parity, and different L = J±1 can be mixed by the tensor force. Indeed, except S_{12} all permissible NN operators appearing in the most general NN potential, i.e. $\vec{\sigma}_1\vec{\sigma}_2$, $\vec{L}\vec{S}$ and Q_{12}, are diagonal in the LSJ representation (see Table 2).

The states of the NN system labelled in the usual notation $^{2S+1}L_J$ are:

Total isospin I = 1, even singlets: ^1S_o, $^1D_2$, $^1G_4$, ..
I = 1, odd triplets: ^3P_o, $^3P_1$, $^3P_2$ and $^3F_2$ with coupling parameter ε_2, ...
I = 0, even triplets: $^3S_1$ and $^3D_1$ with ε_1, $^3D_2$
$^3D_3$ and $^3G_3$ with ε_3, $^3G_4$,..
I = 0, odd singlets: $^1P_1$, $^1F_3$, $^1H_4$,...

All L=J waves as well as a ^3P_o wave are uncoupled.

Though the inelastic scattering: NN Bremsstrahlung: $N+N \to N+N+\gamma$ is possible at all energies, the Bremsstrahlung cross section is several orders of magnitude smaller than the NN cross section. The inelastic cross section becomes appreciable only above the pion-production threshold (E_{lab}> 290 MeV, p_{lab} > 800 MeV/c). At 1 GeV σ^{Tot} has a maximum of 48 mb and about half of this cross section is σ^{inel} (see Fig.4). The presence of inelastic processes makes the phase shift δ complex $\delta = \delta_R + i\delta_I$ and the S matrix for uncoupled waves becomes $S = e^{-2\delta_I} e^{2i\delta_R} = \eta e^{2i\delta_R}$ with $\eta < 1$. One can write an S matrix in terms of a K matrix

$$S = \frac{1-K_I+i\ K_R}{1+K_I-i\ K_R}, \qquad \text{where } K_R = \tan\delta_R,\ K_I = \tan^2\rho$$

$$\eta^2 = \frac{1+K_R^2+K_I^2-2K_I}{1+K_R^2+K_I^2+2K_I} \qquad \text{and}$$

$$\delta_R = \frac{1}{2}[\tan^{-1}\frac{K_R}{I-K_I} + \tan^{-1}\frac{K_R}{I+K_I}] ; \qquad (17)$$

when inelastic cross sections are negligible, then $\rho=0$, $K_I = 0$

Table 2

$\vec{\sigma}_1\vec{\sigma}_2 = 2\vec{S}^2-3$	spin-spin
$\vec{L}\vec{S} = \frac{1}{2}(\vec{J}^2-\vec{L}^2-\vec{S}^2)$	spin-orbit
$S_{12} = 3(\vec{\sigma}_1\vec{r}_o)(\vec{\sigma}_2\vec{r}_o)-\vec{\sigma}_1\vec{\sigma}_2$	tensor
$Q_{12} = \frac{1}{2}[(\vec{\sigma}_1\vec{L})(\vec{\sigma}_2\vec{L}) + (\vec{\sigma}_2\vec{L})(\vec{\sigma}_1\vec{L})]$	quadratic spin-orbit

$$\langle L'SJ|S_{12}|LSJ\rangle = \begin{cases} 0 & \text{if } S=0 \\ A^J_{LL'} & \phantom{\text{if }} S=1 \end{cases}$$

and

$$A^J_{JJ} = 2 \qquad A^J_{J-1,J+1} = \frac{6}{2J+1}\sqrt{J(J+1)}$$

$$A^J_{J-1,J-1} = -2\,\frac{J-1}{2J+1} \qquad A^J_{J+1,J+1} = -2\,\frac{J+2}{2J+1}\,.$$

For S=0 matrix elements of $\vec{L}\vec{S}$ and S_{12} vanish and of Q_{12} and $\vec{\sigma}_1\vec{\sigma}_2$ is a constant

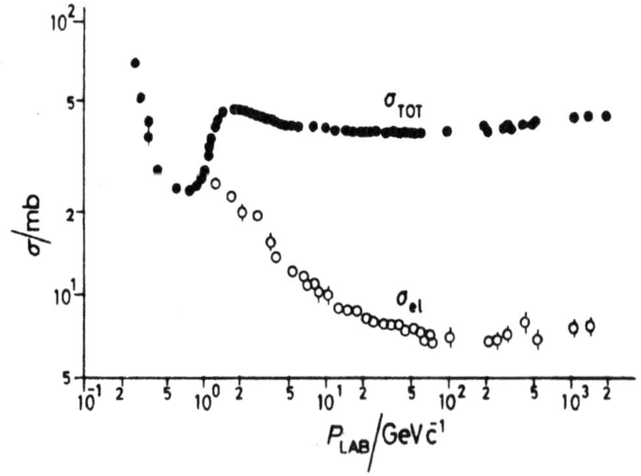

Fig. 4.: Total and elastic pp cross sections.

and $\delta = \delta_R$ and $\eta = 1$.

For coupled waves

$S = \frac{1+iK}{1-iK}$ and K is a 2x2 matrix

$$K = \begin{bmatrix} K_{R_-} & K_{R_o} \\ K_{R_o} & K_{R_+} \end{bmatrix} + i \begin{bmatrix} \tan^2\rho_- & \tan\rho_- \tan\rho_+ \cos\phi \\ \tan\rho_- \tan\rho_+ \cos\phi & \tan^2\rho_+ \end{bmatrix}, \quad (18)$$

where

$$K_{R_\pm} = \frac{\sin(\delta_{J+1}+\delta_{J-1})+\cos 2\varepsilon-\sin(\delta_{J+1}-\delta_{J-1})}{\cos(\delta_{J+1}+\delta_{J-1})+\cos 2\varepsilon-\cos(\delta_{J+1}-\delta_{J-1})}$$

$$K_{R_o} = \frac{\sin 2\varepsilon}{\cos(\delta_{J+1}+\delta_{J-1})+\cos 2\varepsilon-\cos(\delta_{J+1}-\delta_{J-1})}$$

This representation [16]in terms of phase shifts and coupling parameters is the same as the Stapp [17] representation at low energies:

$$S = \begin{bmatrix} \cos 2\varepsilon\; e^{2i\delta_{J-1}} & i \sin 2\varepsilon\; e^{i(\delta_{J+1}+\delta_{J-1})} \\ i \sin 2\varepsilon\; e^{i(\delta_{J+1}+\delta_{J-1})} & \cos 2\varepsilon\; e^{2i\delta_{J+1}} \end{bmatrix},$$

but it does not correspond to any previous representation, when inelasticity occurs. Shortcomings of various parametrizations have been discussed and a new method has been proposed [18] . The procedure consists in introducing two dummy absortive channels and then parametrizing the 4x4 matrix. The elastic but nonunitary 2x2 S matrix describing elastic NN scattering above pion threshold is then recovered as a submatrix of this 4x4 matrix.

2.2 Scattering Matrix

For unpolarized beam on an unpolarized target one can measure $\sigma(\theta)$, three components of the polarization of the scattered particle, three components of the polarization of the recoil particle, and 9 correlations C_{ij} between the directions of polarization of the two nucleons. This is the total of 16 measurements.

However, either the beam or the target or both can be polarized yielding 3x16, 3x16, and 9x16 different measurements and altogether 16x16 = 256 measurements at each angle and at each energy. These measurements are not independent as one can readily see from the most general form of the S matrix satisfying symmetry constraints. The scattering matrix in the spin space is a 4x4 matrix, which depends on the spins σ_1 and σ_2 of two nucleons and on the initial $\vec{k}_i$ and final $\vec{k}_f$ relative momenta. Instead of $\vec{k}_i$ and $\vec{k}_f$ we will use $\vec{p} = (\vec{k}_i+\vec{k}_f)/2$; $\vec{q} = \vec{k}_f-\vec{k}_i$ and $\vec{n} = \vec{k}_i \times \vec{k}_f$. The rotational invariance requires M to be a scalar. The effect of space reflection and time reversal are shown in Table 3. and one sees that the most general M matrix on the energy shell is:

$$M(\vec{k}_f,\vec{k}_i,\vec{\sigma}_1,\vec{\sigma}_2) = A+B(\vec{\sigma}_1\vec{n})(\vec{\sigma}_2\vec{n})+C(\vec{\sigma}_1+\vec{\sigma}_2)\vec{n}+D(\vec{\sigma}_1-\vec{\sigma}_2)\vec{n}+ + E(\vec{\sigma}_1\vec{q})(\vec{\sigma}_2\vec{q})+F(\vec{\sigma}_1\vec{p})(\vec{\sigma}_2\vec{p}) . \tag{19}$$

If M should be symmetric in labels 1 and 2, then D must vanish. The coefficients A, B, C, E and F are functions of $\vec{p}$ and $\vec{q}$, which at a fixed energy and angle are just complex numbers. From equ. (19) it follows that the pp or nn M matrix is completely determined by five complex numbers. However, even these 10 numbers are not independent, if measurements are made at all angles of the same energy. Using the unitary condition in the form of the optical theorem (16) one obtains five conditions:

Table 3

Observables	$\vec{k}_i$	$\vec{k}_f$	$\vec{p}$	$\vec{q}$	$\vec{n}$	$\vec{\sigma}$	$\vec{L}$	$\vec{r}$
parity	$-\vec{k}_i$	$-\vec{k}_f$	$-\vec{p}$	$-\vec{q}$	$\vec{n}$	$\vec{\sigma}$	$\vec{L}$	$-\vec{r}$
time reversal	$-\vec{k}_f$	$-\vec{k}_i$	$-\vec{p}$	$\vec{q}$	$-\vec{n}$	$-\vec{\sigma}$	$-\vec{L}$	$\vec{r}$
interchange 1-2	$-\vec{k}_i$	$-\vec{k}_f$	$-\vec{p}$	$-\vec{q}$	$\vec{n}$	$\sigma_1 \rightleftarrows \sigma_2$	$\vec{L}$	$-\vec{r}$

$$\frac{4\pi}{k}\begin{bmatrix}\mathrm{Im}\,A\\ \mathrm{Im}\,B\\ 2\mathrm{Re}\,C\\ \mathrm{Im}\,E\\ \mathrm{Im}\,F\end{bmatrix} = \int M^*(\vec{k},\vec{k}'') \begin{bmatrix}1\\ (\vec{\sigma}_1\vec{n})(\vec{\sigma}_2\vec{n})\\ (\vec{\sigma}_1+\vec{\sigma}_2)\vec{n}\\ (\vec{\sigma}_1\vec{p})(\vec{\sigma}_2\vec{p})\\ (\vec{\sigma}_1\vec{q})(\vec{\sigma}_2\vec{q})\end{bmatrix} M(\vec{k}'',\vec{k}')\,d^3k''$$

and thus, only 5 independent experiments have to be performed at each energy. Rodebaugh and Bonner showed that these 5 measurements do not have to involve triple-spin parameters H_{ijk} [19].

If parity is not conserved, M also contains terms

$$R(\vec{\sigma}_1-\vec{\sigma}_2)\vec{p} \qquad \text{and } S[(\vec{\sigma}_1\vec{n})(\vec{\sigma}_2\vec{p})-(\vec{\sigma}_1\vec{p})(\vec{\sigma}_2\vec{n})].$$

If time reversal does not hold, one should add $T\,[(\vec{\sigma}_1\vec{p})(\vec{\sigma}_2\vec{q})-(\vec{\sigma}_1\vec{q})(\vec{\sigma}_2\vec{p})]$ and, if both parity and time-reversal invariance are violated, there are two additional terms:

$$P(\vec{\sigma}_1-\vec{\sigma}_2)\vec{q} \qquad \text{and } Q[(\vec{\sigma}_1\vec{q})(\vec{\sigma}_2\vec{n}) - (\vec{\sigma}_1\vec{n})(\vec{\sigma}_2\vec{q})].$$

2.3 Density Matrix Formalism

We define the spin-density matrix

$$\rho = \sum_n |n>w_n<n| \,, \tag{20}$$

where w_n is the probability of finding the system in the spin state $|n>$. The normalized expectation value of an NN spin operator O_μ:

$$O_\mu = 1,\ 1\,\sigma_1,\ \sigma_2\,1,\ \sigma_1\,\sigma_2 \qquad (\mu=1,\dots 16)$$

is

$$\langle O_\mu \rangle = \frac{\sum_n w_n \langle n|O_\mu|n\rangle}{\sum_n w_n} = \frac{Tr(O_\mu \rho)}{Tr\rho} . \tag{21}$$

Expanding $\rho = \sum_{\mu=1}^{16} c_\mu O_\mu$ and using the orthogonality relation $Tr(O_\mu O_\nu) = 4\delta_{\mu\nu}$ one finds $c_\mu = \frac{1}{4} Tr(O_\mu \rho) = \frac{1}{4}\langle O_\mu\rangle Tr\rho$.

Density matrices of the final and initial states are connected through

$$\rho_f = M \rho_i M^* . \tag{22}$$

Thus, the expectation value of the NN spin operators O_μ in the final state can be expressed as

$$\langle O_\mu \rangle_f = \frac{Tr\rho_i}{4Tr\rho_f} \sum_{\nu=1}^{16} \langle O_\nu \rangle_i \, Tr(MO_\nu M^* O_\mu) . \tag{23}$$

We will illustrate how one calculates the observables from equ. (23) by two examples:

i) differential cross section

$$\sigma(\theta) = \frac{Tr\rho_f}{Tr\rho_i} = \frac{1}{4} Tr(MM^*)$$

ii) triple-scattering experiments determine, how the second scattering changes direction and magnitude of the polarization of the incident particle. First scattering (in old times) served as a polarizer and the third scattering as an analyzer:

$$\sigma(\theta)\langle\sigma_1^\alpha\rangle_f = \frac{1}{4} Tr(MM^*\sigma_1{}^\alpha) + \sum_{\beta=1}^{3} \frac{1}{4} Tr(M\sigma_1{}^\beta M^*\sigma_1{}^\alpha)\langle\sigma_1{}^\beta\rangle_i \tag{24}$$

with

$$\sigma(\theta) = \frac{1}{4}Tr(MM^*) + \sum_{\beta=1}^{3} \frac{1}{4} Tr(M\sigma_1{}^\beta M^*)\langle\sigma_2{}^\beta\rangle_i .$$

Denoting the momenta of the incident and scattered particle in the second scattering by $\vec{k}_2$ and $\vec{k}'_2$, respectively and defining two coordinate systems:

one, $\vec{n}_2 = \vec{k}_2 \times \vec{k}'_2$, $\vec{k}_2$ and $\vec{x}_2 = \vec{n}_2 \times \vec{k}_2$ and

the other, $\vec{n}_2, \vec{k}'$, $\vec{s}'_2 = \vec{n}_2 \times \vec{k}'_2$, equ. (26) can be written as:

$$\sigma(\theta)\langle\vec{\sigma}_1\rangle_f = \sigma_o([P_2 + D\langle\vec{\sigma}\rangle_i \vec{n}_2]n_2 + [A\langle\vec{\sigma}\rangle_i \vec{k}_2 +$$

$$+ R\langle\vec{\sigma}\rangle_i (\vec{n}_2 \times \vec{k}_2)]\vec{s}'_2 + [A'\langle\vec{\sigma}\rangle_i \vec{k}_2 + R'\langle\vec{\sigma}\rangle_i (\vec{n}_2 \times \vec{k}_2)]\vec{k}'_2).$$

The parameters D, depolarization, and R, R', A and A', spin-rotation parameters, are called Wolfenstein parameters.

Though we have pointed out that only five measurements are needed at each angle and at each energy, at least one among them should be with nonparallel planes, e.g. $\sigma(\theta)$, $P(\theta)$, $D(\theta)$, C_{NN} and $R(\theta)$ (the analyzing plane is perpendicular to the second scattering plane).

2.4 Phase-Parameter Analysis

Owing to the short range of NN force, it is possible to describe the scattering in terms of a rather small number of phase parameters. The semiclassical relation $\ell\, h = b\, p$ where b is the impact parameter shows that phase parameters with $\ell > \ell_c$ can be generated from the tail of the NN potential. Since we now know that the longest-range NN force is mediated by the exchange of one pion, all phase parameters with $\ell > \ell_c$ are derived from one-pion exchange potential (OPEP). This procedure has been introduced by Moravcsik [20] and it is now a standard approach called a modified phase shift analysis.

The OPEP tail is included by writing the NN scattering amplitude as:

$$f(\theta) = \sum_{\ell=0}^{\ell_{max}} \frac{1}{2i} [(e^{2i\delta_\ell}-1)-(e^{2i\delta_\ell^{OPE}}-1)]P_\ell(\theta)+f_{OPE}(\theta), \tag{25}$$

where δ_ℓ^{OPE} is the phase shift from OPE alone and f_{OPE} is the total OPE amplitude,which can be expressed in a closed form.

Similarly, one can separate electromagnetic (em) effects from the purely nuclear effects:

$$f(\theta) = f_{em}(\theta) + f_N(\theta)$$

$$f_{em}(\theta) = f_c(\theta) + \Delta f_{em}(\theta),$$

where it is convenient to distinguish a point Coulomb contribution

$$f_c(\theta) = -\frac{\eta}{2k \sin^2 \frac{\theta}{2}} \exp[-2i\eta \ln \sin \frac{\theta}{2}]$$

$$\eta = \frac{e^2}{\hbar v}, \quad v = \text{relative velocity}$$

and an additional contribution Δf_{em} due to vaccum polarization, finite charge and magnetic moment. These additional effects are short-range and may be characterized by the phase shift τ_ℓ.

The nuclear phase shift σ_ℓ is now defined relative to the total em phase shift $\sigma_\ell-\sigma_o+\tau_\ell$, where $\sigma_\ell = \arg\Gamma(\ell+1+i\eta)$, and the purely nuclear scattering amplitude is

$$f_N(\theta) = \frac{1}{2ik} \sum_\ell (2\ell+1)e^{2i(\sigma_\ell-\sigma_o)} e^{2i\tau_\ell} (e^{2i\delta_\ell}-1)P_\ell(\theta).$$

For identical-particles scattering (pp or nn) we must also antisymmetrize the spin-singlet amplitude:

$f_s(\theta) + f_s(\pi-\theta)$ and similarly for the spin triplet

$f_t(\theta) - f_t(\pi-\theta)$.

The phase-shift analysis is performed by determining phase parameters in such a way that the expression

$$\chi^2 = \sum_{i=1}^{n} \left|\frac{Y_i(\delta)-Y_i(\exp)}{\Delta Y_i(\exp)}\right|^2 \tag{26}$$

for n experimental data $Y_i(\exp)\pm\Delta Y_i$ has a minimum. The expression $Y_i(\delta)$ is the value of the corresponding observable calculated from the given phase parameters. The search can be performed at one energy (energy-independent analysis) or simultaneously at many energies under the additional constraint that the phase parameters smoothly vary with the energy (energy-dependent analysis).

The experimental accuracy determines to some extent the number of phases, e.g. better angular resolution requires more phase shifts.

Comprehensive phase-shift analyses have been performed covering the energy region of up to 1 GeV [21-23]. We present here the results of the phase-shift analysis of Arndt et al. [21], which uses the data base of 5207 pp data below 1.2 GeV and 5283 np data below 1.1 GeV. Fig.5 illustrates the kinematical distribution of the used data grouped into so-called old (pre 1978) and new data (1978-83). Still there are large gaps in np data above 500 MeV and for $\theta<90^\circ$, scarce np P data above 800 MeV and very few other np spin observable data. The energy-dependent phase-shift analysis and 20 energy-independent analyses [21] are summarized in Fig.6. The χ^2/data are 1.77 and 1.72 for 0-1 GeV pp and np, respectively

Fig. 5.: Kinematical distributions of the data used in the phase-shift analysis (ref.21) grouped into pre-1978 data (points) and zero data (1978-83): a-f.

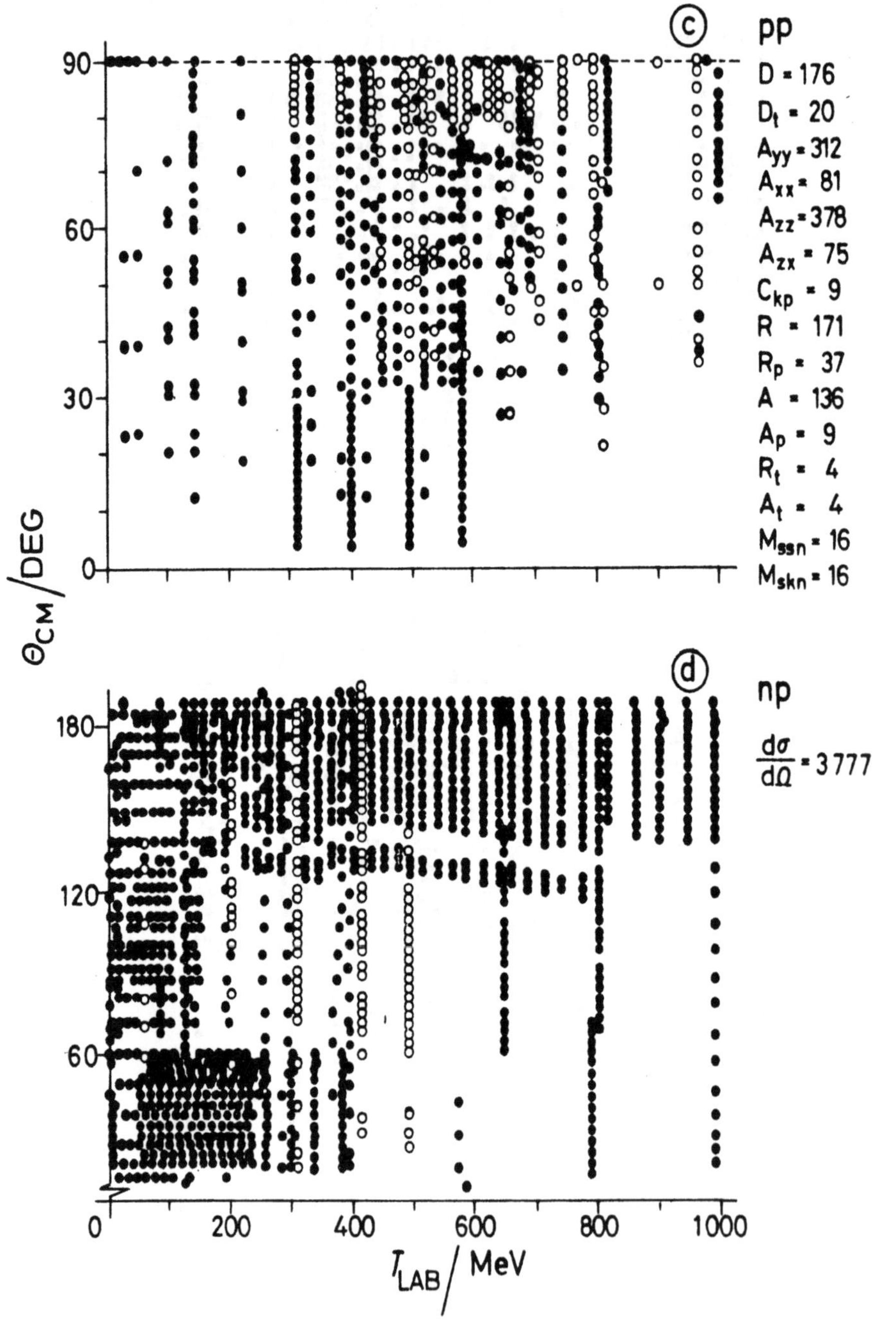

c
pp
D = 176
D_t = 20
A_{yy} = 312
A_{xx} = 81
A_{zz} = 378
A_{zx} = 75
C_{kp} = 9
R = 171
R_p = 37
A = 136
A_p = 9
R_t = 4
A_t = 4
M_{ssn} = 16
M_{skn} = 16
90
60
30
0
Θ_{CM}/DEG
d
np
$\frac{d\sigma}{d\Omega}$ = 3777
180
120
60
0
200
400
600
800
1000
T_{LAB}/ MeV

Fig. 5.: ctd. 1

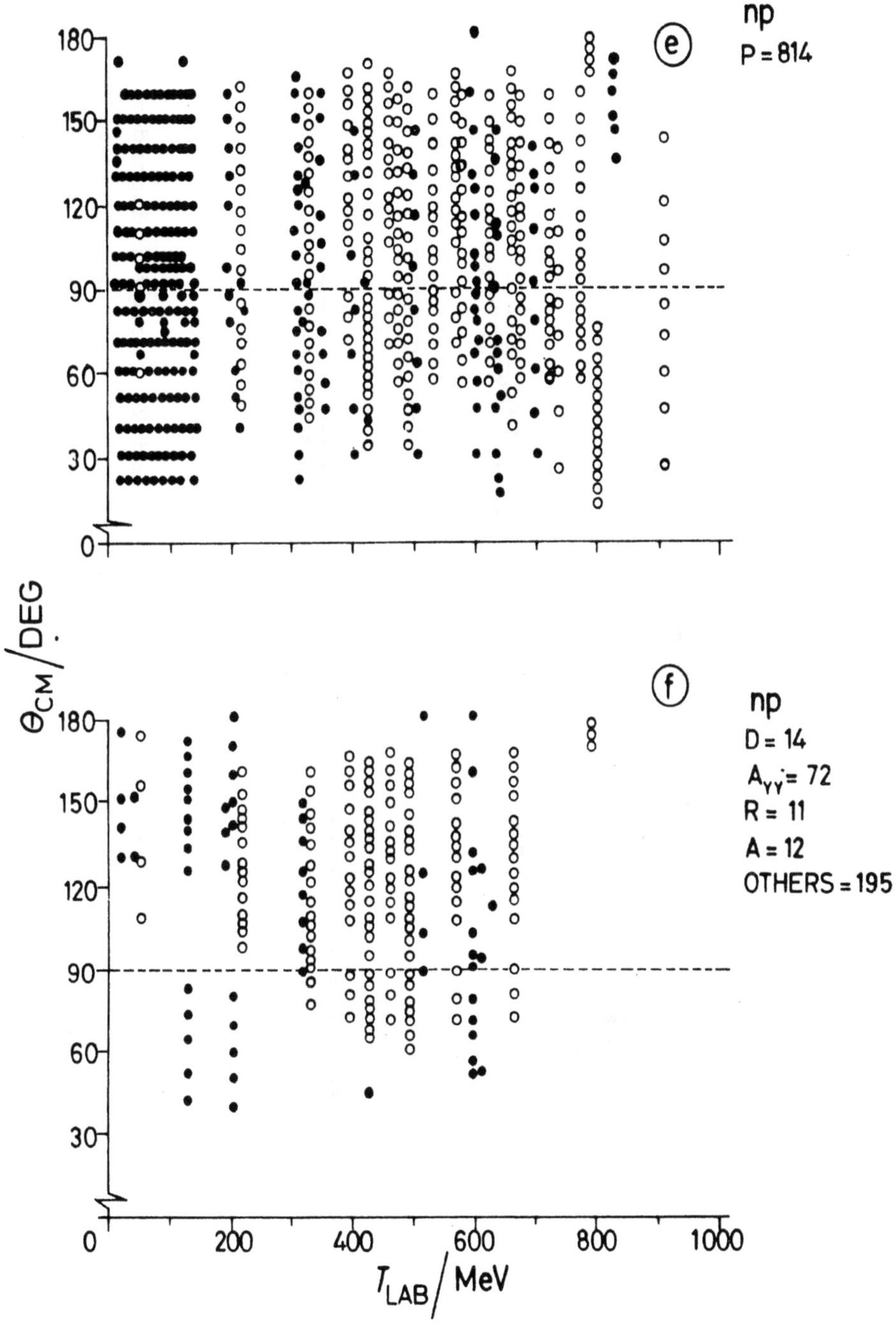
np
P = 814
e
180
150
120
90
60
30
0
Θ_{CM}/DEG
f
np
D = 14
A_{YY} = 72
R = 11
A = 12
OTHERS = 195
180
150
120
90
60
30
0
200
400
600
800
1000
T_{LAB}/MeV

Fig. 5.: ctd. 2

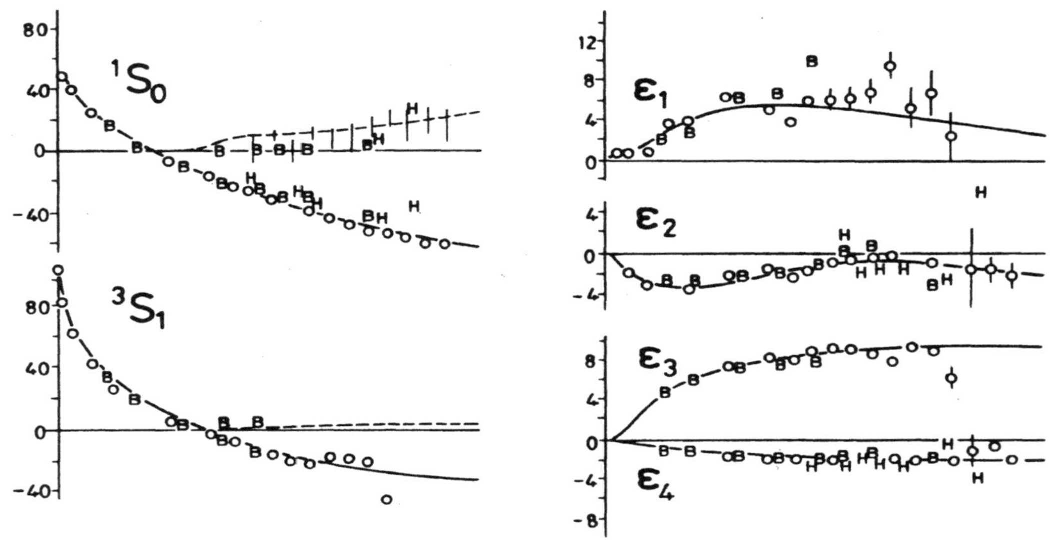

Fig. 6.: The results of the energy-dependent (curves) and 20 energy-independent (points with error bars) phase-shift analyses from ref.21, from ref.22 (B), and ref.23 (H). The results from ref.23 for I = 0 states differs appreciably and are not shown.

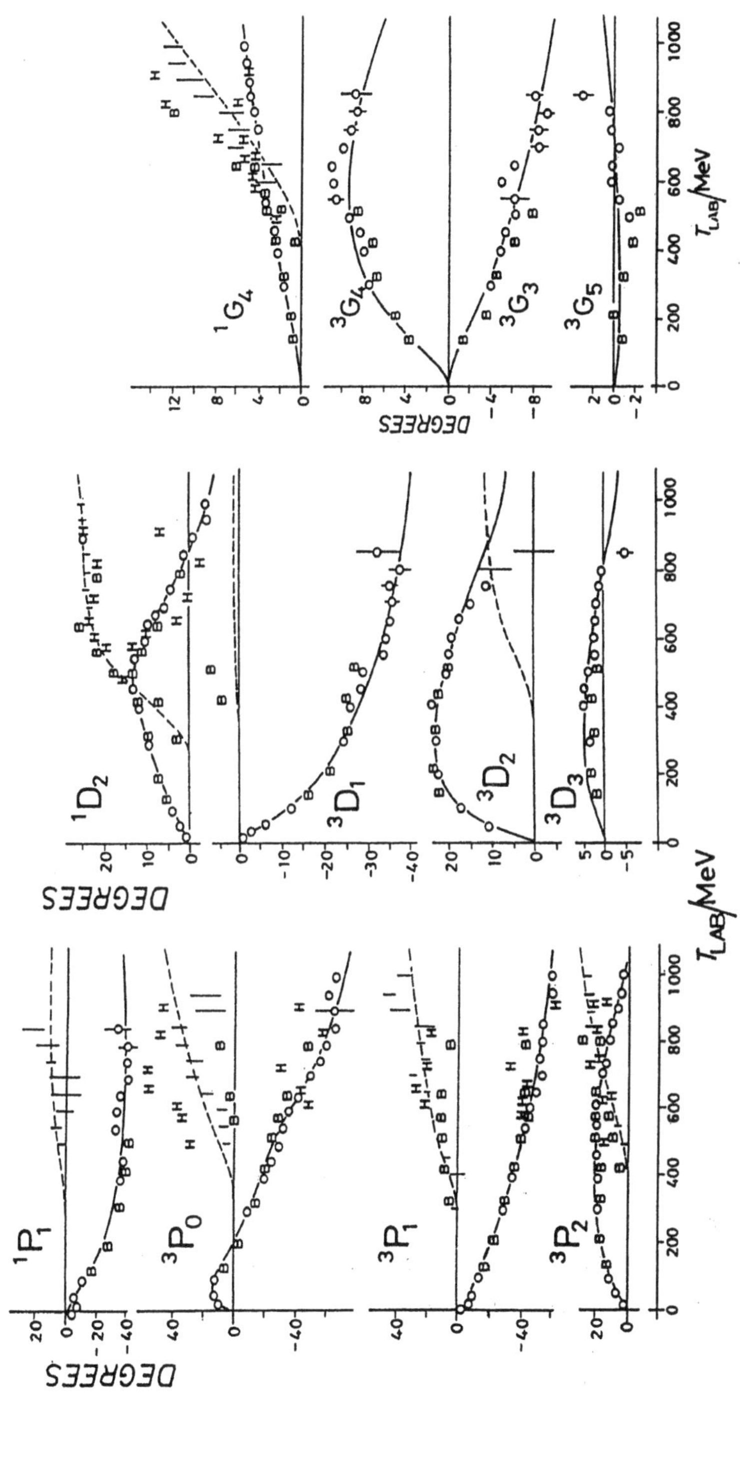

DEGREES
$^1P_1$
$^3P_0$
$^3P_1$
$^3P_2$
$^1D_2$
$^3D_1$
$^3D_2$
$^3D_3$
$^1G_4$
$^3G_4$
$^3G_3$
$^3G_5$
T_{LAB}/MeV

Fig. 6.: ctd.

The χ^2/data at lower energies are smaller, e.g. at 400 MeV for pp χ^2/data is 1.06. The results of the phase shift analyses by Dubois et al.[22] are also shown in Fig.6 and the agreement between the two results are satisfactory except for ρ of $^3P_0$ and $^3P_1$ and δ of $^1G_4$. The results by Hashimoto et al.[23] for I=0 states differs appreciably and it is not shown in Fig.6. The solutions of Arndt et al. [21] can be accessed through a computer interactive dial-in system (scattering analyses interactive dial-in: SAID) at Virginia Polytechnic Institute. The usage level of SAID is quite high sometimes more than 10 dial-ins per day. SAID can also be sent by tape to other sites and this has been done for several centres in North America and Europe.

Several measurements have been performed since 1983 and some are summarized as follows [24]:

i) extensive measurements of the pp spin observable at 448, 494, 515, 536, 560 and 578 MeV. They agree with the phase-shift analysis of ref.21.

ii) p-p $D_{NN}(90^o)$ in 380-800 MeV region, and P and Wolfenstein parameters at 699 and 750 MeV, also agree with prediction of ref.21.

The pp phase shifts up to 800 MeV are quite accurate and well known. The status of np I=0 phase shifts is satisfactory in the 140-515 region, but at lower energies the ε_1 and $^1P_1$ are questionable and there is a crying need for A_{LL} and A_{SS} at 650 and 800 MeV (planned at LAMPF).

From NN scattering observables and phase parameters one concludes:

The $^1S_0$ pp phase shift becomes negative at 250 MeV. It continues to decrease monotonically even up to 2 GeV suggesting that the repulsive core exists up to these energies. The repulsive core was introduced in 1955 by Jastrow [25] in order to explain an almost isotropic angular distribution of the p-p differential cross section. The explanation that this isotropy is the result of only S wave interaction can be ruled out on the basis that the experimentally measured total eleastic cross section for p-p nuclear scattering at 400 MeV is 38 mb, whereas the maximum S wave total cross section is $\frac{4\pi}{k^2}$ = 14 mb. In fact, at these energies one can conclude both from the total elastic cross section and from the impact parameter argument that P and D waves contribute. Therefore the differential cross section is:

$$\sigma(\theta) \sim \sin^2\delta_o + 25 \sin^2\delta_2[P_2(\theta)]^2 + 10 \sin\delta_o \sin\delta_2 \cdot$$

$$\cdot \cos(\delta_o-\delta_2)P_2(\theta) + \text{contribution involving P waves.}$$

In order to obtain an isotropic angular distribution it is necessary that sign δ_o = -sign δ_2. Since at E = 100-400 MeV the D waves are still dominated by the outside attraction, their phase shift is positive. It follows that δ_o has to be negative and that the ^1S_o wave potential has to be repulsive at these energies.

The contribution of P waves and in general of odd waves is appreciably smaller as one can conclude from the np angular distribution (Fig.7), which are symmetric with respect to 90°. Since any odd-L Legendre polynomial will give an asymmetry about 90°, it implies that odd-L contributions are small. This suggests that the spin averaged np potential is weak in odd-L states and this led Serber to propose an idealization that V_{np} acts only in even L states:

$$V_{np}(\vec{r}) = V_o(1+P_r)/2 ,$$

where P_r is a space (or Majorana) exchange operator equal +1 in even and -1 in odd states. Indeed, we see in Fig.6

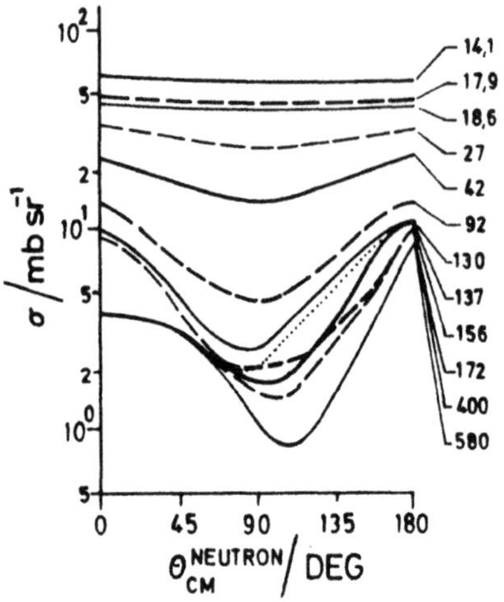

Fig. 7.: The np differential cross sections from 10 to 580 MeV.

that the $^1P_1$ phase is negative and $^1D_2$ is positive.

The downward bending of $^1D_2$ phase at about 500 MeV cannot be explained just by the central force. Since in the singlet states the tensor and spin-orbit term vanish one has to require a term which is L dependent so that the phases for ^1S_o and $^1D_2$ (and similarly $^1G_4$) can be fitted.

At 150 MeV a backward np scattering becomes very pronounced. The scattering amplitude calculated in the Born approximation for the square-well non-exchange force (Wigner force) is

$$f(\theta) \sim \int e^{-i\vec{k}_f\vec{r}}\, V(\vec{r})\, e^{i\vec{k}_i\vec{r}}\, d\vec{r} = \int_0^b V_o r \sin qr dr \, , \tag{27}$$

where b is the range of the square well and $q = 2k \sin\frac{\theta}{2}$. The amplitude is appreciable only for angles satisfying

$\sin\frac{\theta}{2} < \frac{1}{2kb}$, i.e. for $\theta < 30^o$ at 100 MeV. The explanation of the backward peak is in the mechanism which makes the system "believe" that the backward direction is a forward direction. This mechanism is the jump of a proton charge from the target to the forward emitted neutron, which thus becomes a proton. This charge-exchange process is represented by a charge-exchange operator $P_\tau = \frac{1}{2}(1+\vec{\tau}_1\vec{\tau}_2)$. In addition to the charge exchange one can also exchange the spin, i.e. $P_\tau P_\sigma$, where $P_\sigma = \frac{1}{2}(1+\vec{\sigma}_1\vec{\sigma}_2)$. Since nucleons are fermions the complete exchange of particles, i.e. the exchange of charge, spin and space causes the NN wave to change sign:

$$P_\tau P_\sigma P_r = -1 \text{ and } P_\tau P_\sigma = -P_r$$

and the large backward peak implies that the Majorana force is stronger than the Wigner force.

The $^3S_1$ phase shift becomes negative at 315 MeV. The NN force is repulsive in both S waves, but it is more repulsive in the ^1S_o wave. The repulsive core influences all partial waves but at progressively higher energies, e.g. the impact parameters for D and F waves are 0.4 fm at 3.2 and 6.2 GeV, respectively.

We see from Fig.6 that $^3P_2$ is positive, $^3P_1$ is negative and ^3P_o becomes negative at 210 MeV. Such behaviour can be understood, if the triplet phase shifts are written as

$$\delta(^3P_2) = \delta_C - \frac{2}{5}\delta_T + \delta_{LS}$$

$$\delta(^3P_1) = \delta_C + 2\delta_T - \delta_{LS}$$

$$\delta(^3P_o) = \delta_C - 4\delta_T - 2\delta_{LS} \, , \qquad (28)$$

where δ_i are the effective phases of type i, which at high energies are just the Born approximation phase shifts from the respective potentials. Putting in [28] experimental phase shifts and calculating δ_C, δ_T and δ_{LS} we conclude that δ_C is small, δ_T is large and negative and δ_{LS} is larger and positive. This implies that the triplet-odd tensor force is repulsive and the spin-orbit is attractive.

Similar analysis can be done for ^3D waves:

$$\delta_C = \frac{1}{15}[7\delta(^3D_3) + 5\delta\ (^3D_2) + 3\delta\ (^3D_1)]$$

$$\delta_T = \frac{7}{120}[-2\delta(^3D_3) + 5\delta\ (^3D_2) - 3\delta\ (^3D_1)]$$

$$\delta_{LS} = \frac{1}{60}[14\delta(^3D_3) - 5\delta\ (^3D_2) - 9\delta\ (^3D_1)] \tag{29}$$

implying that central and tensor triplet-even potentials are strong and attractive and that the triplet-even spin-orbit is weaker and repulsive.

At energies below 10 MeV S wave scattering is dominant. The S wave function u(k,r) asymptotically behaves as

$$v(k,r) = \frac{\sin(kr+\delta_o)}{\sin \delta_o} = \cos kr + \cot \delta_o \sin kr .$$

The Taylor expansion of $k \cot\delta_o$ in k^2 is called the effective-range expansion.

$$k \cot\delta_o = -\frac{1}{a}\ r_o k^2 + P r_o^{\,3} k^4 + Q r_o^{\,5} k^6 + \ldots \tag{30}$$

$$\text{at } k = 0:\ k \cot\delta_o = -\frac{1}{a} . \tag{31}$$

The quantity a is called the scattering length. At $k = 0$ $v(0,r) = 1 - \frac{r}{a}$ and u(0,r) curves from 0 at $r = 0$ to v(0,r) at very large distances (Fig.8). The asymptotic slope of u(0,r) is given by the slope of v(0,r). This slope is positive for

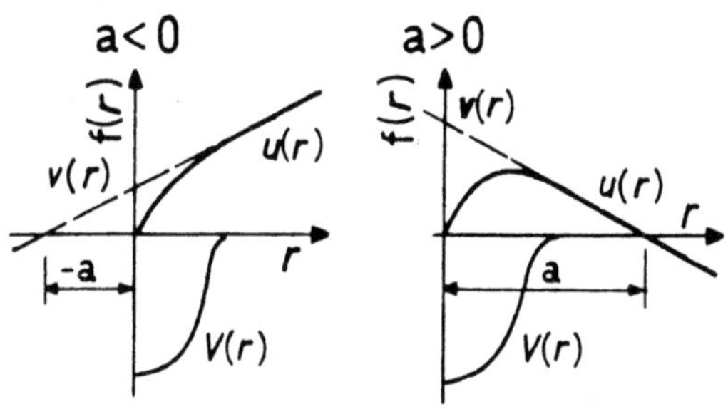

Fig. 8.: The definition and the meaning of the scattering length a.

negative a, zero for a=∞, and negative for positive a. If the slope is negative, it can be matched to an exponentially decreasing function characteristic of bound states. Therefore, an attractive potential giving a wave function u(0,r), which has a positive scattering length, is strong enough to have a bound state. As the strength of the potential decreases the slope of the asymptotic wave function tends to zero and the binding energy tends to zero. For even weaker potentials there is no bound state and a becomes negative. In the vicinity of a = ∞, the change in a is a magnifying glass on the change in V:

$$\frac{\Delta a}{a} = (100\text{-}1000)\ \frac{\Delta V}{V} \ . \tag{32}$$

From (31) one sees that negative a is related to $0<\delta_o<90^o$, and positive a to $90^o<\delta_o<180^o$.

The wave function for the bound state is $e^{-r/R}$, and the ratio $\frac{u'(r)}{u(r)}$ is $-\frac{1}{R}$. Relating these expressions with those for a scattering wave e^{ikr} and $\frac{v'(0,r)}{v(0,r)} = k \cot\delta$, it follows that (30) using only two terms on the right-hand side transforms into:

$$-\frac{1}{R} = -\frac{1}{a} - \frac{1}{2}\frac{r_o}{R^2} \ .$$

Wherefrom we see that the total elastic cross section at zero energy becomes

$$\frac{4\pi\sin^2\delta}{k^2} = \frac{4\pi}{k^2+k^2\cot^2\delta} = 4\pi a^2 = \frac{4\pi R^2}{(1-\frac{r_o}{2R})^2} \ .$$

This is the expression we have already used in (1) and (2).

The quantity r_o in (30) is called an effective range. It is related to the effective-range integral $\rho(E_1,E_2)$ which measures the amount of distortion of u(r) compared to v(r):

$$\rho(E_1,E_2) = 2\int_0^\infty (v_1v_2 - u_1u_2)\,dr$$

and $r_o = \rho(0,0)$.

The OPE branch-cut causes the series (30) to diverge for E_{lab} beyond 9.71 MeV. If the OPE cut is taken into account explicitly the range of validity is extended to 38 MeV. A very good approximation is provided [26] by the modified effective-range expansion

$$k \cot\delta = -\frac{1}{a} + \frac{1}{2} r_o k^2 - \frac{pk^4}{1+qk^2} \; . \tag{33}$$

The pp scattering is influenced by em effects and thus the left-hand side of (33) has to be modified into

$$\frac{c^2 k}{1-\phi_o} [(1+\lambda_o)\cot\delta_o - \tan\tau_o] + 2\eta k \, [h(\eta) + \ell_o(\eta)]$$

where c^2, η and $h(\eta)$ are parameters which arise in a pure Coulomb problem, while λ_o, ϕ_o, τ_o and ℓ_o are non-Coulomb em parameters [27].

The effective-range parameters for ^1S_o and $^3S_1$ states are summarized in Table 4. The procedure to determine a purely nuclear ^1S_o a_{pp} depends on the knowledge of the entire p-p wave function, which in turn implies the knowledge of the NN force at all distances. Since we do not know the NN force at very short distances [28] (say $r<0.8$ fm), the uncertainty introduced by this lack of knowledge is ±0,7 fm, as shown in the value for a^N_{pp} in Table 4. For a_{nn} we use the value extracted from the reaction $D(\pi^-,\gamma)2n$, since we argue that the values obtained from the reaction D(n,2n)p are influenced by the three-nucleon force [29].

Table 4

	a (fm)	r (fm)
^1S_o np [30]	-23.749(009)	2.78(05)
nn [29]	-18.45(46)	2.76(11)
pp [31]	-7.844(003)	2.859(006)
pp^N[32]	-17.9(8)	2.84(03)
$^3S_1$ np [30]	+5.423(004)	1.76(003)

Expression (33) has been used to determine a and r_o for np and pp systems. Errors are given in parentheses.

3. SYMMETRIES

Herman Weyl formulated an excellent operational definition of what is meant by symmetry: an object is symmetric, if there is something you can do to it such that when you have finished, it looks the same as it did before.

Similar definition holds for the symmetry of laws of nature. The higher the degree of symmetries, the more tightly the laws are circumscribed. Symmetries are related to conservation principles.

3.1 Parity Nonconservation

A small parity-violating component in the nuclear interaction stems from hadronic weak interaction. The parity-nonconserving (PNC) effect is expected to be of the order of 10^{-7}, but in many nucleon systems it can sometimes be amplified, if the parity-allowed transition is strongly hindered. A large enhancement of PNC effects has been observed in the asymmetry A_γ of the decay of polarized metastable ^{180}Hf: $A_\gamma = -0.017 \pm 0.002$ [33]. The present status of these investigations is not to qualitatively establish PNC effect, but rather to understand the quantitative aspects of the NN PNC force. Indeed, parity violation is the only way to study strangeness-conserving purely hadronic weak interaction. Therefore, one has to study systems for which one has a good understanding of a nuclear wave function, e.g. two-nucleon system and light nuclei.

The weak interaction does not conserve isospin and thus the weak hadronic force H_{PNC} will contain $\Delta I \neq 0$ components. For NN systems at low energy the only important PNC transitions are those connecting L=0 and L=1 states (Table 5). Though we have a successful standard theory of the electroweak interaction of quarks and leptons and the PNC [31] NN interaction is driven by both charged $W^{\pm}$ and neutral

Z^o currents, there is yet no reliable way to predict the weak interaction between nucleons (of the size of 1 fm) in terms of the exchanges of $W^{\pm}$ and Z^o (their wavelength being 3×10^{-3} fm). Thus the PNC NN force is represented in terms of meson exchanges between nucleons (Table 5) and the true physics of $W^{\pm}$, Z^o exchange between quarks is hidden inside the meson-nucleon vertices [34]. Theoretical predictions [35] based on this simple picture of π, ρ, ω exchanges between nucleons with coupling determined in the standard quark model agree with the low-energy data, but fail to fit the high-energy p-p A_L [36,37] . However, at high energies this picture is inadequate and approximate treatment of the strong force in terms of one-gluon exchange seems to predict a correct value for A_L [38]. It is argued [39] that an additional and at high energies large contribution to PNC arises from the nucleon wave function containing a small negative-parity component due to its self-interaction with the weak bosons. There is still some controversy concerning the employed couplings which influence the agreement with the low energy data [40,41].

Table 5

Transition	Isospin structure	System where it occurs	Dominant meson exchange
$^3S_1 - {}^1P_1$	I=0 - I=0, ΔI=0	np	ρ ω
$^3S_1 - {}^3P_1$	I=0 - I=1, ΔI=1	np	$\pi^{\pm}$
$^1S_o - {}^3P_o$	I=1 - I=1, ΔI=0	nn np pp	ρ ω
	1	nn pp	ρ ω
	2	nn np pp	ρ

The only pseudoscalar observable measured in NN scattering is the longitudinal analyzing power

$$A_L = \frac{\sigma_+ - \sigma_-}{\sigma_+ + \sigma_-}$$

(where σ_+ and σ_- denote the cross section for positive- and negative-helicity proton beams). The observable A_L is expressed in terms of the PNC components of the M matrix (see Section 2) as:

$$2\sigma_o A_L(\theta) = [\mathrm{Re}(A-F)^* R - \mathrm{Re}\, C^* S]\cos\frac{\theta}{2} + $$
$$+ [\mathrm{Im}\,(-B-F)^* S + \mathrm{Im}\, C^* R]\sin\frac{\theta}{2} ,$$

where σ_o is a differential cross section for the unpolarized particles. The results of the measurements and of the theoretical calculations [38,41] are given in Table 6.

Table 6

Energy (MeV)	A_L(exp) 10^{-7}	Ref.	A_L(theo) 10^{-7}	Ref.
15	-(1.7±0.8)	42	-1.3	35
45	-(2.31±0.89)	43	-2.3	35
46	-(1.3±2.3)	44		
47	-(4±3)	45		
800	+(6.6±3.2)	36	+12.0	41
5130	+(26.5±6)	37	+20.0	41

At low energies A_L is mainly sensitive to $^1S_o - {}^3P_o$ $\Delta I = 0$, 1 and 2 PNC amplitudes, but at higher energies $^3P_2 - {}^1D_2$ and $^3F_4 - {}^1G_4$ PNC amplitudes become important.

The circular polarization P_γ of the 2.2 MeV γ ray from thermal n capture by p comes from the interference between dominant $^1S_o \rightarrow {}^3S_1$ M1 amplitude and a PNC E1 amplitude. Thus, P_γ is sensitive to $^1S_o - {}^3P_o$ and $^3S_1 - {}^3P_1$ PNC mixing. The old measurement $P_\gamma = (-1.3\pm0.45)\times10^{-6}$, which was larger than the theoretical value of below 10^{-7}, is now recalled and a new result [46] is $P_\gamma = (1.8\pm1.8)\times10^{-7}$. The analyzing power using polarized cold neutrons is [47] $A = (0.6\pm2.1)\times10^{-7}$.

The study of parity-violating effects in light nuclei is summarized [34,48,49] in Table 7. In the alpha decay of $^{16}O(2-)$ to $^{12}C+\alpha$ one detects the very small $\Gamma_\alpha = 10^{-10}$eV width. Measurements of the asymmetry of γ decays of polarized ^{19}F formed by the nuclear reaction induced by polarized p and the measurements of circular polarization of γ rays from unpolarized ^{18}F and ^{21}Ne provide examples of the study of PNC mixing of two levels of different parity and same J and almost the same energy. The asymmetry in the photon angular distribution of $nd \rightarrow t\gamma$ can be a sensitive test for PNC, since PC part is suppressed: σ_{nd}/σ_{np} capture is 3×10^{-3}.

3.2 Time-Reversal Noninvariance

It is experimentally well established [50] that the time-reversal invariance (TRI) is broken in the neutral K decay. Many theories have been developed to explain this datum: quark mixing, superweak $\Delta S = 2$ interaction, exchange of multiple Higgs bosons and right-handed weak current [34].

Table 7

System	Observable	Exp.	Theo.	Ref.
^{16}O	$\Gamma_\alpha(10^{10}eV)$	1.03±0.28	0.6	34
^{18}F	$P_\gamma(10^3)$	-0.42±0.7	±1.56	48
^{19}F	$A_\gamma(10^5)$	-6.83±1.75	-6.6	77,48
^{21}Ne	$A_\gamma(10^3)$	0.8±1.4	±1.16	48
$p\alpha$	$A_L(10^7)$	-3.34±0.93	-2.49	48
$nd\to t\gamma$	A	7.8±3.4	1.0	49

These theories predict very different values for other TRI violating observables. For instance, Higgs-exchange theories predict the neutron electric dipole to be 10^{-24} e cm, and quark-mixing theories 10^{-28} e cm. Other models predict values as low as 10^{-33} e cm or as high as 10^{-19} e cm.

It is easy to see that the neutron electric dipole is a direct proof of TRI violation: the neutron is composed of charged quarks and it is possible for that charge to spread out to form a dipole: positive at one end and negative at the other. The neutron has a spin and so only a separation of charge along the spin axis can be observed. If we run a film of spinning n backwards, the particle will appear to spin in the opposite direction, but its dipole will still be pointing in the same direction. Thus if a n would have an electric dipole we would be able to distinguish between real n and those on the backward moving film.

If the particle has an electric and a magnetic dipole, then its energy in the electric and magnetic fields is

$$E = -\vec{\mu}_e \vec{E} - \vec{\mu}_m \vec{B} = -g_e \vec{\sigma}\vec{E} - g_m \vec{\sigma}\vec{B} .$$

Under parity $\vec{E}$ is odd and $\vec{\sigma}$ and $\vec{B}$ are even, and under TR $\vec{E}$ is even and $\vec{\sigma}$ and $\vec{B}$ are odd. Thus, $g_e \neq 0$ implies both parity and TRI to be violated.

Experiments using ultra-cold neutrons have established an upper limit for the n electric dipole:

$$d_e < (0.3 \pm 4.8) \times 10^{-25} \text{ e cm } [51]$$

and

$$d_e < (2.3 \pm 2.3) \times 10^{-25} \text{ e cm } [52]$$

(this value for an electric dipole can be visualized by expanding a neutron to the size of the Earth. Then, $d_e = 10^{-25}$ e cm corresponds to a bulge of 20 microns). Measurements with ultra cold neutrons are continuing and in a year one might expect a sensitivity of 10^{-27} e cm, which should be enough to distinguish between various theories.

Measurements of correlations in γ and β decays can also provide a test for TRI violation.

The correlation of the γ emission of the form $J(k\times E)(J\cdot k)(J\cdot E)$, where E is the electric-field vector of the γ ray is odd under TR and even under parity. The mixing ratio of two competing multipoles is real for TRI. Any imaginary component is a signature for TRI violation. The measurement of $\frac{A(E2)}{A(M1)} = |\delta|\, e^{i\eta}$ in 129 keV transition in ^{191}Ir gives [53] $\eta = (-4.8\pm0.2)\times10^{-3}$, but the value due to γ-electron final-state interaction is $(-4.3\pm0.4)\times10^{-3}$. Thus, no conclusive evidence for TRI violation.

Decay rate of polarized J=1/2 beta emitter can be expressed as $A+D\,\vec{J}\,(\vec{p}\times\vec{q})$, where $\vec{J}$ is the spin of the initial nucleus, and p and q are β and ν momenta. The term D is odd under TRI. The measurement [54] gave an upper limit $D = (0.7\pm6.2)\times10^{-4}$.

TRI relates the cross section for the reaction between unpolarized particles and its inverse. This is the detailed balance theorem. Still unsettled disagreement between $p\,d \rightleftarrows {}^3He\gamma$ probably is not related to TRI. The measurement of the detailed balance $p + {}^{27}Al \rightleftarrows \alpha + {}^{24}Mg$ led [55] to the T_{odd}/T_{even} amplitude $<5\times10^{-4}$.

If TRI is valid, then polarizing and analyzing powers should be equal: P = A (Fig.9). Measurements [56] of the A and P in the reactions ${}^9Be(p,\,{}^3He){}^7Li$ and ${}^{11}B(p,\,{}^3He){}^9Be$ showed a very large difference between A and P. However, additional measurements [57] of the same reactions demonstrate

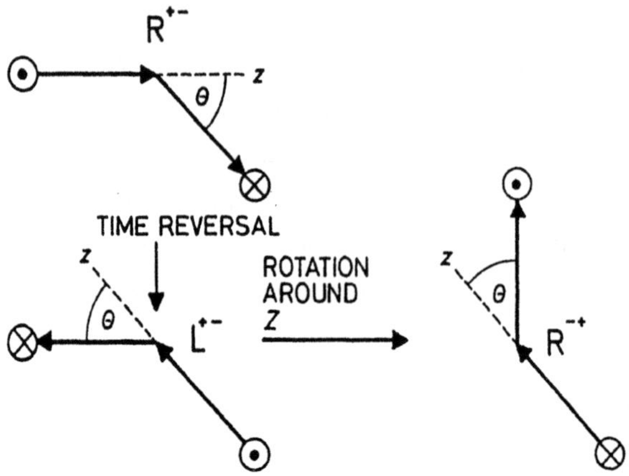

Fig. 9.: L^{+-} is the probability for spin-up particle to scatter to the left and to have in the final state the spin down. R^{+-}, R^{-+}, L^{--} etc. are defined analogously. TR implies $L^{+-} = R^{+-}$ and TR and rotation around z-axis implies $R^{+-} = R^{-+}$. Polarization $P = (R^{++} + R^{-+}) - (R^{+-} + R^{--})$ and analyzing power $A = (R^{++} + R^{+-}) - (L^{++} + L^{+-})$. Rotational invariance implies:
$A = (R^{++} + R^{+-}) - (R^{--} + R^{-+})$ and
$P - A = 2(R^{-+} - R^{+-})$ which is = 0 if TRI.

A = P. P-A is related to the depolarization D

$$D = \frac{\sigma^{++} - \sigma^{+-} + \sigma^{--} - \sigma^{-+}}{\sigma}$$

through

$$P\text{-}A = (1-D)\,\frac{\sigma^{-+} - \sigma^{+-}}{\sigma^{-+} + \sigma^{+-}}$$

and thus P = A if D = 1. Unfortunately several tests of TRI have been done at the conditions were D was close to 1.

Aprile et al.[58] measured 16 polarization parameters at 579 MeV at six cm angles between 66° and 86° and reconstructed the M matrix allowing also for the TRI violating T term. The upper limit on T^2 was $0.0059\times\sigma$.

Bhatia et al.[59] measured P-A in n-p scattering at 800 MeV (1-D is greater than 1) and found P-A = 0.011 ± 0.019 [60]. The difference between P and A, and more generally, nonzero values for observables,which are supported to the zero under TRI, need not indicate for TRNI,but it can also be the consequence of the violation of the rotational invariance [61] (also, see Fig.9). Additional measurements are required to determine TRNI.

3.3 Isotropy of Space [62]

The only measure of the validity of rotational invariance in strong interaction comes from the fact that the partial wave expansion works (which depends on total-angular-momentum conservation). However, this is not a very good limit. The rotational invariance in weak interaction is valid [63] to 10^{-6} and the upper limit [64] on the anisotropy of inertial mass is $\frac{\Delta m}{m} < 10^{-20}$.

Tests of the isotropy of space in strong interaction are almost identical to tests of parity conservation, as can be seen by an example of a reaction involving spin 0 and 1/2 particles:

$$0 + 1/2 \rightarrow 0 - 1/2$$

If the rotational invariance holds, then M matrix is $M = a_0 + a_1\vec{\sigma}\vec{q}_1 + a_2\,\vec{\sigma}(\vec{q}_1 x\vec{q}_2) + a_3\vec{\sigma}\vec{q}_2$, where $\vec{q}_1$ and $\vec{q}_2$ are initial and final cm momenta. a's are functions of scalars q_1^2, q_2^2 and $\vec{q}_1 x\vec{q}_2$. If parity is conserved, then $a_1 = a_3 = 0$.

Let us assume that the rotational invariance is violated and that there is a unit vector $\vec{u}$ pointing in a distinguished direction in space. The M matrix can now be constructed of four vectors $\vec{u}$, $\vec{\sigma}$, $\vec{q}_1$ and $\vec{q}_2$, and since $\vec{u}$ can be expressed in terms of $\vec{q}_1, \vec{q}_2$ and $\vec{q}_1 x \vec{q}_2$, M is:

$$M = a_0 + (a_1 + \omega_1)\vec{\sigma}\vec{q}_1 + (a_2 + \omega_2)\vec{\sigma}(\vec{q}_1 \times \vec{q}_2) + (a_3 + \omega_3)\vec{\sigma}\vec{q}_2 ,$$

where terms ω depend on the absolute orientation of space. In addition a_i depend also on terms $\vec{q}_1\vec{u}$, $\vec{q}_2\vec{u}$, $\vec{q}_1(\vec{q}_2 \times \vec{u})$ and regardless whether $\vec{u}$ is a vector or a pseudovector a_i include scalars and pseudoscalars. Requiring parity conservation does not reduce the M matrix to two terms as it did before:

$$M = a_0^S + (a_2^S + \omega_2^S)\vec{\sigma}(\vec{q}_1 \times \vec{q}_2) \ / \ (a_1^{PS} + \omega_1^{PS})\vec{\sigma}\vec{q}_1 + (a_3^{PS} + \omega_3^{PS})\vec{\sigma}\vec{q}_2.$$

S and PS denote scalars and pseudoscalars.

Polarization in the direction $\vec{q}_i$ is:

$$q_1^2 \mathrm{Re}[a_0^S(a_1^{PS} + {}_1^{PS})^*] + (q_1 q_2)\mathrm{Re}\ a_0^S(a_3^{PS} + {}_3^{PS})^*) +$$

$$+ (q_1 x\ q_2)^2 \mathrm{Im}[(a_2^S + \omega_2^S)(a_3^{PS} + \omega_3^{PS})^*] .$$

This is precisely A_L measured in the test of parity nonconservation. Indeed, the structure of M does not distinguish, whether parity is violated and rotational invariance holds or vice versa. The way to distinguish between them is to perform two measurements: one corresponding to space inversion. Namely, for PNC and rotational invariance all amplitudes a_i are scalars, while for rotational noninvariance amplitudes contain pseudoscalars.

3.4 Charge-Nonconserving Decay

Lower limits on the electron lifetime are set [65] at $10^{22} - 10^{23}$ y. The search for the charge nonconserving decay $^{87}Rb \rightarrow {}^{87m}Sr$ + neutrals gave [66] a lower limit of 7.5×10^{19} y which sets an upper limit on the ratio

$$\frac{\Gamma(n \rightarrow p + \nu_e + \tilde{\nu}_e)}{\Gamma(n \rightarrow p + e^- + \tilde{\nu}_e)} \text{ of } 7.9 \times 10^{-21} .$$

3.5 Charge Dependence

Nuclear phenomena strongly suggest that nuclear forces are to a large extent charge independent CI, i.e. $[H,T^2]_-=0$ or in other words that in a given state nn, np and pp nuclear forces are identical. Charge symmetry CS is a weaker concept implying that the strong interaction satisfies $[H,e^{i\pi T^Y}]_- = 0$, which for nucleons means that the observables are unaffected by changing n into p and p into n. Charge symmetry implies the equality of pp and nn nuclear forces, but it also has consequences for the np system. Henley and Miller classify [67] NN interaction into four groups:

i) charge independent $V_I = V(1 + \vec{\tau}_1\vec{\tau}_2)$

ii) charge symmetric but charge dependent

$$V_{II} = V(\tau_1^Z\tau_2^Z - \frac{1}{3}\vec{\tau}_1\vec{\tau}_2)$$

iii) charge-symmetry violating and charge independent

$$V_{III} = V(\tau_1^Z + \tau_2^Z)$$

Since $[T^Z, T^2] = 0$, i.e. $[V_{III}, T^2]_- = 0$ it does not cause isospin mixing

iv) charge-symmetry violating and charge dependent

$$V_{IV} = V\{E(\tau_1^Z - \tau_2^Z) + F(\vec{\tau}_1 \times \vec{\tau}_2)_Z\} ,$$

$$E = (\vec{\sigma}_1 - \vec{\sigma}_2).\vec{L}\cdot F(r^2L^2); \quad F = (\vec{\sigma}_1 \times \vec{\sigma}_2).\vec{L}.G(r^2L^2).$$

This interaction vanishes for L = 0 states. It causes isospin mixing. (The term $(\vec{\tau}_1 \times \vec{\tau}_2)_Z$ is odd under TR, because it contains τ^Y linearly).

The consequence of V_{IV} is the difference between $A_n(\theta)$ and $A_p(\pi-\theta)$ in n-p elastic scattering. Two experiments measuring $A_n(\theta) - A_p(\pi-\theta)$ are in progress at TRIUMF with 480 MeV neutrons and at IUCF with 200 MeV neutrons. Meson picture of nuclear force predicts CS and CI breakdown (see Fig. 10 for some CS breaking diagrams). Equivalently, difference between u and d quarks masses implies CS and CI breaking.

Experimental evidence for charge dependence and charge asymmetry:

i) a_{np}, a_{nn} and a_{pp} (Table 4 in Section 2). There is a very clear evidence for charge dependence and the difference $a_{nn}-a_{np}$ = 5.3±0.5 fm has been explained [68] in terms of OPE and TPE as well as $\gamma\pi$ exchanges:

(3.6±0.15) (OPE) + (0.88±0.1)(TPE) - (1.6±0.2) (πN coupling constant in OPE) + (1.6±0.2) (coup. const. in TPE) + (1.1±0.4) ($\gamma\pi$) = 5.6±0.5 fm.

The difference $a_{pp} - a_{nn}$ = 0.55±0.95 fm is consistent with the explanation [69] in terms of ρ-ω and π-η isospin mixing, TPE and $\gamma\pi^0$:

0.74 (ρ-ω) - 0.03 (π-η) + 0.54 (2π) + 0.19 ($\gamma\pi^0$) = 1.44 fm.

ii) Differences between the binding energies of mirror nuclei compared to the differences in their electromagnetic energies demonstrate that the nn force is more attractive than the pp force (Table 8). The ^3H-^3He binding-energy difference is largely model independent [70]. The

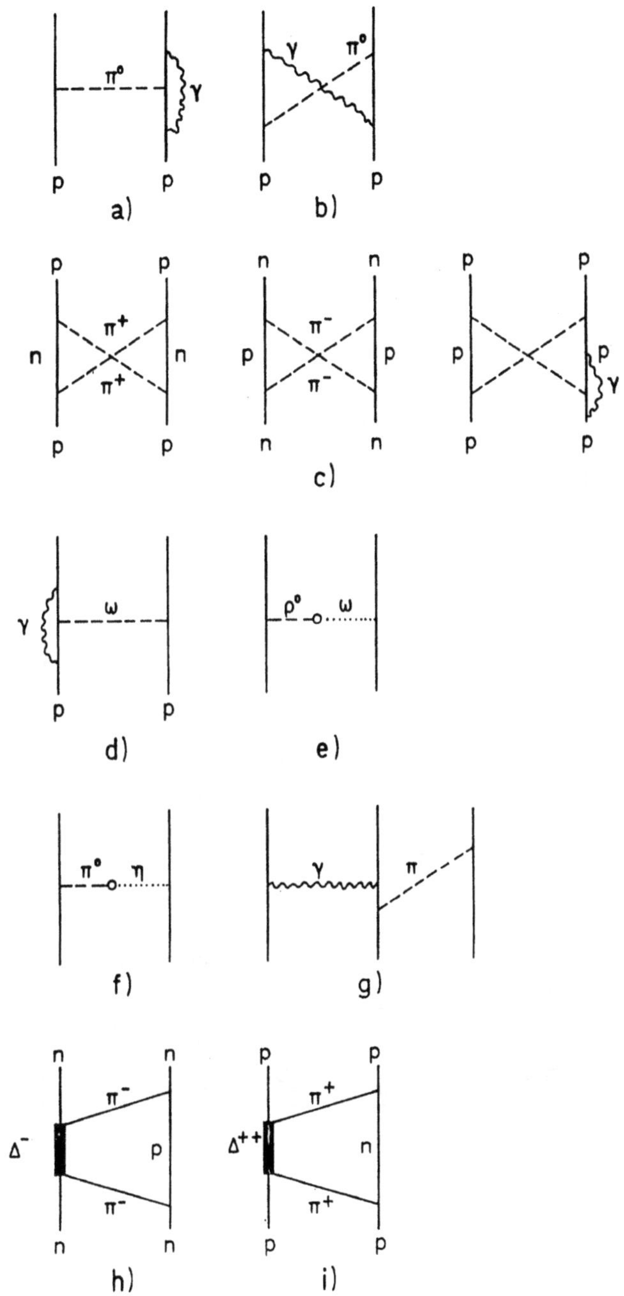

Fig. 10.: Several diagrams describing CS breaking.

Table 8

Mirror Nuclei	D_{em}	$D=\|E(A)-E(B)\|$	$D-D_{em}$ (keV)
^{3}H - ^{3}He	683±19 model independent	763.84±0.26	81±19
^{15}N - ^{15}O	3320±90	3542	222±90
^{17}O - ^{17}F	3300±90	3542	242±90
^{39}K - ^{39}Ca	6900±140	7305	405±140
^{41}Ca - ^{41}Sc	6700±130	7277	537±130

The situation for heavier mirror nuclei is more complicated and indeed the CS breaking potential which explains ^3H-^3He is too small to amount for ^{41}Ca-^{41}Sc. This is explained by the excess of p in the surface due to the Coulomb force. The p outside sees less attraction than the n at the same place, because the average np force is stronger than the average pp force [71].

The effective-range parameters and binding-energy differences are consistent and reveal charge-symmetry breaking due to class III force (though class IV could contribute to energy differences involving heavier mirror nuclei).

iii) A number of other phenomena has been investigated:

a) measurement of the angular asymmetry about 90° of deuterons produced in $np \rightarrow d\pi^o$. The final state is I=1 and thus the np system in the initial state has to be also I =1. For I=1 np system the angular distribution has to be symmetric with respect to 90°. Asymmetry would be caused by class IV force, because any asymmetry comes from an isospin mixing of I=0 and I=1. At 800 MeV the asymmetry is below ±0.55 % [72]. (Fig. 11) The theoretical estimate for the asymmetry due to class IV force is -0.11 % to -0.16 % [72].

b) measurement of the total and differential elastic scattering [75] reveal appreciable difference between π^+d and π^-d, but it is at this stage impossible to conclude anything about charge-symmetry breaking [74] (Fig. 12).

c) measurement [75] of the elastic scattering of π^- and π^+ from ^3He and ^3H at 180 MeV has been used to construct a superratio

$$R = \frac{d\sigma(\pi^{+3}H)/d\sigma(\pi^{-3}He)}{d\sigma(\pi^{+3}He)/d\sigma(\pi^{-3}He)} \quad .$$

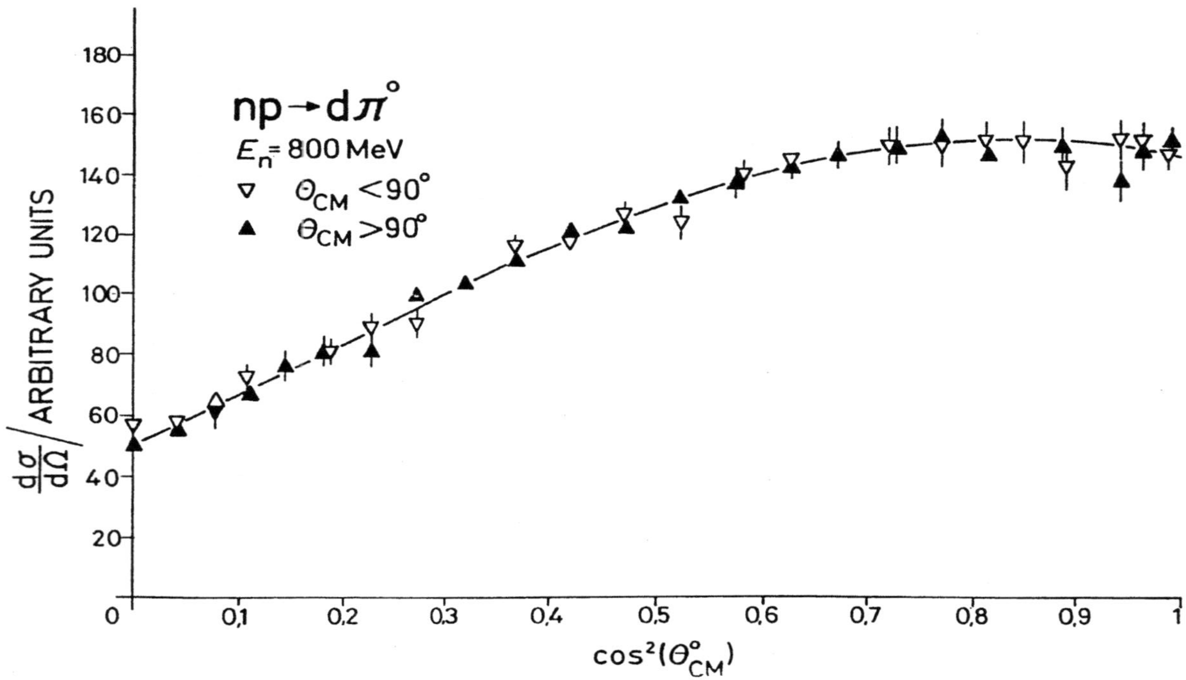

Fig. 11.: Angular distribution of the reaction $np \rightarrow d\pi^0$. Solid wave is the Legendre-polynomial fit using only even terms:

$\frac{A_2}{A_0} = 0.827 \pm 0.12$, $\frac{A_4}{A_0} = -0.247 \pm 0.01$ and

$A_6/A_0 = -0.071 \pm 0.02$ [72].

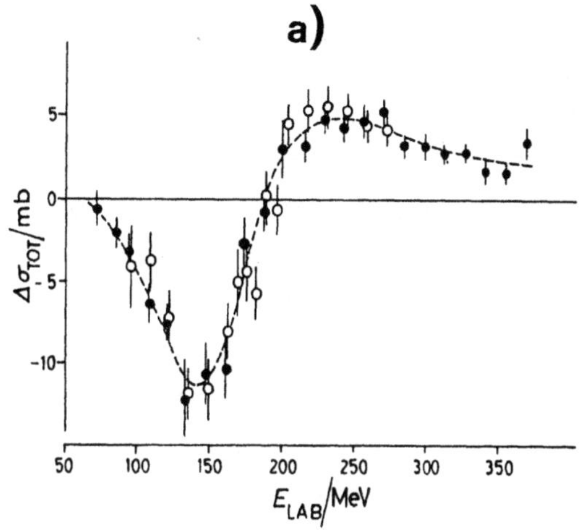

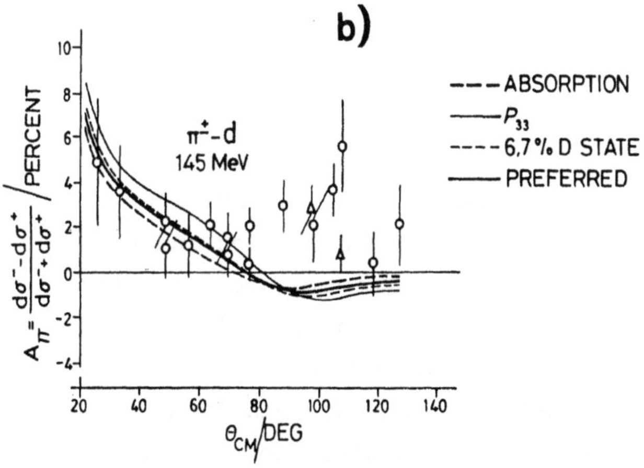

Fig. 12: a) Difference $\Delta\sigma$ between the total π^-d and π^+d cross sections: 10 cm long target (circles) and 20 cm (points). Solid curve is a fit parametrized in terms of masses and widths of nucleons and isobars [73]

b) Angular distribution measured at the energy where $\Delta\sigma$ has a maximum. Curves are various calculations [73]

If charge symmetry is valid, one expects R=1. Experimental R varies with angle reaching at 66° a value of 1.31±0.09. Kim suggests that the effect can be due to multiquark $(q^{10}\tilde{q})$ compound resonance (MQCR) in the interacting hadronic systems and suggest that a test for such a MQCR would be measuring the energy dependence [76].

d) measurements of the strength of E1 transitions in conjugate nuclei, the study of vector and tensor polarization in D(d,p)t and D(d,n)^{3}He reaction and the investigation of Barshay-Temmer theorem are certainly rewarding studies, but so far they did not provide conclusive evidence for charge dependence.

4. THE TWO-NUCLEON POTENTIAL

We have shown that the NN force is invariant under rotation, time reversal, and space reflection and that the charge independence and charge symmetry are only slightly broken. Electric charge is conserved. Certainly the Hamiltonian should be hermitian and invariant under space translation, time translation, and Galileian transformation. We will construct now the most general nonrelativistic NN potential satisfying all these symmetries and being invariant under the interchange of particles 1 and 2 [78].

Nucleons are characterized by their coordinates $\vec{r}_1$ and $\vec{r}_2$, momenta $\vec{p}_1$ and $\vec{p}_2$, spins $\vec{\sigma}_1$ and $\vec{\sigma}_2$, and isospins $\vec{\tau}_1$ and $\vec{\tau}_2$. Instead of $\vec{r}_1$, $\vec{r}_2$, $\vec{p}_1$ and $\vec{p}_2$ we use the relative coordinate $\vec{r} = \vec{r}_2 - \vec{r}_1$, c.m coordinate $\vec{R} = (\vec{r}_1+\vec{r}_2)/2$ and their canonically conjugate momenta $\vec{p} = \vec{p}_2-\vec{p}_1$ and $\vec{P} = \vec{p}_1+\vec{p}_2$. Therefore, the potential is a function $V(\vec{r},\vec{p},\vec{R},\vec{P},\vec{\sigma}_1,\vec{\sigma}_2,\vec{\tau}_1,\vec{\tau}_2)$, since the invariance under time translation requires that t does not appear explicitly.

Charge independence means that the Hamiltonian is invariant under rotation in the isospin space. Since any polynomial expression in $\vec{\tau}_1$ (or $\vec{\tau}_2$) can be reduced to linear expression because of the relations:

$$\tau_1^j \tau_1^k = i \cdot \varepsilon_{jkl} \tau_1^l \quad (j \neq k)$$

$$(\tau_1^j)^2 = 1 , \tag{34}$$

the only scalars in isospin space are 1 and $\vec{\tau}_1\vec{\tau}_2$. (Specifically, $(\vec{\tau}_1\vec{\tau}_2)^2 = 3-3(\vec{\tau}_1\vec{\tau}_2)$). Thus, the most general NN potential is reduced to two terms

$$V_1(\vec{r},\vec{R},\vec{p},\vec{P},\vec{\sigma}_1,\vec{\sigma}_2) + \vec{\tau}_1\vec{\tau}_2 V_2 (\vec{r},\vec{R},\vec{p},\vec{P},\vec{\sigma}_1\vec{\sigma}_2).$$

Since all other symmetry transformations commute with the isospin operator,we will consider the implication of invariance on each term V_i separately.

Translational invariance implies that V_i do not depend on $\vec{R}$ and Galileian invariance that they do not depend on $\vec{P}$.

V_i can contain only scalars because of rotational invariance and parity conservation.

V_i can be written as a sum of a $\vec{\sigma}$-independent term V_{io}, a term V_{i1} linear in $\vec{\sigma}_1$ and $\vec{\sigma}_2$ and a term V_{i2} linear in $\vec{\sigma}_1$ and $\vec{\sigma}_2$. Since for σ there are relations identical to (34), there is no term linear in $\vec{\sigma}_1$ (or $\vec{\sigma}_2$).

The term V_{io} can depend on $\vec{r}$ and $\vec{p}$. However, rotational invariance requires that it must be a function of r^2, p^2, $(\vec{r} \times \vec{p}) = L^2$ and $\vec{p}\vec{r} + \vec{r}\vec{p}$. Time-reversal invariance implies that the last term must appear as $(\vec{p}\vec{r} + \vec{r}\vec{p})^2$. The relation

$$(\vec{p}\vec{r} + \vec{r}\vec{p})^2 = 2(r^2p^2 + p^2r^2) - 4L^2 + 3$$

shows that $\vec{p}\vec{r} + \vec{r}\vec{p}$ depend on r^2, p^2 and L^2 and therefore $V_{io} = V_{io}(r^2,p^2,L^2)$

We will show now that the p^2 dependence is redundant for describing on-shell NN scattering: The potential V_{io} contains terms as $f(r^2)$, $p^2g(r^2) + g(r^2)p^2$ and $L^2h(r^2)+h(r^2)L^2$. Their matrix elements between plane waves $e^{i\vec{k}_i\vec{r}}$ and $e^{i\vec{k}_f\vec{r}}$ introducing $\vec{q} = \vec{k}_i - \vec{k}_f$ and $\vec{K} = \vec{k}_i + \vec{k}_f$ are

$$f(q);\ (k_f^2 + k_i^2)g(q) = \frac{1}{2}(q^2 + K^2)g(q) \text{ and}$$

$$(k_i \times k_f)^2h(q) = \frac{1}{4}(\vec{q} + \vec{K})\times(\vec{K} - \vec{q})h(q) = \frac{1}{4}[q^2K^2-(\vec{q}\vec{K})^2]h(q). \tag{35}$$

For on-shell NN scattering $\vec{q}\,\vec{K} = k_i^2 - k_f^2 = 0$ and three terms in (35) are not independent. Thus, it is possible to express the p^2 dependence in terms of r^2 and L^2 dependence. Off the energy shell $k_i^2 \neq k_f^2$ and $q\,K \neq 0$ and all three terms are needed. This also shows that the elastic NN scattering cannot determine off-shell NN force.

The term linear in $\vec{\sigma}_1$ and $\vec{\sigma}_2$ must contain them in the symmetric combination $\vec{\sigma}_1 + \vec{\sigma}_2 = \vec{S}$, and since $\vec{S}\vec{r}$ and $\vec{S}\vec{p}$ violate parity, the only allowed term is $\vec{L}\,\vec{S}$:

$$V_{i1} = \vec{L}\,\vec{S}\,V_{i1}\,(r^2,L^2) \ : \text{spin orbit force}$$

The bilinear combinations are: $\vec{\sigma}_1\vec{\sigma}_2$, $(\vec{\sigma}_1\vec{r})(\vec{\sigma}_2\vec{r})$, $(\vec{\sigma}_1\vec{p})(\vec{\sigma}_2\vec{p})$ and $(\vec{\sigma}_1\vec{L})(\vec{\sigma}_2\vec{L}) + (\vec{\sigma}_2\vec{L})(\vec{\sigma}_1\vec{L})$, the last one is called a quadratic spin-orbit operator $= 2Q_{12}$. Again, on the energy shell four terms are not independent, but off the energy shell all four terms are independent. The term $(\vec{\sigma}_1\vec{r})(\vec{\sigma}_2\vec{r})$ can be expressed through the tensor-force operator S_{12}.

The most general NN potential is thus:

$$V = (V_{10}(r^2,p^2,L^2) + \vec{L}\,\vec{S}\;V_{11}(r^2,p^2,L^2) + S_{12}V_{12}(r^2,p^2,L^2)$$
$$+ (\vec{\sigma}_1\vec{\sigma}_2)V_{13}(r^2,p^2,L^2) + Q_{12}V_{14}(r^2,p^2,L^2) +$$
$$+ (\vec{\sigma}_1\vec{p})(\vec{\sigma}_2\vec{p})V_{15}(r^2,p^2,L^2) + \vec{\tau}_1\vec{\tau}_2\;V_{20}(r^2,p^2,L^2) +$$
$$+ (\vec{\tau}_1\vec{\tau}_2)\vec{L}\,\vec{S}\;V_{21}(r^2,p^2,L^2) + (\vec{\tau}_1\vec{\tau}_2)S_{12}V_{22}(r^2,p^2,L^2) +$$
$$+ (\vec{\tau}_1\vec{\tau}_2)(\vec{\sigma}_1\vec{\sigma}_2)V_{23}(r^2,p^2,L^2) + (\vec{\tau}_1\vec{\tau}_2)Q_{12}V_{24}(r^2,p^2,L^2) +$$
$$+ (\vec{\tau}_1\vec{\tau}_2)(\vec{\sigma}_1p)(\vec{\sigma}_2p)V_{25}(r^2,p^2,L^2). \qquad (36)$$

On the energy shell p^2 dependence in V_{ij} and terms V_{15} and V_{25} are redundant.

There is strong experimental evidence for each term in equ.(36):

i) deuteron quadrupole moment requires the tensor force;

ii) difference between singlet and triplet scattering call for the spin-spin force;

iii) the explanation of ^1S_o, $^1D_2$, $^1G_4$ phase shifts needs a quadratic spin-orbit term.

iv) experimentally observed polarization cannot be explained just in terms of central + tensor potentials. The effect of the $\vec{L}\,\vec{S}$ potential can be easily visualized in Fig.13;while a repulsive potential has a defocussing and an attractive one a focussing effect, a spin-orbit potential acting on spin-up particles is attractive on one side and repulsive on the other because of the change of sign of the orbital angular momentum. While spin-up particles are preferentially scattered to one side, spin-down particles are scattered to the other.

Therefore, we conclude that the NN potential is as complex as the basic symmetry considerations allow.

We will say that the NN potential is realistic, if it provides a satisfactory fit to the existing NN scattering data up to about 1 GeV and to the properties of the deuteron. The history of realistic NN potentials starts in 1957, when the extensive measurements of pp polarization and Wolfenstein parameters at 310 MeV allowed a first successful phase-shift analysis. Gammel, Christian, and Thaler tried but failed to fit NN data with a static local potential = central + tensor. In 1958 Gammel and Thaler and independently Signell and Marshak added a spin-orbit force and introduced the hard core. Subsequently several potentials were developed. Hamada-Johnston and Yale have a hard core and the pathological change in the potential around the hard-core radius. This called for modification. Soft core of the Yukawa type and finite core were introduced; e.g.

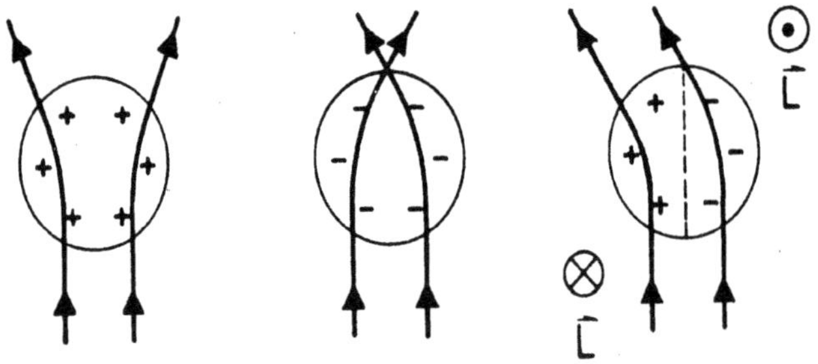

Fig. 13.: The defocussing effect of a repulsive potential, the focussing effect of an attractive potential, and the effect of a spin-orbit potential.

Reid soft-core, Bressel-Kerman-Rouben finite-core, Sprung's super-soft-core. Some of these potentials were constructed in such a way as to facilitate calculation and to some of them the primary aim was to fit nuclear matter and nuclear data, e.g. Baranger et al. potential. A particularly useful potential is a separable nonlocal potential $V(r,r') = V_0 g(r)g(r')$ or in the momentum representation $\lambda g(p)g(p')$. Most of the calculations of the three particle-problem using the Faddeev framework have been done with separable nonlocal potentials. The validity of the assumption of locality of nuclear force depends to what extent the relative coordinate of the nucleons does not change, while the interaction takes place. Since the nuclear force is mediated by exchanged hadrons, for large r the form is local because the mass of the exchanged pion is much smaller than the mass of the nucleon. However, at small r the interaction becomes nonlocal, since the mass of the nucleon is comparable to the mass of exchanged hadrons. Simpler older forms of the separable nonlocal potential $V_0 g(r)g(r')$ become more nonlocal at large r: $\int d^3r' V_0 g(r)g(r')\psi(r') = V_0 g(r)\int d^3r' g(r')\psi(r')$ and the integral has its major contribution around $r'=b$ (range of force) which is more different from r, the larger r is. Fairly sophisticated separable potentials have been developed, notably Graz potentials and Doleschall's potential. These potentials contain several terms adequately describing all the features of the potential (36). Fig. 14 shows matrix elements for two Doleschall's potential. Which differ on- and off-shell.

Our understanding of hadronic structure and interaction has significantly advanced. This knowledge can be and should be included in the formulation of NN potentials. Even some older phenomenological potentials have an OPE tail. Similarly, the understanding of the exchange of uncorrelated and correlated pions has to be incorporated

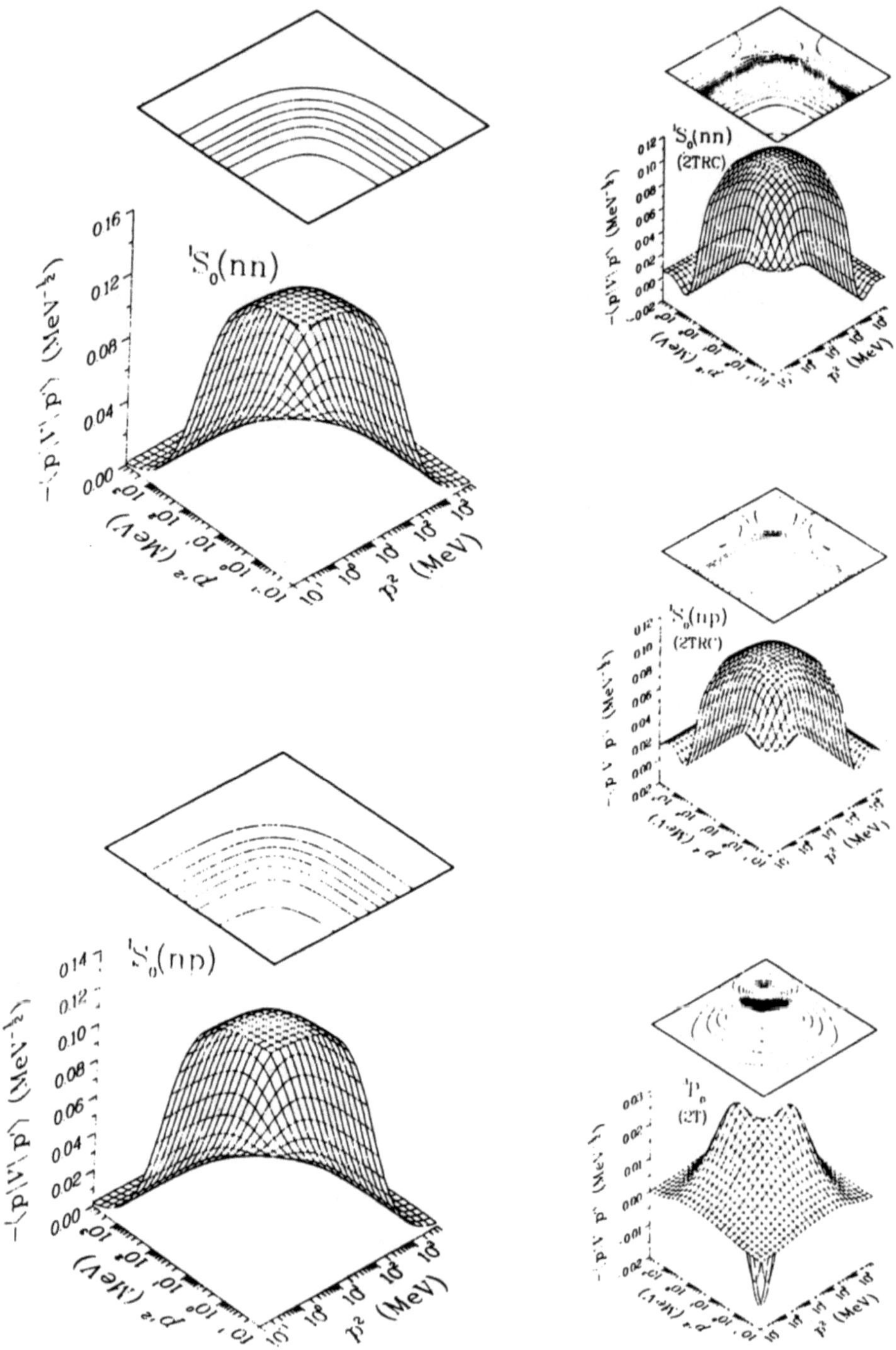

Fig. 14.: Matrix elements for two Doleschall's potentials: ^1S_o(nn) and ^1S_o(np) (Fig. 14a) and 2TRC (Fig.14b).

in the realistic potentials. Such potentials are based on more sound theoretical grounds, though they still contain some phenomenological parameters, e.g. parameters describing the radius characterizing the domain, where subhadronic degrees of freedom: quarks and gluons dominate.

Among these theoretically-based phenomenological potentials we will briefly discuss one-boson-exchange potential (OBEP) [79,80], boundary-condition model [81] and Paris potential [82].

A) OBEP is a generalization of OPEP. It assumes that the interaction is represented by the exchange of various bosons ("resonances"): $\rho,\omega,\eta,\phi,\delta$ and σ. The inputs are the masses of these bosons and their coupling constants to nucleons. Some of these parameters are fixed by other than NN scattering experiments. The shortcomings of the OBEP are [85]:

i) all exchanged particles are considered stable, i.e. zero width. Experimental widths are of the order of 100 MeV;

ii) need of a scalar-isoscalar sigma meson which has not been found experimentally. The mass of sigma is 500 MeV, while there is evidence for a 600 MeV broad resonance at 1 GeV;

iii) it neglects two-pion exchange (TPE) and consequently there is no natural way for excitation of isobars which are obviously very important.

Two outstanding OBEP are Bonn [79] and Nijmegen [80] potentials.

B) Boundary-condition-model of Lomon and Feshbach [81] applies a separate energy-independent logarithmic derivative to each partial-wave radial function at a boundary r_B common to all waves. The wave function inside r_B is taken to be zero. The potential outside r_B contains central and tensor OBE and TPE arguing that $\vec{L}\vec{S}$ and Q_{12} are contained in the boundary condition.

C) Paris potential [82] describes the long and intermediate domain ($r > 0.8$ fm) by OPE, TPE and the ω and A_1 exchange. The TPE is calculated via dispersion relations. It automatically includes all πN isobars and $\pi\pi$ resonances in the S (ε meson) and P waves (ρ meson).

The comparisons of various potentials to fixed sets of data are given in Tables 9-11.

These best realistic NN potentials still fail to predict properties of nuclear matter and light nuclei correctly (see Table 12 and Fig.15): ^3H and ^4He are underbound, ^3He and ^4He charge form factors are not correctly described and theoretical predictions for nuclear matter and ^{16}O form bands which miss the experimental values (potentials in Fig.15 are HJ = Hamada - Johnston, MRSC = modified Reid soft-core, MRBJ = modified Bethe-Johnston, UG = Ueda-Green, BS = Bryan-Scott, LP = Lagaris - Pandharipande, W = Wiringa, SSC = supersoft-core de Tourreil - Sprung, CPW = Carlson et al. with a three-nucleon force). It seems that most of these observables can be described much better, if in addition to the best two-nucleon force one includes also three-nucleon forces and meson-exchange currents. Calculations have been done [84] in a relativistic BHF. The inclusion of the Δ produces saturation points A', B' and C'. The potential A with the Δ, labelled A' gives a good fit [84] to the experimental value. Since in the TPE nucleon force the $\Delta(1232)$ isobar plays an important role and since Δ is in general important for TPE processes, a phenomenological potential is constructed with explicit $\Delta(1232)$ degrees of freedom. Argonne group has formulated two phenomenological potentials [85] v_{14} and v_{28}.

The potential v_{14} is similar to the Urbana potential [86] and it is a conventional NN potential with 14 terms: 8 are of the same form as in equ.(3) and instead of Q_{12} and $(\vec{\sigma}\vec{p})$ terms they have $\vec{L}^2, \vec{L}^2(\vec{\tau}_1\vec{\tau}_2), L^2(\vec{\sigma}_1\vec{\sigma}_2), \vec{L}^2(\vec{\sigma}_1\vec{\sigma}_2)(\vec{\tau}_1\vec{\tau}_2)$,

Table 9: Potentials compared with the same 648 pp data from 10-330 MeV [7]

Potential	χ^2/data	r_o(fm)
Brueckner, Gammel, Thaler	106	0.4
Hamada, Johnston	2.98	0.4855
Reid hard-core	2.72	0.423
Bressel, Kerman, Rouben	2.13	0.7
Boundary Condition	2.69	0.75

Table 10: Potentials compared to the phase-shift analyses [8]

Potential	Bonn	Nijmegen	Reid soft-core	Paris
χ^2/data	2.77	2.17	2.4	2.5 pp 2.7 np

Table 11

	χ^2/data pp	χ^2/data np
Reid soft-core	4.26	9.99
Paris	1.99	2.17
Arndt phase shift	1.33	1.80

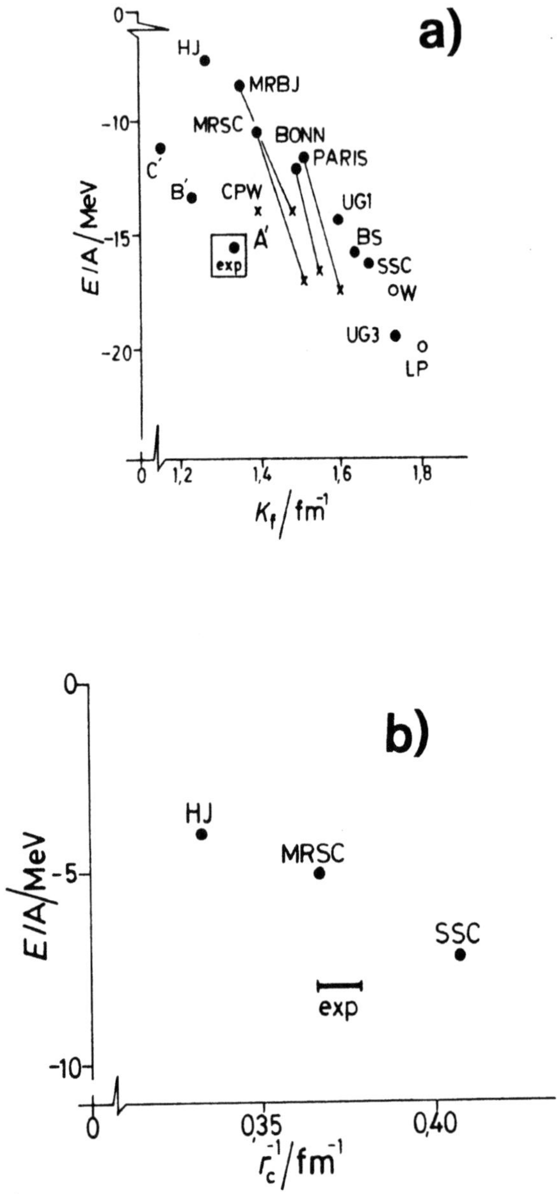

Fig. 15.: Theoretical predictions and the experimental value for the nuclear-matter binding energy E/A and density $\rho = 2k_F^3/3\pi^2$. Potentials are described in the text. The arrows show the shift when higher-order coupled-cluster terms are included (a); and for ^{16}O (b).

Table 12: V_3 is the three-body force: Tucson potential is used for the long and intermediate range and the short range is phenomenological repulsion.

		V_2^{LP}(Urbana)	V_2(Paris)	$V_2^{LP}+V_3$	Exp.
^3H	E(MeV)	-7	-7.4	-8.1	-8.48
	r_c(fm)	1.85	2.02	1.76	1.70±0.05
^4He	E(MeV)	-22.9		-29.1	-28.3
	r_c(fm)	2.63		1.64	1.67±0.03
NM	E/A(MeV)	-20.0		-14.4	-15.5±0.5
	k_F(fm^{-1})	1.78		1.4	1.33±0.04

Table 13

	v_{14}	v_{28}	Paris	Experimental
E_d (MeV)	-2.225	-2.225	-2.2249	-2.22463(3)
Q_d (fm^2)	0.286	0.286	0.279	0.2860(15)
η_d=D/S asympt.	0.0266	0.0265	0.0261	0.0265(5)
μ_d (μ_N)	0.845	0.846	0.853	0.857441(2)
P_D (%)	6.08	6.13	5.77	4-7
P_Δ (%)	0	0.52	0	?

$(\vec{L}\vec{S})^2$, and $(\vec{L}\vec{S})^2(\vec{\tau}_1\vec{\tau}_2)$. The radial dependence is represented by the long-range terms for the spin-spin and tensor potentials, which are

$$\frac{f^2_{\pi NN}}{4\pi}\frac{m_\pi}{3}\frac{e^{-\mu r}}{\mu r}\quad(1 - e^{-cr^2})\ \text{and}$$

$$\frac{f^2_{\pi NN}}{4}\frac{m_\pi}{3}\ (1 + \frac{3}{\mu r} + \frac{3}{(\mu r)^2})\ \frac{e^{-\mu r}}{\mu r}\ (1 - e^{-cr^2})^2 = 3.7268\ T_\pi(r)$$

The smooth Gaussian cutoffs make them vanish at $r = 0$. The intermediate parts are assumed to come from TPE and their radial form is $T^2_\pi(r)$, while the short range is given by a Wood-Saxon shape. The coefficients multiplying intermediate and short range radial function are to be determined by fitting the data. Thus, there are 14 x 2 = 28 free phenomenological parameters. There are additional parameters, i.e. c and Wood-Saxon parameters R and a, but they are kept fixed.

The potential v_{28} has 14 additional operators including transition operators for all possible $\pi N\Delta$ and $\pi\Delta\Delta$ couplings.

The comparison between the SAID 1981 data (solid curves and points) to Paris potential (crosses), v_{14} (dashed-dot), and v_{28} (dashed curves) is shown in Fig. 16. The calculated properties of the deuteron ground state are given in Table 13. The total Δ-state probability P_Δ breaks into $P_\Delta(^3S_1) = 0.04\%$, $P_\Delta(^3D_1) = 0.02\%$, $P_\Delta(^7D_1) =$ $= 0.42\ \%$ and $P_\Delta(^7G_1) = 0.04\ \%$. The deuteron wave functions for v_{28} (solid), v_{14} (dashed), and Paris potentials (dotted) are shown in Fig. 17, where u_{LS} and $\tilde{u}_{LS}$ denote the NN and $\Delta\Delta$ components of the wave function. The agreement between all potentials and the data is remarkably good. The only exception is μ_d for v_{14} and v_{28} but relativistic and meson

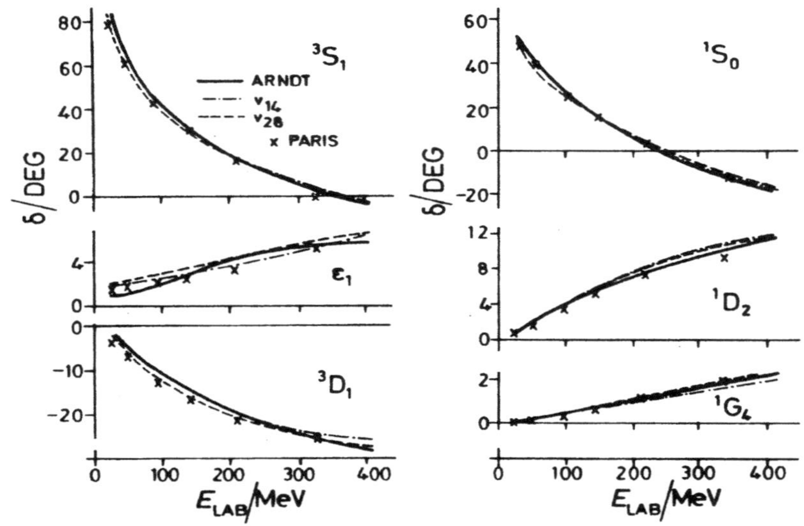

Fig. 16.: Comparison between the phase-shift analysis [21] (solid), v_{14} (dash-dotted), v_{28} (dashed) [86] and Paris [82] potentials predictions.

exchange current effects might influence all observables.

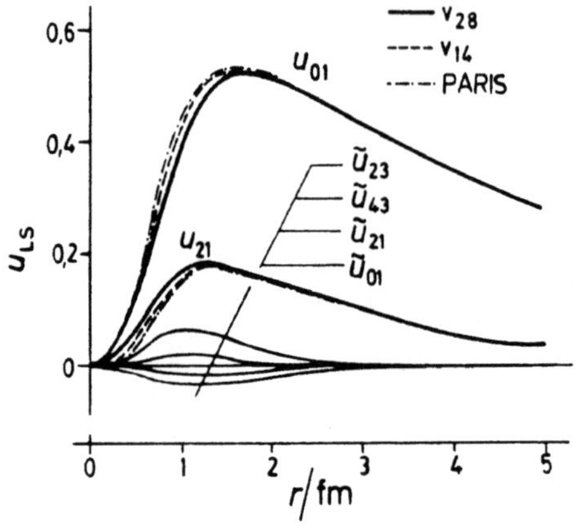

Fig. 17.: The deuteron wave functions for v_{28} (solid), v_{14} (dashed), and Paris (dash-dotted) potentials, where u_{LS} and $\tilde{u}_{LS}$ denote the NN and ΔΔ components [86].

4.1 Off-the-Energy Shell Interaction

The two-particle T matrix is given by the equation

$$T(Z) = V + V \frac{1}{Z-H} V ,$$

where H is the Hamiltonian and Z ist the energy. Three kinds of matrix elements can be distinguished:

$\langle k' | T(Z) | k \rangle$ - fully off-shell

$\langle k' | T(k^2) | k \rangle$ - half off-shell

$$\langle k|T(k^2)|k\rangle \text{ - on-shell.} \quad (37)$$

The elastic scattering process lasts a very long time and therefore the energy Z must be equal to the initial and final states energies:

$$k_{in}^2 = Z = k_{fin}^2 .$$

The elastic scattering gives information only about on-energy-shell matrix elements.
If the interaction is a part of a larger diagram (as in NN Bremsstrahlung, Fig. 18) then it occurs within a finite

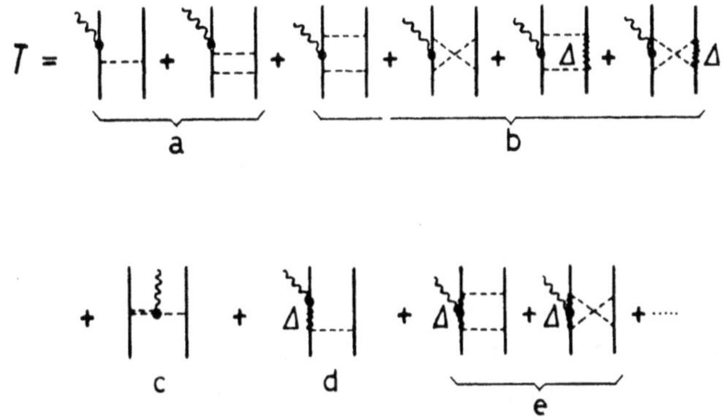

Fig. 18.: Diagrams representing NN Bremsstrahlung: a) external, b) rescattering, c) meson current, d) radiative isobar decay, e) isobar current

time, the energy does not have to be conserved and T matrix elements are off-shell. In all systems composed of more than two nucleons, there are off-shell T matrix elements.

The knowledge of phase parameters $<k|T(k^2)|k>$ does not fully determine the potential. Actually, there is an infinite set of potentials which are different at short distance and still produce identical phase parameters. Such potentials are called phase-equivalent potentials and they can be generated by a unitary transformation U. For example:

$$<r|U|r'> = \delta(r-r') - 2 <r|\Lambda|r'>$$

Since $U^+U = 1$, it follows $\Lambda^+ + \Lambda = 2\Lambda^+\Lambda$

and the transformed $\tilde{T}$ is related to the original T through

$$<k'|\tilde{T}(k^2+i0)|k> = <k'|T(k^2 + i0)|k> + 2(k'^2 - k^2)<k'|\Lambda|\psi_k^+>$$

and it is obvious that $\tilde{T}$ and T are on-shell equivalent.

The off-shell M matrix differs from the on-shell in two ways [87]: i) each term depends on an addditional scalar variable for each off-shell particle and ii) though in the nonrelativistic scattering theory approach the off-shell M matrix, which is only Lorentz invariant contains the same number of terms as the on-shell M matrix, the imposition of TRI imposes no condition on the off-shell M matrix, while it constrains the on-shell M matrix.

It is not possible to phenomenologically determine the off-shell amplitude embedded in a larger process in a model-independent way. Our knowledge of the embedding process is more reliable, if the remaining part is due to soft-photon emission or weak interaction, but then the off-shell effects appear in general in the third-order term [87].

Though information about off-energy-shell interaction can be in principle obtained from any more complex nuclear system, it is necessary to choose a system as simple as possible so that other effects, e.g. three-nucleon force, do not interfere and so that the theoretical framework is adequately reliable. We will briefly summarize three processes:

4.1.1 NN Bremsstrahlung

The process is described by diagrams shown in Fig.18. For low photon energies k the T matrix can be explained as:

$$T = \frac{A}{k} + B + Ck + Dk^2 + \ldots$$

Amplitudes A and B contain only on-shell information and it is necessary to measure the Bremsstrahlung at kinematical conditions corresponding to relatively large k, so that at least the term Ck is significant. The results of the pp Bremsstrahlung measurement at 200 MeV with two photons detected at symmetric angles of 16.3^o is shown in Fig.19 together with the predictions of the theoretical calculations using the BKR potential and the phase-equivalent potentials labeled by α= 12 and 14 appreciably differ for r <1 fm, the difference in predicted cross section is comparable to the experimental errors. It is disturbing that the soft-photon approximation, SPA: $T = \frac{A}{k} + B$ adequately fits the data. Fortunately, the measurement of A_y at 10° would discriminate between phase-equivalent potentials.

So far no information about off-energy-shell interaction has been obtained from pp or np Bremsstrahlung studies.

4.1.2 Electron-Deuteron Tensor Polarization [88]

Electron-deuteron elastic scattering observables depend on the charge $F_C(q)$, quadrupole $F_Q(q)$ and magnetic $F_M(q)$ structure functions of the deuterons. These functions depend

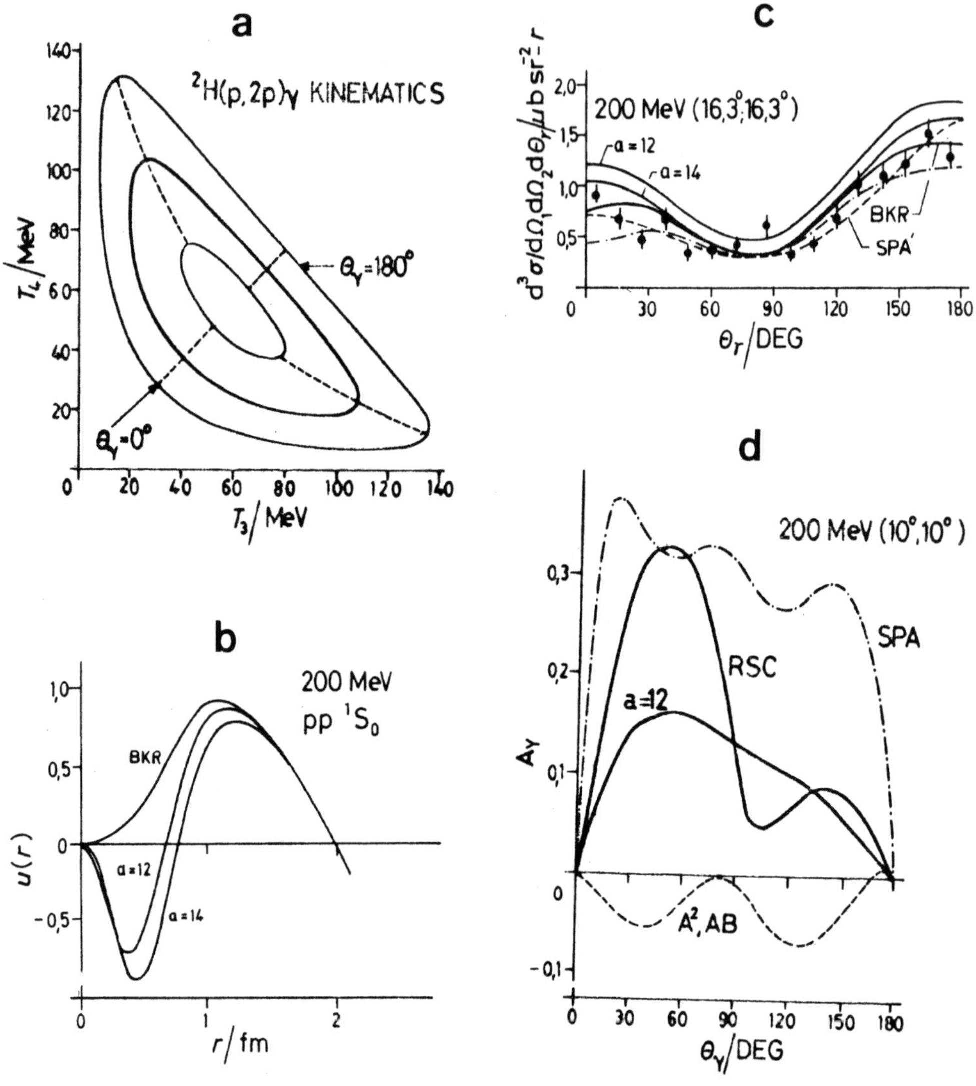

Fig. 19.: a) Kinematics of the ppγ process at T_p = 200 MeV, $\theta_3 = \theta_4 = 16.3^o$ taking into account finite geometry; b) ^1S_o wave function for BKR potential and two phase-equivalent potentials; c) ppγ data and calculations using BKR and two phase-equivalent potentials (α= 12 and 14, respectively) as well as the SPA; d) predictions for A_y for pp at 200 MeV, $\theta_3 = \theta_4 = 10^o$ for RSC potential and a phase-equivalent potential (α=12), together with the SPA and the calculation taking only A^2 and AB terms.

on the S and D deuteron wave functions. The calculation of e-d elastic scattering observables has been done for several potentials that differ in their tensor force and in their off-shell properties. The calculations include the pair-meson exchange-current corrections to the impulse approximation. Fig.20 shows that potentials with different off-shell properties (Paris, Bonn, Nijmegen, Reid as opposed to Graz and Doleschall's) have different F_C and give different t_{20}. Also potentials with different tensor forces (Paris vs. Graz) give different F_Q and t_{21}. The measurement of t_{20} was performed [89]at Bates linear accelerator center at two values of momentum transfer $q = 1.74\ fm^{-1}$ (incident electron energy $E = 310$ MeV) and $2.03\ fm^{-1}$ ($E = 371$ MeV). The polarimeter is the same one that has been used for the measurement of $\pi d\ t_{20}$ at LAMPF. The measurement (Fig.21) clearly favours Paris potential over Graz.

4.1.3 Three-Nucleon Observables

Observables dominated by S wave interaction provide a test on the off-shell force, since the interior region is better probed by S waves. Aside from the differential cross section only second-order polarizations, e.g. $K_y^{y'}$ and $C_{y'y}$ do not vanish, if only S wave interactions are used [90]. The experimental $K_y^{y'}$ at 10 MeV [91] has been compared with the predictions of two Doleschall's potential and evidence for off-shell effects is claimed. However, the two potentials differ both on and off-shell. Calculation [92] of $K_y^{y'}$ has also been done using PEST1 and phase-shift equivalent potential Y(E) (Fig.22). The potential PEST1 is a separable representation of the Paris potential using the Ernst-Shakin-Thaler method. PEST potentials also reproduce the Paris potential Noyes-Kowalski half-shell function

$$f(p,k) = \frac{T(p,k,k^2+io)}{T(k,k,k^2+io)} .$$

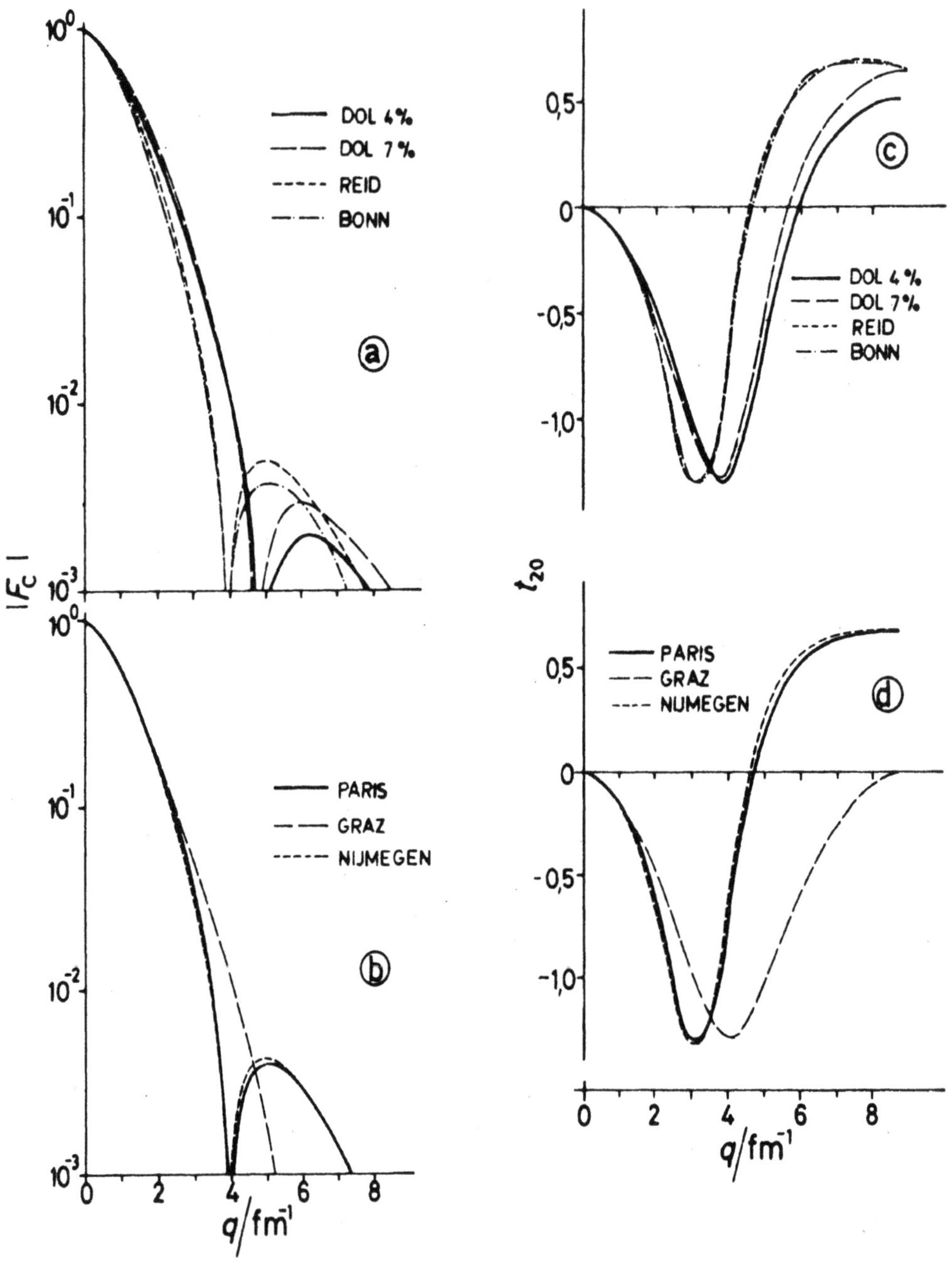

Fig. 20.: Deuteron charge structure function $|F_C(q)|$ and t_{20} for several NN potentials [88].

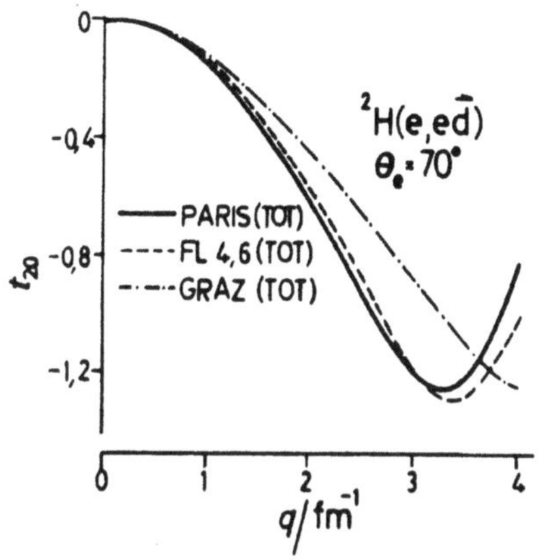

Fig. 21.: The results of the measurement [89] of t_{20} compared with predictions for various potentials.

The potential Y(E) is the Yamaguchi-type equivalent on-shell to PEST 1. Off-shell they differ as shown in Fig.22. The dashed-dotted curve in Fig.22 is the PEST1 including approximate Coulomb correction. The data favour PEST over Y(E).

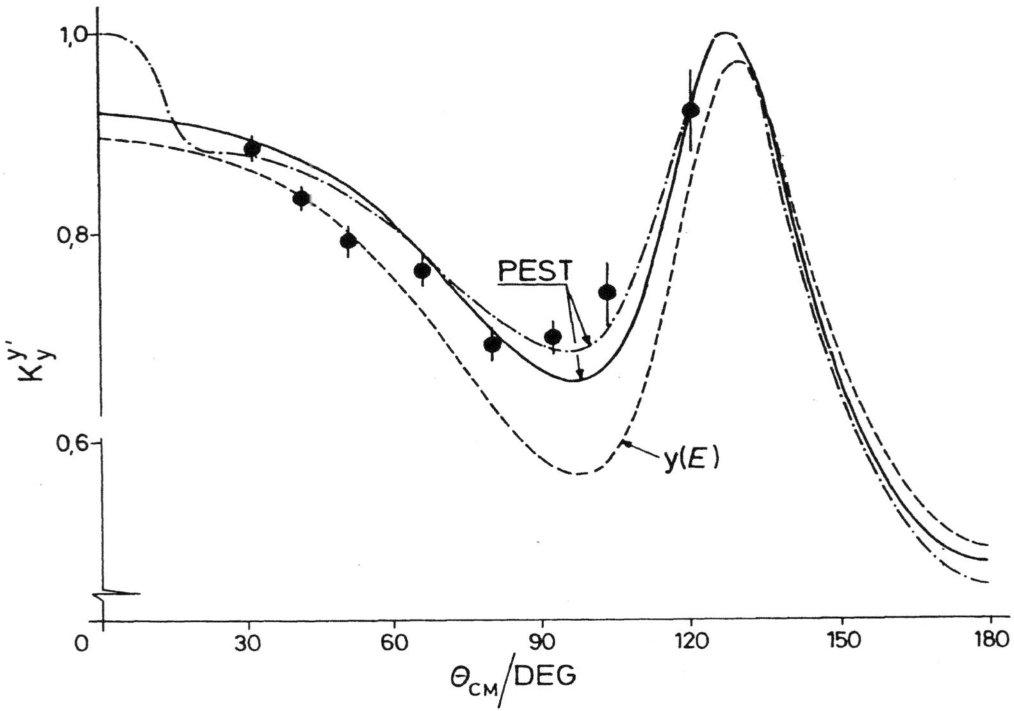

Fig. 22.: The prediction for $K_y^{y'}$ using the separable representation of the Paris potential (PEST) with (dash-dotted) and without (solid) approximate Coulomb correction.

5. DIBARYONS

All NN phase-shift analyses [21-23,93,94] agree that the $^1D_2$ partial wave amplitude has a counterclockwise behaviour in the Argand plot thus, indicating a resonance at 2.15 GeV. Evidence for a $^3F_3$ resonance is found in the work of Arndt [21] and Hoshizaki [93], but not in that of Dubois [22] and Bystricki [94]. The observables $\Delta\sigma_L$, $\Delta\sigma_T$ and A_{NN} show structure around 800 MeV (Figs. 2 and 23). The suggested dibaryon candidates (Table 14) are strongly inelastic and there should be evidence for them in pionic channels:

i) pp→NN channel: successful fit to the $\Delta\sigma_L$ data can be obtained only by including $^3F_3$ resonance (Fig. 24). However, the same success is not achieved for A_y even when including both $^1D_2$ and $^3F_3$ resonances [98].

ii) πd elastic channel: though the cross section is a smooth function of energy and angle, all theories have been unable to fit the large-angle 256 and 292 MeV data [96]. Faddeev calculation of iT_{11} shows a peak shift of about 15°. Including $^1D_2$ and $^1G_4$ resonances the theory fits the data (Fig. 25) [99]. It seems that the two measurements [100,101] of T_{20} are in disagreement. Specially, in one measurement [100] there is a narrow ∿10 MeV structure at 134 MeV in the excitation curve of T_{20} at θ_{cm} = 150°. (Fig.26) Also the angular distribution at 134 MeV shows pronounced oscillations. Three other experiments also report a narrow structures:

a) p(^{3}He,d)X and ^{3}He(p,d)X missing-mass spectra show [102] a 15-18 MeV wide resonance at M = 2.32-2.45 GeV.

b) the pp missing-mass spectrum from the reaction γd→ppπ reveals [103] a 15 MeV wide structure at M = 2.02 GeV.

c) 50 MeV wide resonances have been observed at 2.0 to 2.2 GeV in the reactions pd→pnp and pd→pnnπ$^+$ [104]. However, the same experiment done at KEK does not confirm this result [105].

States with narrow width are rather unlikely in a conventional nucleons, deltas and mesons picture [96]. However, several QCD models [106] predict fairly narrow six-quarks states particularly interesting because of their exotic origin (hidden color).

Measurements of A_{LL} and A_{LS} for $\vec{p}\vec{p}\rightarrow\pi d$ between 500 and 800 MeV appreciably disagree with the theoretical calculation, but the explicit inclusion of dibaryons did not fit the data [111].

Table 14: First experimental evidence for dibaryons was found in a Λp invariant mass [108], first theoretical discussion done by Oakes [109]. Besides those dibaryon listed below there are possible resonances in other states, e.g. in ^3P_o. In addition there are strange-dibaryon states: S= -1 in $^3S_1$, an SU(3) partner of the deuteron[110]:

	$^1D_2$	$^3F_3$	$^1G_4$
Mass (MeV)	2120 - 2190	2155 - 2390	-2480
Γ (MeV)	50 - 140	50 - 210	- 150

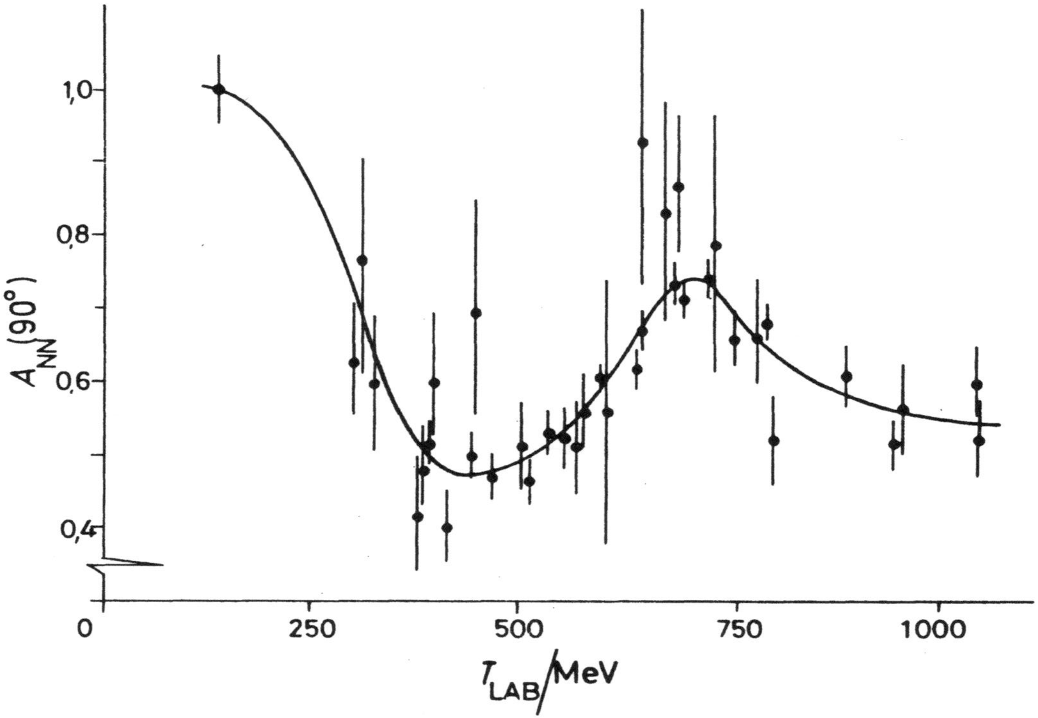

Fig. 23.: The compilation of A_{NN} (90°) measurements [95,96].

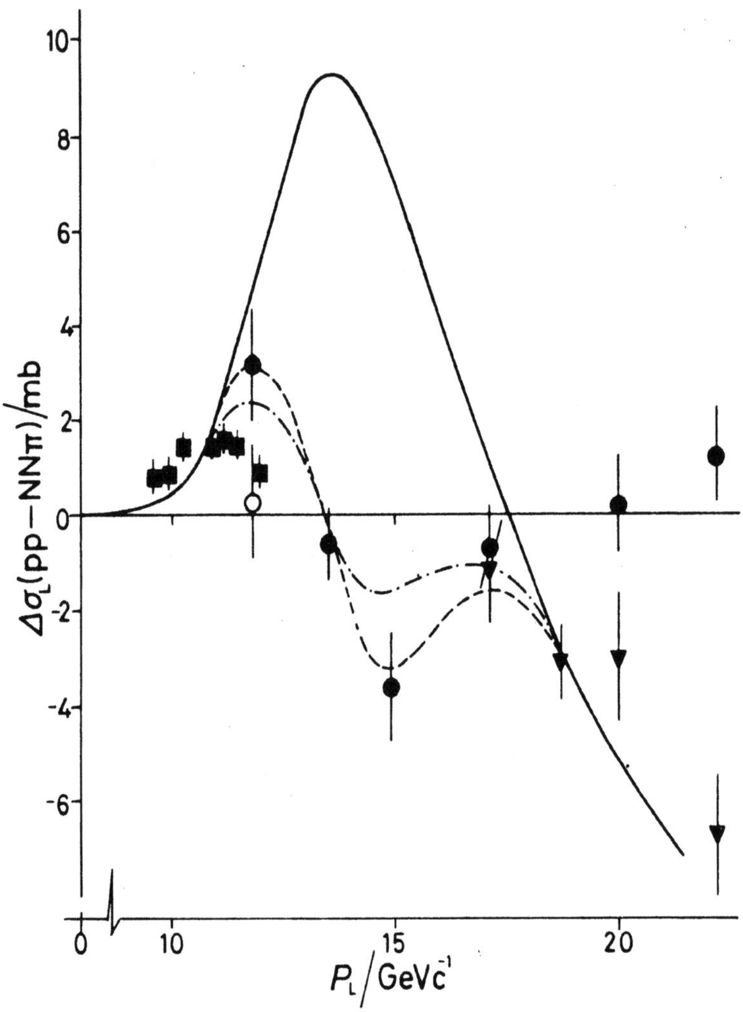

Fig. 24.: The $\Delta\sigma_L$: data and Deck model prediction with I=1, JP = 3^- admixture [96,97]; M_R = 2236 MeV, Γ_{tot} = 120 MeV, Γ_{el} = 26 MeV, β = -60° (dashed) and M_R = 2230 MeV, Γ_{tot} = 150 MeV, Γ_{el} = 30 MeV, β = 135 (dash-dotted).

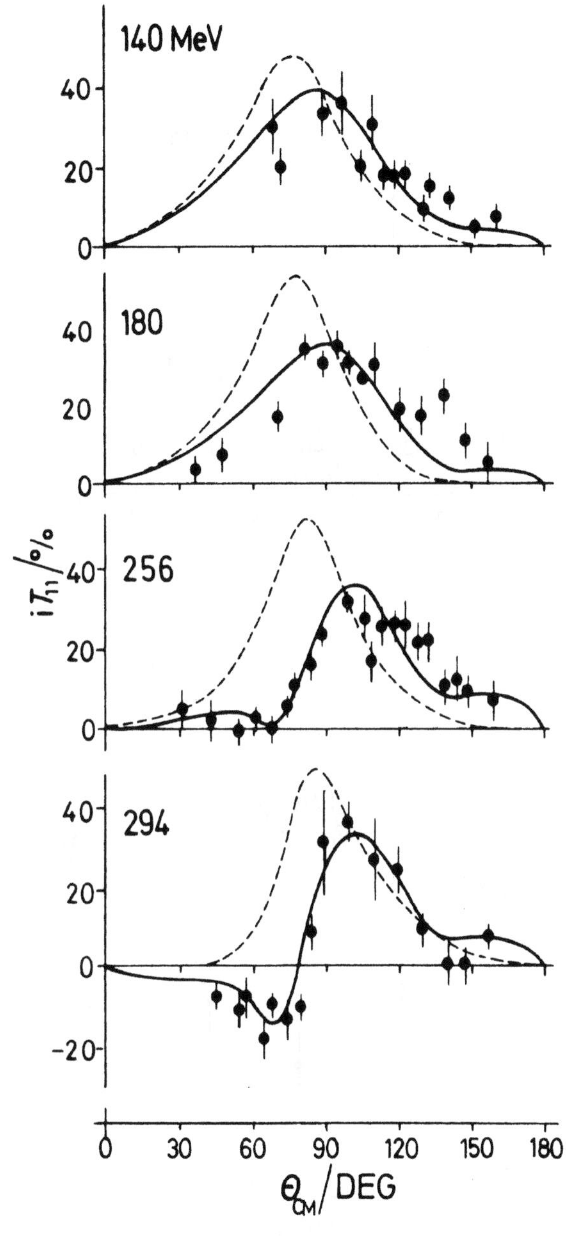

Fig. 25.: Vector-analyzing power in the πd elastic scattering from incident pion energy of 140 MeV to 294 MeV. Curves are Faddeev-type predictions without (dashed) and with (solid) $^1D_2$ and $^1G_4$ dibaryons [99].

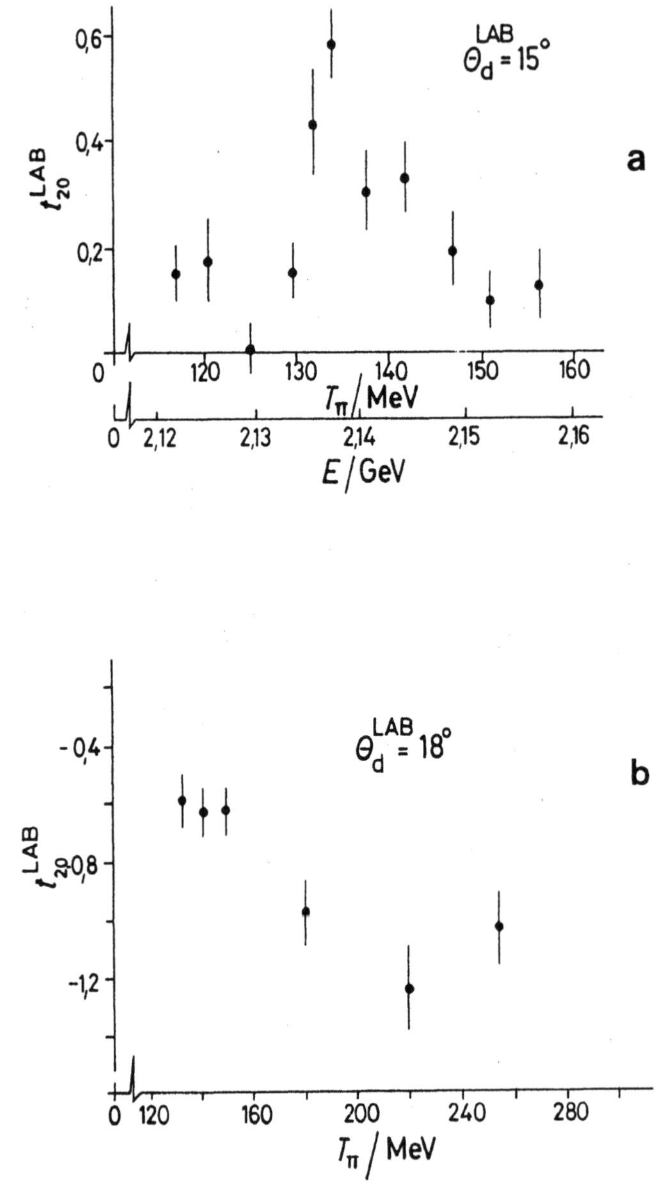

Fig. 26.: The excitation curves of the t_{20} in πd elastic scattering at $\theta_d^{LAB} = 15^o$ (a) [100] and $\theta_d^{LAB} = 18^o$ (b) [101].

REFERENCES

1. H. Yukawa, Proc. Phys. Math. Soc. 17 (1935) 48.
2. see "Mesons in Nuclei" eds. M. Rho and D.H. Wilkinson, North Holland Publ. Comp. (1979).
3. D. Koltun, Comments Nucl. Part. Phys. 11 (1983) 171.
4. H. Taketani et al., Prog. Theor. Phys. 6 (1951) 581.
5. E. Hadjimichael and D.P. Saylor, Phys. Rev. Lett. 45 (1980) 1776.
6. Faessler et al., Phys. Lett. 112B (1982) 201.
7. M. Wakamatsu et al., Phys. Lett. 146B (1984) 148 and 153.
8. E.L. Lomon, Inv. talk given at the Third LAMPF-II Workshop, July 1983.
9. T. Goldman and G.J. Stephenson Jr., Phys. Lett. 146B (1984) 143.
10. E. Eichten et al., Rev. Mod. Phys. 56 (1984) 579.
11. D. Wilkinson in "The Nature of Matter", ed. J.H. Mulvey, Clarendon Press, Oxford 1981 , 1.
12. E.P. Wigner, never published this work. It has been first mentioned in H. Bethe and R. Bacher, Rev. Mod. Phys. 8 (1936) 193.
13. J. Bystricki et al., Saclay DPhPE (1984) submitted to Phys. Lett.
14. A. Yokosawa, Phys. Report 64 (1980) 47.
15. E. Aprile et al., Phys. Rev. D28 (1983) 21.
16. R.A. Arndt and L.D. Roger, Phys. Rev. D25 (1982) 2011.
17. H.P. Stapp et al., Phys. Rev. 105 (1957) 302.
18. R. Bryan, Phys. Rev. C30 (1984) 305, ibid. C24 (1981) 2659.
19. R.F. Rodebaugh and B.E. Bonner, in Few-Body Problems in Physics, Vol. II, ed. B. Zeitnitz, NHPC (1984) 43.
20. M.J. Moravcsik, UCRL-5317-T (1958) unpubl.
21. R.A. Arndt et al., Phys. Rev. D28 (1983) 97.
22. R. Dubois et al., Nucl. Phys. A377 (1982) 554.
23. K. Hashimoto et al., Prog. Theor. Phys. 64 (1980) 1678 and 1693.
24. D.V. Bugg, in Few-Body Problems in Physics, Nucl. Phys. A416 (1984) 227c.

25. R. Jastrow, Phys. Rev. 81 (1951) 165.
26. M. Cini et al., Phys. Rev. 114 (1959) 1633; H.P. Noyes and D.Y. Wong, Phys. Rev. Lett. 3 (1959) 191.
27. L. Heller, Rev. Mod. Phys. 39 (1967) 584.
28. e.g.,P.U. Sauer, Phys. Rev. C11 (1975) 1786, but also J.M. Allen and H. Fiedeldey, Nucl. Phys. A260 (1976) 213; C.W. Wong et al., Nucl. Phys. A253 (1975) 96.
29. I. Slaus et al., Phys. Rev. Lett. 48 (1982) 993.
30. W. Dilg, Phys. Rev. C11 (1975) 103.
31. L. Mathelitsch and B.J. VerWest, preprint 1983.
32. O. Dumbrajs et al., 1982 Data Compilation, Nucl. Phys. B216 (1983) 277.
33. K.S. Krane et al., Phys. Rev. C5 (1972) 1663.
34. E.G. Adelberger, in Proc. Int. Conf. Nucl. Phys. (1983) Tipogr. Compos. Bologna, eds. P. Blasi, R.A. Ricci, p.499.
35. B. Desplanques et al., Ann. Phys. 124 (1980) 449.
36. D.E. Nagle et al., in High-Energy Spin Physics (1983) ed. G.M. Buna,AIP Conf. Proc. No.95, p.150.
37. N. Lockyer et al., Phys. Rev. Lett. 45 (1980) 1821.
38. T. Goldman and D. Preston, LA-UR 1285 (1982).
39. G. Nordulli and G. Preparata, Phys. Lett. 117B (1982) 445.
40. J.F. Donoghue and B.R. Holstein, Phys. Lett. 125B (1983) 509.
41. G. Nordulli and G. Preparata, Phys. Lett. 137B (1984) 111.
42. J.M. Potter et al., Phys. Rev. Lett 33 (1974) 1307.
43. R. Balzer et al., Phys. Rev. C30 (1984) 1409.
44. P. von Rossen et al., in Polarization Phenomena in Nucl. Phys. (1980) AIP Conf. Proc. No.69 eds. G.G. Ohlsen et al.,p.1442.
45. D.M. Tanner et al., in Proc. Int. Conf. Nucl. Phys. (1983) Tipogr. Compos. Bologna, eds. P. Blasi, R.A. Ricci, p.F1.
46. V.A. Knyazhov et al., Nucl. Phys. A417 (1984) 209, V.M. Lobashov et al., Nucl. Phys. A197 (1972) 241.

47. J.F. Cariagnac et al., Phys. Lett. 67B (1977) 148.
48. J. Lang et al., Phys. Rev. Lett. 54 (1985) 170.
49. M. Avenier et al., Phys. Lett. 137B (1984) 125.
50. C.R. Christenson et al., Phys. Rev. Lett. 13 (1964) 138.
51. J.M. Pendlebury et al., Phys. Lett. 136B (1984) 327.
52. I.S. Altarev et al., Phys. Lett. 102B (1981) 13.
53. J.L. Gimlett et al., Phys. Rev. C25 (1982) 1567.
54. A.L. Hallin et al., preprint.
55. E. Blanke et al., Phys. Rev. Lett. 51 (1983) 355.
56. R.J. Slobodrian et al., Phys. Rev. Lett. 47 (1981) 1803.
57. R.A. Hardekopf et al., Phys. Rev. C25 (1982) 1090; C.R. Meitzer et al., Phys. Rev. C30 (1984) 1105.
58. E. Aprile et al., Phys. Rev. Lett. 47 (1981) 1360.
59. T.S. Bhatia et al., Phys. Rev. Lett. 48 (1982) 227.
60. See also J. Binstock et al., Ann. Phys. 133 (1981) 355.
61. M.J. Moravcsik, Phys. Rev. Lett. 48 (1982) 1718.
62. G.R. Goldstein and M.J. Moravcsik, Phys. Rev. D25 (1982) 2934.
63. R. Newman and S. Wiesner, Phys. Rev. D14 (1976) 1.
64. V.W. Hughes et al., Phys. Rev. Lett. 4 (1960) 342.
65. R.I. Steinberg et al., Phys. Rev. D12 (1975) 2582.
66. S.C. Vaidya et al., Phys. Rev. D27 (1983) 486.
67. E.M. Henley and G.A. Miller, in Meson in Nuclei, ed. M. Rho and D.H. Wilkinson (NHPC 1979) Vol.I.
68. T.E.O. Ericson and G.A. Miller, Phys. Lett. 132B (1983) 32.
69. S.A. Coon and M.D. Scadron, Phys. Rev. C26 (1982) 2402.
70. R.A. Brandenburg et al., Nucl. Phys. A294 (1978) 305.
71. S. Shlomo and D.O. Riska, Nucl. Phys. A254 (1975) 281; C.W. Wong et al., Nucl. Phys. A280 (1976) 498; E.S. Auerbach et al., Phys. Rev. Lett. 23 (1969) 1253.
72. C.L. Hollas et al., Phys. Rev. C24 (1981) 1561.
73. E. Pedroni et al., Nucl. Phys. A300 (1978) 321, T.G. Masterson et al., Phys. Rev. C25 (1982) 2091.
74. A.S. Rinat and Y. Alexander, Nucl. Phys. A404 (1983) 467.
75. B.M.K. Nefkens et al., Phys. Rev. Lett. 52 (1984) 735.

76. Y.E. Kim, Phys. Rev. Lett. 53 (1984) 1508.
77. K. Elsener et al., preprint (1985).
78. L. Eisenbud and E.P. Wigner, Proc. Natl. Ac. Sci. US 27 (1941) 781; S. Okubo and R.E. Marshak, Ann. Phys. 4 (1968) 166.
79. K. Holinde and R. Machleidt, Nucl. Phys. A256 (1976) 479.
80. M.M. Nagels et al., Phys. Rev. D17 (1978) 768.
81. E.L. Lomon and H. Feshbach, Ann. Phys. 48 (1968) 94.
82. M. Lacombe et al., Phys. Rev. C21 (1980) 861 and references therein.
83. R. Vinh Mau, Proc. Int. Conf. Nucl. Phys. (1983), Tip. Compos. Bologna, eds. P. Blasi, R.A. Ricci) Vol.2, 61.
84. R. Brockmann, this conference.
85. R.B. Wiringa et al., Phys. Rev. C29 (1984) 1207.
86. I.E. Lagaris and V.R. Pandharipande, Nucl. Phys. A359
87. (1981) 331.
87. H. Fearing et al., Phys. Rev. D29 (1984) 2612; M.J. Moravcsik, Phys. Lett. 65B (1976) 409.
88. M.I. Haftel et al., Phys. Rev. C22 (1980) 1285.
89. M.E. Shulze et al., Phys. Rev. Lett. 52 (1984) 597.
90. D.P. Saylor and F.D. Rad, Phys. Rev. C8 (1973) 507.
91. F. Sperisen et al., Phys. Lett. 102B (1981) 9.
92. J. Haidenbauer and W. Plessas, Phys. Rev. C30 (1984) 1822; H. Zankel et al., Phys. Rev. C28 (1983) 538.
93. N. Hoshizaki et al., Prog. Theor. Phys. 61 (1979) 129.
94. J. Bystricki et al., Saclay DPhPE 82-12, revised Feb.1984.
95. I.P. Auer et al., Phys. Rev. D29 (1984) 2435.
96. A. Svarc, invited talk at PANIC 1984, SIN, PR-84-11.
97. W. Jauch et al., Wuppertal preprint WK B 84-7 (1984).
98. A.D. Hancock et al., Phys. Rev. C27 (1983) 2742, E.A. Umland et al., in Polarization Phenomena in Nucl. Physics, AIP Conf. Proc. No. 69 (1980).
99. G.R. Smith et al., Phys. Rev. C30 (1984) 980.
100. V. König et al., J. Phys. G: Nucl. Phys. 9 (1983) L211.
101. R.J. Holt, LAMPF Users Group Proc., Nov. 1983, p.76.

102. B. Tatischeff et al., Phys. Rev. Lett. 52 (1984) 2022.
103. W.J. Schwille, Pion Few-Nucleon Reaction and Dibaryons, Workshop Leibnitz-Haus, Hannover 1984.
104. T. Siemiarczuk et al., Phys. Lett. 137B (1984) 434.
105. N. Katayama et al., Univ. Tokyo preprint, UT-He-83/14 (1983).
106. A. Aerts et al., Phys. Rev. D21 (1980) 2653.
107. Particle Data Group, Rev. Mod. Phys. 56 (1984) 5288.
108. O.I. Dahl et al., Phys. Rev. Lett. 6 (1961) 142.
109. R.J. Oakes, Phys. Rev. 131 (1963) 2239.
110. R.H. Dalitz, Nucl. Phys. A354 (1981) 101c.
111. G. Glass et al., Phys. Rev. C31 (1985) 288; T. Mizutani et al., Phys. Lett. 107B (1981) 177; and H. Kamo and W. Watari, Prog. Theor. Phys. 64 (1980) 338.

Acta Physica Austriaca, Suppl. XXVII, 91–126 (1985)

SKYRME SOLITONS AND LOW-ENERGY HADRONIC PHYSICS+

by

R. VINH MAU
Division de Physique Théorique*,
Institut de Physique Nucléaire,
91406 Orsay and LPTHE,
Université Pierre et Marie Curie, T16 E1,
75230 Paris Cedex 05, France.

CONTENTS

+) Lectures given at the XXIV. Internationale Universitätswochen für Kernphysik, Schladming, Austria, February 20-March 1, 1985.

*) Laboratorie Associé au C.N.R.S.

The following notes correspond to the second part of my lectures. Written versions of the first part which deals with the NN and N$\bar{N}$ interactions can be found in:
R. Vinh Mau, Proceedings of the International Conference on Nuclear Physics, Editors P. Blasi and R.A. Ricci, Tipografia Compositori-Bologna Vol.2, p.61 (1983), AIP Conference Proceedings No.110, p.187 (1984).

1. INTRODUCTION

QCD is believed to be the theory of hadronic physics. However, one is faced with the difficulty related to the running coupling constant $\alpha(q^2)$. For large momentum transfer q, as in the region of lepton-hadron deep-inelastic scattering, $\alpha(q^2)$ is small. One can then use perturbation theory to make theoretical predictions which are in good agreement with experiments. For problems related to the structure of hadrons, on the other hand, (hadron spectroscopy, static properties of hadrons and low-energy nuclear physics in general) the length scale is typically 1 fm, corresponding to small momentum transfer q^2. The running coupling constant $\alpha(q^2)$ is then large and perturbation theory is useless. In this strong-coupling regime, non-perturbative methods are indispensable. So far, not much success in this direction has been achieved. One is then led to modelling non-perturbative QCD. For example, in bag models, non perturbative effects are simulated by the pressure on the bag, in the string models they are described by the tension, etc...

Another theoretical route was advocated by 't Hooft [1]. He proposed to look for a parameter, other than the coupling "constant" $\alpha(q^2)$, in terms of which one can perform an expansion. He was then led to the $1/N_c$ expansion, N_c being the number of colours. On can note, in passing, that there is nothing fundamental in the choice $N_c = 3$, required by phenomenologcal considerations such as the $\pi^o \to 2\gamma$ rate or the ratio $\sigma_{tot}(e^+e^-)/\sigma(e^+e^- \to \mu^+\mu^-)$. In a detailed analysis of QCD diagrams in the $1/N_c$ expansion, Witten [2] arrived at the following conjecture: in the large-N_c limit and for low energies, QCD is equivalent to effective field theories involving only MESONS. One of the important properties of QCD is its chiral symmetry. This invariance must therefore be preserved in the corresponding effective field theories.

2. THE σ MODEL

The simplest chirally invariant effective field theory is the σ model [3]. Its version <u>without fermions</u> can be considered as a candidate for the class of effective field theories envisaged by Witten. The σ model without fermions can be formulated in the following way. Consider a 2x2 matrix field:

$$U(x) = \frac{1}{f_\pi}\begin{bmatrix} \sigma + i\pi_3 & i\pi_1 + \pi_2 \\ i\pi_1 - \pi_2 & \sigma - i\pi_3 \end{bmatrix} = \frac{1}{f_\pi}[\sigma(x) + i\vec{\tau}.\vec{\pi}(x)],$$

where $\sigma(x)$ is a scalar and isoscalar field and $\vec{\pi}(x)$ is the pion field. Then the Lagrangian density for the σ model involving only mesons is

$$L_2(x) = \frac{f_\pi^2}{4}\,\mathrm{Tr}[\partial_\mu U(x)\partial^\mu U^+(x)] . \qquad (1)$$

In the non-linear realization of the σ model,

$$\sigma^2 + \vec{\pi}^2 = f_\pi^2 \ , \tag{2}$$

which implies

$$U^+U = 1 \ . \tag{3}$$

Defining

$$\sigma = f_\pi \cos\theta(r) \quad \text{and} \quad \vec{\pi} = f_\pi \hat{\pi} \sin\theta(r) \ , \tag{4}$$

we can write

$$U = e^{i\vec{\tau}.\hat{\pi}\theta(r)} \quad . \tag{5}$$

In equations (1-4) f_π is the pion decay constant.

The difficulty with the non-linear σ model is that it leads to energetically unstable solutions. By dimensional arguments, one can see that the energy is proportional to R where R is the size of the system and it vanishes when the system shrinks to zero size.

3. THE SKYRME MODEL

To remedy this shortcoming, Skyrme [4] proposed the addition of higher-order terms in the derivative ∂_μ to the Lagrangian density $L_2(x)$. He explicitly proposed adding to $L_2(x)$ a term of the form

$$L_4(x) = \frac{1}{32e^2} \mathrm{Tr}[(\partial_\mu U)U^+, (\partial_\nu U)U^+]^2. \tag{6}$$

The resulting Lagrangian density

$$L(x) = L_2(x) + L_4(x) \tag{7}$$

leads to a non-linear Euler-Lagrange equation which is difficult to solve. However, if one makes the ansatz that the pion field is directed radially in configuration space, namely $\hat{\pi} = \hat{r}$ ($\hat{r}$: unit vector), the U field is of the "hedgehog" form [5]

$$U = e^{i\vec{\tau}\cdot\hat{r}\theta(|\vec{r}|)} . \tag{8}$$

In this case, the Lagrangian is

$$L = \int d^3x\, L(x)$$

$$= \frac{\pi}{2}\frac{F_\pi}{e}\int_0^\infty d\tilde{r}\{\tilde{r}^2(\dot{\theta}^2 + \frac{2\sin^2\theta}{\tilde{r}^2}) + 4\sin^2\theta(\frac{\sin^2\theta}{\tilde{r}^2} + 2\dot{\theta}^2)\} \tag{9}$$

with the choice of length scale $\vec{\tilde{r}} = e\,F_\pi\,\vec{r}$, $\dot{\theta} = d\theta/d\tilde{r}$ and $F_\pi = 2f_\pi$. Minimizing L with respect to θ, one gets the Euler-Lagrange equation

$$\ddot{\theta}(\frac{\tilde{r}^2}{4} + 2\sin^2\theta) + \tilde{r}\,\frac{\dot{\theta}}{2} - \frac{\sin 2\theta}{4} + \dot{\theta}^2\sin 2\theta - \frac{\sin^2\theta\sin 2\theta}{\tilde{r}^2} = 0. \tag{10}$$

With the boundary conditions $\theta(\infty) = 0$ and $\theta(0) = n\pi$, a soliton solution can be found [4]. This solution is now energetically stable, since the contribution of the term $L_4(x)$ to the energy is proportional to $1/R$ and the total energy given by equ.(7) can be minimized to get a finite value. It can be shown [4] that the solutions characterized by these boundary conditions are topological solitons whose winding number is n.

4. THE BARYONIC CURRENT

Besides the usual Noether currents such as V_μ and A_μ associated with symmetries of the Lagrangian $L(x)$, Skyrme showed that one can construct an "anomalous" current

$$B_\mu = \frac{1}{24\pi^2} \varepsilon_{\mu\alpha\beta\gamma} \mathrm{Tr}[(U^+\partial_\alpha U)(U^+\partial_\beta U)(U^+\partial_\gamma U)], \tag{11}$$

which is automatically conserved. Therefore, a new quantum number can be associated to B_μ. Now, for the hedgehog solution, we have

$$B_o = \frac{1}{4\pi^2 r^2} (\cos 2\theta - 1) \frac{d\theta}{dr} \tag{12}$$

and hence

$$B = \int d^3x\, B_o(r) = \frac{1}{\pi} [\theta(0) - \theta(\infty) - \frac{1}{2} \sin 2\theta(0) + \frac{1}{2} \sin 2\theta(\infty)]$$

$$= n, \text{since } \theta(0) = n \text{ and } \theta(\infty) = 0,$$

where n is the winding number characterizing the soliton solution U. At this point Skyrme conjectured that n is the baryon number, or equivalently B_μ is the baryonic current. Baryons therefore emerge from the model as topological solitons.

5. QUANTIZATION

So far, the Lagrangian as well as the Euler-Lagrange equation are classical. U is a classical field and therefore has no definite spin and isospin values. A simple quantization method starting from the classical solution $U_o = e^{i\vec{\tau}.\hat{r}\theta}$ was proposed by Adkins et al.[6]. These authors introduce collective coordinates through the unitary transformation

$$A = a_4 + i\vec{\tau}.\vec{a}\,, \tag{13}$$

where the a_μ 's are independent of $\vec{r}$ but can depend on time t. The a_μ's also satisfy the constraint

$$\sum_{\mu=1}^{4} a_\mu^2 = 1\,, \tag{14}$$

since $A^+A = 1$. Under this transformation, U_o becomes $U = A(t)U_oA^+(t)$. Substituting U in the Lagrangian one gets

$$L = \int d^3x\, \mathcal{L}(U = A^+U_oA) \tag{15}$$

$$= -\, M + \lambda\ \mathrm{Tr}\, (\frac{\partial A}{\partial t}\frac{\partial A^+}{\partial t}) \tag{16}$$

$$= -M + 2\lambda \sum_{\mu=1}^{4} (\dot{a}_\mu)^2\,, \tag{17}$$

where M is given by equation (9) and

$$\lambda = \frac{2\pi}{3F_\pi e^3}\int_0^\infty d\tilde{r}\, \tilde{r}^2 \sin^2\theta[1 + 4(\frac{d\theta}{d\tilde{r}}^2 + \frac{\sin^2\theta}{\tilde{r}^2})]. \tag{18}$$

The Hamiltonian is

$$H = \Pi_\mu\, \dot{a}_\mu - L, \tag{19}$$

where the Π_μ's are the conjugate momenta

$$\Pi_\mu = \frac{\partial L}{\partial a_\mu} = 4\lambda\dot{a}_\mu\ . \tag{20}$$

Substitution of (17) and (20) in equation (19) yields

$$H = M + \frac{1}{8\lambda} \sum_{\mu=1}^{4} \Pi_\mu^2 \; . \tag{21}$$

With the usual canocial quantization procedure

$$\Pi_\mu = - i \frac{\partial}{\partial a_\mu} \quad , \tag{22}$$

one obtains

$$H = M + \frac{1}{8\lambda} \sum_{\mu=1}^{4} \left(- \frac{\partial^2}{\partial a_\mu^2}\right) \tag{23}$$

with the constraint $\sum_\mu a_\mu^2 = 1$. Because of this constraint the term

$$\sum_\mu \frac{\partial^2}{\partial a_\mu^2}$$

is to be interpreted as the Laplacian on the 3-dimensional sphere. The wave functions are polynomials in a_μ.

6. THE SPIN AND ISOSPIN OPERATORS

Under a rotation about the z-axis by a small angle δ, it is easy to show that

$$e^{i\vec{\tau}.\vec{r}} \rightarrow e^{i\frac{\delta}{2}\tau_z} e^{i\vec{\tau}.\vec{r}} e^{-i\frac{\delta}{2}\tau_z} \; . \tag{24}$$

Hence, under this rotation the field configuration

$$A \, e^{i\vec{\tau}.\hat{r}\theta(r)} A^+$$

transforms into a new one with

$$A \rightarrow A\, e^{i\frac{\delta}{2}\tau_2} \tag{25}$$

or

$$(a_4 + i\vec{\tau}.\vec{a}) \rightarrow (a_4 + i\vec{\tau}.\vec{a})(1 + i\,\frac{\delta}{2}\tau_z). \tag{26}$$

This gives after some algebra

$$a_4 \rightarrow a_4 - \frac{\delta}{2}\,a_z$$

$$a_z \rightarrow a_z + \frac{\delta}{2}\,a_4$$

$$a_x \rightarrow a_x - \frac{\delta}{2}\,a_y$$

$$a_y \rightarrow a_y + \frac{\delta}{2}\,a_x, \tag{27}$$

i.e. $a_\mu \rightarrow (1 + i\delta J_z)a_\mu$

with $J_z = \frac{i}{2}[-a_4\,\frac{\partial}{\partial a_z} + a_z\,\frac{\partial}{\partial a_4} - a_x\,\frac{\partial}{\partial a_y} + a_y\,\frac{\partial}{\partial a_x}].$ (28)

Under a rotation about the z-axis of isospin by a small angle δ

$$A \rightarrow e^{-i\frac{\delta}{2}\tau_z A} \tag{29}$$

and the same calculation as before leads to

$$I_z = \frac{i}{2}[a_4\,\frac{\partial}{\partial a_z} - a_z\,\frac{\partial}{\partial a_4} - a_x\,\frac{\partial}{\partial a_y} + a_y\,\frac{\partial}{\partial a_x}]. \tag{30}$$

One can then easily verify that the spin-isospin wave functions for baryons are:

$$|p\uparrow\rangle = \frac{1}{\pi}(a_x + ia_y) \,, \qquad |p\downarrow\rangle = -\frac{i}{\pi}(a_4 - ia_z)$$

$$|n\uparrow\rangle = \frac{i}{\pi}(a_4 + ia_z) \,, \qquad |n\downarrow\rangle = -\frac{1}{\pi}(a_x - ia_y)$$

$$|\Delta^+, s_z = \tfrac{1}{2}\rangle = -\frac{\sqrt{2}}{\pi}(a_x + ia_y)\,[1 - 3(a_4^2 + a_z^2)]$$

$$|\Delta^{++}, s_z = \tfrac{3}{2}\rangle = \frac{\sqrt{2}}{\pi}(a_x + ia_y)^3 \tag{31}$$

etc....

The calculation of observables consists in taking appropriate matrix elements of operators (constructed from the field configuration) with these spin-isospin wave functions.

For example, the proton mass is

$$\langle p\uparrow|H|p\uparrow\rangle = \frac{1}{\pi^2}\int d^4a\,\delta(\sum_\mu a_\mu^2 - 1)(a_x - ia_y)H(a_x + ia_y) \tag{32}$$

$$= M + \frac{3}{8\lambda} \tag{33}$$

the delta mass is

$$\langle\Delta^{++}, s_z = \tfrac{3}{2}\,|H|\,\Delta^{++}, s_z = \tfrac{3}{2}\rangle = M + \frac{15}{8\lambda} \,, \tag{34}$$

where the expression (23) for $H(U = AU_oA^+)$ has been used.

7. RESULTS FOR THE STATIC PROPERTIES OF BARYONS

The previous procedure has been applied by Adkins et al. [6] to the calculation of baryon masses (nucleon and delta), charge radii, magnetic moments and coupling constants. The results are listed in Table 1. These authors adjust F_π and e to fit the nucleon and Δ masses obtaining F_π = 129 MeV and e = 5.45. The other quantities, except g_A, are in qualitative agreement with experiment (within 30 %).

Quantity	Prediction	Experiment
M_N	input	939 MeV
M_Δ	input	1232 MeV
F_π	129 MeV	186 MeV
$\langle r^2 \rangle^{1/2}_{I=0}$	0.59 fm	0.72 fm
$\langle r^2 \rangle^{1/2}_{M,I=0}$	0.92 fm	0.81 fm
μ_p	1.87	2.79
μ_n	-1.31	-1.91
$\left\|\frac{\mu_p}{\mu_n}\right\|$	1.43	1.46
g_A	0.61	1.23
$g_{\pi NN}$	8.9	13.5
$g_{\pi N\Delta}$	13.2	20.3
$\mu_{N\Delta}$	2.3	3.3

Table 1

8. THE BARYON-BARYON INTERACTION

Another field of application where the model can be confronted by experiment is the baryon-baryon interaction and especially the low-energy nucleon-nucleon interaction about which so much is known.

With the hedgehog form (8) and the Lagrangian density (7) it is not difficult to derive the static interaction between two solitons, if one makes the approximation that the field configuration, when the solitons are at a distance $\vec{R}$ apart, is a simple product of the form

$$U(\vec{r}) = U_o(\vec{r} + \frac{\vec{R}}{2})U_o(\vec{r} - \frac{\vec{R}}{2}), \qquad (35)$$

where $U_o(\vec{r}) = e^{i\vec{\tau}.\hat{r}\theta(|\vec{r}|)}$ is the classical solution. However, this field configuration is again spin and isospin degenerate. One has to map out the spin and isospin content of the two-soliton interaction, if one wants to identify the baryon-baryon interaction of interest. For the nucleon-nucleon interaction this problem has been tackled by Jackson et al. [7], but in an indirect way via the hedgehog quark model. These authors rotate one soliton with respect to the other in configuration space through Euler angles (α,β,γ) and compare this angular dependence with that obtained in the quark hedgehog model under the same rotation. The comparison, made numerically, enables them to identify the nucleon-nucleon potential. A more direct way of lifting the degeneracy without any appeal to the quark hedgehog model is to use the procedure described in section 5 and 6. This is done in ref.8 and in the following, I shall describe the methods used in this reference for the derivation of the baryon-baryon interaction.

The two solitons are rotated independently in SU(2) space

$$U_o(\vec{r} + \frac{\vec{R}}{2}) \rightarrow A\, U_o(\vec{r} + \frac{\vec{R}}{2})\, A^+$$

$$U_o(\vec{r} - \frac{\vec{R}}{2}) \rightarrow B\, U_o(\vec{r} - \frac{\vec{R}}{2})\, B^+, \qquad (36)$$

where A and B are unitary matrices:

$$A = a_4 + i\,\vec{\tau}.\vec{a} \qquad B = b_4 + i\,\vec{\tau}.\vec{b}\,; \qquad (37)$$

the components a_μ and b_μ lie on the surfaces of four-spheres,

$$\sum_{\mu=1}^{4} a_\mu^{\;2} = \sum_{\mu=1}^{4} b_\mu^{\;2} = 1\ .$$

<u>The soliton-soliton interaction</u>

Under the rotations (36) the field configuration (35) for the two solitons becomes

$$U(\vec{r}) = AU_o(\vec{r} + \frac{\vec{R}}{2})\, A^+B\, U_o(\vec{r} - \frac{\vec{R}}{2})\, B^+. \qquad (38)$$

Actually, only the relative rotation of the two solitons is relevant and the energy in the field of equ.(38) is the same as in

$$U(\vec{r}) = U_o(\vec{r} + \frac{\vec{R}}{2})\, C\, U_o(\vec{r} - \frac{\vec{R}}{2})\, C^+, \qquad (39)$$

where $C = A^+B = c_4 + i\,\vec{\tau}.\vec{c}$ and

$$c_4 = a_4 b_4 + \vec{a}.\vec{b},\ \vec{c} = a_4\vec{b} - b_4\vec{a} + \vec{a} \wedge \vec{b}. \qquad (40)$$

The total energy, which is obtained by integrating the energy density over all space $\vec{r}$, can be expressed as the sum of two parts. The first comes from terms in which the derivatives all act upon the same soliton $U_o(\vec{r} + \vec{R}/2)$ or $U_o(\vec{r} - \vec{R}/2)$. This term is independent of C and of $\vec{R}$ and is simply twice the static baryon mass. The second part, which we denote by $v(\vec{R},C)$ involves only mixed derivatives. It is a function of both $\vec{R}$ and the variables c_μ and is identified as the baryon-baryon potential.

By a choice of length scale $\tilde{r} = eF_\pi\vec{r}$, $\tilde{R} = eF_\pi\vec{R}$ the two parameters in the theory are made into simple scales of length and of energy and we find that

$$v(\tilde{R},C) = v_2(\tilde{R},C) + v_4(\tilde{R},C), \tag{41}$$

where

$$v_2(\tilde{R},C) = \frac{F_\pi}{8e} \int d^3\tilde{r}\, \rho_2(\tilde{R},\tilde{r},C) \tag{42}$$

and

$$\begin{aligned} \rho_2 = \mathrm{Tr}[&-X_i C Y_i C^+ + C^+ X_i C(Y_j Y_j Y_i + Y_i Y_j Y_j - 2Y_j Y_i Y_j) \\ &+ C Y_i C^+ (X_j X_j X_i + X_i X_j X_j - 2X_j X_i X_j) \\ &+ C\, Y_i Y_i C^+\, X_j X_j + C Y_i Y_j C^+ X_j X_i \\ &- 2C\, Y_i Y_j\, C^+ X_i X_j]\,. \end{aligned} \tag{43}$$

For calculational purpose it is convenient not to immediately impose the constraints $\sum_\mu c_\mu^2 = 1$. Before this constraint is imposed v_2 and ρ_2 are quadratic forms in the variables c_μ.

$$v_4(\tilde{R},C) = \frac{F_\pi}{8e} \int d^3\tilde{r}\, \rho_4(\tilde{R},\tilde{r},C) \tag{44}$$

and

$$\rho_4 = \frac{1}{2}\,\mathrm{Tr}[X_i C Y_j C^+ (X_j C Y_i C^+ - X_i C Y_j C^+)$$

$$+ C Y_i C^+ X_j (C Y_j C^+ X_i - C Y_i C^+ X_j)] \tag{45}$$

and

$$X_i = U_o^+(\tilde{r} + \frac{\tilde{R}}{2})\,\partial_i\, U_o(\tilde{r} + \frac{\tilde{R}}{2})$$

$$Y_i = \partial_i\, U_o(\tilde{r} - \frac{\tilde{R}}{2})\, U_o^+(\tilde{r} - \frac{\tilde{R}}{2}). \tag{46}$$

∂_i is the partial derivative with respect to $\tilde{r}$.

The first term of equ.(43) which is quadratic in X_i and Y_i is the contribution of the quadratic term in the Lagrangian (7), all the remaining terms in equs.(43) and (45), which are quartic in X_i and Y_j, are from the quartic piece of the Lagrangian.

The expectation value of $v(\tilde{R},C)$ for a general baryon number B = 2 state $\phi(a,b)$ is

$$\langle v(\tilde{R},C)\rangle = \iint dadb\; \phi^*(a,b)\, v(\tilde{R},c)\,\phi(a,b), \tag{47}$$

where the integrals are over the surface of the unit four-sphere. The c_μ are related to a_μ and b_μ by equ.(40) and a_μ act upon the soliton centered at $\tilde{r} = \tilde{R}/2$. These functions of a_μ and b_μ act like spin and isospin operators, but can also transform one baryon state into an other and in particular they mix the nucleon and the delta states. The baryon number B = 2 wave functions $\phi(a,b)$ are constructed from single nucleon and delta states and these are given in equs.(31).

It can be shown from equs.(43) and (45) that the general form of $v(\tilde{R},C)$ is

$$v_2(\underset{\sim}{R},C) = \alpha_1(c_4^{\,2} + \vec{c}^{\,2}) + \alpha_2 c_4^{\,2} + \alpha_3 \frac{(\vec{c}.\underset{\sim}{R})^2}{\underset{\sim}{R}^2} \tag{48}$$

and

$$v_4(\underset{\sim}{R},C) = (\vec{c}^{\,2} + c_4^{\,2})\,[\beta_1(\vec{c}^{\,2} + c_4^{\,2}) + \beta_2 c_4^{\,2} + \beta_3 \frac{(\vec{c}.\underset{\sim}{R})^2}{\underset{\sim}{R}^2}]$$

$$+ \beta_4 c_4^{\,4} + \beta_5 c_4^{\,2} \frac{(\vec{c}.\underset{\sim}{R})^2}{\underset{\sim}{R}^2} + \beta_6 \frac{(\vec{c}.\underset{\sim}{R})^4}{\underset{\sim}{R}^4}\,, \tag{49}$$

where the α's and β's are functions of $|\underset{\sim}{R}|$ only. They can be calculated by letting the vector $\underset{\sim}{R}$ lie in the z-direction; then for example $\alpha_1(|\underset{\sim}{R}|)$ is the right-hand side of equs. (42) (43) with $C = i\tau_x$, $\beta_1 + \beta_2 + \beta_4$ is the right-hand side of equ. (44) with $C = 1$. Six choices of the matrix C can yield enough independent linear equations to determine the α's and β's.

Imposing the constraint $c_4^2 + \vec{c}^{\,2} = 1$ yields

$$v(\underset{\sim}{R},C) = (\alpha_1 + \beta_1) + (\alpha_2 + \beta_2)c_4^{\,2} + (\alpha_3 + \beta_3) \frac{(\vec{c}.\underset{\sim}{R})^2}{\underset{\sim}{R}^2}$$

$$+ \beta_4\, c_4^{\,4} + \beta_5\, c_4^{\,2} \frac{(\vec{c}.\underset{\sim}{R})^2}{\underset{\sim}{R}^2} + \beta_6 \frac{(\vec{c}.\underset{\sim}{R})^4}{\underset{\sim}{R}^4}\,. \tag{50}$$

The static nucleon-nucleon potential

Let us now consider the contribution of $v(\underset{\sim}{R},C)$ to the static nucleon-nucleon potential. This potential, in both isospin states, is the sum of central spin-spin and tensor parts, and can be written as

$$V(\underset{\sim}{R}) = V_c^+(|\underset{\sim}{R}|) + \vec{\tau}_1.\vec{\tau}_2 V_c^-(|\underset{\sim}{R}|) + \vec{\sigma}_1.\vec{\sigma}_2[V_{ss}^+(|\underset{\sim}{R}|) + \vec{\tau}_1.\vec{\tau}_2\,.$$

$$\cdot V_{ss}^{-}(|\tilde{R}|)] + \left(\frac{3\vec{\sigma}_1\cdot\tilde{R}\vec{\sigma}_2\cdot\tilde{R}}{\tilde{R}^2} - \vec{\sigma}_1\cdot\vec{\sigma}_2\right)[V_T^{+}(|\tilde{R}|) + \vec{\tau}_1\cdot\vec{\tau}_2 V_T^{-}(|\tilde{R}|)]. \quad (51)$$

These conventional potentials can be related to the functions $\alpha_i(\tilde{R})$ and $\beta_i(\tilde{R})$ by taking six matrix elements of $v(\tilde{R},C)$. Equ.(47) is used for calculating the latter with the wave functions of equs.(31) in a_μ for the first soliton and in b_μ for the second. For example,

$$\langle p\uparrow p\uparrow|V|p\uparrow p\uparrow\rangle = \int d^4a\,\delta(\Sigma a_\mu^2-1)\,d^4b\,\delta(\Sigma b_\mu^2-1)\,(a_x^2+a_y^2)(b_x^2+b_y^2)\,v(\tilde{R},C)$$

$$= (\alpha_1 + \beta_1) + \frac{5}{18}(\alpha_2 + \beta_2 + \alpha_3 + \beta_3) +$$

$$+ \frac{7}{48}(\beta_4 + \frac{1}{3}\beta_5 + \beta_6), \quad (52)$$

where $|p\uparrow p\uparrow\rangle$ is the state of two protons both with spin oriented along the z-axis (the direction of $\tilde{R}$). This matrix element can be equated to

$$V_C^{+} + V_C^{-} + V_{ss}^{+} + 2(V_T^{+} + V_T^{-}). \quad (53)$$

By solving six such algebraic equations one finds that independent of the functional forms of the α's and β's

$$V_C^{-} = V_{ss}^{++} = V_T^{+} = 0$$

$$V_C^{-} = V_{ss}^{+} = V_T^{+} = 0$$

$$V_C^{+} = \alpha_1 + \beta_1 + \frac{1}{4}(\alpha_2 + \beta_2 + \alpha_3 + \beta_3) + \frac{1}{8}(\alpha_4 + \beta_6) + \frac{1}{24}\beta_5$$

$$V_{ss}^{-} = \frac{1}{36}(\alpha_2 + \beta_2) - \frac{1}{108}(\alpha_3 + \beta_3) + \frac{1}{48}\beta_4 + \frac{1}{432}\beta_5 - \frac{1}{144}\beta_6$$

$$V_T^{-} = \frac{1}{54}(\alpha_3 + \beta_3) + \frac{1}{432}\beta_5 + \frac{1}{72}\beta_6. \qquad (54)$$

We also find the following identities in the functions α's and β's

$$8\beta_3 + \beta_5 + 6\beta_6 = 0$$

$$8\beta_2 + \beta_5 + 6\beta_4 = 0. \qquad (55)$$

These are in fact the combinations which occur in V_{ss}^{-} and V_T^{-} so that the quartic terms in C in $v(\tilde{R},C)$ contribute only to the central potential V_c^{+}. We also find that the quadratic term in Lagrangian (7) contribute only to V_{ss}^{-} and V_T^{-}.

The nucleon-delta potential and the transition potential

Let us consider first the transition-matrix elements. Since all matrix elements are invariant under nucleon interchange it is sufficient to consider the case in which one nucleon is transformed into a delta.

The matrix elements have two simplifying features. The first is that only isospin-1 states are involved and the second is that the potential cannot change the total spin component along the direction of the line of centers $\tilde{R}$. Quantizing spin along $\tilde{R}$ it is sufficient to consider three initial spin states of two nucleons of total spin S, projection m,

$$|N,N,S,m\rangle = |N,N,0,0\rangle , \quad |N,N,1,1\rangle \text{ and } |N,N,1,0\rangle \qquad (56)$$

and four final delta-nucleon states,

$$|\Delta,N,S,m\rangle = |\Delta,N,1,0\rangle, \ |\Delta,N,1,1\rangle, \ |\Delta,N,2,0\rangle \text{ and } |\Delta,N,2,1\rangle. \quad (57)$$

Because of conservation of m there are only six independent matrix elements and we find

$$\langle\Delta,N,1,0|\nu|N,N,0,0\rangle = \langle\Delta,N,2,0|\nu|N,N,1,0\rangle = 0. \quad (58)$$

Remarkably the other four are combinations of the α's and β's that occur in the nucleon-nucleon matrix elements. It is therefore possible to express the transition-matrix elements entirely in terms of the nucleon-nucleon potentials,

$$\langle\Delta,N,2,1|v|N,N,1,1\rangle = 3\sqrt{6}\ V_T^-$$

$$\langle\Delta,N,1,0|v|N,N,1,0\rangle = -2\sqrt{2}\ (2V_{ss}^- + V_T^-)$$

$$\langle\Delta,N,1,1|v|N,N,1,1\rangle = \sqrt{2}\ (-4\ V_{ss}^- + V_T^-)$$

$$\langle\Delta,N,2,0|v|N,N,0,0\rangle = -6\sqrt{2}\ V_T^-. \quad (59)$$

A similar simplification occurs in the direct nucleon-delta potential; we find for the isospin-1 states

$$\langle\Delta,N,2,2|v|\Delta,N,2,2\rangle = V_c^+ - 3(V_{ss}^- + 2\ V_T^-)$$

$$\langle\Delta,N,2,1|v|\Delta,N,2,1\rangle = V_c^+ - 3(V_{ss}^- - V_T^-)$$

$$\langle\Delta,N,1,1|v|\Delta,N,1,1\rangle = V_c^+ + 5\ V_{ss}^- + V_T^-$$

$$\langle\Delta,N,1,1|v|\Delta,N,2,1\rangle = \sqrt{3}\ (-V_{ss}^- + 4\ V_T^-)$$

$$\langle\Delta,N,2,0|v|\Delta,N,2,0\rangle = V_c^+ - 3\ V_{ss}^- + 6\ V_T^-$$

$$\langle\Delta,N,1,0|v|\Delta,N,1,0\rangle = V_c^+ + 5\,V_{ss}^- - 2\,V_T^-$$

$$\langle\Delta,N,1,0|v|\Delta,N,2,0\rangle = 0. \tag{60}$$

It would be of interest to investigate the consequence of such simple relationships for the delta-production cross sections in nucleon-nucleon collisions.

Some numerical results

The functions α_i and β_i are culculated numerically for the model of equs. (35) and (36). This model wich neglects nucleon distortion should be at its best at large $\tilde{R}$ where this is small.

At large $\tilde{R}$ the nucleon-nucleon potential is known to be dominated by the one-pion-exchange tensor potential which for massless pions is

$$V_T^- = (4m^2)^{-1}(g^2/4\pi)R^{-3}, \tag{61}$$

where g is the conventional pion-nucleon coupling constant and m is the nucleon mass. We find that at large R this tensor potential does indeed dominate and is of the form

$$V_T^- = (F_\pi/e)25.6/\tilde{R}^3 = (F_\pi^2e^4)^{-1}25.6/R^3. \tag{62}$$

Identifying the long-range Skyrme-model potential with the conventional pion exchange yields the result

$$g/2m = 18/F_\pi e^2 , \tag{63}$$

which is the same as that of ref.6 but by a very different method.

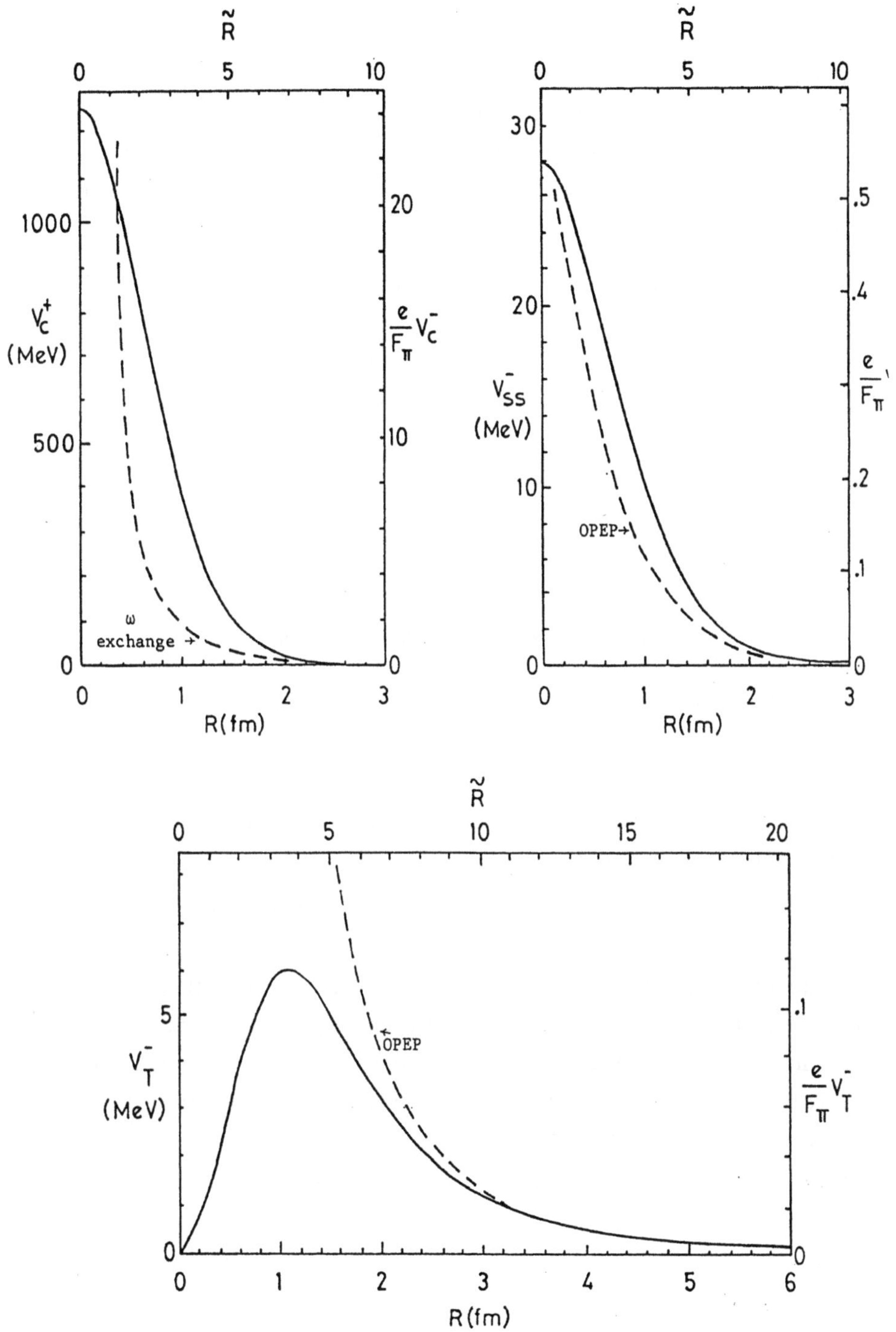

R̃
0
5
10
1000
500
0
20
10
0
V_C^+ (MeV)
$\frac{e}{F_\pi} V_C^-$
ω exchange →
0
1
2
3
R(fm)
R̃
0
5
10
30
20
10
0
.5
.4
.2
.1
0
V_{SS}^- (MeV)
$\frac{e}{F_\pi}$
OPEP→
0
1
2
3
R(fm)
R̃
0
5
10
15
20
5
0
.1
0
V_T^- (MeV)
$\frac{e}{F_\pi} V_T^-$
←OPEP
0
1
2
3
4
5
6
R(fm)

Fig. 1

The numerical results are shown in fig. 1. We give the parameter-free scales $(e/F_\pi)V(\tilde{R})$ as functions of $\tilde{R}$ and also the potentials with the parameter choice $F_\pi = 186$ MeV and $e = 3.6$. With this choice, F_π is given by the pion-decay constant and e is given by the pion-nucleon coupling constant g through equ.(63).

These results are obtained within the approximation that the field configuration for two solitons is a product of the two single soliton fields. This should be a good approximation at large separations, but need not be so when R is small since the field of one soliton can be significantly distorted by the close proximity of another.

9. THE SYMMETRIC QUARTIC TERM

The results of the previous section show that there is no attraction in V_c^+ at intermediate range, i.e. for internucleon distances between one and two fermi. This is a serious shortcoming, since it is the attraction which is responsible for binding nucleons in nuclei.

A modification of the Skyrme model was recently proposed by A. Jackson et al. [9]. These authors suggest a change of sign of the quartic term in the Skyrme model. Such a sign change turns the previous repulsion into attraction but by itself it would destabilize the soliton. In ref.9 this is prevented by the addition of a repulsive sixth-order term. The balance between an attractive 4th-order term and a repulsive 6th-order term can generate a NN central potential, which is attractive at long range and repulsive at short range, but the model then predicts the wrong signs for the pion-pion scattering lengths.

In a recent work, Lacombe et al. [10] point out that a generalization of the Skyrme model that does not introduce higher powers of derivatives can maintain the correct form for the pion-pion scattering lengths whilst providing attractive nucleon-nucleon forces at intermediate range.

The Lagrangian density (7) can be considered to be the first two terms of an expansion of the chiral-invariant effective Lagrangian in powers of derivatives of the meson fields. However, equ.(7) is not the most general expansion which terminates at the quartic level. Another independent quartic term can be chosen to be [11]

$$L_{sym} = \frac{\gamma}{8e^2} [Tr(\partial_\mu U \partial^\mu U^+)]^2. \qquad (64)$$

Contrary to the quartic term appearing in equ.(7), this new term is symmetrical with respect to the derivatives. In the following we shall call it the symmetric quartic term. Adding this term to the original Skyrme Lagrangian and quantizing in the same way as in section 5, one obtains the following expressions for the classical soliton mass M (equ.(9)) and for the moment of inertia of the rotated soliton λ (equ.(18):

$$M = \frac{F_\pi}{e} \frac{\pi}{2} \int_0^\infty d\tilde{r} \{ \tilde{r}^2 (\dot{\theta}^2 + \frac{2\sin^2\theta}{\tilde{r}^2}) + 4\sin^2\theta (\frac{\sin^2\theta}{\tilde{r}^2} + 2\dot{\theta}^2)$$

$$- 4\gamma \tilde{r}^2 (\dot{\theta}^2 + \frac{2\sin^2\theta}{\tilde{r}^2})^2 \} \qquad (65)$$

$$\lambda = \frac{2\pi}{3F_\pi e^3} \int_0^\infty \tilde{r}^2 d\tilde{r} \sin^2\theta [1 + 4(\dot{\theta}^2 + \frac{\sin^2\theta}{\tilde{r}^2}) - 8\gamma(\dot{\theta}^2 + \frac{2\sin^2\theta}{\tilde{r}^2})]. \qquad (66)$$

The chiral angle $\theta(r)$ is now the solution of the Euler-

Lagrange equation:

$$\ddot{\theta}\left(\frac{\tilde{r}^2}{4} + 2(1-2\gamma)\sin^2\theta - 6\gamma\tilde{r}^2\dot{\theta}^2\right) + \tilde{r}\,\frac{\dot{\theta}}{2} - \frac{\sin 2\theta}{4} +$$

$$+ (1-2\gamma)\dot{\theta}^2\sin 2\theta - 4\gamma\tilde{r}\dot{\theta}^3 - (1-4\gamma)\,\frac{\sin^2\theta\,\sin 2\theta}{\tilde{r}^2} = 0. \quad (67)$$

In deriving the equs.(65) - (67) terms proportional to quartic time derivatives are neglected, since they are small, of order $(v^2/c^2)^2$ compared to the static term M.

In principle, the values of the parameters F_π, e, and γ can be determined in the meson sector, for example by fitting the $\pi\pi$ scattering lengths as is done in refs. 11, 12. Fitting the parameters in this way all the low-energy properties of baryons would be predicted. We here choose the parameters to be consistent with the pion-pion scattering lengths, but since we are principally interested in predicting the effects of this new symmetric quartic Lagrangian in the NN interaction we have determined F_π and e by fitting the nucleon and Δ masses. Note that the contributions of the symmetric quartic term to equs.(65) and (66) occur with a negative sign. We therefore expect a local minimum for M only for values of γ smaller than some critical value γ_c.

We solve equation (67) numerically and find that with our numerical code a solution exists for $\gamma \leqslant 0.12$. Whether this upper limit is actually the critical value γ_c, we do not know, but 0.12 is surely less than γ_c. The solution corresponding to $\gamma = 0.12$ gives $m = F_\pi/e\ 28.6$ and $\lambda = 53.6/ /F_\pi e^3$. Fitting the nucleon and Δ masses which are respectively given by

$m_N = M + \frac{3}{8\lambda}$ and $m_\Delta = M + \frac{15}{8\lambda}$ yields $F_\pi = 130$ MeV and $e = 4.32$.

Although the value of γ we found is similar to that of ref.11, the agreement can be fortuituous. We find a sign error in their expression for λ, a discrepancy in their numerical expression for M and an inconsistent normalization factor in their lagrangian L_o (their formulae (10), (12), and (2)).

Contribution of the symmetric quartic term to the nucleon-nucleon potential

The field configuration for the two solitons is taken again as

$$U(\vec{r}) = AU_o(\vec{r} + \frac{\vec{R}}{2})\ A^+BU_o(\vec{r} - \frac{\vec{R}}{2})B^+. \tag{68}$$

The potential-energy density corresponding to the Lagrangian density (64) is then given by

$$v(\vec{R},\vec{r},A,B) = -\frac{\gamma}{8e^2}\{[Tr(\partial_i U\partial^i U^+)]^2 - m_1 - m_2\}, \tag{69}$$

where m_1 and m_2 are, up the factor $\gamma/(8e^2)$, the single soliton-mass densities

$$m_1 = [Tr\ \partial_i U_o(\vec{r} + \frac{\vec{R}}{2})\partial^i U_o^+(\vec{r} + \frac{\vec{R}}{2})]^2 = [Tr(X_i X^{i+})]^2$$

$$m_2 = [Tr\ \partial_i U_o(\vec{r} - \frac{\vec{R}}{2})\partial^i U_o^+(\vec{r} - \frac{\vec{R}}{2})]^2 = [Tr(Y_i Y^{i+})]^2 \tag{70}$$

with

$$X_i = U_o^+(\vec{r} + \frac{\vec{R}}{2})\partial_i U_o(\vec{r} + \frac{\vec{R}}{2}),\ Y_i = \partial_i U_o(\vec{r} - \frac{\vec{R}}{2})U_o^+(\vec{r} - \frac{\vec{R}}{2}). \tag{71}$$

∂_i is the partial derivative with respect to $\vec{r}$.

After some algebra, we find

$$v(\vec{R},\vec{r},C) = -\frac{\gamma}{2e^2}\{[\mathrm{Tr}(X_i C Y^i C^+)]^2 + \mathrm{Tr}(X_i C Y^i C^+)(\sqrt{m_1}+\sqrt{m_2}) + $$

$$+\frac{1}{2}\sqrt{m_1 m_2}\}, \quad (72)$$

where C represents the relative rotation of one soliton with respect to the other as given by equ.(40).

The potential operator is

$$V(\vec{R},C) = \int d^3r\; v(\vec{R},\vec{r},C). \quad (73)$$

Expanding equ.(72) and imposing the condition $c_4^2 + \vec{c}^{\,2} = 1$, we can write the potential as

$$V(\tilde{R},C) = -\frac{\gamma F_\pi}{2e}(\beta_1 + \beta_2 c_4^2 + \beta_3 c_z^2 + \beta_4 c_4^4 + \beta_5 c_4^2 c_z^2 + \beta_6 c_z^4), \quad (74)$$

where

$$\beta_1 = \int d^3\tilde{r}[\tfrac{1}{2}\sqrt{m_1 m_2} - G(\sqrt{m_1}+\sqrt{m_2}-G)],$$

$$\beta_2 = \int d^3\tilde{r}[(G-E)(\sqrt{m_1}+\sqrt{m_2}-2G) + I^2],$$

$$\beta_3 = \int d^3\tilde{r}[(G-F)(\sqrt{m_1}+\sqrt{m_2}-2G) + J^2],$$

$$\beta_4 = \int d^3\tilde{r}[(G-E)^2 - I^2],$$

$$\beta_5 = \int d^3\tilde{r}[2(G-E)(G-F) - I^2 - J^2 + H^2],$$

$$\beta_6 = \int d^3\tilde{r}[(G-F)^2 - J^2], \quad \text{and} \quad (75)$$

$$E = -\mathrm{Tr}(X_i Y^i), \quad F = -\mathrm{Tr}(X_i \tau_3 Y^i \tau_3), \quad G = -\mathrm{Tr}(X_i \tau_1 Y^i \tau_1)$$

$$I = 2i\,\mathrm{Tr}(X_i Y^i \tau_1), \quad J = 2\,\mathrm{Tr}(X_i \tau_1 Y^i \tau_3), \quad H = 2i\,\mathrm{Tr}(X_i Y^i \tau_3)$$

Taking matrix elements of this operator for different two-nucleon states and using the same procedure as in section 8, we obtain

$$V_C^+ = -\frac{\gamma F_\pi}{2e}(\beta_1 + \frac{\beta_2}{4} + \frac{\beta_3}{4} + \frac{\beta_4}{8} + \frac{\beta_5}{24} + \frac{\beta_6}{8})$$

$$V_{SS}^- = -\frac{\gamma F_\pi}{2e}(\frac{\beta_2}{36} - \frac{\beta_3}{108} + \frac{\beta_4}{48} + \frac{\beta_5}{432} - \frac{\beta_6}{144})$$

$$V_T^- = -\frac{\gamma F_\pi}{2e}(\frac{\beta_3}{54} + \frac{\beta_5}{432} + \frac{\beta_6}{72}) \quad ; \quad V_C^- = V_{SS}^+ = V_T^+ = 0 . \qquad (76)$$

We shall here concentrate on the central, isospin-independent potential V_C^+,since this is the dominant component of the NN potential. It is easy to show from equs.(75) and (76) that the contribution of the symmetric quartic term (64) to this component V_C^+ is attractive.

We have calculated numerically the contribution of the symmetric quartic term to V_C^+, using the solution θ (or U_O) corresponding to $\gamma = 0.12$ and using for F_π and e the values 130 MeV and 4.32 which reproduce the nucleon and Δ masses. The results are given in table 2. In the same table we also show the contribution from the Skyrme term calculated with the same values for F_π and e.

It can be seen that the symmetric quartic term, equ. (64), gives rise to a very significant attraction reducing considerably the repulsion due to the Skyrme term. Note, however, that the latter repulsion is unrealistically large at medium range and it cannot be overcome by the attraction corresponding to a value of γ as small as 0.12. We now discuss a possibility of increasing the value of γ such that one can get a net attraction in the medium range.

As will be explained in the next section the next improvement of the effective Lagrangians (7), (64) con-

R	Contribution from the Skyrme term (equ.1)	Contribution from the symmetric quartic term (equ.(2))
0.1 fm	896 MeV	- 532 MeV
1.1 fm	212.7 MeV	- 113 MeV
2.1 fm	8.12 MeV	- 4.54 MeV

Table 2: Contributions to the isospin-independent central potential V_C^+ from the Lagrangians (7) and (64) (with $\gamma = 0.12$, $F_\pi = 130$ MeV, $e = 4.32$).

sists in incorporating vector mesons. The case of the ω meson is particularly interesting and its inclusion to the effective Lagrangian is very simple. It is given by the following Lagrangian density [13]

$$L_\omega = -\frac{1}{4}(\partial_\mu \omega_\nu - \partial_\nu \omega_\mu)^2 + \frac{1}{2} m_\omega^2 \omega_\mu^2 + \beta \omega_\mu B^\mu , \qquad (77)$$

where ω_μ is the ω meson field and

$$B_\mu = \frac{1}{24\pi^2} \varepsilon_{\mu\alpha\beta\gamma} \, Tr(U^+\partial^\alpha U \, U^+\partial^\beta U \, U^+\partial^\gamma U) \qquad (78)$$

is the baryon current.
The contribution of the Lagrangian (77) to the soliton mass M and to the moment of inertia λ are positive and they counterbalance the destabilizing effect of the symmetric quartic term (equ.(64)).

We can therefore, by adding the ω meson term (equ.(77)) to the Lagrangian (7) and (64), increase the values of γ and e, making the contribution of the symmetric quartic term larger (in magnitude) and that of the quartic Skyrme term smaller without spoiling the good fit of the nucleon and Δ masses. Then, for the NN potential V_C^+, the attraction due to the symmetric quartic term can become large enough to overtake the repulsion due to the Skyrme term. On the other hand the contribution of the ω-meson term (equ.(77)) to V_C^+ can be shown to be repulsive and, for reasonable values of β, of shorter range than those of the quartic terms [14]. The effect of the ω-meson term is then to restore the repulsion at short range without overtaking the net attraction due to the quartic terms at intermediate and long ranges. One thus recovers in a natural way the well-known empirical features of the NN interaction.

10. EXTENSION OF THE SKYRME MODEL

The different calculations reviewed in the previous sections can be regarded as part of a more general program attempting to deduce baryon physics from meson physics via solitons [15]. In this context the Skyrme model, taken as an effective Lagrangian, is probably too crude for a simultaneous fit of the meson physics and the baryon physics, since it contains only the pion field. The next improvement consists in incorporating vector mesons to the effective Lagragian. This should be done in a chirally invariant way.

The case of the ω meson which in SU(2) × SU(2) is a chiral singlet is simple and it gives rise to equs.(77) - (78). Note that the interaction term $\omega_\mu B^\mu$ is responsible for the coupling of the ω meson to 3 pions and to baryons in the meson and baryon sectors, respectively.

The isospin-1 vector mesons (ρ and A_1) can be introduced as gauge fields through the transformation [14], [16],

[17]

$$\partial_\mu U \to D_\mu U = \partial_\mu U - ig[A^L_\mu U - UA^R_\mu]. \tag{79}$$

The left-handed vector field A^L_μ and the right-handed vector field A^R_μ are hermitian matrices

$$A^L_\mu = \vec{A}^L_\mu \cdot \vec{\tau} \tag{80}$$

$$A^R_\mu = \vec{A}^R_\mu \cdot \vec{\tau} \tag{81}$$

and they transform under a chiral transformation as

$$A^L_\mu \to e^{i\vec{\tau}.\vec{\alpha}} A^L_\mu e^{-i\vec{\tau}.\vec{\alpha}} \tag{82}$$

$$A^R_\mu \to e^{-i\vec{\tau}.\vec{\beta}} A^R_\mu e^{i\vec{\tau}.\vec{\beta}} . \tag{83}$$

An effective Lagrangian generalizing the Skyrme model which incorporates the isospin-1 vector mesons can then be constructed by replacing $\partial_\mu U$ by $D_\mu U$ in equs.(1), (6), and (64). In the meson sector the resulting Lagrangian accounts for all possible interactions of pions and the gauge fields. Since experiments suggest that multi-vector meson processes are negligible, it is sufficient to retain only terms which are quadratic in the gauge fields. Adding to the resulting Lagrangian kinetic energy and mass terms on obtains [14]

$$\begin{aligned} L(x) = &- \frac{1}{8} Tr[A^L_{\mu\nu}A^{L\mu\nu} + A^R_{\mu\nu}A^{R\mu\nu}] - \frac{1}{4} \omega_{\mu\nu}\omega^{\mu\nu} \\ &+ \frac{1}{4} m^2_\rho Tr[A^L_\mu A^{L\mu} + A^R_\mu A^{R\mu}] + \frac{1}{2} m^2_\omega \omega_\mu\omega^\mu + \beta\omega_\mu B^\mu \\ &+ \frac{f}{16} Tr[D_\mu U D^\mu U^+] + L_4(x) + L_{sym}(x) \\ &- \frac{i\gamma g}{4e^2} Tr(\partial_\mu U \partial^\mu U^+) \, Tr(U\partial_\mu U^+ A^{L\mu} + U^+\partial_\mu U A^{R\mu}) \end{aligned} \tag{84}$$

$$-\frac{ig}{16e^2}\mathrm{Tr}\{([(\partial_\mu U)U^+,(A^L_\nu-UA^R_\nu U^+)]+[(A^L_\mu-UA^R_\mu U^+),(\partial_\nu U)U^+])[(\partial_\mu U)U^+,(\partial_\nu U)U^+]\}$$

In equ.(84),

$$A_{\mu\nu} = \partial_\mu A_\nu - \partial_\nu A_\mu \tag{85}$$

$$U = e^{2i\vec{\tau}.\vec{\phi}/F_\pi} . \tag{86}$$

$L(x)$ is invariant under the chiral transformation

$$U \to e^{i\vec{\tau}.\vec{\alpha}} U e^{i\vec{\tau}.\vec{\beta}}, \quad A^L_\mu \to e^{i\vec{\tau}.\vec{\alpha}} A^L_\mu\, e^{-i\vec{\tau}.\vec{\alpha}}, \quad A^R_\mu \to e^{-i\vec{\tau}.\vec{\beta}} A^R_\mu\, e^{i\vec{\tau}.\vec{\beta}} . \tag{87}$$

Under parity transformation,

$$\vec{\phi} \to -\vec{\phi}, \quad U \to U^+, \quad A^L_i \to A^R_i, \; A^R_i \to A^L_i$$

$$A^L_o \to -A^R_o , \quad A^R_o \to -A^L_o, \quad \omega_i \to -\omega_i , \quad \omega_o \to \omega_o . \tag{88}$$

To identify the physical meson fields, it is sufficient to expand the U field as a power series of the pion field $\vec{\phi}$ and keep only the terms in the Lagrangian (84) that are quadratic in the fields,

$$\begin{aligned} L^{quadratic} = & -\frac{1}{4}\omega^2_{\mu\nu} - \frac{1}{4}(\partial_\mu\vec{A}^L_\nu - \partial_\nu\vec{A}^L_\mu)^2 - \frac{1}{4}(\partial_\mu\vec{A}^R_\nu - \partial_\nu\vec{A}^R_\mu)^2 + \\ & + \frac{f}{2F^2_\pi}(\partial_\mu\vec{\phi})^2 \\ & + \frac{1}{2} m^2_\omega \omega_\mu \omega^\mu + \frac{1}{4} m^2_\rho [(\vec{A}^R_\mu - \vec{A}^L_\mu)^2 + (\vec{A}^R_\mu + \vec{A}^L_\mu)^2] \\ & + \frac{fg^2}{8}(\vec{A}^R_\mu - \vec{A}^L_\mu)^2 + \frac{fg}{2F_\pi}(\vec{A}^R_\mu - \vec{A}^L_\mu) .\partial_\mu\vec{\phi} . \end{aligned} \tag{89}$$

In equ.(89) ω_μ is the field of an isoscalar vector meson of mass m_ω (the ω meson),

$$\vec{\rho}_\mu = \frac{1}{\sqrt{2}} (\vec{A}^R_\mu + \vec{A}^L_\mu)$$

is the field of an isovector meson of mass m_ρ (the ρ meson), $\frac{1}{\sqrt{2}}(\vec{A}^R_\mu - \vec{A}^L_\mu)$ and $\partial_\mu\vec{\phi}$ are mixed.

By defining [17] the A_1 meson field as

$$\vec{A}_{1\mu} = \frac{1}{\sqrt{2}} (\vec{A}^R_\mu - \vec{A}^L_\mu) + \frac{(m_A^2 - m_\rho^2)^{1/2}}{m_A m_\rho} \partial_\mu\vec{\phi} \tag{90}$$

of mass

$$m_A^2 = m_\rho^2 + \frac{fg^2}{2} \tag{91}$$

one can diagonalize $L^{\text{quadratic}}$ and find

$$L^{\text{quadratic}} = \frac{1}{4}[\omega_{\mu\nu}\omega^{\mu\nu} + \vec{\rho}_{\mu\nu}.\vec{\rho}^{\mu\nu} + \vec{A}_{1\mu\nu}.\vec{A}_1^{\mu\nu}] + \frac{1}{2}\partial_\mu\vec{\phi}.\partial^\mu\vec{\phi}$$
$$+ \frac{m_\omega^2}{2}\omega_\mu\omega^\mu + \frac{m_\rho^2}{2}\vec{\rho}_\mu.\vec{\rho}^\mu + \frac{m_A}{2}\vec{A}_{1\mu}.\vec{A}_1^\mu . \tag{92}$$

In equ. (92) the parameters f and g have been determined by the A_1-ρ mass difference (equ. (91)) and by the standard normalization of the pion field,

$$f = F_\pi^2 \frac{m_A^2}{m_\rho^2} \quad \text{and } g^2 = \frac{2m_\rho^2}{F_\pi^2} (1 - \frac{m_\rho^2}{m_A^2}) . \tag{93}$$

The other parameters that occur in the complete Lagrangian (84) are m_ω, m_ρ, m_A, F_π, β, e and γ. Among them only e and γ are really free parameters, since the vector-meson masses and the pion-decay constant F_π can be taken from experiment and β is related to the ω coupling constant to nucleons.

Those free parameters (e and γ) can be adjusted to fit, for example, the low-energy $\pi\pi$ scattering amplitudes. In this way all parameters are fixed and predictions for the baryon properties and the baryon-baryon interaction can be computed. With the hedgehog form of U

$$U_o = e^{i\vec{\tau}\cdot\hat{r}\theta(r)}$$

one can show, from the interaction term $Tr[U\partial_\mu U^+ A_\mu^L + U^+\partial_\mu U A_\mu^R]$ of the Lagrangian (84) that the ρ and A_1 meson fields are of the form

$$\rho_i = \gamma(r)(\vec{\tau} \times \vec{r}).\vec{i}$$

$$A_i = \alpha(r)\vec{\tau}.\vec{i} + \beta(r)(\vec{\tau}.\vec{r})(\vec{r}.\vec{i}), \tag{94}$$

The Euler-Lagrange equations are now coupled equations relating the chiral angle $\theta(r)$ and the vector-meson fields $\alpha(r)$, $\beta(r)$, $\gamma(r)$ and $\omega(r)$. The solution of these equations is in progress [14].

Some numerical results

It is of interest to compute predictions for some three-body coupling constants. For example, in the meson sector it is easy to extract from equ.(84) the $\rho\pi\pi$ interaction. One finds

$$L_{\rho\pi\pi} = \sqrt{2}\, g\, \vec{\rho}_\mu \cdot (\partial_\mu\vec{\phi} \times \vec{\phi}). \tag{95}$$

With value of g given by equ.(93) and the usual of the particle masses one gets the $\rho\pi\pi$ coupling constant

$$G_{\rho\pi\pi} = 6.6 \tag{96}$$

in fair agreement with the experimental value 5.8.

Likewise, in the baryon sector the nucleon $\to$ nucleon $+ \rho$ tensor amplitude can be constructed by considering the matrix element

$$M = \frac{\sqrt{2}}{16} gf \langle p\uparrow | \int d\vec{r} \, \mathrm{Tr}[\rho_i (U^+ \partial_i U + U \partial_i U^+)] | p\uparrow \rangle \tag{97}$$

$$= \frac{\sqrt{2}}{16} gf \, \vec{\varepsilon} . (\vec{\sigma} \wedge \vec{k}) \, \frac{4\pi}{9} \int_0^\infty \sin^2\theta \, r^2 \, dr , \tag{98}$$

where $\vec{\varepsilon}$ is the ρ-meson polarization.
This matrix element is to be compared with the usual ρNN effective coupling

$$\bar{N} \frac{G^T_{NN\rho}}{4m_p} \frac{\partial_\mu \vec{\rho}_\nu - \partial_\nu \vec{\rho}_\mu}{2} . \vec{\tau} \sigma_{\mu\nu} N \to \frac{G^T_{NN\rho}}{4m_p} \vec{\varepsilon} . (\vec{\sigma} \wedge \vec{k}) . \tag{99}$$

Identifying equs. (98) and (99) one obtains the tensor coupling constant of the ρ meson

$$\frac{G^T_{NN\rho}}{4m_p} = \frac{\sqrt{2}}{16} gf \frac{4\pi}{9} \int_0^\infty \sin^2\theta \, r^2 \, dr$$

$$= G_{\rho\pi\pi} \frac{F_\pi^2}{16} \frac{m_A^2}{m_\rho^2} \frac{4\pi}{9} \int_0^\infty \sin^2\theta \, r^2 \, dr . \tag{100}$$

An estimate of $G^T_{NN\rho}$ can be made by taking θ as given by the Skyrme model and $G_{\rho\pi\pi}$ from equ. (96). This gives $G^T_{\rho NN} = 17.3$ in qualitative agreement with the empirical value 16.

These simple and rather curde numerical results lead to the hope that Skyrme solitons emerging from realistic effective Lagrangians can allow a new phenomenology for low-energy hadron physics. Of course, this hope is to be substantiated by more detailed studies.

11. CONCLUDING REMARKS

In these lectures we reviewed the theoretical and phenomenological arguments suggesting that an approximation scheme to non perturbative QCD at low energies can be provided by effective Lagrangians of the type considered by Skyrme. These effective Lagrangians are constructed from meson fields and baryons emerge as solitons. Meson physics and baryon physics are thus intimately related. Predictions can be made not only for the single-baryon properties but also for the baryon-baryon interaction. They are in qualitative agreement with experiment and their accuracy can be improved. An interesting feature concerning the baryon-baryon interaction is that a connection can be made with more conventional approaches like meson-exchange models, which are known to be phenomenologically quite successful.

ACKNOWLEDGEMENTS

Much of the material presented in these lectures results from work done with W.N. Cottingham, M. Lacombe, and B. Loiseau. I would like to thank them sincerely for our pleasurable and fruitful collaboration.

REFERENCES

1. G. 't Hooft, Nucl. Phys. B72 (1974) 461, Nucl. Phys. B75 (1974) 461. See also Erice Lectures.
2. E. Witten, Nucl. Phys. B160 (1979) 57.
3. M. Gell-Mann and M. Lévy, Il Nuovo Cimento 16 (1960) 705.
4. T.H.R. Skyrme, Proc. Roy. Soc. A260 (1961) 127, Nucl. Phys. 31 (1962) 556.
5. T.H.R. Skyrme, ref.4; A. Chodos and C.B. Thorn, Phys. Rev.

D12 (1975) 2733; G.E. Brown and M. Rho, Phys. Lett. 82B (1979) 177.

6. G.S. Adkins, C.R. Nappi, and E. Witten, Nucl. Phys. B228 (1983) 552.
7. A. Jackson, A.D. Jackson, and V. Pasquier, Nucl. Phys. A432 (1985) 567.
8. R. Vinh Mau, M. Lacombe, B. Loiseau, W.N. Cottingham, and P. Lisboa, Phys. Lett. 150B (1985) 259.
9. A. Jackson, A.D. Jackson, A.S. Goldhaber, G.E. Brown, and L.C. Castillejo, Stony Brook preprint (1985) to be published.
10. M. Lacombe, B. Loiseau, R. Vinh Mau, and W.N. Cottingham, Orsay preprint IPNO/TH 85-17 (1985).
11. J.F. Donoghue, E. Golowich and B.R. Holstein, Phys. Rev. Lett. 53 (1984) 747.
12. T.N. Pham and T.N. Truong, Ecole Polytechnique preprint (1985).
13. G.S. Adkins and C.R. Nappi, Phys. Lett. 137B (1984) 251.
14. R. Vinh Mau, Talk at the Meeting on the Skyrme Model held at the Rutherford Appleton Laboratory (Nov. 1984); M. Lacombe, B. Loiseau, R. Vinh Mau, and W.N. Cottingham, in preparation.
15. E. Witten, Proc. Workshop on Solitons in Nuclear and Particle Physics, Lewes Center for Physics (June 1984).
16. Y. Brihaye, N.K. Pak, and P. Rossi, CERN preprint TH-3984/84; O. Kaymakcalan, S. Rajeev, and J. Schechter, Phys. Rev. D30 (1984) 594; H. Gomm, O. Kaymakcalan, and J. Schechter, Phys. Rev. D30 (1984) 2345.
17. J. Schwinger in Particles and Sources, Ed. Gordon and Breach, p.60. (1969).

Acta Physica Austriaca, Suppl. XXVII, 127 (1985)

NUCLEON-ANTINUCLEON EXPERIMENTS AT LEAR ENERGIES

by

K. KILIAN
CERN
Geneva

Prof. K. Kilian reviewed the experimental evidence on the $N\bar{N}$ system at low and medium energies. Particular emphasis was laid on experiments performed at LEAR. We regret Prof.Kilian's lectures to be missing in these Proceedings.

The Editors

Acta Physica Austriaca, Suppl. XXVII, 129–175 (1985)

MECHANISMS FOR $N\bar{N}$ ANNIHILATION+

by

A.M. GREEN
University of Helsinki
Research Institute for Theoretical Physics
Helsinki, Finland

1. INTRODUCTION

The $N\bar{N}$ interaction at low energies is a subject of current interest both experimentally and theoretically.

On the experimental side, the advent of LEAR (Low-Energy Antiproton Ring) at CERN will (or should) revolutionize $N\bar{N}$ physics at low energies - as will be described in the lectures at this school by Dr.Kilian and as can also be found in the proceedings of the two LEAR workshops [12]. The outcome, amongst other things, should be detailed measurements of various $N\bar{N}$ cross sections (with $\bar{p}$ and possibly $\bar{n}$ beams) as a function of energy $0 \lesssim E(lab) \lesssim 500$ MeV and angle, and also the branching ratios for $N\bar{N}$ annihilation into specific meson channels. These should provide very stringent tests for theoretical models. In particular, since annihilation involves the <u>inner</u> part of the $N\bar{N}$ wavefunction

+Lectures given at the XXIV. Internationale Universitätswochen für Kernphysik, Schladming, Austria, February 20-March 1, 1985.

it is less easy to find equivalent descriptions - unlike NN scattering which only depends on the asymptotic form of the wave function.

On the theoretical side, now that the long-range part ($r \gtrsim 1$ fm) of the NN potential is rather well understood [4] the emphasis over the last few years has been on attempts to understand the short-range NN interaction in terms of quark degrees-of-freedom [13] and more recently in terms of Skyrmions [14]. The time is therefore ripe to confront the corresponding $N\bar{N}$ problem.

The theoretical aspects of $N\bar{N}$ physics at low energies have been reviewed in the past e.g. by Shapiro in ref.[15] and Green and Niskanen in ref.[16]. In fact from the latter reference much of the present lectures can be found in an expanded form. Only the more recent developments (post 1983) are now given in some detail.

2. $N\bar{N}$ SCATTERING AT LOW ENERGIES

The framework for describing $N\bar{N}$ scattering and annihilation at low energies is usually based on some form of the non-relativistic Schrödinger equation

$$[T + V(ME) + V(ANN) - E]\psi(N\bar{N}) = 0 , \qquad (2.1)$$

where V(ME) is a real potential based on meson exchanges and V(ANN) is a complex potential to take into account annihilation.

2.1 Forms of V(ME)

In the literature V(ME) has been written down in many forms most of which are based on meson-exchange models of NN scattering. Through the G-parity transformation this type of NN potential instantly yields the corresponding $N\bar{N}$ potential by simply reversing the sign of the π- and ω-exchange contributions. Such an approach dates back to at least 1968 with the Bryan-Phillips potential [1] derived from the OBEP of Bryan-Scott [2]. Later versions employed more sophisticated NN potentials e.g. Dover-Richard [3] start with a simplified form of the Paris potential [4] in which the σ- and ρ-mesons of the OBE approach are replaced by two-pion exchange. More recently the Paris group [5] have proceeded with their full NN potential by retaining its various momentum-dependent contributions. In conclusion it can be said that the best forms of V(ME) can be represented by fig.1.

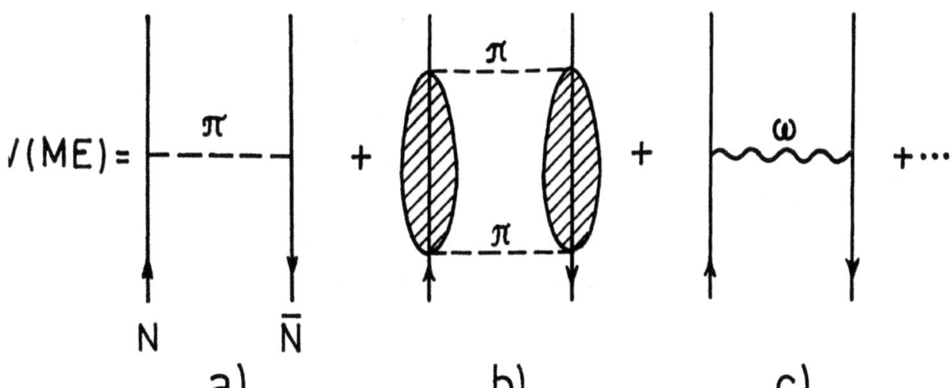

Fig.1: Contribtions to the meson exchange potential V(ME)

a) One-pion exchange (OPEP)

b) Two-pion exchange - the "blobs" represent various intermediate states I in the scattering $\pi N \rightarrow I \rightarrow \pi N$

c) One-omega exchange

It should be added that in some cases it is convenient to use forms of V(ME) that can be partially treated by analytic techniques. The best example of this is, where V(ME) is expressed in terms of separable potentials. Again this can be carried out with varying degrees of complexity. These range from: -

a) the 1-term separable form of the Heidelberg group [6] and the 2-term form of ref.[7], both of which are designed to fit low-energy properties of the Dover-Richard potential up to several MeV.
b) the multi-term forms of the Graz group [8] which simulate the Paris potential [5] over a wide range of energies.

With these separable potentials the equ.(2.1) can to some extent be solved analytically. This also facilitates the introduction of many-body effects when the $N\bar{N}$ interaction is employed in some $\bar{N}$-nucleus situation [7].

For a description of NN scattering at energies E(NN,lab.) $\lesssim$ 400 MeV, it is possible to ignore the structure of the $\pi N \to \pi N$ vertices of fig.1b) - depicted as black "blobs". This structure only plays a direct role at higher energies as inelasticity begins to enter. However, in the $N\bar{N}$ case, inelasticity (mainly annihilation) occurs even at the $N\bar{N}$ threshold and so it is necessary to include explicitly some of the more important components within the $\pi N \to \pi N$ vertex. Usually the only term treated is $\pi N \to \Delta$(1236 MeV) $\to \pi N$. However, as will be discussed later, it is possible that other terms are needed to describe some particular annihilation channels e.g. $\pi N \to N^*$(-ve parity) $\to \pi N$ for the two-body decays $N\bar{N}$ (Stopped) $\to M^S M^S$ where the M^S are s-wave mesons π,η,ρ,ω or η'. When the Δ or N^* are treated explicitly, V(ME) must be modified to avoid double counting their effect. This is illustrated in fig.2 and the appropriate coupled Schrödinger equations to be solved are

$$[T+\tilde{V}(ME)+V(ANN)-E]\psi(N\bar{N}) = V(Tr)\psi(N\bar{\Delta})$$
$$[T+\tilde{V}(\Delta,ME)+V(\Delta,ANN)-E]\psi(N\bar{\Delta}) = V(Tr)\psi(N\bar{N}). \qquad (2.2)$$

$$V(ME) \rightarrow \tilde{V}(ME) = V(ME) \; -$$

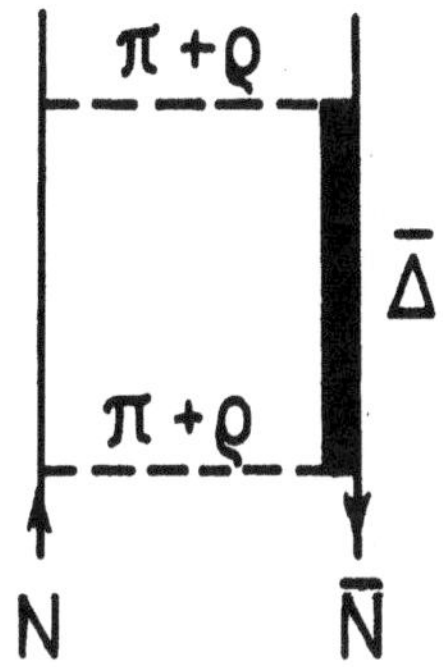

Fig.2: The removal from V(ME) of the Δ-box diagram to avoid double counting in coupled-channels approaches.

Here $\psi(N\bar{\Delta})$ is the wavefunction for the $(N\bar{\Delta} \pm \Delta\bar{N})$ configuration, V(Tr) the transition potential for $N\bar{N} \rightarrow (N\bar{\Delta} \pm \Delta\bar{N})$ and V(Δ,ANN) is the complex potential for the direct annihilation $N\bar{\Delta}\rightarrow$Mesons. There is no reason to believe that V(ANN), which represents $N\bar{N}\rightarrow$Mesons, should be equal to V(Δ,ANN). The term $\tilde{V}(\Delta,ME)$ represents the diagonal interaction $N\bar{\Delta} \rightarrow N\bar{\Delta}$ and is usually set to zero. The conclusion to be made about the explicit introduction of, in particular, the Δ(1236) is that:- "If the N and $\bar{N}$ are sufficiently close to annihilate, then they are certainly sufficiently close to polarize each other into Δ or $\bar{\Delta}$ by the exchange of a pion".

2.2 Forms of V(ANN)

The complex potential V(ANN) in equ.(2.1) is designed to take into account $N\bar{N}$ annihilation. Until recently this has been purely phenomenological with only the range having some theoretical basis. For example, Bryan-Phillips [1] consider V(ANN) to be only imaginary

$$V(\mathrm{ANN,B.P}) = -i8\ \mathrm{GeV}/[1 + \exp(5r)]\,, \tag{2.3}$$

whereas Dover-Richard [3] include a real part

$$V(\mathrm{ANN,D.R}) = (-21-i20)\,\mathrm{GEV}/[1 + \exp(5r)]\,. \tag{2.4}$$

In both cases the short range (1/5 fm) is dictated by the idea that annihilation can be explained in terms of N-exchange, as depicted in fig.3. The more recent potential of Paris [5] parametrizes the imaginary part of V(ANN) as

$$V(\mathrm{ANN,P}) = [g_c(1+f_c\,T_{lab}) + g_{ss}(1 + f_{ss}\,T_{lab})\underline{\sigma}_1\cdot\underline{\sigma}_2 + \\ + g_T\,S_{12} + \frac{g_{LS}}{4M^2}\,\underline{L}\cdot\underline{S}\,\frac{1}{r}\,\frac{d}{dr}]\,\frac{K_o(2Mr)}{r}\,, \tag{2.5}$$

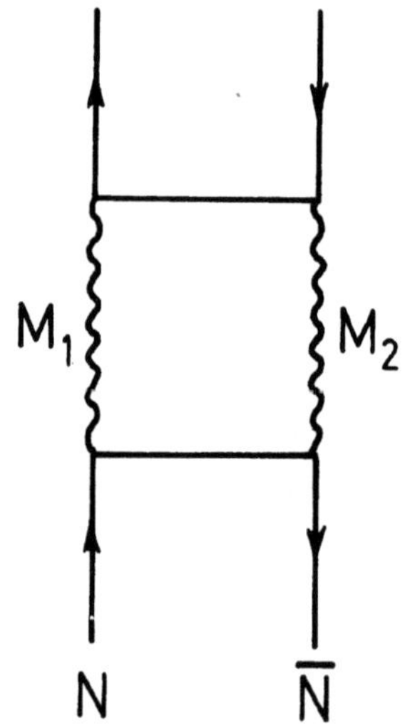

Fig.3: $N\bar{N}$ annihilation into two mesons M_1M_2 treated as N-exchange between a point N and a point $\bar{N}$.

where K_o is the modified Bessel function

$$K_o(2Mr) = \int_{4M^2}^{\infty} dt \exp[-\sqrt{t}\, r]/[t(t-4M^2)]^{1/2}$$

$$\rightarrow \exp[-2Mr]/r^{1/2} \quad \text{as} \quad r \rightarrow \infty$$

and M is the nucleon mass. This radial dependence is now precisely of the form expected from the N-exchange model of fig.3. The main characteristic of the above three forms of ImV(ANN) is that they are all $\simeq$100 MeV for $r \simeq 0.8$ fm - a feature that seems to be necessary to achieve a good fit to experimental data [9]. Such a strong potential at such large values of r is somewhat surprising when it is remembered that the range is only $\simeq \frac{1}{2M} \simeq 0.1$ fm. However, this is arrived at by the use of potential strengths that are enormous being in the 10 - 100 GeV range. In section 3 it will be shown that similar effects can be achieved without the use of these very strong short-ranged potentials.

In addition to those phenomenological approaches that treat annihilation in terms of a complex potential, there are others [10,11] that introduce the effect through the use of coupled channels. These represent in some average way the inelastic channels entering into the problem. The most complete attempt in this direction is by the Nijmegen group [11] with

$$V = \begin{bmatrix} V(N\bar{N}) & V(A) \\ V^+(A) & 0 \end{bmatrix}, \qquad (2.6)$$

where $V(N\bar{N}) = V(ME) + V(PHEN) + V(COULOMB)$ and $V(A) = V_0/[1 + \exp(2.2\, r)]$. Here V(ME) is a theoretical meson-exchange potential and V(PHEN) a phenomenological short-ranged, real, spin- and isospin-dependent potential. The coupling potential V(A) is also phenomenological. Furthermore, it is dependent only on isospin and connects the $N\bar{N}$ to only two "typical" 2-meson channels with masses $2M_1 = 1700$ MeV and $2M_2 = 420$ MeV i.e. $\simeq 2\rho$ (or 2ω) and $\simeq 2\pi$. The outcome is a model giving a fit to experimental data comparable to that of the Paris [5] potential.

3. MICROSCOPIC MECHANISMS FOR $N\bar{N}$ ANNIHILATION

The complex potential V(ANN) of equ.(2.1) can be thought of as arising from a whole series of processes some of which are depicted in fig.4. Most effort has been devoted to the study of the mechanisms in fig.4a) and b) since these lead naturally to threshold effects and do not involve exotic intermediate states containing gluons [as in d)] or $N\bar{N}$ bound states [as in c)]. The problem in a) and b) now reduces to knowing the coupling potentials $N\bar{N} \rightarrow$ mesons - the V(A) of equ.(2.6). This has been, and still is, a source of heated debate. In fig.5 are shown some of the possibilities for these couplings. Most of these lectures will now be devoted to a detailed study of these alternatives.

3.1 Nucleon-exchange mechanisms

A good philosophy in physics is to make developments that are, as far as possible, extensions of well founded ideas without introducing new ones for the particular process under consideration. In $N\bar{N}$ annihilation this means using models based on Nucleon-Nucleon-Meson (NNM) vertices and avoiding microscopic descriptions (e.g. quark models) of the nucleons and mesons. These NNM approaches have a long history and often their general features, such as having a range of $\simeq 1/2M \simeq 0.1$ fm, are imposed on phenomenological potentials (see section 2.2). However, only recently, by

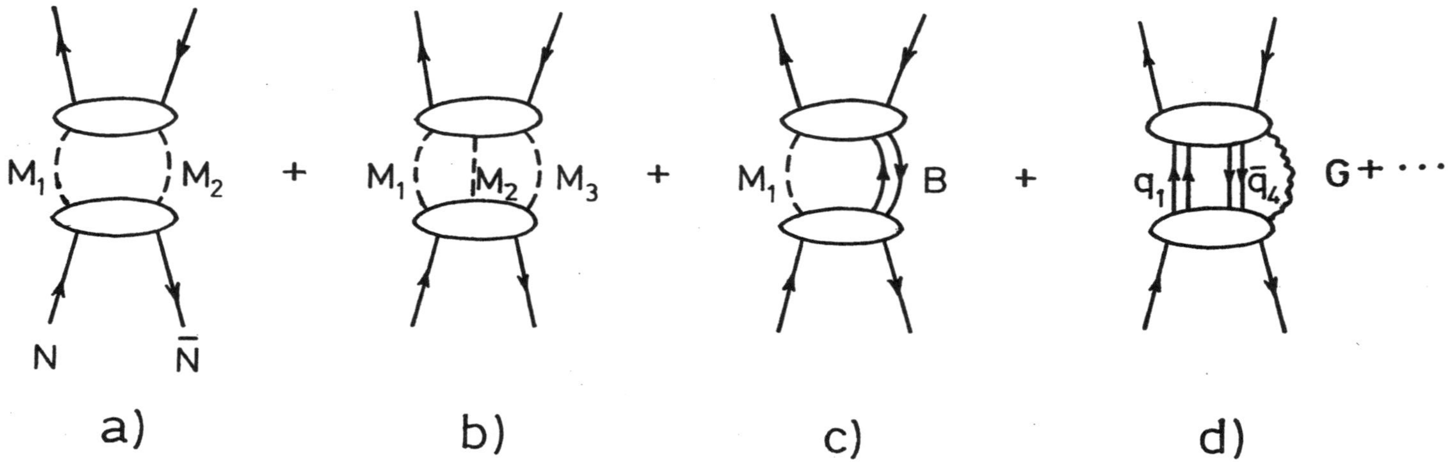

<u>Fig.4:</u> Contributions to V(ANN)

a) Two-meson intermediate states M_1M_2

b) Three-meson intermediate states $M_1M_2M_3$

c) $N\bar{N}$ bound state B <u>plus</u> a single meson M_1

d) Quark-Antiquark-Gluon intermediate state

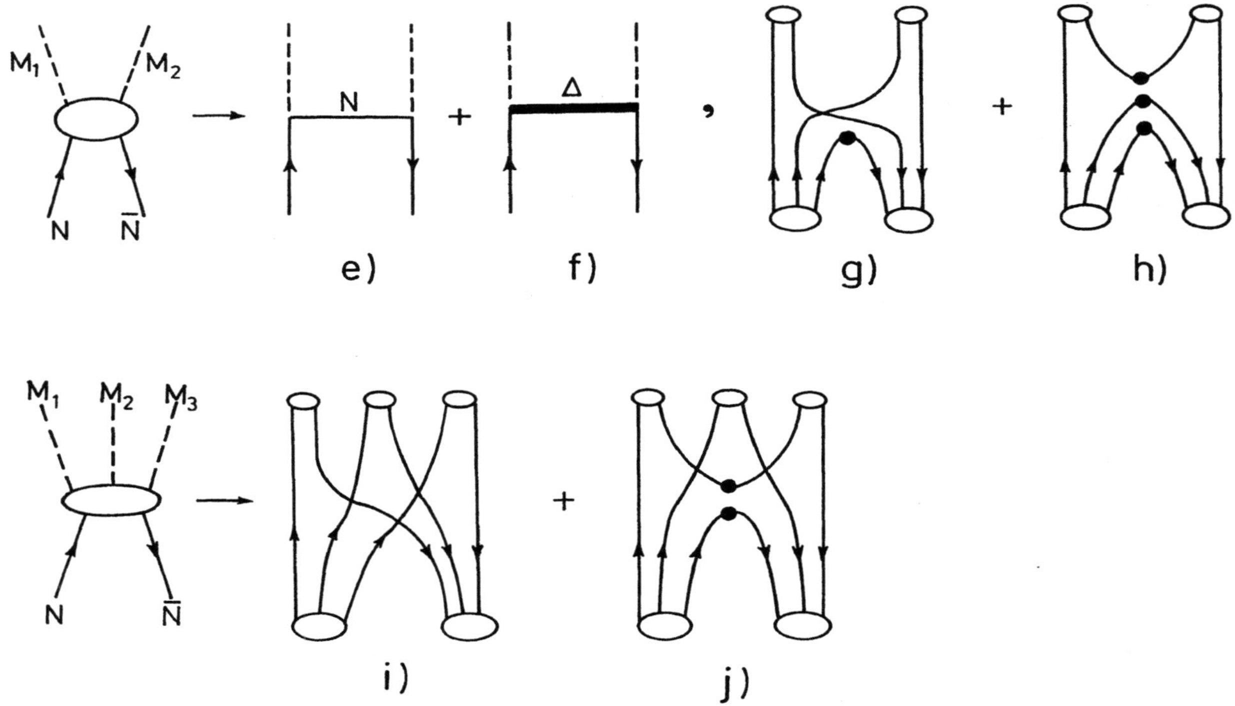

Fig. 5.: Contributions to the coupling $N\bar{N} \rightarrow$ Mesons

e) N-exchange between N and $\bar{N}$

f) Δ-exchange

g) Quark-Antiquark ($q\bar{q}$) annihilation at ● and ($qq\bar{q}\ \bar{q}$) rearrangement into two mesons

h) ($q\bar{q}$) ($q\bar{q}$) annihilation and ($q\bar{q}$) creation to give two mesons

i) ($qqq\ \bar{q}\bar{q}\bar{q}$ rearrangement into three

j) ($q\bar{q}$) annihilation and creation to

Moussallam [17,18], has the approach been carried through in a realistic manner. In ref. 17 he describes, for $T_{lab} \lesssim 400$ MeV, $N\bar{N}$ annihilation in terms of the N-exchange process of fig.5e) and is able to get a reasonable understanding of some specific 2-meson annihilation channels $N\bar{N} \to M_1M_2$. However, the sum of the various cross-sections, i.e. $\sigma(N\bar{N}\to\Sigma_{ij} M_iM_j)$ with $M_{i,j} = \pi,\eta,\rho$ and ω, accounts for less than 20 % of the observed total annihilation cross section. In his later paper, ref. 18 , Moussallam extends his calculations to include the Δ-exchange process of fig.5f) and finds this has a large effect on his earlier work. Of course, this should not have been surprising, since, in spite of the 300 MeV mass difference between the Δ and the N, there is the inequality $f^{*2}\exp(-\Delta r) > f^2\exp(-Mr)$ for $r \lesssim 1$ fm, where $f^{*2}/4\pi$ the $N\Delta\pi$ coupling has a value $\simeq 0.35$ compared to the $NN\pi$ coupling $f^2/4\pi \simeq 0.08$. In his second paper [18] Moussallam has only calculated $N\bar{N} \to \pi^o\pi^o$ and $\pi^+\pi^-$. It will be of interest to see the Δ-exchange effect on other two-meson channels. Possibly the present model is most appropriate only for the $N\bar{N} \to \pi\pi$ reactions, since in this case the pions could be originating from the pion cloud surrounding the quark and antiquark confinement regions of the N and $\bar{N}$. In fact as discussed after equ.(3.8) the emission of pions from the $N\bar{N}$ confinement region is suppressed by a factor of $\simeq 1/2$ because of sizeable $q\bar{q}$ components within the pion [34]. For a more complete calculation Moussallam should also consider the effect of virtual Δ's before the N- or Δ-exchange process takes place - as shown in fig.6. These have an effect way below the $N\bar{\Delta}$ threshold and presumably become very large as this threshold is approached.

The observation that the N- and Δ-exchange mechanisms are very strongly dependent on the NNM and $N\Delta M$ vertex form factors shows that a microscopic description is necessary for these form factors. However, nowadays such descriptions

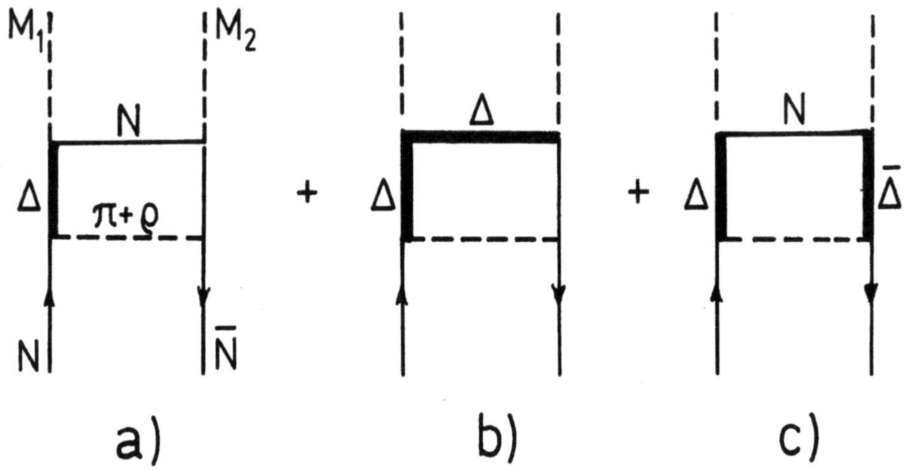

Fig.6: The effect of virtual Δ's
a) $N\bar{N} \to N\bar{\Delta} \pm \Delta\bar{N} \to M_1M_2$ with N-exchange
b) $N\bar{N} \to N\bar{\Delta} \pm \Delta\bar{N} \to M_1M_2$ with Δ-exchange
c) $N\bar{N} \to \Delta\bar{\Delta} \to M_1M_2$ with N-exchange

are normally based on some version of the quark model. Therefore, if the quark model plays so crucial a role in the form factors for the N- and Δ-exchange mechanisms, it would seem natural to abandon these mechanisms and introduce quarks right from the beginning. Even so, it is of interest to complete the programme initiated by Moussallam, since the purely quark approaches of section 3.3 can in some cases be expressed as N- and Δ-exchange with form factors. In this way, a better understanding of these form factors may emerge, since they could be of use in more complicated situations involving many particles where quark approaches may be too complicated to carry out.

For more details on this approach the reader should refer to sect. 3.1 of ref.[16] and also refs.[17-19].

3.2 The effect of possible $N\bar{N}$ bound states on $N\bar{N}$ inelasticity

When antiprotons were stopped in hydrogen the authors of ref. 20 saw a series of essentially monochromatic γ-rays. One possible explanation of these is that the initial $\bar{p}p$ atomic states decay by γ emission to narrow $\bar{N}N$ bound states of nuclear dimensions and binding energies $\simeq$ 105,182,236 and 666 MeV. It should be added that the very existence of these γ-rays is the source of much controversy - see the lectures of Dr. Kilian. Furthermore, the interpretation as $\bar{N}N$ bound states is also very model dependent. With the standard forms of V(ANN) in sect. 2.2 there is no hope of getting such narrow ($\Gamma \lesssim$ 20 MeV) bound states, since the ImV(ANN) are so strong giving $\Gamma \simeq$ 1 GeV [35]. The form of the Paris potential given in equ. (2.5), because of its linear T_{lab} dependence, cannot be used for $T_{lab} \lesssim -$ 40 MeV. Below that energy V(ANN,P) of equ. (2.5) becomes positive. However, the authors of ref. 5 could have parametrized their energy dependence in an exponential form and so avoided this problem [16,28]. But at the time they were only concerned with fitting $N\bar{N}$ scattering data over the energy range $20 \lesssim T_{lab} \lesssim 370$ MeV, where the difference between a linear and exponential T_{lab} dependence was not significant (R. Vinh Mau private communication).

Assuming that there are indeed narrow bound $N\bar{N}$ states B, then it follows [21] that, through the inelastic process $N\bar{N} \rightarrow B\pi \rightarrow N\bar{N}$, there arises a complex contribution to the $N\bar{N}$ potential in addition to the conventional annihilation potential. As discussed in ref. 21 this could account for upto 10 % of the total annihilation cross section for $T_{lab} \lesssim$ 400 MeV.

The effective $N\bar{N}$ potential generated by the above mechanism has a range that is dictated by the wave function of B and takes on the energy-dependent form

$$\mathrm{ImV(ANN.B)} \alpha \sum_i \psi(B_i,r)\psi(B_i,r')[(\tfrac{1}{2}T_{lab}+B_i)^2-m_\pi^2]^{3/2} . \qquad (3.1)$$

Similarly, ReV(ANN.B) has a range $\simeq 1/2\sqrt{B}$ and contributes even below the $N\bar{N} \to B\pi$ threshold. This is no problem in the $N\bar{N}$ system. However, for the NN system the corresponding argument would suggest [22] a contribution to the NN potential arising from the $pp \to d\pi^+$ reaction. Such a potential would have a range $\simeq 1/2\sqrt{B(\text{deuteron})} \simeq 2$ fm which is longer than OPEP. This argument is correct. But it must be remembered that this long-range potential only acts in low partial waves since the initial pp must be in $^1S_0$ to connect significantly with $d(^3S_1)$. Furthermore, the dominant contributions to the $pp \to d\pi^+$ reaction involve Δ-configurations, which have a built-in short-range effect overpowering the long range of the deuteron. Since the model is expected to only apply at energies $T_{lab} >> B$ - otherwise the process in fig. 4c) resembles part of the nucleon mass renormalisation - the question reduces to the possible existence of a long-range NN potential in S-waves for $T_{lab} \gtrsim 50$ MeV. However, the dominance of OPEP at long ranges is only seen in high partial waves - the phase shifts in low partial waves being much different from δ(OPEP) and fitted by the phenomenological parts of the potential.

3.3 Quark-antiquark approaches to annihilation

When confronted with the series of quark vertices in figs. 5g,h,i and j), can any statements be made concerning their relative importance? In the literature two general approaches are advocated. The first [23-26] considers specific annihilation channels and attempts to extract the relative importance of the vertices in fig. 5 by using directly experimental data. Whereas, the second [16,27,28] concentrates on trying to understand the total annihilation cross section by including more and more of the processes in fig. 5.

The first approach can be applied with varying degrees

of sophistication. On the one extreme a specific mechanism can be singled out of fig. 5 and shown to fit a few pieces of data reasonably well or, by parameter adjustment, to give an overall fit to much more data. Examples of this are the works of Genz [23] , who shows that $p\bar{p} \rightarrow \pi^o\pi^o, \pi^o\eta'$ and $\eta\eta'$ can be understood simply in terms of fig. 5h) without the need for fig. 5g), and the works of Maruyama and Ueda [26], who describe $p\bar{p}$ annihilation into a whole series of three-meson states using fig. 5i) by introducing parameters to take into account initial-state interactions and $N,\bar{N}$ form factors. In these cases the reasons for the specific choices from fig. 5 are by no means convincing. In ref. 23a) the author invokes the idea that since fig. 5g) involves crossed quark lines compared with fig. 5h) it is noncoplanar, and so, as shown in the high-energy limit, it is negligible. However, at the energies of interest here, the internal momenta of the quarks and antiquarks in the N and $\bar{N}$ are comparable to the relative momenta between the quarks in N and the antiquarks in $\bar{N}$ - a fact that throws serious doubt on the use of high-energy ideas. A better way of implementing this approach is that of refs. 24,25,32,41. These include both figs. 5g) and h) and attempt to extract their relative importance. Furui et al.[24] find that, in agreement with Genz [23] , fig. 5h) is sufficient, whereas the authors of refs. 25,41 need both mechanisms. For annihilation into three mesons, Furui et al.[24] find that both figs. 5i) and 5j) can explain the data. On the other hand, the authors in ref. 32 find that, for specific decays, the amplitude in fig. 5i) is $\simeq 7$ times that in fig. 5j). However, this whole approach is full of uncertainties some of which are discussed in the conclusion of ref. 28 and in ref. 29. These should now be modified in view of more recent developments. Firstly, the criticism that Genz [23] uses N and $\bar{N}$ wave functions with incorrect symmetry [25] is denied by him in a recent preprint [30]. Secondly, it appears that Furui et al. [24] did not include the possibility of spin flip in their $q\bar{q}$ annihilation vertex. As known for ordinary meson decays - in particular $B\rightarrow\pi\omega$ - this is not a good approximation [39]. This has been corrected in Furui's recent preprint [31].

The second approach for including the mechanisms of fig. 5 is much more ambitious and is the one being pursued by the Helsinki group. The first step is to attempt an understanding of the total annihilation cross section for $T_{lab} \lesssim 400$ MeV. This amounts to a theoretical description of the V(ANN) in equ. (2.1). The input is again a selection of the processes in fig. 5 and eventually the corresponding annihilation branching ratios will be calculated with the appropriate initial-state $N\bar{N}$ interaction automatically included in a self-consistent manner. This avoids, for example, the adjustable parameters needed in ref. 26 to simulate these initial $N\bar{N}$ interactions. In the first papers [27,28] by the Helsinki group only the purely rearrangement mechanism of fig. 5i) was included. This choice was guided by several considerations:

1) The earlier work of Maruyama and Ueda [26] made the same assumption with the claim that experimentally >80 % of all $N\bar{N}$ annihilation seemed to be directly into three mesons. However, this is apparently far from being a well accepted experimental fact (C. Amsler private communication).

2) The analyses of ref. 32 indicated that fig. 5i) dominated over fig. 5j)

3) Since the early attempts in ref. 27,28 concentrate on the total cross sections and not specific annihilation channels, the final results are not biassed by the adjustment of any free parameters. In fact, as will be seen later, the mechanism in fig. 5i) leads to a <u>separable</u> $N\bar{N}$ potential. Such a potential has the interesting feature that it can only give up to a certain maximum cross section independent of the strength of the potential.

4) The model is easily extended to include other terms e.g. in ref. 33 fig. 5g) is now being incorporated. This choice of fig. 5g) over that of 5h) is based on ref. 32 and the hope that the $(q\bar{q})$ annihilation vertices introduce a small parameter into the model. A similar hope is also made in the lectures of Dr. Isgur for his discussion of $q^2\bar{q}^2$ states. In the near future it is expected that fig. 5h) will also be included in the Helsinki model.

So far no statement has been made about the form of the $(q\bar{q})$ annihilation and creation vertices in fig. 5. In practice two alternatives are widely used - the $^{13}P_0$ and $^{13}S_1$ models. The $^{13}P_0$ model admits that $q\bar{q}$ vertices are presumably complicated objects involving the exchange of many soft gluons with the neighbouring quarks and antiquarks. This situation is therefore simulated by an effective coupling of the $q\bar{q}$ directly into the vacuum - see equ. (3.12) for the explicit form. With this assumption and using non-relativistic quark wave functions a whole series of meson decays and baryon vertices can be understood with a single coupling strength γ[38]. Recently Furui [42] has shown, how the $^{13}P_0$ model and essentially the same value of γ can be extracted using relativistic wave functions. The main advantage of the $^{13}P_0$ model is that the interaction $q\bar{q} \to$ Vacuum is expressed as a one-body operator. This makes it easy to deal with many-body systems. On the other hand the operator in the $^{13}S_1$ model is of two-body form and is guided by single-gluon exchange. However, it is hard to see, how this can be anything other than an effective interaction again, since the $^{13}S_1$ model as applied in low-energy $N\bar{N}$ annihilation involves situations containing many soft gluons. In ref. 43 this point is to some extent confirmed by the attempts to fit the total annihilation cross section with this model. There it is found that the qqG coupling needed is $\simeq 10$ the usually accepted QCD value. However, it should be added that a fit to a selection of elastic and charge-exchange data can be achieved with a qqG coupling that is more in line with the QCD value [44]. Furthermore, the authors of ref. 46 find that the explicit gluon processes in fig. 7 - the same as considered in ref. 41 - appear to lead to some interesting selection rules. In particular they find that, unlike the $^{13}P_0$ model, $p\bar{p}$ annihilation into $\pi^{\pm}\rho^{+}$ is inhibited from $\bar{p}p$ $^{31}S_0$ and $^{33}P_J$ states at threshold - an observation that is in agreement with the experimental data [47]. However, as pointed out in ref. 56, for decay processes relativistic effects can be very important especially for those decays that would be forbidden in the non-relativistic limit. To clarify this issue one of the experiments now being performed at LEAR [37] hopes to see, whether it is the $^{13}P_0$ - or $^{13}S_1$-model dominating the reaction $p\bar{p} \rightarrow \Lambda\bar{\Lambda}$ at the $\Lambda\bar{\Lambda}$ threshold.

It should be added that the Cornell group (eg., ref. 48) has an alternative $q\bar{q}$ vertex in which the annihilation takes place in a $q\bar{q}$ $^{1}S_0$ state - N.B. a spin singlet not triplet as suggested by a single-gluon vertex. This model has been restricted to treating heavy-meson decays for which it has been successful.

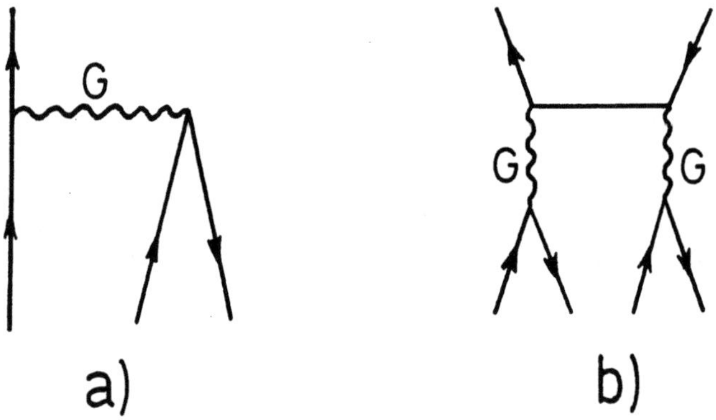

Fig. 7: Interpretation of figs. 5g) and h) in terms of a) single-and b) double-gluon exchange between quarks.

3.4 The Helsinki model

In principle, $N\bar{N}$ scattering can be described by the series of coupled equations

$$[T(N\bar{N}) + V(ME)-E]\psi(N\bar{N}) = \sum_i V_i\psi_i(M's)$$

$$[T_1(M's)+\Delta M_1-E]\psi_1(M's) = V_1\psi(N\bar{N})$$

$$[T_2(M's)+\Delta M_2-E]\psi_2(M's) = V_2\psi(N\bar{N}) \quad \text{etc.,} \tag{3.2}$$

where $\psi(N\bar{N})$ is the $N\bar{N}$ wave function, the $\psi_i(M's)$ are the various mesonic annihilation-channel wave functions and ΔM_i is the mass difference between the mesons and $N\bar{N}$. Equ. (3.2) is assumed to also contain the effect of $N\bar{N} \rightarrow N\bar{\Delta} \pm \Delta\bar{N}$ and $N\bar{N} \rightarrow \Delta\bar{\Delta}$ configurations with double counting removed as shown in fig. 2 and equ. (2.2). If, as in equ. (3.2), there are no diagonal interactions between the mesons, then the problem can be reduced without approximations to the form of equ. (2.1) involving only $\psi(N\bar{N})$ and

$$V(ANN) = - \sum_i V_i \frac{1}{T_i(M's)+\Delta M_i-E} V_i \ . \tag{3.3}$$

The radial form of V_i depends on which mechanism is being considered in fig. 5.

3.4.1 The coupling potentials V_i

In the first papers [16,27,28] of the Helsinki group only fig. 5i) was treated. As shown in these papers V_i can be expressed as the simple overlap

$$V_i \propto \langle N\bar{N}|M_1M_2M_3\rangle, \tag{3.4}$$

where

$$\psi(N) = N_1 \exp - [\frac{\alpha}{2} \sum_{i=1}^{3} (\underline{r}_i - \frac{\underline{r}}{2} - \underline{R})^2]$$

$$\psi(\bar{N}) = N_1 \exp[- \frac{\alpha}{2} \sum_{i=4}^{6} (\underline{r}_i + \frac{\underline{r}}{2} - \underline{R})^2]$$

$$\psi(M_1) = N_2 \exp[- \frac{\beta}{2} \sum_{i=1,4} (\underline{r}_i - \frac{\underline{r}'}{2} - \frac{\underline{\rho}'}{3} - \underline{R}')^2] \quad \text{etc.,} \qquad (3.5)$$

are the $N, \bar{N}$ and meson wave functions expressed in terms of quark and antiquark coordinates $\underline{r}_1$ - - - $\underline{r}_6$. The coordinates $\underline{r}, \underline{R}$ and $\underline{r}'$, $\underline{\rho}'$, $\underline{R}'$ describe the relative and centre-of-mass positions of the $N, \bar{N}$ and $M_1 M_2 M_3$ and are related to the $\underline{r}_i$ by a series of δ-functions, e.g.

$$\delta(\frac{\underline{r}_1+\underline{r}_2+\underline{r}_3}{3} - \frac{\underline{r}_4+\underline{r}_5+\underline{r}_6}{3} - \underline{r}) \text{ and}$$

$$\delta(\frac{\underline{r}_1+\underline{r}_4}{2} - \frac{\underline{r}'}{2} - \frac{\underline{\rho}'}{3} - \underline{R}') .$$

Integrating over the $\underline{r}_i$ results in

$$V_i \propto \exp[- \frac{\alpha}{2}(r'^2 + \frac{4}{3}\rho'^2) - \frac{3}{4} \beta r^2] , \qquad (3.6)$$

where $\alpha = 1/b^2$ and $\beta = \sqrt{\frac{4}{3}} \alpha$ - with b being the proton rms charge radius. In this way V(ANN) from equ. (3.3) can be expressed as

$$V(ANN) = - \frac{\hbar^2}{M} \lambda \, I(E) \exp[- \frac{3}{4} \beta (r^2+r''^2)] \qquad (3.7)$$

- a separable interaction acting only in S-waves. The overall factor I(E) is both energy and state dependent and has the form

$$I(E,TS) = \sum_{m=M_1M_2M_3} A_m\, C(TS,m)\, I(E,m). \qquad (3.8)$$

The C(TS,m) are SU(4) recoupling coefficients and are tabulated in several places, e.g. in ref. 26 for the $N\bar{N}$ system and ref. 27 for the $N\bar{N}$, $N\bar{\Delta}$ and $\Delta\bar{\Delta}$ systems. The A_m allows for the possibility that a particular meson is not a pure $q\bar{q}$ system made up from u, d quarks. This means that each η or η', with ideal mixing, introduces a factor of 1/2 into A_m and each pion another factor of $\simeq 1/2$ to take into account the presence of their more complex $q^n\bar{q}^n$ configurations [34]. In equ. (3.8) I(E,m) is a phase-space integral

$$I(E,m) = \iint_0^\infty \frac{dk_1 dk_2 k_1^2\ k_2^2 \exp[-\frac{1}{\alpha}(k_1^2+\frac{3}{4}k_2^2)]}{\frac{k_1^2}{2\mu_{12}} + \frac{k_2^2}{2\mu_3} - E_m - i\varepsilon}\ , \qquad (3.9)$$

where $\mu_{12} = \frac{M_1M_2}{M_1+M_2}$ and $\mu_3 = \frac{(M_1+M_2)M_3}{M_1+M_2+M_3}$. The $E_m = E(cm) + 2M - \sum_{j=1,2,3}(M_j - i\frac{\Gamma_j}{2})$, where the presence of the meson widths Γ_j tends to smooth out the various threshold effects. These widths play a crucial role in annihilation, since, in particular, the ρ-meson would otherwise often not contribute at low energies. In some cases the C(TS,m) in equ. (3.8) can be very large for configurations involving ρ-mesons, so that non-negligible effects arise from the ρ-width at energies far below the nominal threshold. In ref. 28 various forms for the energy dependences of the meson widths are discussed. However, the different possibilities tend to give essentially the same results for $0 \lesssim T_{lab} \lesssim 400$ MeV. Only in the prediction of deeply bound $N\bar{N}$ states do the various choices lead to different estimates for the bound state widths [35].

In ref. 16,27,28 the λ in equ. (3.7) is essentially a free parameter with the only theoretical estimate being a very crude one in the appendix of ref. 27. However, for a better estimate of V_i as

$$V_i = \langle N\bar{N}|H-E|M_1M_2M_3\rangle \tag{3.10}$$

the form in equ. (3.4) is misleading. Conceptually it is better to consider that only the interaction terms in H - the total Hamiltonian - give rise to coupling, with the kinetic-energy terms and E not contributing. This amounts to assuming $\langle N\bar{N}|M_1M_2M_3\rangle = 0$ with the orthogonality being guaranteed by some property of the $|N\bar{N}\rangle$ or $|M_1M_2M_3\rangle$ that is not explicitly given in equs. (3.5). Such a property could be that of the N and $\bar{N}$ each having a "junction" [36] or "winding number" [14] which must be annihilated before mesons are produced. Furthermore, since at low energies the $N\bar{N}$ annihilation and elastic cross sections are comparable and the short-range part of the real component of the $N\bar{N}$ potential is thought to be of a non-perturbative origin [14], then it follows that the imaginary component of the $N\bar{N}$ potential should also be due to a non-perturbative mechanism. In H the only candidates to fulfill this last requirement are:

1) the quark and antiquark confining potential, which is of the form

$$U = K \sum_{i,j} \underline{\lambda}_i \cdot \underline{\lambda}_j |\underline{r}_i - \underline{r}_j|^2 , \qquad K = \frac{\hbar^2}{M_q} \frac{1}{(2b)^4} \tag{3.11}$$

to be consistent with equs. (3.5) and

2) the $q\bar{q}$ annihilation vertex [40] of the $^{13}P_o$ model

$$U(^{13}P_o) = i\, 3\gamma\sqrt{\tfrac{2}{3}}\, \left(\tfrac{3}{4\pi}\right)^{1/2} [2\underline{\nabla}_x \chi(q\bar{q},\ S = 1)]^{J(q\bar{q})=0}\, \delta(\underline{x}) \tag{3.12}$$

- a vertex simulating multigluon effects and found to give a satisfactory understanding of many-meson decays $M_1 \to M_2M_3$ and baryon vertices $B_1B_2M_3$ [38] with $\gamma = 3.4$.

The use of simply the U of equ. (3.11) as the coupling potential between configurations with different topologies is, to some extent, justified in the lectures at this school of Dr. Isgur.

The requirement that the annihilation should be non-perturbative makes it hard to understand how the $^{13}S_1$ $q\bar{q}$ model, which is based on the idea that a $q\bar{q}$ annihilate into a <u>single</u> gluon, could play a large role in $N\bar{N}$ annihilation.

Assuming that U in equ. (3.11) is the coupling interaction [6,33] leads to <u>precisely</u> the expressions for V(ANN) in equ. (3.7) with $\lambda = 8.32$ (MeV fm^2)/b (fm) $\simeq$ 15 MeV·fm for the usually accepted value of $b \simeq 0.6$ fm. Similar forms can be obtained for other situations, e.g. with $v_1(r) = \exp(-\frac{3}{4}\beta r^2)$

$$V[N\bar{N}(S) \to M_1^pM_2^pM_3 \to N\bar{N}(S)]$$

$$= -\frac{\hbar^2}{M}\lambda\frac{\beta^2}{3} I(pps)v_2(r)v_2(r'),$$

where $v_2(r) = [r^2 - \frac{8/3}{\alpha+\beta}]v_1(r)$

$$V[N\bar{N}(P) \to M_1^pM_2M_3 \to N\bar{N}(P)]$$

$$= -\frac{\hbar^2}{M}\frac{\lambda\beta}{3} I(pss)v_3(r)v_3(r'),$$

where $v_3(r) = rv_1(r)$

$$V[N\bar{N}(D) \to M_1^dM_2M_3 \to N\bar{N}(D)]$$

$$= -\frac{\hbar^2}{M}\frac{\lambda\beta^2}{15} I(dss)v_4(r)v_4(r'),$$

where $v_4(r) = r^2v_1(r)$

$$V[N\bar{N}(D) \rightarrow M_1^p M_2^p M_3 \rightarrow N\bar{N}(D)]$$

$$= -\frac{\hbar^2}{M} \lambda \frac{2}{15} \beta^2 \; I(pps) v_5(r) v_5(r'),$$

where $v_5(r) = v_4(r)$. (3.13)

Here M^p and M^d are p- and d-wave mesons e.g. B,H,A, -- and ω(1670) ..., and I(pps) etc. have definitions similar to the I(sss) of equ. (3.8). It should be added that in ref. 28 the second interaction - the one proportional to I(pps) - did not contain the factor of $\frac{1}{3}$ in the overall constant. However, this does not alter the conclusions of ref. 28.

The interaction of equ. (3.11) only leads to three-meson final states, whereas that of equ. (3.12) gives two-meson final states and, with $v'(r) = \exp[-\frac{3}{8}(\alpha+3\beta)r^2]$,

$$V[N\bar{N}(S) \rightarrow M_1^p M_2 \rightarrow N\bar{N}(S)]$$

$$= -\frac{\hbar^2}{M} C \frac{9}{4} \beta I(ps) v_6(r) v_6(r'),$$

where $v_6(r) = [r^2 - \frac{4/3}{\alpha+\beta}] v'(r)$

$$V[N\bar{N}(P) \rightarrow M_1 M_2 \rightarrow N\bar{N}(P)]$$

$$= -\frac{\hbar^2}{M} C \; I(ss) v_7(r) v_7(r'),$$

where $v_7(r) = r\, v'(r)$

$$V[N\bar{N}(D) \rightarrow M_1^p M_2 \rightarrow N\bar{N}(D)]$$

$$= -\frac{\hbar^2}{M} C \frac{9}{10} \beta I(ps) v_8(r) v_8(r'), \quad (3.14)$$

where $v_8(r) = r^2 v'(r)$.

Here $C = 0.023\ \beta^3/(\alpha+\beta)$ and I(ps), I(ss) have definitions similar to equ. (3.9). Unfortunately, the problem with using the interaction of equ. (2.12) is that its strength γ is extracted from situations [$M_1 \to M_2M_3$ and $B_1B_2M_3$] which do not involve the annihilation of junctions or winding numbers. Therefore, a more correct treatment could be a two-step process [29,33] in which first the interaction in equ. (3.11) annihilates the junctions and produces three mesons one of which is in a $^{13}P_0$ state - the ε meson. Then the interaction of equ. (3.12) annihilates the ε-meson into the vacuum. This is depicted in fig. 8.

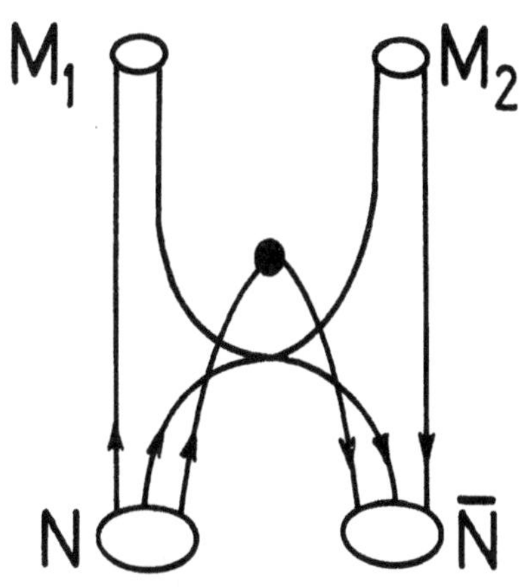

Fig.8: $N\bar{N}$ (rearrangement) $\to M_1M_2\varepsilon$ followed by $q\bar{q}$ annihilation to give M_1M_2.

When $N\bar{N}$ annihilation involves p- or d-wave mesons, then the SU(4) recoupling coefficients C(TS,m) are not those tabulated in refs. 26,27. Such a complete tabulation would cover many pages. However, in ref. 28 a special selection is given for $N\bar{N}(P) \to M_1M_2$ where M_i are π,η,ρ or ω. Of course, these now depend on the $q\bar{q}$ annihilation model considered - in this case $^{13}P_o$. These have been tabulated - for the same $^{13}P_o$ model - elsewhere in the literature but showing, for some unknown reason, different results. However, as proved in ref. 33 there is a simple relationship between these two-meson coefficients and the three-meson coefficients tabulated in ref. 26,27 - namely

$$C(2,J) = \frac{1}{q}\frac{2S+1}{2J+1} C(3,J) . \tag{3.15}$$

Here C(2,J) is the two-meson coefficient in the state P_J and C(3,J) is the three-meson coefficient for $\omega[M_2M_3]^J$ in the state S. The one complication in applying equ. (3.15) is that in refs. 26,27 only $\sum_J C(3,J)$ is quoted when M_2 and M_3 are both vector mesons.

3.4.2 Results

At short distances the meson-exchange potential V(ME) of equs. (2.1) and (3.2) and also the transition potentials V(Tr) of equ. (2.2) are cut off by the form factor

$$F_c(r) = (r/r_c)^{10}/[1+(r/r_c)^{10}] , \tag{3.16}$$

where $r_c \simeq 0.6 - 0.8$ fm. The function $F_c(r)$ drops rapidly to zero for $r < r_c$ and so removes the contributions from meson exchange for those values of r where interquark effects are expected to give a better description of the overall interaction.

For the latest version of the Helsinki model [33] the main results are as follows:

a) Including only the s-wave mesons π,η,ρ,ω and η' in the rearrangement process of fig. 5i) i.e. $N\bar{N}(S) \rightarrow M_1M_2M_3$ gives a contribution of $\simeq .10$ mb to the $N\bar{N}$

S-wave annihilation cross section at E(lab) = 100 MeV. This should be compared with the unitarity limit of 26 mb and the experimental cross section of ≃100 mb for the sum of all partial waves. This S-wave result is changed by at most 5 % when p-wave mesons are included in the configurations $M_1^pM_2^pM_3$.

b) Configurations containing a single p-wave meson ($M_1^pM_2M_3$) contribute at 100 MeV up to 15 mb to the $N\bar{N}$ P-wave annihilation cross section.

c) Configurations containing a single d-wave meson ($M_1^dM_2M_3$) contribute at 100 MeV ≃ 5 mb i.e. about 40 % of the theoretical value. Again the $M_1^pM_2^pM_3$ configurations contribute very little.

d) In all, the three-meson states $M_1M_2M_3$, $M_1^pM_2^pM_3$, $M_1^pM_2M_3$ and $M_1^dM_2M_3$ can yield up to ≃40 % of the observed cross section, the precise number depending on the λ in equs. (3.13) and the r_c in equ. (3.16).

The large difference compared with the earlier result of ≃70 % in ref. 28 arises from two sources. The presence of the 1/3 factor in the $V[N\bar{N}(P) \rightarrow M_1^pM_2M_3 \rightarrow N\bar{N}(P)]$ of equ. (3.13) now makes the S,P and D wave contributions peak at <u>different</u> values of λ. Furthermore, the problem of double counting $N\bar{\Delta} \pm \Delta\bar{N}$ and $\Delta\bar{\Delta}$ effects, as illustrated in fig. 2, was not corrected with sufficient accuracy.

e) The effect of including two-meson configurations via the $^{13}P_o$ model and the mechanism of fig. 5g) is large in all three partial waves considered, raising the calculated cross section up to ≃ 60 % of the observed value at 100 MeV. However, the resultant branching ratios for $N\bar{N} \rightarrow M_1M_2$ and $M_1^pM_2$ show some disturbing features e.g. 1) $N\bar{N}$(S-waves) → $B\pi$ is ≃ 5-10 %, whereas experimentally it is 0.7 ± 0.1 %, 2) $N\bar{N}(P) \rightarrow \pi^{\pm}\rho^{\mp}$ is dominated by the $^{33}P_1$ channel compared with the experimental indications [46,47] that it is $^{11}P_1$.

f) $N\bar{\Delta} \pm \Delta\bar{N}$ and $\Delta\bar{\Delta}$ configurations can be important in both scattering and specific branching ratios as seen in figs. 9 and 10.

g) As seen in fig. 11 P-wave annihilation dominates over S-wave for $E_{lab} \gtrsim 10$ MeV. Even at 1 MeV the P-wave contribution is ≈10 % of the S-wave.

h) As seen in fig. 9 the effect of including the meson widths in the denominator of equ. (3.3) and (3.9) is important. As discussed in ref. 28 this can be done in several ways - with differing energy dependences. However, for $0 \lesssim E(lab) \lesssim 400$ MeV they lead to essentially the same results. Only in the discussion of deeply bound $N\bar{N}$ states [35] do differences arise.

3.4.3 Improvements

In section 3.4.2 the most serious defect of the model seems to be the prediction of a $N\bar{N}(S) \rightarrow B\pi$ branching ratio that is an order of magnitude too large. Theoretical branching ratios and annihilation cross sections that are too small are less worrying, since the argument can always be made that they arise from some mechanism not yet included. Of course, one simple explanation of the $B\pi$ problem is that for stopped antiprotons only a small fraction of the annihilation actually takes place from the $N\bar{N}(^{13}S_1 - {}^{13}D_1)$ state, the one appropriate for the $B\pi$ branching i.e. the annihilation is dominated by other S-wave, and also P-wave, $N\bar{N}$ states. However, it should be added that the model developed in the previous subsections has many deficiencies some of which will now be briefly described.

a) Induced tensor effects.

Part of the $B\pi$ problem is the occurrence of a large $^{13}S_1$ and $^{13}D_1$ components in thewave function even at very low

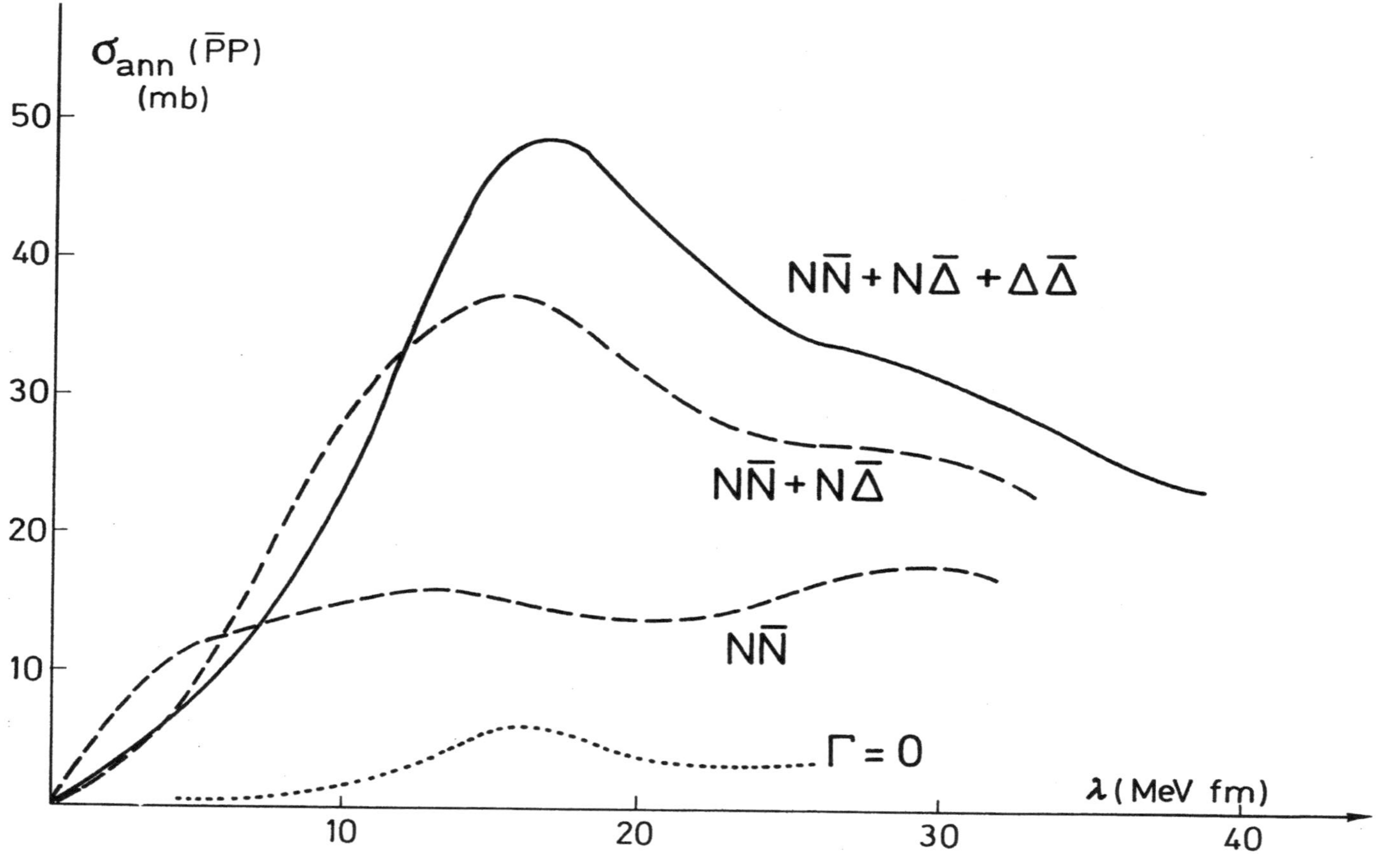

Fig. 9.: The effect of $N\bar{\Delta} \pm \Delta\bar{N}$ and $\Delta\bar{\Delta}$ configurations on the P-wave annihilation cross section at E_{lab} = 100 MeV. The dotted line shows the result of putting all meson widths (Γ) to zero.

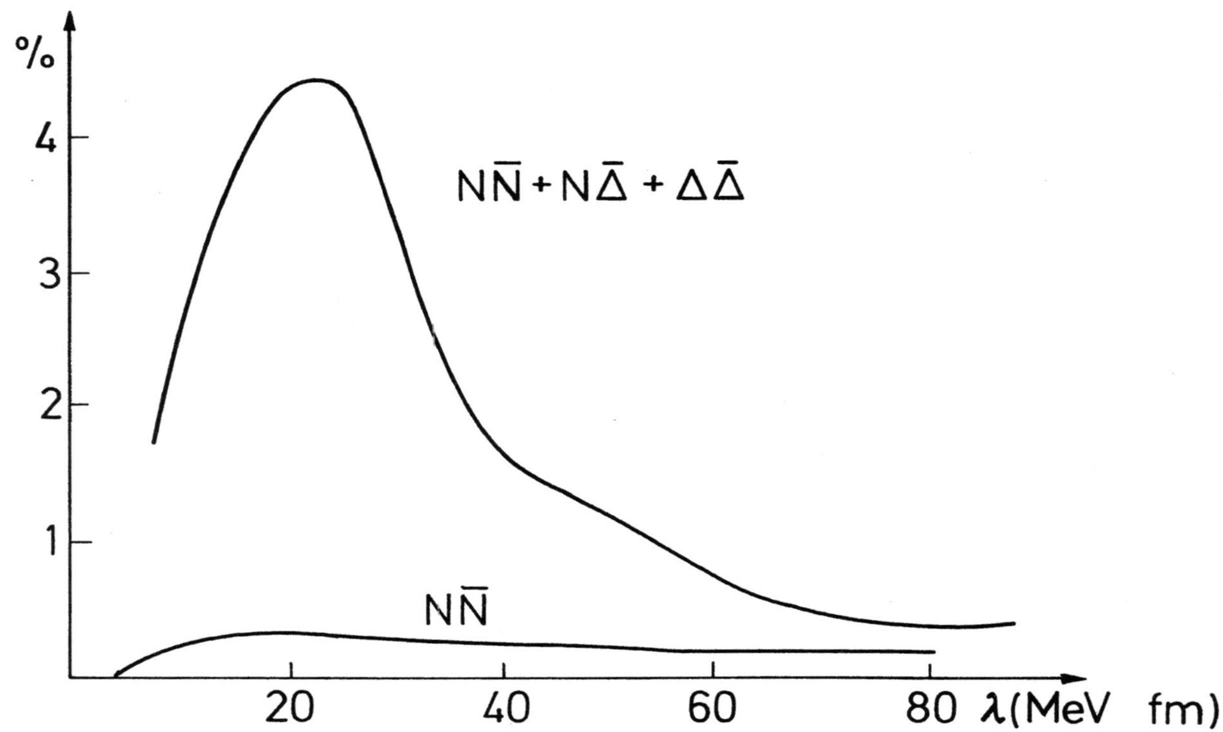

Fig. 10.: The effect of $N\bar{\Delta} \pm \Delta\bar{N}$ and $\Delta\bar{\Delta}$ configurations on the $N\bar{N}(S) \rightarrow \eta\pi^{o}\omega$ branching ratio for stopped $\bar{p}$'s. [54]

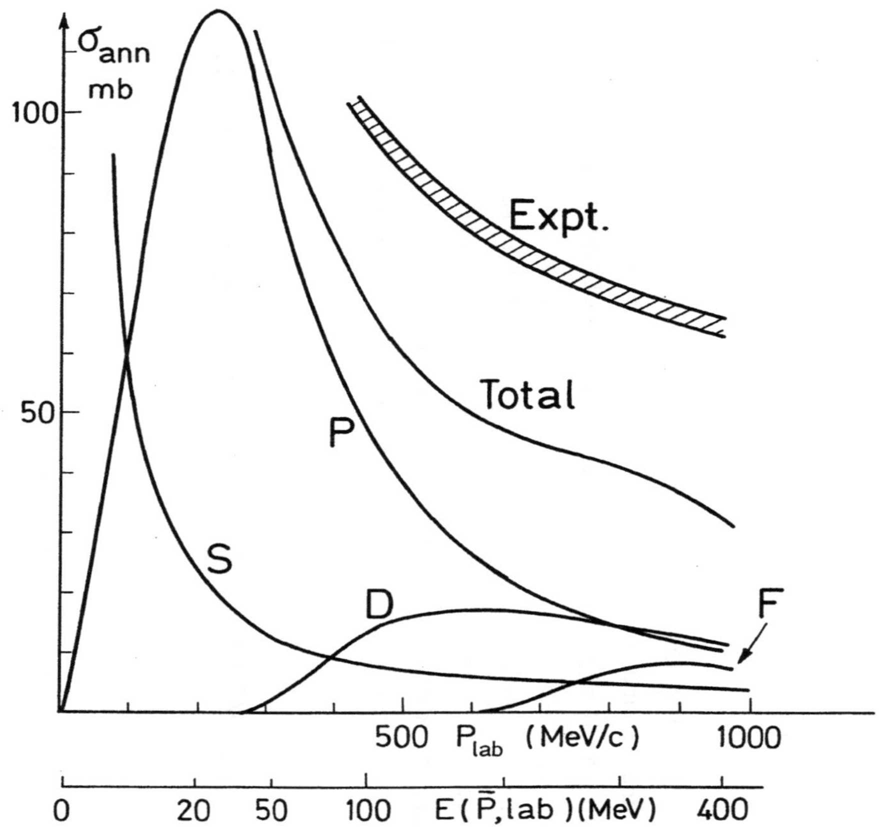

Fig. 11.: Contributions to the total annihilation cross section.

energy - indicating that tensor effects are possibly too strong. One way of reducing this could be the off-diagonal effect of $N\bar{N}(S) \to M_1^p M_2 \to N\bar{N}(D)$, which is non-zero because the various s(p)-wave mesons are not degenerate in mass. This could and, perhaps, should be included in the present framework.

b) The effect of $N\bar{N} \to$ [a single meson] $\to N\bar{N}$. This could be as important as Partovi [45] suggests. In the elementary-particle description of fig. 12a) the contribution to the real part of V(ANN) has the form

$$\mathrm{Re}\, V(\mathrm{ANN},M_1) = 16\pi f^2 \frac{\delta(\underline{r})\delta(\underline{r}')}{(2M+E)^2-\mu^2}, \tag{3.17}$$

where f^2 is the NNM_1 coupling constant (e.g. 14 and $\simeq 5$ for the π and ω) and μ the meson mass. This formula is expected to hold for $E \simeq \mu - 2M$. For the energies of interest ($0 \lesssim E \lesssim 200$ MeV) the quark-model description of fig. 12b) suggests in the standard $^{13}P_0$ model a specific radial dependence. If this is now identified with the form factor that arises when going away from the energy $E \simeq \mu - 2M$ to $E \simeq 0$, then the $\delta(\underline{r})\delta(\underline{r}')$ in equ. (3.17) becomes $F(r)F(r')$, where

$$F(r) = 3\left(\frac{3}{2\pi}\right)^{3/2} \frac{\alpha+\beta}{\alpha}\left(\alpha+\frac{3}{2}\beta\right)^{5/2}\left[r^2-\frac{2}{3(\alpha+\beta)}\right]\exp\left[-\frac{3}{2}r^2\left(\alpha+\frac{3}{2}\beta\right)\right]. \tag{3.18}$$

Here $\int d\underline{r}F(r) = 1$ and α and β are the parameters in the oscillator wave functions for the N, $\bar{N}$ and M_1 - see equ. (3.5). If, however, the philosophy behind fig. 8 is invoked, i.e. first $N\bar{N}$ rearrangement into $M_1\varepsilon\varepsilon$ and then $(q\bar{q})(q\bar{q})$ annihilation into M_1, the appropriate F(r) would be

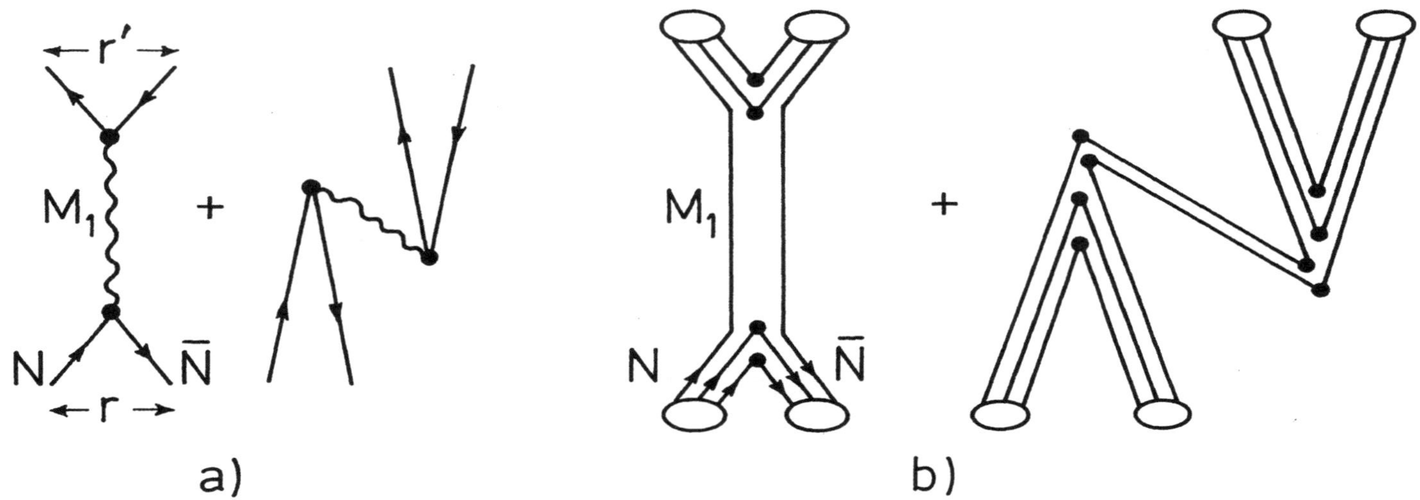

<u>Fig.12:</u> a) $N\bar{N} \rightarrow M_1 \rightarrow N\bar{N}$ as elementary particles
b) $N\bar{N} \rightarrow M_1 \rightarrow N\bar{N}$ in terms of $q\bar{q}$ annihilation and creation vertices.

$$F(r) = \left(\frac{3\beta}{4\pi}\right)^{3/2} \exp\left(-\frac{3}{4}\beta r^2\right). \tag{3.19}$$

This form has the attractive feature of becoming $\delta(\underline{r})$ as $\delta \to \infty$, i.e. the meson becomes point-like. On the other hand, the F(r) in equ. (3.18) tends to $\alpha^{5/2} r^2 \exp[-\frac{15}{4}\alpha r^2]$. This latter r^2 factor is unavoidable when the two $^3P_0$ $q\bar{q}$ vertices act directly on the initial $N\bar{N}$ state. Replacing the $\delta(\underline{r})\delta(\underline{r}')$ in equ. (3.17) by $F(r)F(r')$ gives a separable $N\bar{N}$ potential appropriate for $E \simeq 0$. A local equivalent potential can be formed as

$$\begin{aligned} V(M_1,\text{local}) &= \int d\underline{r}' \ \text{Re}\ V(ANN,M_1) \\ &= 16\pi f^2 \frac{F(r)}{(2M)^2-\mu^2}\,. \end{aligned} \tag{3.20}$$

Using the F(r) in equ. (3.19) this results in the spin-averaged $p\bar{p}$ potential

$$\bar{V}(M_1 = \pi+\eta+\rho+\omega+\eta') = 600 \exp(-2.4 r^2)\,\text{MeV}, \tag{3.21}$$

which is comparable, but of opposite sign, to the spin-averaged $p\bar{p}$ potentials due to the configurations $M_1M_2M_3$ and $M_1^pM_2M_3$ - see ref. 28 equs. (3.7) and (4.3), namely,

$$\text{Re}\ U(\text{S-wave},\ M_1M_2M_3) = -\ 950 \exp(-2.1\ r^2)\,\text{MeV}$$

$$\text{Re}\ U(\text{P-wave},\ M_1^pM_2M_3) = -\ 1350 \exp(-2.1\ r^2)\,\text{MeV}.$$

However, it should be remembered that these local equivalent potential arguments can be misleading when there is a possibility for nodes in the $N\bar{N}$ wavefunction. Here this could indeed be the case, since there have been indications for $N\bar{N}$ bound states [20]. Advocates for a strong policy of duality might be worried by the inclusion of mesons in the t-channel [as in V(ME)] and, at the same time, also in the s-channel [as in fig. 4]. However, it should be added that in situations as complicated as the multimeson exchanges considered in fig. 4, proofs of duality do not exist. On the other hand, for the single-meson case discussed in this section the duality violation is more blatant. In this case the double-counting problem, if any, is minimized by cutting off V(ME) inside $r_c \simeq 0.6 - 0.8$ fm by the $F_c(r)$ in equ. (3.16). Since the mechanisms in fig. 4 only contribute at short ranges, the t- and s-channel effects at least do not double count in coordinate space.

c) Rearrangement before $(q\bar{q})$ annihilation. As mentioned after equ. (3.14) and depicted in fig. 8, it may be necessary to treat fig. 5g) for the $N\bar{N} \to M_1M_2$ as a two-step process. First the $N(q^3)\bar{N}(\bar{q}^3)$ rearrange as in fig. 5i) into $M_1M_2\ \varepsilon$, where ε has the vacuum quantum numbers. This will be proportional to $\sqrt{\lambda}$, where λ is given in equ. (3.13), and removes the "junctions" or "winding numbers" that characterize baryons compared with mesons [36]-see the discussion after equ. (3.10). The ε now propagates as $1/2\ m_q$, with m_q being the mass of an u,d quark, and then annihilates into the vacuum by means of the operator in equ. (3.12). In this way the process in fig. 5g) becomes proportional to $\sqrt{\lambda}\ \gamma/2m_q \exp[-\frac{3}{4}\beta r^2]$ compared with $\gamma \exp[-\frac{3}{2}\beta r^2]$. However, in spite of the longer range, preliminary calculations indicate that the overall strength is considerably weaker.

d) $^{13}P_0$ model inadequate. As discussed at the end of sect. 3.3 there are indications both experimental [47] and theoretical [46] that, unlike the $^{13}P_0$-model, the single-gluon exchange model - $^{13}S_1$ model - can account for various selection rules for meson branching ratios.

e) More "meson physics" should be included. In NN physics it is now a well established fact that the transition NN → NΔ (both virtual and real) has important consequences [49]. As seen from figs. 9 and 10 the indications are that $N\bar{\Delta}$ and $\Delta\bar{\Delta}$ configurations could be equally important - if not more so - in $N\bar{N}$ physics. The question then arises concerning the possible effects of resonances other than the Δ(1236). At present, in NN physics, experimentally there does not seem to be any overwhelming need for other resonances. However, recently [50] the analysis of some spin correlations in the $pp \to d\pi^+$ reaction does seem to indicate effects that can be explained by $\pi N \to N^* \to \pi N$ amplitudes characteristic of N^* [S_{11}(≃ 1600 MeV) or S_{31}]. On the theoretical side, these odd-parity resonances appear to be much more important than often realized [51]. This is illustrated in fig. 13 where it is seen that much of the effect of the NΔ diagrams (fig. 14a,b) is cancelled by the corresponding diagrams of figs. 14c) and d) which contain an intermediate N^*. In ref. 51 the N^* is considered to be the resonances P_{11} - the Roper - and D_{13}. But the effects of these two are not separated in ref. 51. However, the indications are that it is the D_{13} that gives the largest correction. Furthermore, it is clearly fig. 14d) that dominates over fig. 14c), since the latter is purely attractive. The N^* giving rise to mainly repulsion is also consistent with the $pp \to d\pi^+$ data mentioned above. (D. Bugg private communication)

What has this got to do with $N\bar{N}$ physics? The answer is the large $N\bar{N}(S) \to \pi^\pm\rho^\mp$ branching ratio of 5.8 ± 0.3 %. For the $^{13}S_0$ model fig. 5g) gives this branching ratio as zero. It should be added that in the single-gluon exchange model of ref. 46 there is no problem for this decay. However, an alternative is the two-step process shown in fig. 15 in which the initial S-wave $N\bar{N}$ state is converted into a P-wave $N^*\bar{N}$ state. This relative P-wave along with the internal P-wave of the N^* enables within the $^{13}P_0$ model the creation of a $\pi^\pm\rho^\mp$ state also in a relative P-wave. In

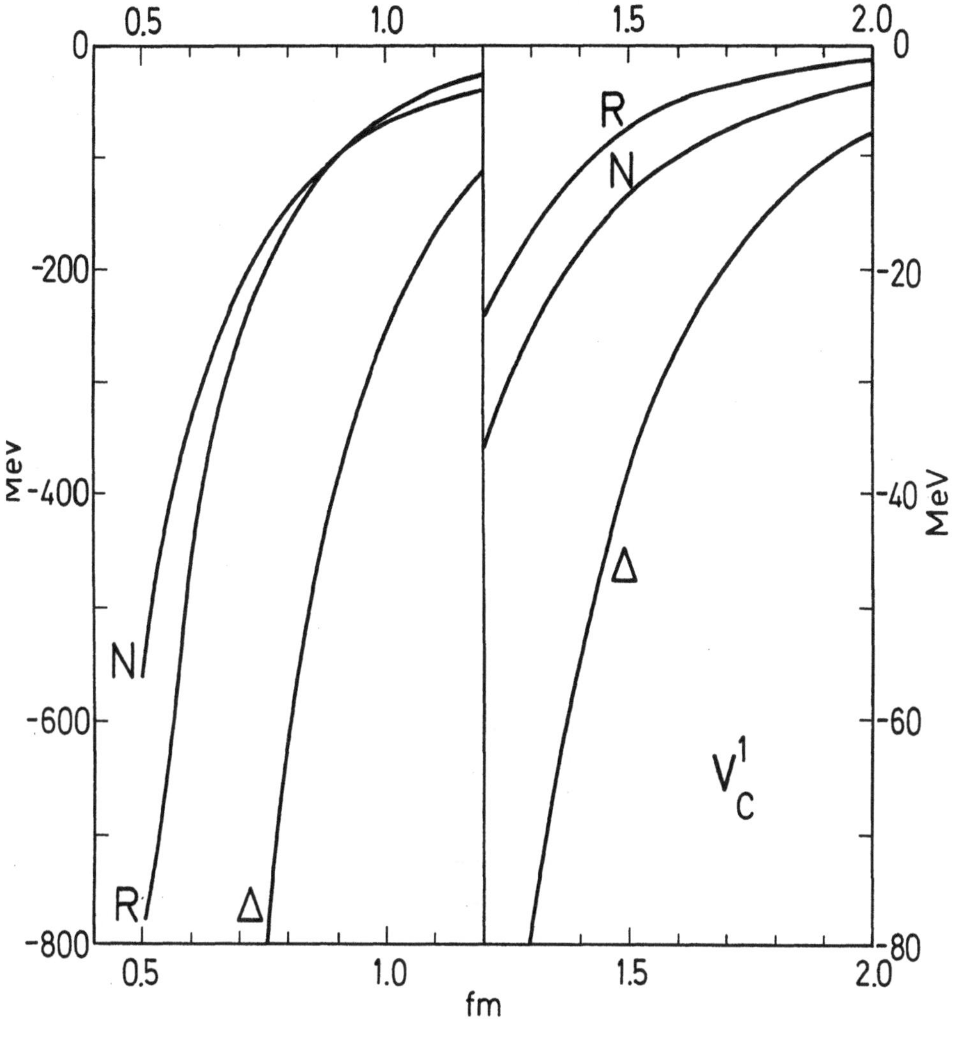

Fig.13: The T = 1 central contribution from the processes in fig. 14 - N is for the all-nucleon-box plus cross-pion process - Δ the effect of including figs. 14a,b) - R the effect of including further figs. 14c,d) in curve Δ.

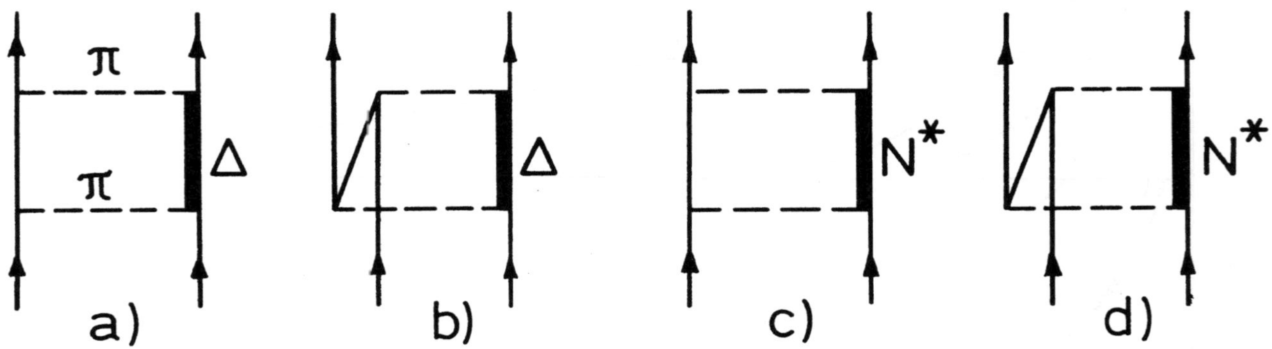

<u>Fig.14:</u> Contributions to the NN potential.

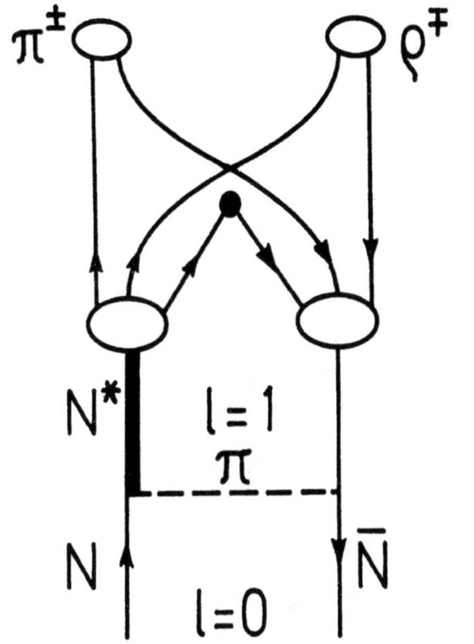

Fig.15: The decay $N\bar{N}(S) \rightarrow N^*\bar{N}(P) \rightarrow \pi^{\pm}\rho^{\mp}$

these arguments only the corresponding $N\bar{\Delta}$ and $N\bar{N}^*$ box diagrams [figs. 14a) and c)] contribute to the annihilation. Whereas, the cross-pion diagrams - here interpreted as the virtual excitation of another $\bar{N}$ in figs. 14b) and d) - do not contribute owing to their intermediate mass being 2M larger than their partners in figs. 14a) and c).

Another example of where "meson physics" before "quarks" could be important is in annihilation processes resulting in K-mesons. Usually decays of the type $N\bar{N} \rightarrow KK$ are given as processes needing fig. 5h) with its three $(q\bar{q})$ vertices, and ruling out the rearrangement process of fig. 5g) - with its single $(q\bar{q})$ vertex. The reason for this is simply that strange quarks need to be created and so two $(q\bar{q})$ pairs have to be annihilated, However, with the two-stage mechanism depicted in fig. 16, first the $N\bar{N}$ exchange a K(493 MeV), K^*(892 MeV)

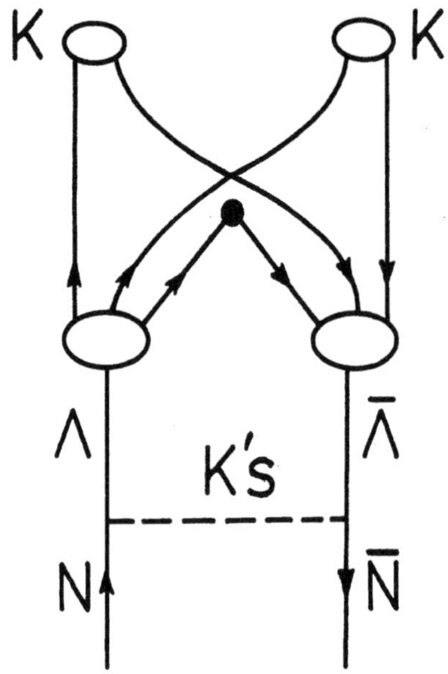

Fig.16: The annihilation $N\bar{N} \to KK$ as a two step process involving an intermediate $\Lambda\bar{\Lambda}$ configuration [55].

and possibly [53] the tensor meson K^{**}(1430 MeV) to generate a $\Lambda\bar{\Lambda}$ configuration, which is then followed by the rearrangement process of fig. 5g). In fact, some support for this mechanism comes from its ability to explain in a very direct way [52] the observation that $N\bar{N} \to K_S K_S$ is inhibited by about two orders of magnitude compared with $N\bar{N} \to K_L K_S$. This can be understood when it is noted that the $K_S K_S$ state has purely isospin T = 1 whereas $K_L K_S$ is 50 % T = 0. Since $\Lambda\bar{\Lambda}$ has also purely T = 0, it can not decay to $K_S K_S$. Furthermore, the $\Lambda\bar{\Sigma}$ and $\Sigma\bar{\Sigma}$ states, which do have T = 1 components, only couple weakly to $N\bar{N}$ since the $N\Sigma K$ coupling constant is much weaker (by about an order of magnitude) than the corresponding $N\Lambda K$ coupling.

f) The short-range $N\bar{N}$ potential. The present model and its suggested improvements contain several mechanisms which contribute to the real part of the $N\bar{N}$ potential at short distances,e.g. all the processes in fig. 4. However, none of these correspond to the short-range NN potential of refs. 13 due to quark-quark interactions or of ref. 14 due to the Skyrmion-Skyrmion interaction. In the NN case, the "baryon property" of junction [36] or winding number [14] is preserved throughout the interaction, unlike the mechanisms in fig. 4 in which the intermediate states contain no such "baryon property". This is illustrated in fig. 17a) versus b) where the dotted line represents the "baryon property". In addition, for $N\bar{N}$ there are other intermediate states such as the exotic states of figs. 17c) and d), where the "baryon property" is preserved. However, these can be viewed as contributions described by single- and double-meson exchange, and as such appear in V(ME). Possibly, at shorter distances the Skyrmion-Skyrmion approaches of ref. 14 include also such effects and so give an estimate of not only fig. 17a) by also in addition c) and d).

4. CONCLUSION

Hopefully these lectures have shown that the study of $N\bar{N}$ annihilation mechanisms presents challenging problems in a field where much rapid progress is currently occuring both experimentally and theoretically. In addition, I hope it could be seen that these studies are still very much in their infancy. At present it is not even clear which processes dominate and what (if any) is the small expansion

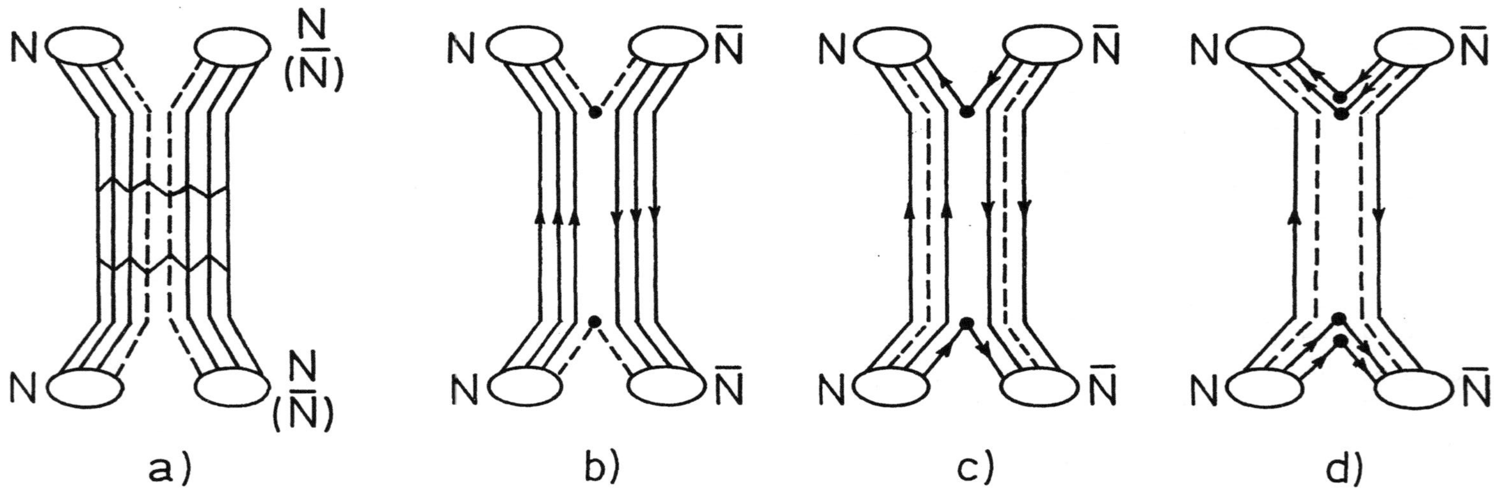

Fig.17: The description of NN(a) and $N\bar{N}$(a,b,c,d) scattering including the "baryon property". In a) ∿∿ implies quark and gluon exchanges.

parameter suppressing other processes. Of course, purists will say that since annihilation presumably involves soft gluons the search for a microscopic theory based on QCD is premature. This means that "effective interactions" are probably necessary and these must be adjusted to fit as much data as possible not only in the $N\bar{N}$ system but preferably in other situations. Probably the best example of this is the γ of the $^{13}P_0$ model - see equ. (3.12). Another could be the oscillator strength K in equ. (3.11). However, as pointed out at the end of the last section, "meson physics" effects such as the excitation of $N\bar{\Delta}$, and $\Lambda\bar{\Lambda}$ configurations should not be forgotten. In fact, probably the most consistent approach is to first include the "meson physics", since these are usually on a much firmer foundation having been often tested and found to be successful in the corresponding NN situations. Of course, the importance of "meson physics" should not come as a surprise to quark-only advocates, since even they usually restrict their quarks to a well or bag so that the quark contribution to the proton charge R.M.S. radius is only ≈ 0.6 fm. The remaining 0.2 fm, to give the experimental value of ≈ 0.8 fm, is then attributed to the meson cloud outside the quark region. However, this passive role of the cloud should now be modified to include the excitation of any neighbouring baryons before annihilation.

Much of these lectures are based on a joint effort of past and current research with Doc. J.A. Niskanen at the Research Institute for Theoretical Physics in the University of Helsinki. The author also wishes to thank V. Kuikka for many useful comments and computations incorporated here without specific acknowledgement to him.

REFERENCES

1. R.A. Bryan and R.J.N. Phillips, Nucl. Phys. B5 (1968) 201; R.J.N. Phillips, Rev. Mod. Phys. 39 (1967) 681.
2. R.A. Bryan and B.L. Scott, Phys. Rev. 177 (1969) 1435.
3. C.B. Dover and J.M. Richard, Phys. Rev. C21 (1980) 1466.
4. M. Lacombe et al., Phys. Rev. C21 (1980) 861.
5. J. Côté et al., Phys. Rev. Lett. 48 (1982) 1319.
6. G. Ihle, J. Pirner, and J.M. Richard - private communication.
7. A.M. Green and S. Wycech, Nucl. Phys. A377 (1982) 441; A.M. Green, W. Stepien-Rudzka and S. Wycech, Nucl. Phys. A399 (1983) 307.
8. J. Haidenbauer and W. Plessas, Phys. Rev. C27 (1983) 63, C30 (1984) 1822; W. Schweiger, J. Haidenbauer and W. Plessas Graz preprint UNIGRAZ-UTP 01/85, to appear in Phys. Rev. C.
9. T. Ueda, Prog. Theor. Phys. 62 (1979) 1670, 63 (1980) 195.
10. A.M. Badalyan et al., Phys. Repts. 82C (1982) 31.
11. D.H. Timmers, W.A. Van der Sanden, and J.J. De Swart, Phys. Rev. D29 (1984) 1928.
12. Physics at LEAR (Erice Workshop) - Ettore Majorana International Science Series Vol. 17, eds. U. Gastaldi and R. Klapisch (Plenum, New York 1984); Physics at LEAR (Tignes Workshop) - (Editions Frontieres 1985).
13. D.A. Liberman, Phys. Rev. D16 (1977) 1542; C.E. de Tar, Phys. Rev. D17 (1978) 323; C.S. Warke and R. Shanker, Phys. Rev. C21 (1980) 2643; M. Harvey, Nucl. Phys. A352 (1981) 301, 326; S. Ohta et al., Phys. Lett. 119B (1982) 35 Y. Suzuki and K.T. Hecht, Phys. Rev. C27 (1983) 299, C28 (1983) 1458; A. Faessler and F. Fernandez, J. Phys. G: Nucl. Phys. 9 (1983) L39; K. Maltman and N. Isgur, Phys. Rev. Lett. 50 (1983) 1827; J.E.F.T. Ribeiro, Z. Phys. C5 (1980) 27.
14. M. Rho, G.E. Brown, and A.S. Goldhaber, Phys. Rev. Lett. 51 (1983) 747; A. Jackson, A.D. Jackson, and V. Pasquier, Nucl. Phys. A432 (1985) 567; R. Vinh Mau et al., Stony Brook preprint 1984.

15. I. Shapiro, Phys. Repts. 35C (1978) 129.
16. A.M. Green and J.A. Niskanen, to be published in International Review of Nuclear Physics, vol. 1, ed. T.T.S. Kuo (World Scientific, Singapore, 1985).
17. B. Moussallam, Nucl. Phys. A407 (1983) 413.
18. B. Moussallam, Nucl. Phys. A429 (1984) 429.
19. F. Hupe, D. Schütte and K. Bleuler, Phys. Lett. 106B (1981) 264.
20. P. Pavlopoulos et al., Phys. Lett. 72B (1978) 415; B. Richter et al., Phys. Lett. 126B (1983) 284.
21. A.M. Green and M.E. Sainio, J. Phys. G: Nucl. Phys. 6 (1980) 1375; A.M. Green, J. Phys. G: Nucl. Phys. 8 (1982) 485.
22. R. Vinh Mau, Proceedings of the 1983 Indiana University Nuclear Phys. Workshop on "Manifestations of Hadron Substructure in Nuclear Physics".
23. a) H. Genz, Phys. Rev. D28 (1983) 1094
    b) U. Hartmann, E. Klempt and J.C. Korner, Mainz preprint MZ-TH/85-01.
24. S. Furui, A. Faessler and S.B. Khadkikar, Nucl. Phys. A424 (1984) 495.
25. C. Dover and P. Fishbane, Nucl. Phys. B244 (1984) 349.
26. M. Maruyama and T. Ueda, Nucl. Phys. A364 (1981) 297, Phys. Lett. 124B (1983) 121.
27. A.M. Green and J.A. Niskanen, Nucl. Phys. A412 (1984) 448.
28. A.M. Green and J.A. Niskanen, Nucl. Phys. A430 (1984) 605.
29. A.M. Green, Proc. VIIth European Symp. on Antiproton Interactions, Durham, U.K. 1984.
30. H. Genz, Karlsruhe preprint "Comment concerning annihilation diagrams ..."
31. S. Furui, Orsay preprint, December 1984 "$p\bar{p}$ Annihilation into two mesons in the quark model".
32. T. Fields, Phys. Rev. Lett. 36 (1976) 489; R.K. Logan, S. Kogitz and S. Tanaka, Can. J. Phys. 55 (1977) 2059.
33. A.M. Green, J.A. Niskanen and V. Kuikka, in preparation.

34. R. Brockman, W. Weise and E. Werner, Phys. Lett. 122B (1983) 201.
35. J.A. Niskanen and A.M. Green, Nucl. Phys. A431 (1984) 593.
36. G.C. Rossi and G. Veneziano, Nucl. Phys. B123 (1977) 507, Phys. Repts. 63C (1980) 149.
37. R.A. Eisenstein for PS185, Physics at LEAR - Ettore Majorana Int. Science Series Vol. 17, eds. U. Gastaldi and R. Klapisch (Plenum, New York 1984) 477.
38. A. Le Yaouanc et al., Phys. Rev. D8 (1973) 2223, D11 (1975) 1272, and Introduction to the Quark Model of Elementary Particles, Vol. 2, to be edited by D. Flamm. M. Chaichian and R. Kögerler, Ann. Physics 124 (1980) 61; J.P. Ader, B. Bonnier and S. Sood, Nuovo Cim. 68A (1982) 1.
39. F.E. Close - An Introduction to Quarks and Partons (Academic Press, 1979) p. 123.
40. A.M. Green, J.A. Niskanen and S. Wycech, Phys. Lett. 139B (1984) 15.
41. F. Ando, Proc. Int. Summer School on NN Interaction and Nuclear Many-Body Problems, Changchun, China 1983, eds. S.S. Wu and T.T.S. Kuo (World Scientific 1984) p. 689; M. Kohno and W. Weise, University of Regensburg preprint TPR-84-12.
42. S. Furui, Orsay preprint, Dec. 1984 "A $^3P_0$ model with a relativistic quark wave function".
43. R.A. Freedman, W.Y.P. Hwang, and L. Wilets, Phys. Rev. D23 (1981) 1103; M.A. Alberg et al., Phys. Rev. D27 (1983) 536.
44. R. Tegen, T. Mizutani and F. Myhrer, University of South Carolina preprint.
45. H. Partovi, Phys. Rev. D18 (1978) 2861.
46. J.A. Niskanen and F. Myhrer, Helsinki preprint HU-TFT-85-4.
47. M. Foster et al., Nucl. Phys. B6 (1968) 107; S. Ahmad et al., Proc. VIIth European Symp. on Antiproton Interactions, Durham, U.K. 1984.

48. E. Eichten et al., Phys. Rev. D17 (1978) 3090.
49. A.M. Green, Repts. Prog. Phys. 39 (1976) 1109; J.A. Niskanen, Nucl. Phys. A298 (1978) 417.
50. D. Bugg, Queen Mary College preprint 1984.
51. M. Chemtob, J. Durso and D. Riska, Nucl. Phys. B38 (1972) 141.
52. J.A. Niskanen, Helsinki preprint HU-TFT-85-3.
53. F. Tabakin and R.A. Eisenstein, University of Illinois preprint P/85/1/12.
54. J.A. Niskanen, V. Kuikka and A.M. Green, Helsinki preprint HU-TFT-84-33 and 85-6.
55. J.A. Niskanen, A.M. Green and V. Kuikka, in preparation.
56. C. Hayne and N. Isgur, Phys. Rev. D25 (1982) 1944.

Acta Physica Austriaca, Suppl. XXVII, 177–266 (1985)

NUCLEAR PHYSICS FROM THE QUARK MODEL WITH CHROMODYNAMICS+

by

N. ISGUR
Department of Physics
University of Toronto
Toronto, Canada M5S 1A7

It now seems nearly certain that nuclear physics can be derived from quantum chromodynamics. In these lectures I describe the ideas and methods underlying a possible route to such a derivation through the quark model, and some of the progress that has already been made along this path.

PREFACE

The ideas and work described in these lectures are not presented in chronological order. In fact, if they have any value, it will be because the opportunity to lecture on the topics discussed here in La Troisieme Cycle de la Physique en Suisse Romande in the summer of 1984, at the Schladming Winter School in 1985, and at the Los Alamos Summer School in 1985 forced me to recast these developments in a more logical and pedagogical form. I offer my thanks

+) Lectures given at the XXIV. Internationale Universitätswochen für Kernphysik, Schladming, Austria, February 20-March 1, 1985.

to the organizers of these schools (Pierre Extermann at Geneva, Heinrich Mitter and Willibald Plessas at Graz, and Mikkel Johnson at Los Alamos) for their interest and encouragement in this task.

Since these lectures are mainly a reordering and integration of published work, I would like to explicitly acknowledge at the outset my principal collaborators in the areas described here. In chronological order these are Gabriel Karl (baryons), Roman Koniuk (decays), John Weinstein (the $qq\bar{q}\bar{q}$ system), Kim Maltman (the 6q system), Steve Godfrey and Simon Capstick (the relativised quark model for mesons and baryons), and Jack Paton (the flux tube model).

Finally, I apologise in advance for the fact that these lectures will concentrate on developments as seen from Toronto; they do not do justice to the work of many others who have contributed to this field, although occasional references to very closely related work are given. For more complete references the interested student should refer to the works cited.

Nathan Isgur
Toronto, March 1985

1. INTRODUCTION AND OUTLINE

It now seems nearly certain that nuclear physics can be derived from quantum chromodynamics (QCD). In these lectures I will describe a possible route to this derivation via the quark model, and some preliminary (but very encouraging) results which have already been obtained.

I will begin at the beginning: with QCD. From these foundations I will discuss a flux tube model for hadrons based on the strong coupling limit of Hamiltonian lattice QCD. This model will provide a framework which guides (and may eventually justify) the remainder of our discussion. The first stop along our route is to consider the nature and validity of the quark model in QCD. These considerations naturally lead to a recollection of the great successes of the quark model in helping us to understand mesons and baryons; I will take this opportunity to point out some recent progress in removing the derogatory "non-relativistic" prefix usually attached to the model. The remainder of the lectures will be devoted to the extension of the quark model to multiquark systems. These systems are characterized by the fact that they allow many internal colour couplings, in contrast to $q\bar{q}$ and qqq where this coupling is unique. We will first discuss such systems in general, using the simplest multiquark system $qq\bar{q}\bar{q}$ as an example occasionally. With this preparation we will finally turn to the 6q system, from which we will see that we can begin to understand many of the basic features of nuclear physics from the quark model.

2. QCD AND THE QUARK MODEL

While it now seems likely that quantum chromodynamics (QCD) is the correct dynamics underlying the strong interactions, it has not been possible to put this theory to many very severe tests. The most rigorous tests to date have been those in which, via asymptotic freedom, perturbative techniques can be applied, but even in such "simple" processes as deep inelastic scattering these tests remain imprecise. When we turn to phenomena in which the strong interactions are really strong, then aside from some very general features we have almost no rigorous tests of this theory: the calculation of hadronic masses, intrinsic moments, and couplings is in a very rudimentary stage of development. Even many qualitative features of this strong interaction problem remain unclear: for example, do pure glue states exist, and if so are they narrow or broad?

Some progress in the strong interaction problem has been made, however. On one front, exact analytic relations can be derived in the form of sum rules which, with suitable techniques, can reveal many features of the physics of QCD. However, at the moment the greatest hope for eventually extracting precise predictions from QCD seems to lie with methods for solving the theory numerically via lattice techniques. Lattice simulations have, for example, already provided us with convincing (if not conclusive) evidence that QCD is a confining theory. Also, numerous attempts have been made to numerically calculate on the lattice the low-lying hadronic spectrum, with some qualitatively encouraging results. It is very unclear, however, how much further these ab initio calculations can be taken with foreseeable computer capacity.

By contrast, the quark model has had considerable success in systematizing many of the details of meson and baryon spectra, intrinsic moments, and couplings, especially after the naive quark model was supplemented with certain simple dynamical features characteristic of QCD.

In this first lecture we will discuss a model for hadrons based on ideas gleaned from the strong coupling lattice Hamiltonian formulation of QCD. The model contains the ordinary quark model as a special case, but includes as well pure glue states, hybrids (which have both quark and gluonic degrees of freedom in evidence), and other exotics. This model is not QCD but it may be useful as a guide in the present period when rigorous solutions for QCD are unavailable; even later, given the inevitable complexity, it is possible that the model will continue to be useful as a simple qualitative picture of the dyanmics of QCD.

A. A review of QCD

QCD is a non-abelian generalization of QED in which local U(1) phase invariance is replaced by a local SU(3) invariance acting on a new three dimensional internal degree of freedom called colour [1]. If $\psi_\alpha(x)$ ($\alpha = 1,2,3$) is a Dirac field, then under a local (i.e., x-dependent) SU(3) transformation,

$$\psi_\alpha(x) \to \psi'_\alpha(x) = \{\exp(-ig\sum_{i=1}^{8} \theta^i(x) \frac{\lambda^i}{2})\}_{\alpha\beta}\psi_\beta(x) \tag{1}$$

where the λ^i are the usual eight 3x3 generators of SU(3) and $\theta^i(x)$ are "phases". If the Dirac Lagrangian is to be invariant under such transformations, then there must be a set of eight minimally coupled compensating vector fields G^μ_1 (i = 1,...8) (the gluons); by this reasoning one is led to

$$L_{QCD} = \sum_{\text{flavour}} \bar{q}\{\gamma_\mu[i\partial^\mu - g\frac{\lambda^i}{2} G^\mu_i]-m_q\}q - \frac{1}{4} F_{i\mu\nu}F_i^{\mu\nu} \tag{2}$$

where $F^{\mu\nu}_i$ is the QCD field strength tensor

$$F^{\mu\nu}_i = \partial^\mu G^\nu_i - \partial^\nu G^\mu_i - gf_{ijk}G^\mu_j G^\nu_k \tag{3}$$

in which the f_{ijk} are defined via

$$[\frac{\lambda^i}{2}, \frac{\lambda^j}{2}] = if_{ijk} \frac{\lambda^k}{2} \tag{4}$$

i.e., they are the structure constants of SU(3). Notice that the "extra term" in (3) arises here and not in QED because the gauge group is non-abelian. Thus under the gauge transformation which leads to (1) the gluon fields are both "shifted" and "rotated" so that the QED gauge shift $A^\mu(x) \to A'^\mu(x) = A^\mu(x) + \partial^\mu\theta(x)$ becomes here (in the form appropriate to differential phases $\delta\theta^i(x)$)

$$G_i^\mu(x) \to G_i'^\mu(x) = G_i^\mu + \partial^\mu\delta\theta^i + gf_{ijk}\delta\theta_j G_k . \qquad (5)$$

It is this extra non-linear term which is responsible for the remarkable properties of confinement and asymptotic freedom; it is also this innocent-looking term which makes QCD so difficult to solve.

Although it is not the regime in which we are most interested here, we briefly review the phenomenon of asymptotic freedom, which has allowed QCD to be applied with success using perturbative techniques to certain processes involving short-distance physics. At first sight it seems paradoxical that a theory of strong interactions, in which, naively, g would seem required to be large, could ever be treated by perturbation theory. The first step in understanding this is to realize that the coupling constant g appearing in (2) is the "bare" coupling constant and, as such, is not necessarily simply related to the couplings one will observe in various physical processes. Quantum chromodynamics, like any quantum field theory, only makes sense if it is defined (regulated) in terms of some renormalization scheme since the point-like operators appearing in L_{QCD} are highly divergent in character. In the next section we will define QCD in terms of a lattice regularization scheme, but it is more normal to cut the theory off at some very large mass scale M. A renormalizable theory, like QCD, has the property that its predictions for processes with intrinsic scales much less than M are independent of M. Consider, in this light, the quark-gluon coupling at some scale $Q^2 \ll M^2$. We would calculate

$$\alpha_s(Q^2) = f(g,Q^2,M^2) \equiv \alpha_s(g,Q^2,M^2) \tag{6}$$

where f is, for example, obtained by the sum of all relevant Feynman diagrams with integrations cut off at the scale M. Now for a given bare g and cut off M we see that we can calculate $\alpha_s(Q^2)$ at any Q^2; consequently, any measurement of α_s can be used to fix the bare value of g. This choice of g will, however, depend on M since, crudely speaking, g is acting not as the bare coupling of the point-like L_{QCD} but rather as the bare coupling of the M-regulated theory. Thus, given an observed $\alpha_s(Q^2)$, we must choose $g = g(M^2)$; it is this observation that leads to the renormalization group equations. In QED such an analysis indicates that the bare electric charge $e(M^2)$ must be increased with increasing M^2 to maintain α at its observed value of 1/137 at low Q^2; this fact can be attributed to the screening of the bare charge by vacuum polarization, much as an ordinary charge is screened in a dielectric medium. In QCD one finds that the bare colour charge $g(M^2)$ must be decreased with increasing M^2 in order to maintain the physical coupling $\alpha_s(Q^2)$ at fixed values. This observation then leads to the conclusion that for sufficiently small scale phenomena, QCD will behave like a theory with a small coupling and so may be studied by perburbation theory. Conversely, analysis leads to the conclusion that $\alpha_s(Q^2)$ inevitably grows as Q^2 is decreased, until at some characteristic scale Λ_{QCD} one finds $\alpha_s(\Lambda^2_{QCD}) \sim O(1)$, at which point the perturbative analysis fails and along with its failure the possibility that QCD confines emerges.

In the perturbative regime the colour analogs of electrons and photons, namely quarks and gluons, are the obvious degrees of freedom to employ in calculations, and all of the familiar tools of perturbative quantum field theory rather naturally apply. In the low Q^2, strong coupling regime it turns out to be useful to abandon these familiar elements in favour of others.

B. Strong coupling Hamiltonian lattice QCD

Most numerical studies of QCD have been done using a four dimensional lattice on which Feynman path integral techniques can be employed. We are interested here not in numerical but rather in conceptual developments and so will concentrate instead on a Hamiltonian version of QCD in which only space and not time is discretized [2]. In this formulation the spatial lattice spacing a will play the role of the regulator mass M in standard quantum field theory. Latticizing the theory also has another advantage: it will allow us to set up a strong coupling perturbation expansion in which the expansion parameter for lattice QCD is $1/g$ instead of g. We may expect to be able to learn more about the strongly coupled regime of the theory in terms of such an expansion, and indeed this seems to be the case: for example, confinement is an automatic property of the $g\to\infty$ limit of lattice QCD. Moreover, the natural degrees of freedom of the strong coupling regime are not quarks and gluons, but rather quarks and flux tubes, the latter being more in accord with various qualitative ideas on the nature of confinement in QCD. Of course, space is not coarse-grained (at least not on the scale of 10^{-15} metres), so that to relate lattice QCD to real QCD we must consider the limit $a\to 0$. In this limit, as we have discussed, $g\to\infty$ so that a strong coupling expansion must fail; this is just the other side of the failure of the weak coupling expansion for small Q^2. If, however, it can be shown that the two regimes "match" around $g = 1$, thereby proving that lattice QCD as $a\to 0$ is QCD, then one would, nevertheless expect the strong coupling expansion to be useful in many situations where large scales dominate, just as the weak coupling expansion is useful for short distance physics.

A simple analogy may be useful. Consider approximating a continuous one dimensional harmonic oscillator by a particle hopping along a one dimensional lattice of points $x = na$ ($n = \ldots,-2,-1,0,1,2,\ldots$) with lattice spacing a. The lattice Hamiltonian could be chosen to be

$$H_{mn} = \left(\frac{1}{ma^2} + \frac{1}{2} ka^2n^2\right)\delta_{mn} - \frac{1}{2ma^2}(\delta_{m,n+1} + \delta_{m,n-1}) \tag{7}$$

since then the Schrödinger equation

$$i \frac{\partial\psi_m}{\partial t}(t) = H_{mn}\psi_n(t) \tag{8}$$

becomes

$$i \frac{\partial\psi(x,t)}{t} = \left[-\frac{1}{2m}\frac{\partial^2}{\partial x^2} + \frac{1}{2}kx^2\right]\psi(x,t) \tag{9}$$

as $a \to 0$. Now for $a \to \infty$ with k and m fixed, the potential energy term $\frac{1}{2}ka^2n^2\delta_{mn}$ dominates and the eigenstates correspond to the particle sitting on single lattice sites; corrections to this limit are of relative order $\chi = 1/kma^4$ and one can proceed to systematically do perturbation theory in this hopping strength. Since the characteristic scale of the harmonic oscillator is $\alpha^{-1} = (km)^{-1/4}$, one will not get realistic wave functions or eigenenergies for the harmonic oscillator for $a \gg \alpha^{-1}$ where lowest order perturbation theory applies, but for $\chi \sim 1$ one begins to get good approximations to the solutions of the continuum problem if one works to sufficiently high order in χ. By contrast, starting with free particle solutions to the continuum Hamiltonian and treating $\frac{1}{2}kx^2$ as a perturbation is hopeless. (The difference, of course, is that the hopping parameter expansion for the ground state, for example, will be accurate, if a matrix of dimension of order $\frac{1}{a\alpha}$ is diagonalized).

We now turn to the formulation of QCD on a (cubic) spatial lattice. In this formulation the quark degrees of freedom of the theory "live" on the lattice sites, while the gluonic degrees of freedom "live" on the links between these sites (see Fig.1). Let's consider first the theory without quarks: we describe this theory in terms of link variables U_ℓ which (before quantization) are 3x3 SU(3) group elements. The pure gauge field Hamiltonian is then the sum of two parts, one involving only the U's and one which has non-trivial commutation relations with the U's:

$$H_{glue} = \frac{g^2}{2a} \sum_\ell C_\ell^2 + \frac{1}{ag^2} \sum_p \mathrm{Tr}[2-(U_{\ell 4} U_{\ell 3} U_{\ell 2} U_{\ell 1} + \mathrm{h.c.})] \qquad (10)$$

with a the lattice spacing and g the corresponding coupling constant. Here C_ℓ^2 is defined in terms of the eight generators $E^a_{\ell\pm}$ of SU(3) transformations of U_ℓ at the beginning (-) or the end (+) of the link ℓ

$$[E^a_{\ell\,+}, U_\ell] = -\frac{\lambda^a}{2} U_\ell \qquad (11)$$

$$[E^a_{\ell\,-}, U_\ell] = +U_\ell \frac{\lambda^a}{2} \qquad (12)$$

by $C_\ell^2 = \Sigma_a (E^a_{\ell+})^2 = \Sigma_a (E^a_{\ell-})^2$. In the second term the product of the U's is taken in order around the plaquettes p. To complete lattice QCD one simply adds to (10) a lattice Hamiltonian H_{quark} for the quarks interacting with the glue. With the quark fields as site variables we have

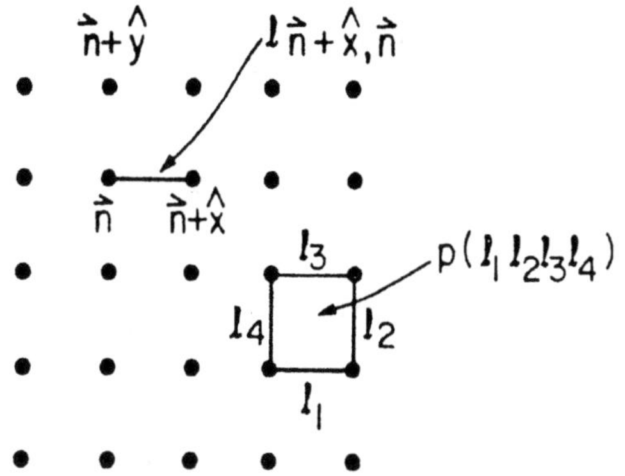

Fig. 1.: A two dimensional (x,y) slice of the lattice showing a typical lattice point $\vec{n} = (n_x, n_y, n_z)$, a typical link $\ell_{\vec{n}+\hat{x},\vec{n}}$ from $\vec{n}$ to $\vec{n}+\hat{x}$, and a typical plaquette $p(\ell_1\ell_2\ell_3\ell_4)$

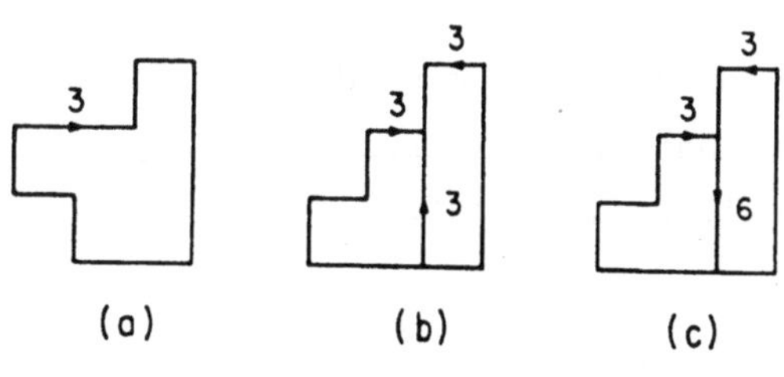

Fig. 2: Some primitive pure glue states

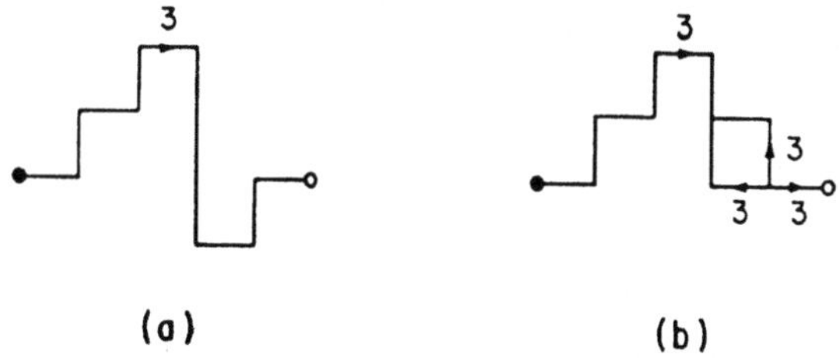

Fig. 3.: Some primitive meson states

$$H_{quark} = \sum_{flavour} m_q \sum_{\vec{n}} q^{\dagger}_{\vec{n}} q_{\vec{n}} + \frac{1}{a} \sum_{flavour} q^{\dagger}_j U_{\ell_{ji}} \alpha_{\ell_{ji}} q_i \tag{13}$$

where $\alpha_{\ell_{ji}}$ is the Dirac matrix in the direction of the link ℓ_{ji}.

Our complete Hamiltonian $H^{lattice}_{QCD} = H_{glue} + H_{quark}$ has H_{QCD} in $A^0 = 0$ gauge as its naive continuum limit; it is, furthermore, invariant under arbitrary gauge transformation at the lattice sites. Gauss' law takes the form of a constraint in the theory that the only physically relevant states are those which are gauge invariant. Finally, we mention that the quark Hamiltonian as written suffers from what is known as the "species doubling problem", but there are several methods for correcting this problem and this technical point needn't concern us here.

C. The spectrum of QCD

We are now ready to consider the properties of $H^{lattice}_{QCD}$. We note first that in the strong coupling limit where a (and as we shall see, therefore g) is large

$$H^{lattice}_{QCD} \to H_{sc} = \frac{g^2}{2a} \sum_{\ell} C^2_{\ell} + \sum_{flavour,\vec{n}} m_q q^{\dagger}_{\vec{n}} q_{\vec{n}} \ . \tag{14}$$

The eigenvalues of C^2_{ℓ} are just those of the square Casimir of SU(3): 0 for the singlet, 4/3 for 3 or $\bar{3}$, 10/3 for 6 or $\bar{6}$, 3 for the octet, etc. The quark part of H_{sc} is, on the other hand, diagonalized by an arbitrary number of quarks and antiquarks at arbitrary lattice sites. Since, however, the only physically relevant eigenstates are those which are gauge invariant, the strong coupling eigenstates may be classified as follows:

1) The strong coupled vacuum:

In this case all links are unoccupied ($C_\ell^2 = 0$) and there are no fermions; the total energy is zero.

2) The pure glue sector:

There are still no quarks, but links are excited in such a way that gauge invariant states are produced. The simplest such pure glue states ("glueloops") have a closed path of links in the 3 (or $\bar{3}$) representation. These have energy $2g^2L/3a^2$ where L is the length of the path; the simplest such state just has the links around the perimeter of an elementary plaquette excited: $\mathrm{Tr}\,[U_{\ell 4}U_{\ell 3}U_{\ell 2}U_{\ell 1}]|0\rangle$, where $|0\rangle$ is the vacuum. Of course more complicated configurations are allowed, including those with non-triplet flux and those with more complicated topology. For example, three flux links can emerge from a single lattice site since a gauge invariant combination can be formed with the ε_{ijk} invariant tensor. See Fig.2.

3) The meson sector:

The simplest quark-containing state consists of a quark and antiquark on the lattice joined by a path of flux links (for gauge invariance). These will have energy $m_q + m_{\bar{q}} + 2g^2L/3a^2$ so that we automatically have quark confinement in strong coupling. See Fig.3.

4) The baryon sector:

The next quark-containing state consists of three quarks connected by an ε_{ijk}-type flux junction. Such quarks will also be confined. See fig.4.

5) Multiquark sectors:

When there are more quarks than those required for a meson or baryon, then in general the system will not be completely confined. The simplest such system consists of two quarks and two antiquarks. See fig.5.

With these examples, the general structure of the eigenstates of the strong coupling limit is clear: it consists of "frozen" gauge invariant configurations of quarks and flux lines. Of course these are not the eigenstates of QCD, but they do form a complete basis (in the limit $a \to 0$) for the expansion of the true strong interaction eigenstates.

The full eigenstates of QCD can be found (in principle!) by considering corrections to the strong coupling limit from the terms we have neglected so far. These terms can induce a variety of effects. Consider first of all the $q^{\dagger}U\alpha q$ term. It can, among other things,

1) annihilate a quark at one point and recreate it at a neighbouring point with an appropriate flux link (Fig.6a),

2) break a 3-flux line and create a pair (Fig.6b). This term thus plays a role analogous to both the usual quark kinetic energy term and the quark-gluon coupling term of the weak coupled theory. Next consider the $\mathrm{Tr}[2-(U_{\ell_4}U_{\ell_3}U_{\ell_2}U_{\ell_1} + \mathrm{h.c.})]$ term. It can, among other things,

1) allow flux to hop across plaquettes (Fig.7a)

2) change flux topology (Fig.7b)

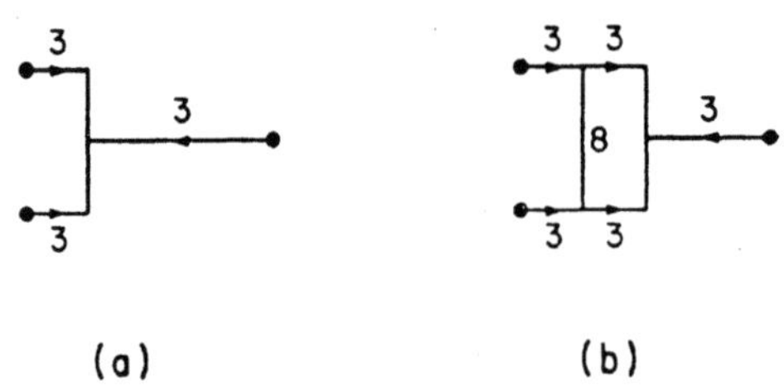

Fig. 4.: Some primitive baryon states

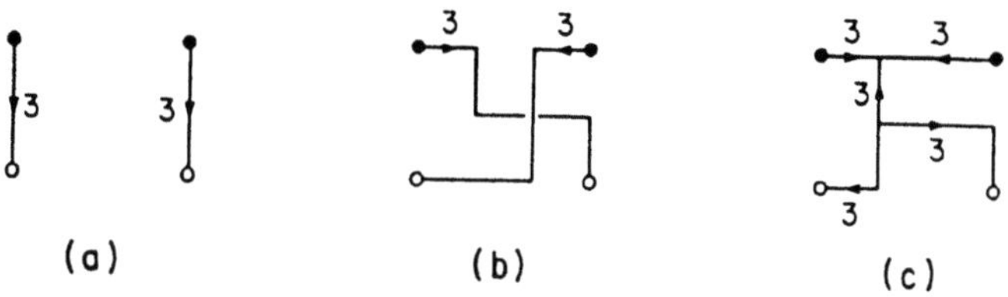

Fig. 5.: Some primitive $qq\bar{q}\bar{q}$ states

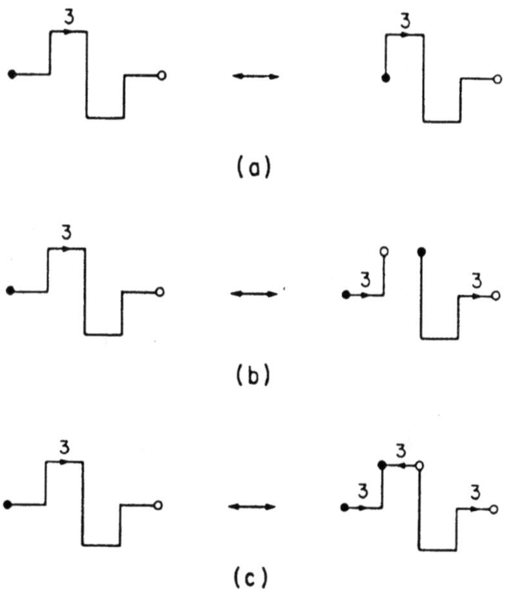

Fig. 6.: Some effects of $q^{\dagger}U\alpha q$: (a) quark hopping,
(b) flux breaking pair creation,
(c) $q\bar{q}$ "seeding"

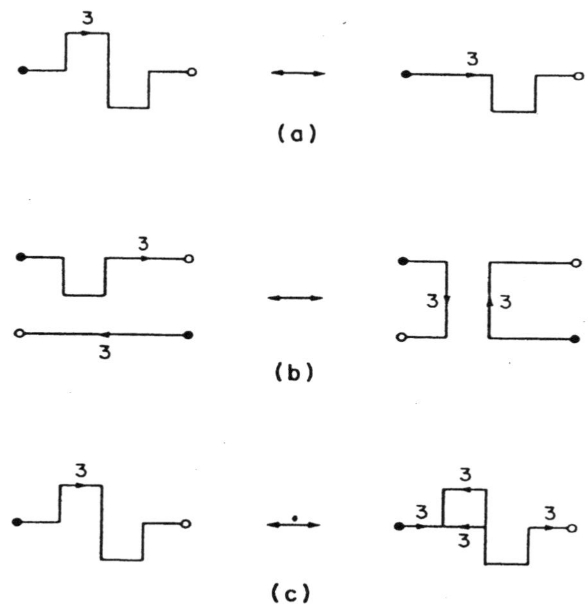

Fig. 7.: Some effects of Tr[2-(UUUU+h.c.)]: (a) flux tube hopping,
(b) flux tube topological mixing by rearrangement,
(c) flux tube topological mixing by "bubble formation"

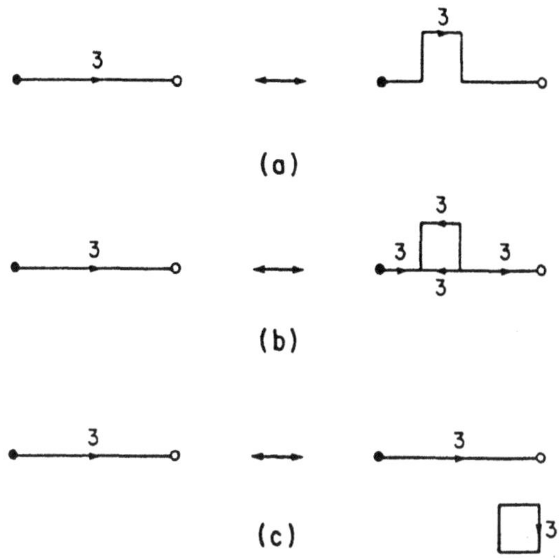

Fig. 8.: Some lowest order corrections to the string tension

We now consider an instructive calculation on the departures from the strong coupling limit: we will examine the effects of such departures on quark confinement by considering two infinitely massive quarks separated by N lattice sites along an axis of the lattice. In the strong-coupled limit, ignoring the constant $2m_Q$,

$$E_{Q\bar{Q}} - E_{vac} = \frac{2g^2}{3a^2} L = \frac{2g^2}{3a} N \ . \tag{15}$$

Since we wish to associate this energy with the linear potential bL of quark models (to be discussed below), we would require that $2q^2/3a^2 = b$. This in turn shows that the bare coupling g must be chosen to depend on a: $g^2(a) = \frac{3}{2} ba^2$, and we see already that as we let a decrease so that we can recover the continuum limit, we will eventually encounter values of g that will require consideration of corrections to the strong coupling limit. In the case at hand the lowest order corrections will consist of

1) corrections to E_{vac} via mixing between the strong

coupled vaccum and states with single excited plaquettes

2) corrections to $E_{Q\bar{Q}}$ via mixing between the straight line of flux and flux configurations with a) kinks, b) bubbles of flux, and c) disconnected excited plaquettes (Figs. 8a,b, and c).

The result of such a second order calculation is that

$$E_{Q\bar{Q}}^{(2)} - E_{vac}^{(2)} = \frac{2g^2}{3a^2} L \left[1 - \frac{11\chi^2}{204}\right] \tag{16}$$

where $\chi = 2/g^4$ is the strong coupling expansion parameter. We see that we still have a linear confinement potential with

$$b^{(2)} = \frac{2g^2}{3a^2} \left[1 - \frac{11}{51g^8}\right] . \tag{17}$$

Note that to maintain the same physical string tension we must now choose

$$g^2(a) = \frac{3}{2} ba^2 \left[1 + \frac{176}{4131b^4a^8}\right] . \tag{18}$$

Clearly this calculation can be carried out in principle to arbitrarily high order. When one does this one will find

$$b = \frac{1}{a^2} f(g^2) \tag{19}$$

where $f(g^2)$ is a pure power series expansion in $1/g^2$. If QCD is to confine then we see that as $a \to 0$ we must be able to choose $g(a)$ in such a way that b remains fixed at its physical value. This requirement, along with the requirement that, as $a \to 0$, $g(a)$ must behave as deduced from weak coupling, shows that confinement, while natural, is not trivial in lattice QCD. Incidentally, from the fact that $db/da = 0$ and from knowledge of $g(a)$ from weak coupling, one can deduce that as $a \to 0$

$$f(g^2) = \frac{b}{\Lambda^2} [\beta_0 g^2]^{-\beta_1/\beta_0^2} \exp(-1/\beta_0 g^2) \tag{20}$$

where Λ is a constant, and where $\beta_0 = 11/16\pi^2$ and $\beta_1 = 102/256\pi^4$ are the first two coefficients in the expansion of the perturbative "beta function" of asymptotic freedom frame Note that $f(g^2)$ has no power series expansion near $g^2 = 0$ proving that confinement cannot be seen in QCD via perturbation theory about weak coupling. This clearly makes the strong coupling expansion the more attractive starting point for understanding hadron physics.

D. The flux tube model

Even after successful numerical calculations within QCD are possible, it will still be useful to have models which summarize the very complex structure of this theory. Many such models have already been proposed, including potential models, bag models, string models, et cetera. We will base our discussion instead on a simple model [3] which emerges from the considerations of the previous sections on Hamiltonian lattice QCD.

To illustrate this model in the simplest possible context, consider first a heavy $Q\bar{Q}$ system and ignore light quarks. Departures from the strong coupling limit will have a number of effects, but among them will be flux hopping terms which will convert the frozen flux lines of the strong coupling limit into dynamical strings which can sustain transverse oscillations. (The conversion from frozen to fluid strings corresponds to the "roughening transition" seen in numerical lattice work). We accordingly first imagine summing all such terms so that we can deal with the $Q\bar{Q}$ pair connected by a dynamical string. Next consider departures from strong coupling that involve "topological mixing" either to other connected string states or to states with disconnected vacuum fluctuations. If L is large we can hope to approximate the state of the system in terms of a lattice of scale a with $L \gg a \sim \lambda_0$, where λ_0 is the scale where g = 1 and topological mixing becomes important. Our model for this system is thus that of a discrete string[4]. The ground state of this system for infinitely heavy quarks will consist of the $Q\bar{Q}$ pair with a ground state string stretched between them, the first excited state will be doubly degenerate with either a right-handed or left-handed phonon excited in the lowest string mode, et cetera. As the distance R between the Q and $\bar{Q}$ is varied slowly, the eigenenergy of the string eigenstate S will trace out an adiabatic potential $V^{(S)}(R)$; we associate such potentials with an adiabatic approximation to the physics of mesons. When the $Q\bar{Q}$ pair moves in the adiabatic potential bR of the ground state of this QCD string, one recovers the usual spectrum of mesons in the quark model. When the pair moves in the adiabatic potential of an excited string, the resulting hadrons correspond to a new species not contained in the usual quark model: hybrid mesons with both quark and gluonic degrees of freedom in evidence.

The baryon sector, while more complicated, is analogous to the mesons: the ordinary baryons of the quark model correspond to three quarks moving in the adiabatic potential of the ground state of the three junction (Y) string, while excited strings will lead to hybrid baryons. The phenomenological implications of hybrid mesons and baryons will be discussed later.

In multiquark systems we are forced for the first

time to go beyond the simple vibrating string picture to consider topological mixing. This is because in such systems adiabatic surfaces will always cross in the absence of mixing, as can be seen by considering Fig.7(b) for the case when the $qq\bar{q}\bar{q}$ system is arranged at the corners of a square. Such systems are consequently considerably more complicated than those we have already considered. They are the main subjects of the last two Chapters.

Of course the simple classification scheme we have presented here is both incomplete and inaccurate. For example, one must expect to find (at high masses) "topological meson hybrids" in which the $Q\bar{Q}$ system moves in the potential of more complicated string topologies. There will also be mixing between topologies via the $\mathrm{Tr}[U_{\ell_1}U_{\ell_2}U_{\ell_3}U_{\ell_4} + \mathrm{h.c.}]$ and $q^{\dagger}U\alpha q$ terms and deviations from the adiabatic limit that mix states with different degrees of phonon excitation. While all such effects can be described within the context of this model, its utility will depend on the degree to which they can be ignored, at least for low-lying states. Perhaps the very success of the naive quark model for mesons and baryons provides some evidence in favour of the viability of such a separation.

E. The interface between strong and weak coupling

While the model just presented may be useful for large distance QCD, it must fail at short distances. Consider the $Q\bar{Q}$ system once again, but now for distances $R = O(\lambda_0)$. To describe the physics of this regime will clearly require a lattice spacing $a < \lambda_0$ where g is small so that a strong coupling expansion becomes, at best, less useful. In particular, at these distance scales, the vacuum will be full of fluctuations and one must expect strong topological mixing as well. It is, therefore, natural to expect that at such scales the string will be mixed by new physics beyond simple flux hopping so that it is plausible that the linear potential bR will be pushed to lower energies in the small R regime. This picture fits with what we know must happen: at short distance the $Q\bar{Q}$ potential must become the Coulombic potential $-4\alpha_s/3R$. It is thus reasonable to supplement our string potentials for mesons by the effects of one gluon exchange (as we do), although it is possible that the transition between these two regimes is significantly more complicated than this simple picture suggests. A similar argument applies to the short distance interquark potential in a baryon, but unfortunately there is no analogous argument to suggest the exact form of the short-distance potential in the pure gluon sector.

Since all presently known hadrons straddle these two regimes (although some high orbital angular momentum light quark mesons are mainly in the confining regime and the lowest-lying $b\bar{b}$ states are mainly in the Coulombic region), we spend a moment here on the effective potential from one gluon exchange.

Consider $q\bar{q}$ scattering in the centre-of-mass frame from momenta $\vec{p}_q = -\vec{p}_{\bar{q}} = \vec{p}$ to $\vec{p}\,'$. With one gluon exchange we get a certain transition amplitude. If we now require a potential V_{eff} which gives exactly the same on-shell transition amplitude, then we get

$$\chi^{\dagger}_{s'}\chi^{\dagger}_{\bar{s}'}V_{eff}\chi_s\chi_{\bar{s}} = \frac{1}{(2\pi)^3}\int d^3Q e^{i\vec{Q}\cdot\vec{r}}\bar{u}(\vec{p}\,'s')\bar{v}(-\vec{p}\bar{s})G(Q^2)$$

$$\times\ \gamma^{\mu}_{q}\gamma^{\bar{q}}_{\mu}u(\vec{p}s)v(-\vec{p}\,'\bar{s}') \tag{21}$$

with $\vec{Q} = \vec{p}\,' - \vec{p}$ and

$$G(Q^2) = \frac{4\pi\alpha_s(Q^2)}{3Q^2}\ \vec{F}_q\cdot\vec{F}_{\bar{q}} \tag{22}$$

where $F_q^i = \frac{\lambda^i}{2}$, $F_{\bar{q}}^i = \frac{\lambda^{c^i}}{2} = \frac{-\lambda^{i^*}}{2}$ are the q and $\bar{q}$ colour matrices. Multiplying out the spinors and defining $\vec{P} = (\vec{p}+\vec{p}')/2$ then gives

$$V_{eff}\ (\vec{P},\vec{r}) = \frac{1}{(2\pi)^3} \int d^3Q e^{i\vec{Q}.\vec{r}} G(Q^2) \Big(\Big[1 - \frac{Q^2}{4E(E+m)} + \frac{i\vec{Q}\times\vec{P}.\vec{S}_q}{E(E+m)}\Big] \times$$

$$\times\Big[1 - \frac{Q^2}{4\bar{E}(\bar{E}+\bar{m})} + \frac{i\vec{Q}\times\vec{P}.\vec{S}_{\bar{q}}}{\bar{E}(\bar{E}+\bar{m})}\Big]$$

$$+\Big[\frac{\vec{P}-2i\vec{Q}\times\vec{S}_q}{2E} \cdot \frac{\vec{P}-2i\vec{Q}\times\vec{S}_{\bar{q}}}{2\bar{E}}\Big]\Big) \qquad (23)$$

where m and $\bar{m}$ are the quark and antiquark masses, $E = (p^2 + m^2)^{1/2}$ and $\bar{E} = (p^2 + m^2)^{1/2}$ their energies, and $\vec{S}_q$ and $\vec{S}_{\bar{q}}$ their spins.

In the non-relativistic limit this expression gives the familiar Breit-Fermi interaction:

$$V_{eff}^{ij} = V_{coul}^{ij} + V_{hyp}^{ij} + V_{so}^{ij} \qquad (24)$$

where

$$V_{coul}^{ij} = \frac{\alpha_s(\vec{r}_{ij})}{r_{ij}} \vec{F}_i \cdot \vec{F}_j \qquad (25)$$

is the colour Coulomb interaction,

$$V_{hyp}^{ij} = \frac{\alpha_s(r_{ij})}{m_i m_j} \Big(\frac{8\pi}{3}\vec{S}_i.\vec{S}_j \delta^3(\vec{r}_{ij}) + \frac{1}{r_{ij}^3}\Big[\frac{3\vec{S}_i.\vec{r}_{ij}\vec{S}_j.\vec{r}_{ij}}{r_{ij}^2} - \vec{S}_i.\vec{S}_j\Big]\Big)\vec{F}_i.\vec{F}_j$$

$$(26)$$

is the hyperfine interaction, and where

$$V^{ij}_{so} \equiv V^{ij}_{so(cm)} + V^{ij}_{so(tp)} \tag{27}$$

is the spin-orbit interaction with

$$V^{ij}_{so(cm)} = - \frac{\alpha_s(r_{ij})}{r^3_{ij}}\left(\frac{1}{m_i} + \frac{1}{m_j}\right)\left[\frac{\vec{S}_i}{m_i} + \frac{\vec{S}_j}{m_j}\right]. \vec{L}_{ij}\ \vec{F}_i.\vec{F}_j \tag{28}$$

its colour magnetic piece and with

$$V^{ij}_{so(tp)} = - \frac{1}{2r_{ij}} \frac{\partial V^{ij}_{coul}}{\partial r_{ij}}\left[\frac{\vec{S}_i}{m^2_i} + \frac{\vec{S}_j}{m^2_j}\right]. \vec{L}_{ij} \tag{29}$$

being the Thomas precession term.

These interactions, especially the colour hyperfine interaction, play a very important role, particularly in the physics of light quark systems.

REFERENCES FOR CHAPTER 2

1. The idea of colour was introduced by O.W. Greenberg, Phys. Rev. Lett. 13 (1964) 598. The idea that the strong interactions might be described by a non-Abelian gauge theory of colour originates in H. Fritzsch and M. Gell-Mann, Proc. XVI Int. Conf. on High Energy Physics (National Accelerator Laboratory, 1972) p. 135; H. Fritzsch, M. Gell-Mann, and H. Leutwyler, Phys. Lett. 74B (1973) 365; S. Weinberg, Phys. Rev. Lett. 31 (1973) 494.
2. For reviews of lattice gauge theories and calculations, see, for example, M. Bander, Physics Reports 75 (1981) 206; J. Kogut, Rev. Mod. Phys. 55 (1983) 775; L. Susskind, in Weak and Electromagnetic Interactions at High Energies,

editors R. Balian and C.H. Llewellyn Smith (North Holland, Amsterdam, 1977). Hamiltonian lattice gauge theories were introduced via the SU(2) theory by J. Kogut and L. Susskind, Phys. Rev. D11 (1975) 395.
3. The flux tube picture described here was introduced in N. Isgur and J. Paton, Phys. Lett. 124B (1983) 247. A more complete discussion and references to related models may be found in N. Isgur and J. Paton, "A Flux Tube Model for Hadrons in QCD", to appear in Phys. Rev. D.
4. String theory has a venerable history: see, e.g., S. Mandelstam, Physics Reports 13C (1974) 261.

3. MESONS AND BARYONS

A. Heavy quarkonia: the hydrogen atoms of QCD

Although, as we have seen, progress is being made in solving strongly coupled QCD, quantum chromodynamics is only understood with any precision in its application to short distance phenomena. One place where this understanding can be applied to hadron spectroscopy is in the study of the low-lying states of a heavy $Q\bar{Q}$ (heavy quarkonium) system: for sufficiently heavy quarks such states are so small that the effects of confinement become unimportant. In this situation the system is dominated by the one gluon exchange potential of Section II.E. and so, apart from some colour factors, becomes almost exactly analogous to the e^+e^- (positronium) system.

Fig.9 compares the spectrum of positronium (m_e = = 0.511 MeV, α = 1/137) with a hypothetical heavy quarkonium (m_Q = 0.511 TeV, α_s = 1/13.7). Note that the Bohr levels differ by a factor of 10^8, while fine and hyperfine effects differ by a factor of 10^{10}. Since the same forces

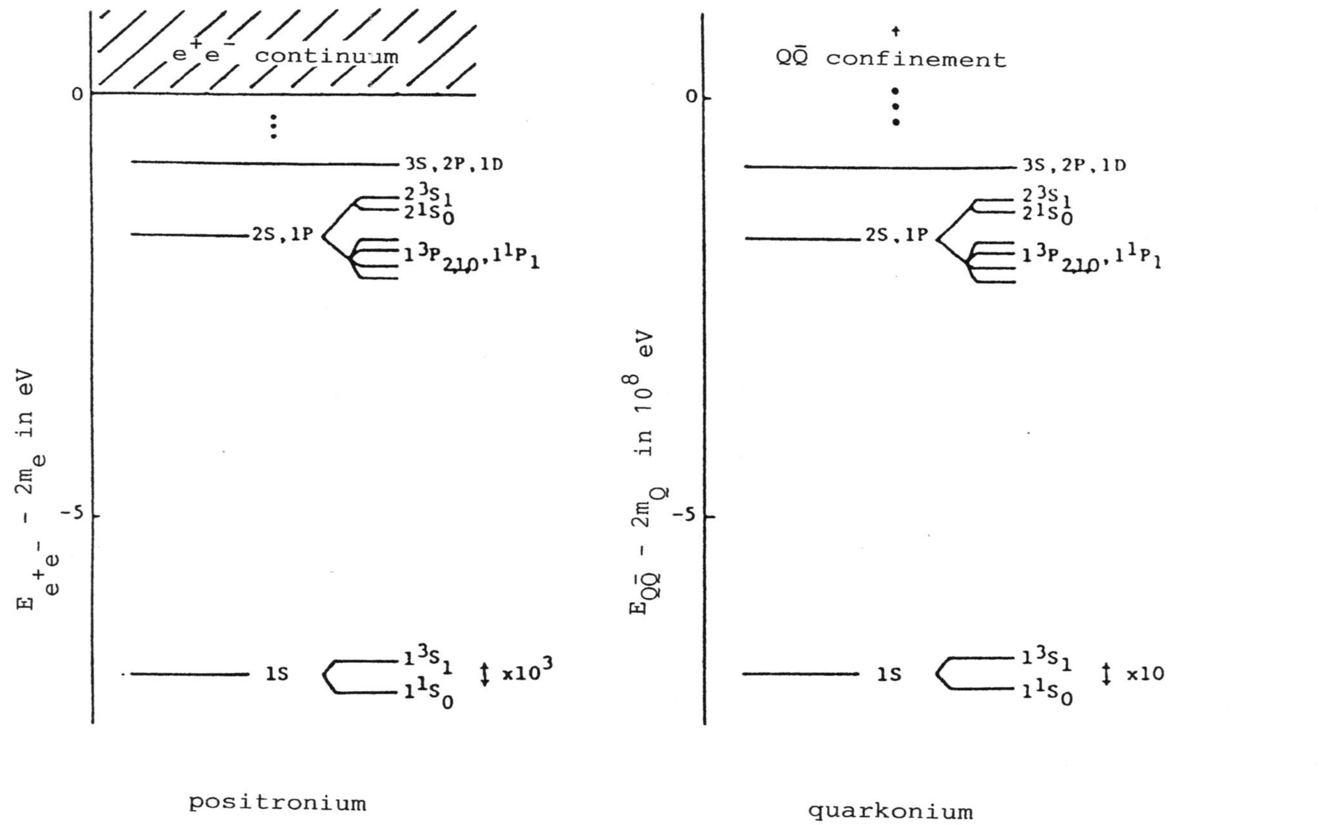

Fig. 9.: A comparison of the spectra of positronium and a hypothetical heavy quarkonium with $\frac{4}{3}\alpha_s = 10\alpha$ and $m_Q = 10^6\ m_e$; Bohr levels are shown to scale, but the internal splittings of these levels are only shown very schematically

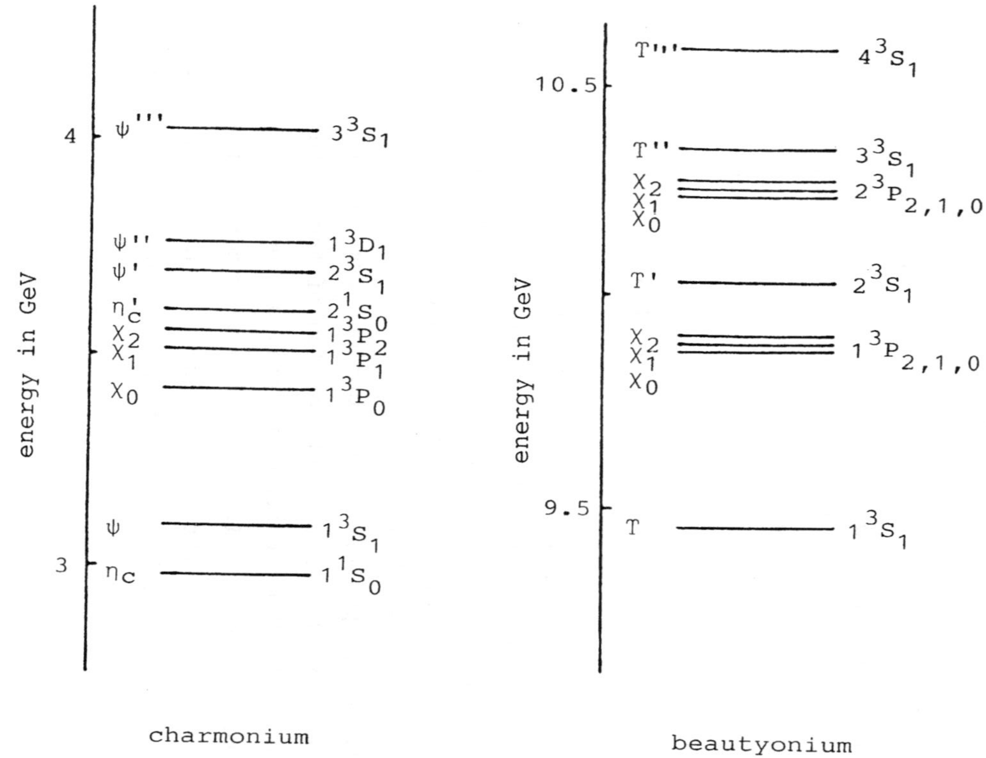

Fig. 10.: The experimental spectra of charmonium ($c\bar{c}$) and of beautyonium ($b\bar{b}$); levels are labelled by both their standard names and their quarkonium interpretation

at work here prove very important throughout hadron spectroscopy, we briefly review the origin of various energy levels:

1) Bohr levels: these levels result from the Coulombic potential (25) and correspond to a Bohr radius of $(\frac{2}{3} m_Q \alpha_s)^{-1} \simeq 10^{-2}$ fm.

2) $^3S_1 - {}^1S_0$ splittings: these splittings arise from the $\vec{S}_i \cdot \vec{S}_j$ part (the "Fermi contact term") of the hyperfine interaction (26) and are the analogs of the 21 cm line of atomic hydrogen; the remainder of the hyperfine interaction (the "tensor interaction") averages to zero in S-waves since it is a tensor of rank two in space.

3) the $^3L_{L+1}$, ^3L_L, ^1L_L, $^3L_{L-1}$ splittings for $L \neq 0$: these splittings arise from two sources, the now-operative tensor force of the hyperfine interaction (26) and the spin-orbit interaction (27).

In addition to the above real analogies between these two systems there is also one apparent analogy which is spurious: the 2S-1P splitting. In positronium this "Lamb shift" receives two contributions: a small negative contribution from vacuum polarization (in QED the effective coupling seen by 2S is stronger than that seen by the more spread out 1P state, so its energy is lowered slightly) and a much larger positive contribution from infrared effects associated with the interaction of the system with soft photons. In QCD the same ordering of the states has a quite different origin: now the vacuum polarization makes a positive contribution and is dominant, while since gluons are confined the analog of the strong infrared mixing of $e^+e^-\gamma$ is suppressed. We mention this minor point since the perturbative 2S-1P splitting of heavy quarkonium provides an important link to our next topic: light quarkonia.

As the quark mass m_q is decreased, the "Bohr radius" of the quarkonium system varies as $(\frac{2}{3} m_q \alpha_s)^{-1}$ (note that both m_q and α_s will vary in this expression since as m_q decreases the running coupling constant will increase). Examining equations (25), (26) and (27) we can see that the ratio of hyperfine and spin-orbit splittings to Bohr splittings will vary like α_s^2 and so as we decrease m_q to regions where $\alpha_s \sim O(1)$ we can expect spectra in which, for example, the $^3S_1$ - $^1S_0$ splitting has become comparable with the 1P-1S splitting. However, more significant than these quantitative changes is the fact that the small 2S-1P splitting, which we think of as characteristic of a Coulombic system (and which is normally in the ratio α_s^3 to the Bohr splittings), will become comparable to the 2S-1S splitting, signalling a large departure of the potential from Coulombic form. This is of course just the perturbative signal for the onset of confinement: in a linear (quadratic) potential the level splittings 1P-1S and 2S-1S are in the ratio of roughly 2 to 3 (exactly 1 to 2).

B. The extension to light quarkonia

As already implied in Section II.E., since all known quarkonium systems straddle the one gluon exchange and confinement regimes, their treatment is at the moment dependent on models. We do not attempt to do justice here to the many other alternatives for such models --- foremost among which is the bag model --- but rather stick to the picture we have built up in the previous sections and discuss quarkonium in the confinement regime in terms of a $q\bar{q}$ pair moving in the lowest adiabatic potential of a string. It is then natural to treat quarkonium by considering its full potential to be a two-component potential: the sum of a one gluon exchange potential ($-\frac{4}{3}\frac{\alpha_s}{r}$ which will dominate at small r) and the ground state string potential (br, which will dominate at large r). Indeed, this approach has proved to be immensely successful; the main topic of this chapter will be to briefly review these successful applications. [1]

Fig.10 shows the observed spectra of the two heaviest quarkonium systems: beautyonium ($b\bar{b}$) and charmonium ($c\bar{c}$). Comparison with Fig.9, in light of our comments on the variation of 2S-1P and of hyperfine and spin-orbit splittings with m_q, reveals immediately the qualitative success of this picture. In fact, the potential model has proved to be much more than a qualitative tool for these systems: it predicts energy levels, total decay rates, decay rates to lepton pairs, and photon transition rates between levels with typically ±20 % accuracy. There are many problems associated with our understanding of the physics of these systems (some of which will be discussed below) but it seems safe to conclude that the correct basic picture for them is in hand and that future understanding will lead to refinements of this picture.

When we turn our attention to systems which contain light (u,d, or s) quarks, the situation is less clear. One of the main reasons for this obscurity is that the non-relativistic treatment appropriate to $b\bar{b}$ and $c\bar{c}$ becomes itself a serious source of error when light quarks are involved so that it becomes difficult to distinguish inadequacies of the model from inadequacies in the treatment of relativistic effects. Nevertheless, Fig.11 offers what might be considered to be strong evidence that the qualitative physics of quarkonium systems persists even for light quarkonia: the smooth transition of the analogous energy levels from heavy to light quarkonia certainly argues for continuity of the underlying physics. We see first of all from this figure that the 1^3P_2 - 1^3S_1 "Bohr level" splitting remains rather constant from the heaviest to the lightest quarkonia. In a Coulombic potential such a splitting would increase in direct proportion to the reduced mass μ while in a linear potential it would decrease like $\mu^{-1/3}$; the approximate constancy in the case at hand emerges automatically as compromise between these two extremes from the mixture of Coulomb and linear potentials present in the two-component model. Also interesting is the evolution of the 1^3S_1 - 1^1S_0 splitting which, as expected, increases smoothly from a strength relative to 2S - 1S of order $\alpha_s^2 << 1$ in $b\bar{b}$ to a strength of order unity in light quark systems where the scales are such that $\alpha_s \sim O(1)$. In addition to these qualitative successes, quarkonium models of mesons containing light quarks have, in studies of both spectra and decays, had considerable quantitative success. Although less accurate than when applied to $b\bar{b}$ and $c\bar{c}$, such models once again seem to give

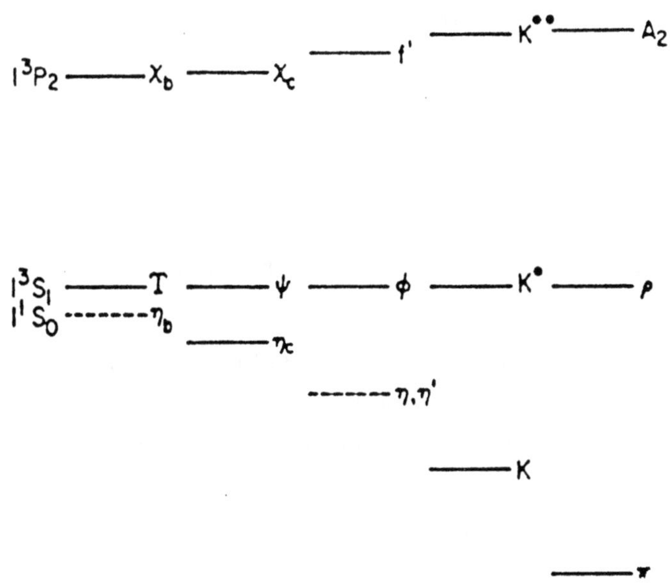

Fig. 11.: A graphic illustration of the universality of meson dynamics from the π to the Υ, showing the splittings of 1^3P_2 and 1^3S_1 in the $b\bar{b}$, $c\bar{c}$, $s\bar{s}$, $u\bar{s}$, and $u\bar{d}$ families

a fairly good account of the masses, decay rates, and static properties of such mesons.

The success of the non-relativistic quark model for mesons (and, as we will soon see, baryons) can probably be understood: while relativistic effects are important in these systems, it seems that they can for the most part be hidden (at least for low-lying states) in the choice of such potential model parameters as the constituent quark mass. When all the dust settles, apparently it is "better to have the right degrees of freedom moving at the wrong speed than the wrong degrees of freedom moving at the right speed".

Of course it would be even better to have the right degrees of freedom moving at the right speed. We have recently shown that such a relativization of the quark model for mesons can be carried out[2]. Some of the features of such a model are:

1) all mesons from the π to the $b\bar{b}$ are treated in the same framework,
2) relativistic kinematics,
3) a semi-relativistic potential including realistic momentum dependences, non-localities, and coordinate smearing.

The resulting picture of mesons is very satisfactory: it provides in a unified framework a good description of the spectroscopy and decays of essentially all known mesons.

Nevertheless, let me conclude this section by mentioning briefly some of the problems of the potential picture which we temporarily "swept under the rug" in our account of its successes. On the theoretical side these problems include:

1) The transition region is unlikely to be simple enough to strictly allow an addition of the two limiting

potentials.

2) Stating that the long range potential is linear is insufficient: we must also specify its Lorentz properties or, alternatively, its momentum dependent properties. These are often taken to be those of a Lorentz scalar interaction in which case a calculation analogous to that leading to (23) leads (with the replacement of $\gamma^{\mu}\times\gamma_{\mu}$ by 1×1 and $G(Q^2)$ by an $S(Q^2)$ corresponding to a linear potential) to only one spin-dependent term: a Thomas precession spin-orbit term which tends to cancel the one-gluon-exchange spin-orbit effect. While this prescription leads to reasonable results, it remains controversial.

3) Two kinds of Fock-space mixing have been neglected: from $q\bar{q}$ to $q\bar{q}q\bar{q}$ via pair creation (including mixing with real and virtual decay channels) and non-adiabatic mixing from the string ground state to excited string states. Preliminary studies of both types of effects indicate that they probably do not change things very drastically, but this remains to be established.

4) The relationship of quarkonium models to the picture of the pseudoscalar mesons as the almost-Goldstone-bosons of broken chiral symmetry remains obscure (although some progress has been made in this direction).

On the phenomenological side there are no major problems with the $b\bar{b}$ and $c\bar{c}$ systems, but for light quarks there may be several. Some examples include:

1) The lightest scalar ($J^P = 0^+$) mesons, the $S^*(980)$ with $I = 0$ and the $\delta(980)$ with $I = 1$, do not seem to be 1^3P_0 $q\bar{q}$ states. One possibility, to be mentioned in Chapter IV, is that they are really $qq\bar{q}\bar{q}$ states (for example, $K\bar{K}$ bound states), but the whole scalar meson sector is both experimentally and theoretically confused at the moment.

2) The $J^P = 1^-$ sector, which shows the expected 1^3S_1, 1^3S_1 and 1^3D_1 states in the $c\bar{c}$ sector, is confused for light quarks: two nonets of such states corresponding

to 1^3S_1 and 1^3D_1 should have nonstrange members around 1.6 GeV, but at present the signals from various experiments are often apparently inconsistent both theoretically and experimentally.

3) There are at the moment reports of various sorts of unusual states: e.g., the $\iota(1440)$, the $\theta(1690)$, and the $\xi(2200)$. Whether these are $q\bar{q}$ mesons or more exotic objects like meson hybrids or glueballs remains very controversial.

Despite these shortcomings and uncertainties, it is reasonable to consider the potential model for mesons an outstanding success. With it we can make sense of (almost?) all of the hundreds of known mesons and in most cases the model can go well beyond a qualitative description of these states to produce detailed and reasonably accurate predictions for masses, decay rates, and static properties for mesons from the lightest to the heaviest known.

C. Baryons with light quarks

It is ironic that the more complex qqq baryon system is better understood experimentally, and for light quarks possibly even theoretically, than the mesons. On the experimental side the reason for this is clear: non-strange and strangeness -1 (S = 0 and -1) baryons can be formed as resonances in πN and $\bar{K}N$ scattering, respectively, while no such s-channel resonance production is possible for mesons (except in $J^{PC} = 1^{--}$ via e^+e^-). Why the simple two component potential models work as well as they do for light quark baryons is unclear, but their effectiveness may be related to the fact that the "multibody" character of these systems makes them somewhat less relativistic than mesons.

In any event, a very simple and analytically solvable

model [3] for qqq has proved quite useful for understanding baryons and so, since an understanding of qqq is obviously essential to our eventual goal of treating 6q, we will devote some time in this section to developing various details of the model. First, however, let us go back to basics.

We expect the lowest-lying baryons to correspond approximately to three quarks moving in the adiabatic potential generated by the ground state of three flux tubes emerging from the quarks and fusing in an ε_{ijk} junction (see Fig.4). When two of the quarks are close together, then the flux tube configuration is essentially identical to that of a meson and so we expect the same linear potential br to be in operation. In terms of an effective two-body potential we would say that in this orientation each of the two nearby quarks is attracting the third with a potential 1/2 br. Although the Y junction certainly has three body components, numerical studies have in fact demonstrated that replacing it by a sum of two body potentials with half the strength of mesons provides a good approximation to the spectrum of excited states, and for now we will adopt this approximation in our discussion. This same factor of 1/2, as it turns out, applies to the one-gluon-exchange part of the interquark potential since the expectation values

$$\left\langle \frac{\vec{\lambda}_i}{2} \cdot \frac{-\vec{\lambda}_j^*}{2} \right\rangle_{q\bar{q} \text{ in a meson}} = -\frac{4}{3} \tag{30}$$

and

$$\left\langle \frac{\vec{\lambda}_i}{2} \cdot \frac{\vec{\lambda}_j}{2} \right\rangle_{qq \text{ in a baryon}} = -\frac{2}{3} \tag{31}$$

have the same ratio, so we can actually apply this "qq=1/2 $q\bar{q}$"

rule uniformly.

Let us now turn to our naive but solvable model: a non-relativistic Hamiltonian with

$$H = H_{si} + H_{sd} \tag{32}$$

where

$$H_{si} = \sum_i \left(m_i + \frac{p_i^2}{2m_i}\right) + \sum_{i<j} \left(\frac{1}{2}br_{ij} + C - \frac{2\alpha_s}{3r_{ij}}\right) \tag{33}$$

is spin-independent and where H_{sd} is a spin-dependent interaction to be introduced later. The quark masses in (33) are the usual non-relativistic constituent quark masses $m_u \simeq m_d \simeq 0.35$ GeV, $m_s \simeq 0.58$ GeV, $m_c \simeq 1.5$ GeV, and $m_b \simeq 5.0$ GeV; the various couplings could in principle be taken from meson spectroscopy.

i) The spin-independent spectrum

We initially concentrate on H_{si} and write

$$H_{si} = \sum_i \left(m_i + \frac{p_i^2}{2m_i}\right) + \sum_{i<j} \left(\frac{1}{2}kr_{ij}^2 + U(r_{ij})\right) \tag{34}$$

$$\equiv H_0 + \sum_{i<j} U(r_{ij}) \tag{35}$$

where $U(r_{ij}) = 1/2\ br_{ij} + C - \frac{2\alpha_s}{3r_{ij}} - 1/2\ kr_{ij}^2$ so that equations (34) and (35) are simply restatements of equ. (33). However, written in this form we can solve exactly for the eigenstates of H_0 and then, choosing k to minimize the perturbation U in the low-lying states, treat U by perturbation theory. Our first step is therefore to find the eigenstates of H_0. It is convenient in doing so to first distinguish between two cases: the case where all three quarks have equal masses (the S = 0 and -3 sectors uuu, uud, ddu, ddd, and sss), and the S = -1 and -2 sectors (uus, uds, dds, ssu, and ssd)) where one quark has a mass different from the other two.

In the case $m_1 = m_2 = m_3 \equiv m$, the change of variables to

$$\vec{R} = \frac{m_1\vec{r}_1+m_2\vec{r}_2+m_3\vec{r}_3}{m_2+m_3+m_3} \tag{36}$$

$$\vec{\rho} = \frac{1}{\sqrt{2}}(\vec{r}_1-\vec{r}_2) \tag{37}$$

$$\vec{\lambda} = \frac{1}{\sqrt{6}}(\vec{r}_1+\vec{r}_2-2\vec{r}_3) \tag{38}$$

converts H_0 to

$$H_0 = \frac{P^2}{2M} + \left(\frac{p_\rho^2}{2m} + \frac{3}{2}k\rho^2\right) + \left(\frac{p_\lambda^2}{2m} + \frac{3}{2}k\lambda^2\right) \tag{39}$$

where M = 3m and $\vec{P}$, $\vec{p}_\rho$, and $\vec{p}_\lambda$ are the momenta conjugate to $\vec{R}, \vec{\rho}$, and $\vec{\lambda}$. Ignoring centre-of-mass motion, the low lying eigenstates of these two degenerate oscillators may be chosen to be of the form $\Psi^\sigma_{LM}=\psi^\sigma_{LM}\ \alpha^3/\pi^{3/2}\exp(-1/2\alpha^2\{\rho^2+\lambda^2\})$ where

$$N = 0 \qquad \psi^S_{00} = 1 \tag{40}$$

$$N = 1 \qquad \psi^\rho_{11} = \alpha\rho_+ \tag{41}$$

$$N = 1 \qquad \Psi^{\lambda}_{11} = \alpha\lambda_+ \tag{42}$$

$$N = 2 \qquad \Psi^{S'}_{00} = \frac{1}{\sqrt{3}}\,\alpha^2(\rho^2+\lambda^2-3\alpha^{-2}) \tag{43}$$

$$N = 2 \qquad \Psi^{\rho}_{00} = \frac{2}{\sqrt{3}}\,\alpha^2\,\vec{\rho}\cdot\vec{\lambda} \tag{44}$$

$$N = 2 \qquad \Psi^{\lambda}_{00} = \frac{1}{\sqrt{3}}\,\alpha^2(\rho^2-\lambda^2) \tag{45}$$

$$N = 2 \qquad \Psi^{S}_{22} = \frac{1}{2}\,\alpha^2(\rho_+^2+\lambda_+^2) \tag{46}$$

$$N = 2 \qquad \Psi^{\rho}_{22} = \alpha^2\rho_+\lambda_+ \tag{47}$$

$$N = 2 \qquad \Psi^{\lambda}_{22} = \frac{1}{2}\,\alpha^2(\rho_+^2-\lambda_+^2) \tag{48}$$

$$N = 2 \qquad \Psi^{A}_{11} = \alpha^2(\rho_+\lambda_z-\rho_z\lambda_+) \tag{49}$$

where we have explicitly displayed only the highest state Ψ^{σ}_{LL} of a multiplet of states Ψ^{σ}_{LM} (the others following by normal Clebsch-Gordanry) and where

$$\alpha \equiv (3km)^{1/4} \tag{50}$$

$$\rho_{\pm} \equiv \rho_1 \pm i\rho_2 \tag{51}$$

$$\lambda_{\pm} \equiv \lambda_1 \pm i\lambda_2\,. \tag{52}$$

Note that the states have energies $(N+3)\omega$ where

$$\omega \equiv (3k/m)^{1/2}. \tag{53}$$

The notation for σ used here is based on the group S_3 of permutations of three objects: S states transform totally symmetrically, A states transform totally antisymmetrically, while pairs of states (M^ρ, M^λ) have mixed symmetry under permutations and transform like the two relative coordinates $(\vec{\rho}, \vec{\lambda})$, e.g.,

$$(213)\ M^\rho = -M^\rho \tag{54}$$

$$(213)\ M^\lambda = +M^\lambda \tag{55}$$

$$(132)\ M^\rho = \frac{1}{2} M^\rho + \frac{\sqrt{3}}{2} M^\lambda \tag{56}$$

et cetera. This classification is particularly useful since from such states one can readily construct product states of definite symmetry. The relevant multiplication rules are

1) SR transforms like R for R = S, A, M
2) AM^ρ and AM^λ transform like M^λ and M^ρ while AA transforms like an S.
3) $\sqrt{1/2}\ (M_1^\rho M_2^\rho + M_1^\lambda M_2^\lambda)$ transforms like S, the pairs, $\sqrt{1/2}\ (M_1^\rho M_2^\lambda + M_1^\lambda M_2^\rho)$, $\sqrt{1/2}\ (M_1^\rho M_2^\rho - M_1^\lambda M_2^\lambda)$ transform like a mixed symmetry pair M^ρ, M^λ and $\sqrt{1/2}(M_1^\rho M_2^\lambda - M_1^\lambda M_2^\rho)$ transforms like A.

These rules are easily verified by explicit calculation. To construct totally antisymmetric baryon wavefunctions we require similarly constructed flavour and spin wavefunctions (the colour wavefunction is automatically the antisymmetric colour singlet $C^A = \sqrt{\frac{1}{6}}(RBY - BRY + BYR - YBR + YRB - RYB)$ so that we need to construct totally <u>symmetric</u> states in flavour × space × spin). As spin wavefunctions we choose

$$S = \frac{3}{2}(\text{with } \sigma=S) : \quad \chi^S_{3/2} = \uparrow\uparrow\uparrow, \text{ etc.} \tag{57}$$

$$S = \frac{1}{2}(\text{with } \sigma=M^\rho) : \quad \chi^\rho_+ = \frac{1}{\sqrt{2}}(\uparrow\downarrow\uparrow - \downarrow\uparrow\uparrow), \text{ etc.} \tag{58}$$

$$S = \frac{1}{2}(\text{with } \sigma=M^{\lambda}): \quad \chi_{+}^{\lambda} = -\frac{1}{\sqrt{6}}(\uparrow\downarrow\uparrow+\downarrow\uparrow\uparrow-2\uparrow\uparrow\downarrow), \text{ etc.} \tag{59}$$

where we once again display only the top state of an angular momentum multiplet. As flavour wavefunctions we take

decuplet

$$\phi_{\Delta^{++}}^{S} = uuu \tag{60}$$

$$\phi_{\Sigma^+}^{S} = \frac{1}{\sqrt{3}}(uus+usu+suu) \tag{61}$$

$$\phi_{\Xi^0}^{S} = \frac{1}{\sqrt{3}}(ssu+sus+uss) \tag{62}$$

$$\phi_{\Omega^-}^{S} = sss \tag{63}$$

octet

$$\phi_{p}^{\rho} = \frac{1}{\sqrt{2}}(udu-duu) \tag{64}$$

$$\phi_{\Lambda}^{\rho} = \frac{1}{\sqrt{12}}(2uds-2dus+usd-dsu-sud+sdu) \tag{65}$$

$$\phi_{\Sigma^+}^{\rho} = \frac{1}{\sqrt{2}}(suu-usu) \tag{66}$$

$$\phi_{\Xi^0}^{\rho} = \frac{1}{\sqrt{2}}(sus-uss) \tag{67}$$

octet

$$\phi_{p}^{\lambda} = \frac{-1}{\sqrt{6}}(udu+duu-2uud) \tag{68}$$

$$\phi_{\Lambda}^{\lambda} = \frac{1}{2}(usd-dsu+sud-sdu) \tag{69}$$

$$\phi_{\Sigma^+}^{\lambda} = \frac{1}{\sqrt{6}}(suu+usu-2uus) \tag{70}$$

$$\phi_{\Xi^0}^{\lambda} = \frac{-1}{\sqrt{6}}(sus+uss-2ssu) \tag{71}$$

singlet

$$\phi_{\Lambda}^{A} = \frac{1}{\sqrt{6}}(uds-dus-usd+dsu+sud-sdu) \tag{72}$$

where in this case we have displayed only the top states of isospin multiplets. We have listed all flavour wavefunctions, not just those associated with S = 0 and -3, for future use.

We are finally ready to construct the baryons, beginning with the S = 0 sector. In the notation $|X^{2S+1}L_\pi J^P>$ where X = flavour, S is the quark spin, L = S,P,D. ... is the orbital angular momentum, π = S,M, or A is the permutational symmetry of the wave function, and J^P is the total angular momentum and parity, we have at N = 0, for S = 0, only the two possibilities

$$|N^2S_S \tfrac{1}{2}^+> \;=\; C_A\ \psi^S_{00}\ \frac{1}{\sqrt{2}}[\chi^\rho_+\phi^\rho_N+\chi^\lambda_+\phi^\lambda_N] \tag{73}$$

$$|\Delta^4S_S \tfrac{3}{2}^+> \;=\; C_A\ \psi^S_{00}\ \chi^S_{3/2}\ \phi^S_\Delta\ . \tag{74}$$

At N = 1 there are more possibilities like

$$|N^4P_M \tfrac{5}{2}^-> \;=\; C_A\ \chi^S{}_{3/2}\ \frac{1}{\sqrt{2}}[\phi^\rho_N\psi^\rho_{11}+\phi^\lambda_N\psi^\lambda_{11}] \tag{75}$$

$$|N^2P_M \tfrac{3}{2}^-> \;=\; C_A\ \tfrac{1}{2}[\chi^\rho_+\phi^\rho_N\psi^\lambda_{11}+\chi^\rho_+\phi^\lambda_N\psi^\rho_{11}+\ \chi^\lambda_+\phi^\rho_N\psi^\rho_{11}\ -\ \chi^\lambda_+\phi^\lambda_N\psi^\lambda_{11}] \tag{76}$$

$$|\Delta^2P_M \tfrac{3}{2}^-> \;=\; C_A\phi^S_\Delta\ \frac{1}{\sqrt{2}}[\chi^\rho_+\psi^\rho_{11}+\chi^\lambda_+\psi^\lambda_{11}] \tag{77}$$

and at N = 2 even more like

$$|\Delta^4D_S \tfrac{7}{2}^+> \;=\; C_A\ \psi^S_{22}\ \chi^S_{3/2}\phi^S_\Delta \tag{78}$$

$$|\Delta^2D_M \tfrac{5}{2}^+> \;=\; C_A\ \phi^S_\Delta\ \frac{1}{\sqrt{2}}[\chi^\rho_+\psi^\rho_{22}+\chi^\lambda_+\psi^\lambda_{22}]\ . \tag{79}$$

Note that in these examples we have chosen for simplicity

states that are the top states of "fully stretched" states of a given L and S; states of lower J from a given L and S may be constructed by the usual Clebsch-Gordan techniques. Of course the S = -3 states are all identical to the S = 0 Δ-like states with the replacement of ϕ_Δ^S by $\phi_{\Omega^-}^S$ and of m_d by m_s.

If we were to neglect the mass difference $m_s - m_d$, then construction of the S = -1 and -2 states would also proceed in exactly parallel fashion to those for S = 0. In fact in this limit the states $|X\,{}^{2S+1}L_\pi J^P\rangle$ we have defined would be in exact correspondence with the usual SU(6) states $|X_\alpha\,{}^{2S+1}[\mu, L^P]J^P\rangle$ where μ = 56,70, or 20 is the SU(6) multiplicity and corresponds respectively to π = S,M, and A, and where α is an SU(3) label to distinguish between Λ_1 and Λ_8, Σ_8 and Σ_{10}, and Ξ_8 and Ξ_{10}. Thus, the wave functions (40), (43), and (46) are of the types S_S, S'_S, and D_S and give the states of the $[56,0^+]$, $[56',0^+]$, and $[56,2^+]$ SU(6) supermultiplets, the pairs ((41), (42)), ((44), (45)), and ((47), (48)) are of the mixed symmetry types P_M, S_M, and D_M and lead to the states of the $[70,1^-]$, $[70,0^+]$, and $[70,2^+]$ supermultiplets, while the lone state (49) is of the type P_A and leads to the $[20,1^+]$ supermultiplet.

In the harmonic limit associated with H_0 and in the SU(3) limit, all of the N = 2 supermultiplets $[56',0^+]$, $[70,0^+]$, $[56,2^+]$, $[70,2^+]$ and $[20,1^+]$ would be degenerate with a spacing ω above the $[70,1^-]$ equal to the spacing of the $[70,1^-]$ above the $[56,0^+]$. We will see below that the anharmonicity potential U has the effect of breaking this high degree of degeneracy of the harmonic limit.

We now complete our construction of baryon states by turning to the S = -1 and -2 sectors. In these cases we have either

$$m_1 = m_2 = m_d < m_3 = m_s$$

or

$m_1 = m_2 = m_s > m_3 = m_d$.

With the same change of variables to $\vec{\rho}, \vec{\lambda}$, and $\vec{R}$, H_0 once again separates to the form

$$H_0 = \frac{P^2}{2M} + \left(\frac{p_\rho^2}{2m_\rho} + \frac{3}{2} k\rho^2\right) + \left(\frac{p_\lambda^2}{2m_\lambda} + \frac{3}{2} k\lambda^2\right) \tag{80}$$

where now

$$M = m_1 + m_2 + m_3 , \qquad m_\rho = m_1 = m_2 , \qquad m_\lambda = \frac{3m_1 m_3}{2m_1+m_3} . \tag{81}$$

The eigenstates of this Hamiltonian are quite distinct from those of the equal mass case, since the ρ-λ degeneracy has been broken:

$$\omega_\rho = (3k/m_\rho)^{1/2} , \qquad \omega_\lambda = (3k/m_\lambda)^{1/2} \tag{82}$$

so that for $S = -1$ $\omega_\lambda < \omega_\rho$ and for $S = -2$ $\omega_\rho < \omega_\lambda$. The low-lying eigenfunctions of H_0 are now, in an obvious notation with $\Psi \equiv \Psi \left(\frac{\alpha_\rho \alpha_\lambda}{\pi}\right)^{3/2} \exp(-\frac{1}{2}\{\alpha_\rho^2 \rho^2 + \alpha_\lambda^2 \lambda^2\})$, given by

$$N_\rho = 0 \quad N_\lambda = 0 \qquad \Psi_{00} = 1 \tag{83}$$

$$N_\rho = 1 \quad N_\lambda = 0 \qquad \Psi_{11}^\rho = \alpha_\rho \rho_+ \tag{84}$$

$$N_\rho = 0 \quad N_\lambda = 1 \qquad \Psi_{11}^\lambda = \alpha_\lambda \lambda_+ \tag{85}$$

$$N_\rho = 0 \quad N_\lambda = 2 \qquad \Psi_{00}^{\lambda\lambda} = \sqrt{2/3}\, \alpha_\lambda^2 (\lambda^2 - \tfrac{3}{2}\alpha_\lambda^{-2}) \tag{86}$$

$$N_\rho = 1 \quad N_\lambda = 1 \qquad \Psi_{00}^{\lambda\lambda} = \sqrt{4/3}\, \alpha_\rho \alpha_\lambda \vec{\rho} \cdot \vec{\lambda} \tag{87}$$

$$N_\rho = 2 \quad N_\lambda = 0 \qquad \psi_{00}^{\rho\rho} = \sqrt{2/3}\ \alpha_\rho^2(\rho^2 - \tfrac{3}{2}\alpha_\rho^{-2}) \tag{88}$$

$$N_\rho = 0 \quad N_\lambda = 2 \qquad \psi_{22}^{\lambda\lambda} = \sqrt{1/2}\ \alpha_\lambda^2\lambda_+^2 \tag{89}$$

$$N_\rho = 1 \quad N_\lambda = 1 \qquad \psi_{22}^{\rho\lambda} = \alpha_\rho\alpha_\lambda\rho_+\lambda_+ \tag{90}$$

$$N_\rho = 2 \quad N_\lambda = 0 \qquad \psi_{22}^{\rho\rho} = \sqrt{1/2}\ \alpha_\rho^2\rho_+^2 \tag{91}$$

$$N_\rho = 1 \quad N_\lambda = 1 \qquad \psi_{11}^{\rho\lambda} = \alpha_\rho\alpha_\lambda(\rho_+\lambda_z-\rho_z\lambda_+) \tag{92}$$

with

$$E = (N_\rho + \tfrac{3}{2})\omega_\rho + (N_\lambda + \tfrac{3}{2})\omega_\lambda \tag{93}$$

so that much of the degeneracy of the harmonic oscillator levels has disappeared. We will see that this has the consequence that (at least in the absence of spin-dependent interactions) SU(6) is maximally violated in these sectors.

The lack of symmetry of the $m_\rho \neq m_\lambda$ wave functions at first seems to pose a paradox: it is impossible using functions to construct totally antisymmetric states as required by the generalized Pauli principle. The way out is simple but perhaps a little surprising: the exclusion principle only requires that identical fermions be antisymmetrized, and so obvious does not apply in the unequal mass case. The confusion arises because in the equal mass case the Hamiltonian has S_3 symmetry so that its wave functions can be chosen to belong to the definite representations S, M = (M^ρ, M^λ), A; if one then elevates flavour to the status of a quark variable, states satisfying the generalized Pauli principle will automatically be eigenstates of not only the energy, but also of flavour SU(3).

In the S = -1 and -2 sectors it is therefore more

appropriate to use flavour wavefunctions like

$$\Phi_\Sigma = \frac{1}{\sqrt{2}} (ud + du)s \tag{94}$$

$$\Phi_\Lambda = \frac{1}{\sqrt{2}} (ud - du)s \tag{95}$$

$$\Phi_\Xi = ssu \tag{96}$$

which single out the third quark which has a different mass, and to retain only the S_2 subgroup of S_3 as a symmetry group. (We shall see below that even this is unnecessary!). By isospin invariance these three flavour wavefunctions are in fact sufficient for S = -1 and -2 spectroscopy and in terms of this "uds basis" for flavour we can construct from the usual colour and spin states and the $m_\rho \neq m_\lambda$ spatial wavefunctions (all of which still have explicit S_2 symmetry) a complete set of states antisymmetric under $1 \leftrightarrow 2$. We label these states by $|Y\,{}^{2S+1}L_\sigma J^P\rangle$, where $Y = \Sigma, \Lambda$, or Ξ, and S,L, and J are as in the equal mass sectors, and where $\sigma = \rho,\lambda$, $\rho\rho,\rho\lambda,\lambda\lambda$ et cetera denotes the level of excitation of the ρ and λ oscillators. For example, for $N_\rho = N_\lambda = 0$ we can construct only

$$|\Lambda\,{}^2S\,\tfrac{1}{2}^+\rangle = C_A\,\Phi_\Lambda\,\chi^\rho_+\,\psi_{00} \tag{97}$$

$$|\Sigma\,{}^2S\,\tfrac{1}{2}^+\rangle = C_A\,\Phi_\Sigma\,\chi^\lambda_+\,\psi_{00} \tag{98}$$

$$|\Xi\,{}^2S\,\tfrac{1}{2}^+\rangle = C_A\,\Phi_\Xi\,\chi^\lambda_+\,\psi_{00} \tag{99}$$

$$|\Sigma\,{}^4S\,\tfrac{1}{2}^+\rangle = C_A\,\Phi_\Sigma\,\chi^S_{3/2}\,\psi_{00} \tag{100}$$

$$|\Xi\,{}^4S\,\tfrac{1}{2}^+\rangle = C_A\,\Phi_\Xi\,\chi^S_{3/2}\,\psi_{00} \tag{101}$$

which correspond to the ordinary Λ, Σ, and Ξ of the octet and the Σ^* and Ξ^* of the decuplet. For $N_\rho + N_\lambda = 1$ we can construct the "fully stretched" Λ states

$$|\Lambda\, {}^2P_\lambda\, \tfrac{3}{2}^-\rangle = C_A\, \Phi_\Lambda\, \chi_+^\rho\, \psi_{11}^\lambda \tag{102}$$

$$|\Lambda\, {}^2P_\rho\, \tfrac{3}{2}^-\rangle = C_A\, \Phi_\Lambda\, \chi_+^\lambda\, \psi_{11}^\rho \tag{103}$$

$$|\Lambda\, {}^4P_\rho\, \tfrac{5}{2}^-\rangle = C_A\, \Phi_\Lambda \chi_{3/2}^S\, \psi_{11}^\rho \tag{104}$$

plus four states with $J<L+S$: $|\Lambda\, {}^2P_\lambda\; 1/2^-\rangle$, $|\Lambda\, {}^2P_\rho\; 1/2^-\rangle$, $|\Lambda\, {}^4P_\rho\; 3/2^-\rangle$, and $|\Lambda\, {}^4P_\rho 1/2^-\rangle$; the "fully stretched" Σ states

$$|\Sigma\, {}^2P_\lambda\, \tfrac{3}{2}^-\rangle = C_A\, \Phi_\Sigma\, \chi_+^\lambda\, \psi_{11}^\lambda \tag{105}$$

$$|\Sigma\, {}^2P_\rho\, \tfrac{3}{2}^-\rangle = C_A\, \Phi_\Sigma\, \chi_+^\rho\, \psi_{11}^\rho \tag{106}$$

$$|\Sigma\, {}^4P_\lambda\, \tfrac{5}{2}^-\rangle = C_A\, \Phi_\Sigma\, \chi_{3/2}^S\, \psi_{11}^\lambda \tag{107}$$

plus the four states $|\Sigma\, {}^2P_\lambda\; 1/2^-\rangle$, $|\Sigma\, {}^2P_\rho\; 1/2^-\rangle$, $|\Sigma\, {}^4P_\lambda\; 3/2^-\rangle$, and $|\Sigma\, {}^4P_\lambda\; 1/2^-\rangle$, with $J<L+S$; and finally the Ξ states which are identical in form to the Σ's (since both Φ_Σ and Φ_Ξ are symmetric under $1\leftrightarrow 2$ interchange) with the replacement $\Phi_\Sigma \to \Phi_\Xi$ and $m_d = m_u \leftrightarrow m_s$. These states correspond exactly in number to the expected $S = -1$ and -2 states of the $[70,1^-]$ SU(6) supermultiplet which has the $[SU(3),J^P]$decomposition

$$[70,1^-] = [1,\tfrac{3}{2}^-]+[1,\tfrac{1}{2}^-]+[8,\tfrac{5}{2}^-]+2[8,\tfrac{3}{2}^-]+2[8,\tfrac{1}{2}^-]+[10,\tfrac{3}{2}^-]+[10,\tfrac{1}{2}^-] \tag{108}$$

leading one to expect 7 states of each of the types Λ,Σ,Ξ along with 5 N's, 2Δ's and 2 Ω's. However, we can now see that the true eigenstates of H_0 are far from being eigenfunctions of SU(6). Consider, for example, the state $|\Lambda^2P_\lambda\ 3/2^->$ of equation (102). In terms of SU(6) states it has the composition (for $\alpha_\rho = \alpha_\lambda$)

$$|\Lambda^2P_\lambda \tfrac{3}{2}^- > = \frac{1}{\sqrt{2}}\{|\Lambda_1{}^2[70,1^-]\tfrac{3}{2}^-> + |\Lambda_8{}^2[70,1^-]\tfrac{3}{2}^->\} \qquad (109)$$

and we see a maximal violation of, in this case, SU(3) symmetry. We will discuss the effects of these violations below. One consequence, which is very transparent in the uds basis, may suffice for now however: the $\Sigma 5/2^-$ and $\Lambda 5/2^-$ which would be expected to be degenerate in the SU(3) limit will (in this approximation) be split by $\omega_\rho-\omega_\lambda \sim 0(50\ \text{MeV})$. Note that this effect is expected to make $\Lambda 5/2^-$ heavier than $\Sigma 5/2^-$ as observed and in reversal to the ground state ordering (which we will see below is a spin dependent effect). The construction of the $N_\rho+N_\lambda = 2$ states proceeds along similar lines and has similar effects. As examples, we list just the possible fully stretched Λ-like states

$$|\Lambda^2S_{\lambda\lambda}\ \tfrac{1}{2}^+> = C_A\ \Phi_\Lambda\ \chi^\rho_+\ \psi^{\lambda\lambda}_{00} \qquad (110)$$

$$|\Lambda^2S_{\rho\lambda}\ \tfrac{1}{2}^+> = C_A\ \Phi_\Lambda\ \chi^\lambda_+\ \psi^{\rho\lambda}_{00} \qquad (111)$$

$$|\Lambda^2S_{\rho\rho}\ \tfrac{1}{2}^+> = C_A\ \Phi_\Lambda\ \chi^\rho_+\ \psi^{\rho\rho}_{00} \qquad (112)$$

$$|\Lambda^2D_{\lambda\lambda}\ \tfrac{5}{2}^+> = C_A\ \Phi_\Lambda\ \chi^\rho_+\ \psi^{\lambda\lambda}_{22} \qquad (113)$$

$$|\Lambda^2D_{\rho\lambda}\ \tfrac{5}{2}^+> = C_A\ \Phi_\Lambda\ \chi^\lambda_+\ \psi^{\rho\lambda}_{22} \qquad (114)$$

$$|\Lambda^2D_{\rho\rho}\ \tfrac{5}{2}^+\rangle = C_A\ \Phi_\Lambda\ \chi_+^\rho\ \psi_{22}^{\rho\rho} \qquad (115)$$

$$|\Lambda^2P_{\rho\lambda}\ \tfrac{3}{2}^+\rangle = C_A\ \Phi_\Lambda\ \chi_+^\lambda\ \psi_{11}^{\lambda\rho} \qquad (116)$$

$$|\Lambda^4S_{\rho\lambda}\ \tfrac{3}{2}^+\rangle = C_A\ \Phi_\Lambda\ \chi_{3/2}^S\ \psi_{00}^{\rho\lambda} \qquad (117)$$

$$|\Lambda^4D_{\rho\lambda}\ \tfrac{7}{2}^+\rangle = C_A\ \Phi_\Lambda\ \chi_{3/2}^S\ \psi_{22}^{\rho\lambda} \qquad (118)$$

$$|\Lambda^4P_{\rho\lambda}\ \tfrac{5}{2}^+\rangle = C_A\ \Phi_\Lambda\ \chi_{3/2}^S\ \psi_{11}^{\rho\lambda} \qquad (119)$$

there are an additional 9 Λ states with $J < L + S$. (In total there are 13N's, 8Δ's, 19Λ's, 21Σ's, 21 Ξ's, and 8Ω's at $N_\rho+N_\lambda=2$). These eigenstates of H_0 not only violate SU(3) but also SU(6); e.g., for $\alpha_\rho = \alpha_\lambda$

$$|\Lambda^2D_{\lambda\lambda}\tfrac{5}{2}^+\rangle = \frac{1}{\sqrt{2}}|\Lambda_8\,{}^2[56,2^+]\tfrac{5}{2}^+\rangle - \frac{1}{2}|\Lambda_8\,{}^2[70,2^+]\tfrac{5}{2}^+\rangle + \frac{1}{2}|\Lambda_1\,{}^2[70,2^+]\tfrac{5}{2}^+ \qquad (120)$$

is a mixture of the SU(6) multiplets $[56,2^+]$ and $[70,2^+]$ and of SU(3) representations within the $[70,2^+]$.

At this point one might wonder why we have introduced SU(6) into this discussion at all. There were two main reasons: 1) in some calculations it is technically useful to use the SU(6) basis, and 2) for historical reasons it is still very common to use SU(6) language in describing many aspects of light baryon physics. Nevertheless, we stress that from the modern point of view SU(6) has very little to do with baryons.

As an amusing and possibly useful aside, we note that it is not usually necessary to use S_3 symmetry at all! Consider, for example, the S = 0 sector. Since by isospin invariance it is sufficient to consider only states of the form uud and since up and down quarks are <u>not</u> identical, we can discuss such states in exact parallelism to the states Ξ = ssu. At $N_\rho + N_\lambda = 0$, for example, the only such states are

$$C_A \text{ (uud) } \chi_+^\lambda \, \psi_{00}^S \tag{121}$$

$$C_A \text{ (uud) } \chi_{3/2}^S \, \psi_{00}^S \tag{122}$$

which we can identify with the proton and the the Δ^+. For $N_\rho + N_\lambda > 0$ this method has the disadvantage that it will only automatically diagonalize isospin if the I = 1/2 and 3/2 states are not degenerate, but this ambiguity is always resolved in practice once one considers spin dependent interactions.

With our zeroth order eigenfunctions established, to complete our discussion of H_{si} it remains to do perturbation theory in U of equation (35). We begin by discussing the equal mass case where in the harmonic limit we have only three equally spaced energy levels. Since H_{si} is in this case S_3 symmetric even when U $\neq$ 0, the degeneracy of all pairs of wavefunctions $(\psi^{M\rho}, \psi^{M\lambda})$ remains unbroken. This leaves us with still a unique ground state and first excited state energy corresponding to the SU(6) supermultiplets $[56,0^+]$ and $[70,1^-]$ respectively, but with in general five independent N=2 levels associated with $\psi_{00}^{S'}$, $(\psi_{00}^{M\rho}, \psi_{00}^{M\lambda})$, ψ_{22}^S, $(\psi_{22}^{M\rho}, \psi_{22}^{M\lambda})$, and ψ_{11}^A, corresponding to the five SU(6) supermultiplets $[56',0^+]$, $[70,0^+]$, $[56,2^+]$, $[70,2^+]$ and $[20,1^+]$, respectively. Given these seven independent energies, one would naturally suppose that without specifying U and doing explicit calculations it would be necessary to introduce these seven energy levels via seven parameters. Since it would be necessary to introduce four parameters --- the common quark mass, b, C, and α_s --- even in a first principles calculation, expanding to seven does not seem too bad. However, it turns out that these seven energy levels are constrained by a remarkable theorem: in first order perturbation theory any potential U will split the harmonic oscillator spectrum into exactly the same pattern, and this pattern can be described in terms of only <u>three</u> parameters. Namely

$$E\,[56,0^+] = E\,[S_S] = E_0 \tag{123}$$

$$E\,[70,1^-] = E\,[P_M] = E_0 + \Omega \tag{124}$$

$$E\,[56',0^+] = E\,[S'_S] = E_0 + 2\Omega - \Delta \tag{125}$$

$$E\,[70,0^+] = E\,[S_M] = E_0 + 2\Omega - \frac{1}{2}\Delta \tag{126}$$

$$E\,[56,2^+] = E[D_S] = E_0 + 2\Omega - \frac{2}{5}\Delta \tag{127}$$

$$E\,[70,2^+] = E\,[D_M] = E_0 + 2\Omega - \frac{1}{5}\Delta \tag{128}$$

$$E\,[20,1^+] = E\,[P_A] = E_0 + 2\Omega \tag{129}$$

as shown in Fig.12. This theorem is easily proved by noting that the expectation value of U in the seven sets of independent wavefunctions can be expressed in terms of the first three (n = 0,1,2) of the moments

$$a_n = \frac{3\alpha^{3+2n}}{\pi^{3/2}} \int d^3\rho\, U(\sqrt{2}\rho)\, e^{-\alpha^2\rho^2}\, \rho^{2n} \tag{130}$$

in terms of which

$$E_0 = 3m + 3\omega + a_0 \tag{131}$$

$$\Omega = \omega - \frac{1}{2} a_0 + \frac{1}{3} a_1 \tag{132}$$

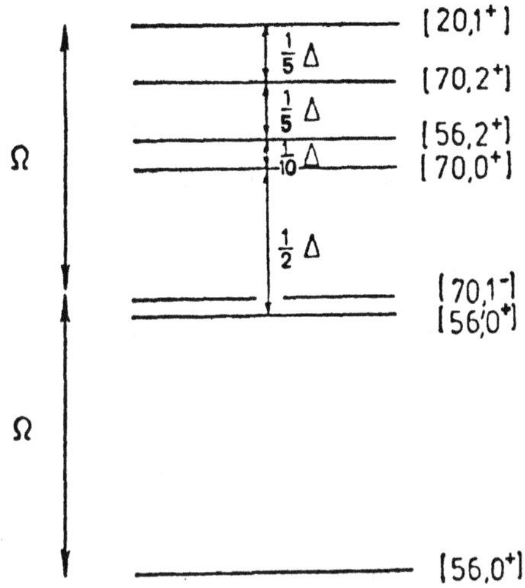

Fig. 12.: The pattern of low-lying SU(6) supermultiplets from first order perturbation theory in the anharmonic potential U

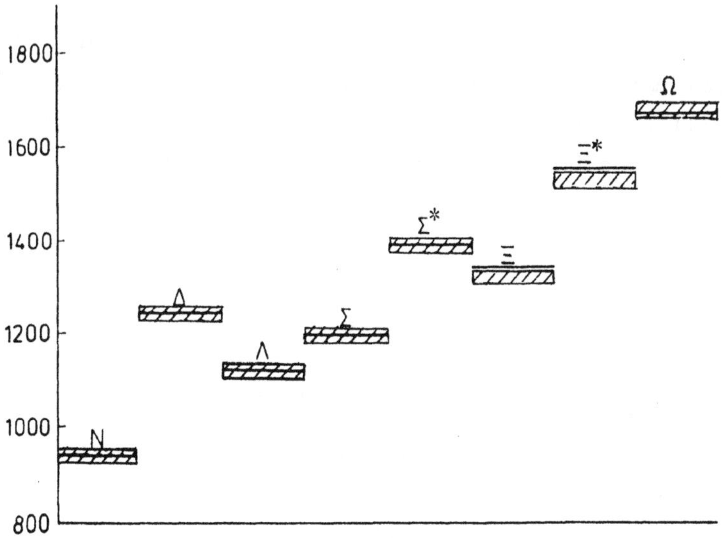

Fig. 13: The experimental spectrum (hatched boxes) of ground state baryons compared to theory (horizontal bars)

$$\Delta = -\frac{5}{4}a_0 + \frac{5}{3}a_1 - \frac{1}{3}a_2 . \tag{133}$$

The actual pattern of levels corresponding to the spin-unperturbed $S = 0$ multiplets seems to be in accord with this theoretical pattern and indicates $E_0 \simeq 1150$ MeV, $\Omega \simeq 440$ MeV, and $\Delta \simeq 400$ MeV. That Δ turns out to be positive is exactly what one expects from the actual form of U: both the attractive Coulomb potential and the fact that the true potential is linear instead of quadratic give positive contributions to Δ. The well-known fact that the $[56',0^+]$ lies far below the $[56,2^+]$ is thus naturally explained. Of course first order perturbation theory can only be expected to be reliable for $\Delta \ll 2\Omega$, but explicit calculations show that 1) this pattern is generally respected for realistic potentials, and 2) the required values for E_0, Ω, and Δ are consistent with the values of the fundamental parameters deduced from meson spectroscopy.

The case $U \neq 0$ is somewhat more complicated in the unequal mass case $m_3 \neq m_1 = m_2$, since now the ρ and λ oscillators behave differently. Of course the energies of the states before turning on U are as previously discussed:

$$\frac{\omega_\lambda}{\omega_\rho} = \left(\frac{m_\rho}{m_\lambda}\right)^{1/2} . \tag{134}$$

Although in the presence of a U perturbation this relationship will change, it should remain a reasonable approximation to the actual situation and can be used to correct the relations (123) to (129). The generality of the important qualitative feature here that modes involving heavy quarks will have a lower frequency than those involving light quarks can most easily be seen in operation in an extreme case like $m_3 >> m_1 = m_2$. In this limit baryons behave like a helium atom with inverted statistics. In the absence of interactions between quarks 1 and 2 this system will have two degenerate negative parity excited spatial wavefunctions above the ground state, namely,

$$|\tilde{\lambda}> \equiv \frac{1}{\sqrt{2}}\{|sp>+|ps>\} \quad (135)$$

$$|\tilde{\rho}> \equiv \frac{1}{\sqrt{2}}\{|sp>-|ps>\} \quad (136)$$

where s and p are the lowest-lying single particle s and p orbitals about quark 3. If one now turns on the residual attractive potential energy of the two light quarks (from any confinement potential), then $|\tilde{\lambda}>$ will end up lying lower than $|\tilde{\rho}>$ since in $|\tilde{\rho}>$ the two quarks are unlikely to be near to each other. Thus though our numerical results will depend on the use of the harmonic oscillator approximations sketched above, our principal conclusions are more general.

This concludes our discussion of the spectrum and eigenstates of H_{si}. There is a considerable body of interesting physics contained just in this spin-independent spectroscopy: we have already mentioned several features of the baryons which are specifically related to the expected form of H_{si}. However, we should also note that the spin-independent eigenstates are sufficient as well to give us a very good zeroth order understanding of several non-spectroscopic features of baryons. The best example is probably the baryon magnetic moments which, assuming Dirac magnetic moments for the usual constituent quark masses, are predicted within $\pm 0.2\mu_N$ in terms of these eigenstates.

In the following section we will see that other physics which is at least as interesting emerges from spin dependent effects.

ii) Spin-dependent effects in baryons

In a heavy $Q\bar{Q}$ or QQQ system the nonrelativistic reduction of one gluon exchange given in Section II. E. should be applicable. However, in light quark systems m/E will be very different from unity and we must expect that spin dependent effects could be significantly modified from this simple form. After all, as we indicated above, such effects are themselves relativistic corrections to the potentials. We accordingly will begin by taking a more inductive approach to these effects in the light baryons than we would in, say, the $b\bar{b}$ system.

We first consider the ground state baryons in which we can completely isolate the hyperfine interaction (26). (Since $\vec{L}_\rho$ and $\vec{L}_\lambda$ annihilate ψ_{00}, the spin-orbit interaction does not even contribute any off-diagonal effects to the ground states). It is completely straightforward to calculate the diagonal hyperfine contribution to these states which comes entirely from the $\vec{S}_i\cdot\vec{S}_j$ contact interaction. Using (31), the fact that

$$\langle\psi_{00}^S|\delta^3(\vec{r}_{ij})|\psi_{00}^S\rangle = \alpha^3/(2\pi)^{3/2}$$

and

$$\langle\vec{S}_i\cdot\vec{S}_j\rangle = \begin{cases} +1/4 & \text{if i and j have spin 1} \\ -3/4 & \text{if i and j have spin 0} \end{cases} \tag{137}$$

one readily obtains all of the ground state diagonal

splittings in terms of a single parameter

$$\delta \equiv \frac{4\alpha_s\alpha^3}{3\sqrt{2}\pi m_d^2} r_{hyp} \tag{138}$$

where r_{hyp} is a factor we introduce into (26) as a crude measure of relativistic corrections to the naive hyperfine interaction. The relations one finds are (ignoring the fact that ($\alpha_\rho \neq \alpha_\lambda$)

$$\delta M_\Delta = +\frac{1}{2}\delta \qquad \delta M_N = -\frac{1}{2}\delta \tag{139}$$

$$\delta M_{\Sigma^*} = [\frac{x}{3} + \frac{1}{6}]\delta \qquad \delta M_\Sigma = [\frac{1}{6} - \frac{2x}{3}]\delta \qquad \delta M_\Lambda = -\frac{1}{2}\delta \tag{140}$$

$$\delta M_{\Xi^*} = [\frac{x}{3} + \frac{x^2}{6}]\delta \qquad \delta M_\Xi = [\frac{x^2}{6} - \frac{2x}{3}]\delta \tag{141}$$

$$\delta M_\Omega = +\frac{x^2}{2}\delta \tag{142}$$

where we have defined $x \equiv m_d/m_s \simeq 2/3$ for brevity. Equation (139) shows that with (26) the Δ is automatically heavier than the N. Another noteworthy feature is that the mass of the Λ is lowered by as much as the N; this SU(3)-like equality comes about since the Λ has a χ^ρ spin wavefunction (see equ. (97)) which makes the suppressed strange quark chromomagnetic moment irrelevant. Since the same is not true for the χ^λ spin function of the Σ, one gets a Σ-Λ splitting even though both have the same quark content. In fact one gets

$$\Sigma - \Lambda = \frac{2}{3}(1-x)[\Delta - N] \tag{143}$$

which works very well indeed. The complete results of this programme, including mixing and $\alpha_\rho \neq \alpha_\lambda$ effects, is shown in Fig.13, in which δ has been adjusted to fit the Δ-N splitting. In addition to this spectroscopic evidence, the is other evidence for the role of the hyperfine interactio in the ground states. Most interesting perhaps is that it gives rise to a charge radius for the neutron. Formally this occurs via an off-diagonal mixing with the N=2 $|N^2S_M\ 1/2^+\rangle$ state, but its physical origin is simply visualized. In the (ddu) flavour basis, the neutron has the spin wavefunction χ^λ. Since the two down quarks therefore have spin 1 they repel (see equ.(137)) each other, but the two up-down pairs of quarks are in an attractive state (recall that the sum of all three pairwise interacti is attractive). Thus the two down quarks are pushed to the periphery of the neutron and the up quark is pulled to its centre, producing a negative charge radius. Detaile calculations automatically produce the observed magnitude of this effect in terms of the same parameter δ. Note that, unlike the pure ground state spectroscopy, this effect is sensitive to the short range nature of H_{hyp}. This is in accord with the evidence from mesons and also with other effects to be mentioned below. In conclusion, there seems to be excellent evidence from the ground states (in support of similar conclusions from elsewhere) for an interaction in light quark baryons very similar in form to (26). Finally, we mention that these techniques have been applied with apparent success to $m_d \neq m_u$ isospin violating effects and to ground state baryons containing a heavy quark.

When we turn to excited baryons we must expect to encounter not only the well-established hyperfine interaction, but also effects due to spin-orbit interactions. Here we meet a surprise which greatly simplifies the description of excited baryons but which is quite mysterious in this context: spin orbit effects are very small. It requires no great subtlety to see this; the situation can be read right out of the Particle Data Group tables via the masses of the four candidate L = 2 states $\Delta 7/2^+$ (1935±25), $\Delta 5/2^+$ (1905±15), $\Delta 3/2^+$ (2010±150), and $\Delta 1/2^+$(1900±50). Spin-orbit effects would, if present, split these states since, roughly speaking, they can all be associated with S = 3/2 coupled with L = 2. A possible explanation of this observation will be discussed below, but for now we note that this situation is not so different from light mesons where the ρ-π hyperfine splitting is 640 MeV, but the P-wave spin-orbit splittings are small: e.g., A_2-A_1 ≃ 45±35 MeV and A_2-B≃85±15 MeV, with some part of these splittings attributable to hyperfine interactions. In the case of mesons, it seems plausible that these small values are due to two effects: i) the radial dependence of the hyperfine interaction is more singular than that of the spin-orbit interactions so that the former is more effective for a given α_s, ii) the full Thomas precession interaction automatically tends to cancel colour magnetic spin-orbit terms, and iii) the relativistic corrections factors for spin orbit effects are samller than their hyperfine analog r_{hyp} in (138). The second of these mechanisms can (probably) be seen at work in charmonium. Thus in terms of mesons spin orbit effects in baryons do not seem so unreasonable: if they were present in the same proportion to the Δ-N splitting as they are in mesons to the ρ-π splitting, then their strength would be only about 30 MeV, well within the "noise" of the light quark models. For now we will therefore adopt the usual approach and ignore spin-orbit forces, postponing to subsection iv) a discussion of the mechanism for this suppression.

With this approximation, one need only calculate hyperfine matrix elements and diagonalize. We will avoid most of the resulting gory details, but use the S=0 negative parity baryons as an example. The P-wave baryons are in a sense a perfect testing ground for hyperfine interactions since they simultaneously have one relative coordinate in an S-wave and one in a P-wave. When two quarks are in a relative S-wave, the contact force can operate but the tensor force vanishes, while when two quarks are in a relative P-wave the contact force vanishes (since there is zero amplitude to find the quarks "in contact") while the tensor force operates. Thus in P-wave baryons the two pieces of the hyperfine interaction operate side by side and their relative strength, as well as the short range character of the contact force, is testable.

Another feature of the P-wave baryons is that there are many states with the same IJ^P quantum numbers which are degenerate in the SU(6) limit; such states will in general be very strongly mixed by the SU(6)-violating terms in our Hamiltonian. Apart from the anomalously large matrix element responsible for the interband mixing of the N=2 level S_M state into the nucleon, the resulting intraband mixing angles, which can be "measured" via an analysis of baryon decays, provide the first real test of whether the model is correctly reproducing the internal structure as well as the spectroscopy of baryons.

The calculation is straightforward and gives, in terms of the same parameter δ,

$$\langle N^2P_M | H_{contact} | N^2P_M \rangle = -\frac{1}{4}\delta \tag{144}$$

$$\langle N^4P_M | H_{contact} | N^4P_M \rangle = \langle \Delta^2P_M | H_{contact} | \Delta^2P_M \rangle = +\frac{1}{4}\delta \tag{145}$$

independent, of course, of J^P and

$$<N\,^4P_M \begin{bmatrix} 5/2^- \\ 3/2^- \\ 1/2^- \end{bmatrix} | H_{tensor} | N\,^4P_M \begin{bmatrix} 5/2^- \\ 3/2^- \\ 1/2^- \end{bmatrix} > \; = \begin{bmatrix} -1/20 \\ +1/5 \\ -1/4 \end{bmatrix} \delta \qquad (146)$$

$$<N\,^4P_M \begin{bmatrix} 3/2^- \\ 1/2^- \end{bmatrix} | H_{tensor} | N\,^2P_M \begin{bmatrix} 3/2^- \\ 1/2^- \end{bmatrix} > = \begin{bmatrix} -1/4\sqrt{10} \\ +1/4 \end{bmatrix} \delta \qquad (147)$$

with all other matrix elements zero. Note that the $^2\Delta$ states have the same contact interaction as the ^4N states because of the short range character of this force. Indeed, at first glance these seven states do seem to split into two bands separated by about $1/2(\Delta-N)$ and with the Δ's in the upper band as predicted (see Fig.14). When tensor forces are considered the SU(6)-like N $1/2^-$ and N $3/2^-$ states become mixed, and we see that the resulting mixing angles are even independent of δ! These mixings can have dramatic effects on the decays of these states and in this way they can be determined experimentally, as we will discuss below. In both sectors the sign and magnitude of the measured mixing angles are in good agreement with those expected.

This is not the place to proceed to a blow by blow description of every sector of baryon spectroscopy. Suffice it to say that when applied to the negative parity hyperons and to the positive parity baryons of the N=2 band, the same simple picture continues to work quite well. This is illustrated in Fig.15 which shows the results for N = 2 N's and Δ's, including not only spectroscopy but also indicating on the basis of an analysis of meson couplings which states ought to have been seen in πN elastic scattering.

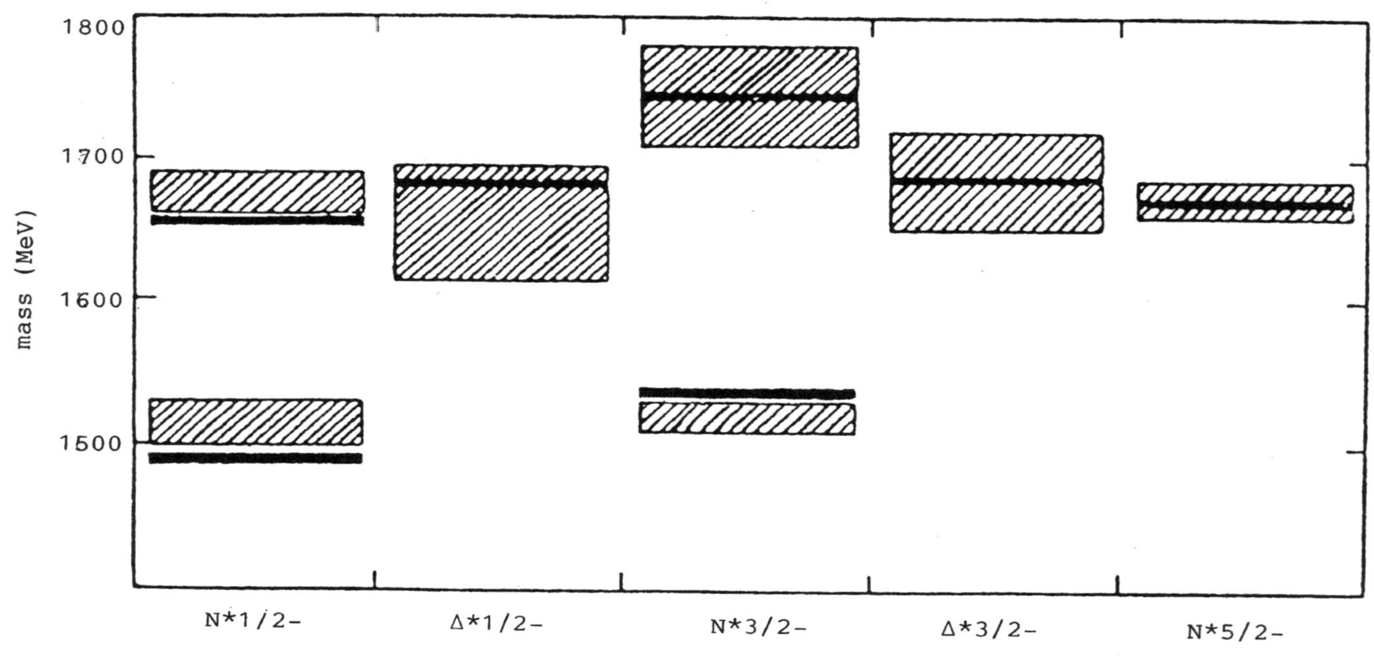

Fig. 14.: The predicted S=O negative parity baryons compared to experiment; the experimentally determined regions in which the masses lie are denoted by shaded boxes, theory is given by the horizontal bars

iii) Baryon couplings

The last comment above leads us to discuss the importance of the study of hadronic couplings as a test of any model of hadron structure. It is clearly insufficient to predict just the masses of hadrons: any complete model should also predict their internal structure. Of course such gross features as the J^P of a state reflect this internal structure but by far the most sensitive indicators of internal composition are decay amplitudes since they can vary dramatically with interferences between different components of a state. Especially when working with a model and with data of limited accuracy, such tests are crucial.

Let me illustrate this point with an example. Notice the two $\Delta 5/2^+$ baryons in Fig.15 at about 1950 MeV. In the SU(6) limit, to mention just two points, each of these states is strongly coupled to πN, and the lower one decays to a $\pi\Delta$ P-wave more strongly than to the F-wave. Tensor forces completely change this state of affairs and the model we have been discussing predicts that only the lower state will be observable in πN elastic scattering and that its $\pi\Delta$ F-wave decay will dominate its P-wave. This is in accord with observation. There has, moreover, been a further very recent development: while the upper state is predicted to be practically "invisible" in $\pi N \to \pi N$, it was predicted that it should be visible in $\pi N \to \rho N$. This previously "missing" upper state now seems to have been found in precisely this channel.

As another example, let us now consider S= -1 baryons. To the extent that we can neglect U and hyperfine perturbations of the ρ-λ excitation structure, we can see trivially that ρ excitations will only couple weakly to the usual $\bar{K} N$ resonance channel. This is illustrated in Fig.16 which shows that, in the spectator approximation,

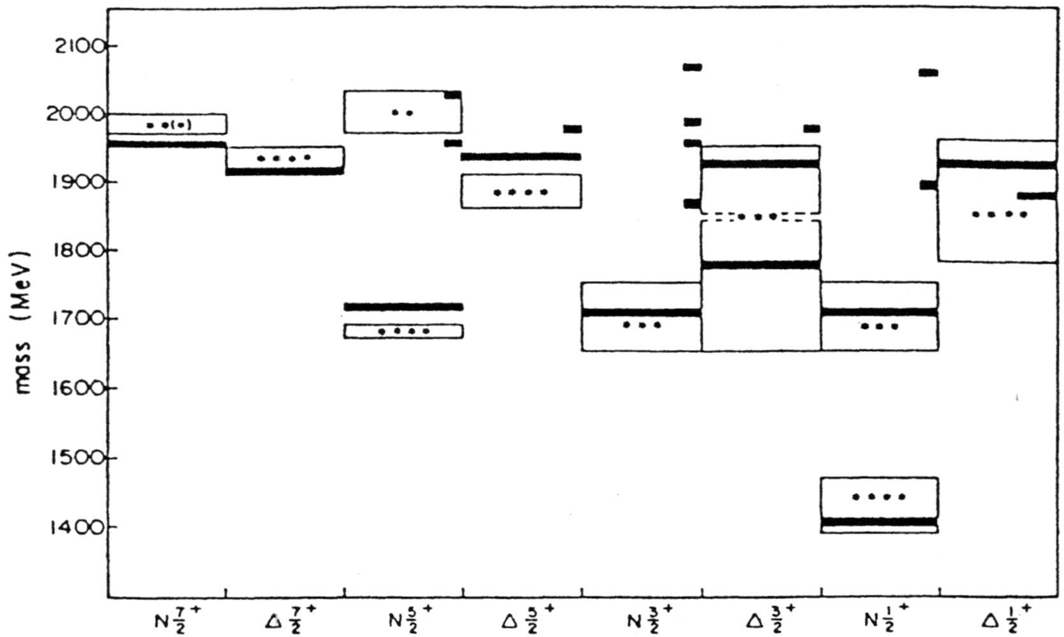

Fig. 15.: The predicted S=O positive parity baryons compared to experiment; the legend is as for Fig.14, with now in addition states that are predicted to decouple from πN being shown as short horizontal bars

$\bar{K}$ emission will leave the ρ oscillator excited and therefore with no overlap with a ground state nucleon. Fig.17 shows the N=2 states with strangeness -1, where the fact that many of the states are missing is attributable to this effect (though a full understanding requires effects like those responsible for decoupling the upper $\Delta\ 5/2^+$ in the previous example).

In concluding this brief sketch of our understanding of the light baryons, we first state that it appears that the simple model we have outlined here is a rather good guide to the physics of baryons with light quarks. It manages to predict masses with an accuracy of roughly 50 MeV and in hundreds of cases it rather consistently predicts the signs and magnitudes of the couplings of baryons. Nevertheless, it is important to remember that the model cannot be right. On the theoretical side, ignoring spin-orbit forces must be just an approximation and one cannot be satisfied with the model until this approximation is better understood. There also seem to be a number of places (like the $\Lambda 3/2^-(1520)$ - $\Lambda 1/2^-(1405)$ splitting, the photocouplings of the $N3/2^+(1720)$, et cetera) where the model seems unacceptably inaccurate. Problems such as these indicate that it is essential to test the model in new regimes (like Ξ^* and Ω^* spectroscopy) in the hope that it will fail and in doing so provide us with clues as to how to proceed to improve it.

iv) Relativization and unification with mesons

As with mesons (see above), the quark model for baryons not only survives relativization, but is improved by it. Recent work [4] has shown that when the baryon problem is solved with the potential parameters and relativistic modifications described above for mesons, some

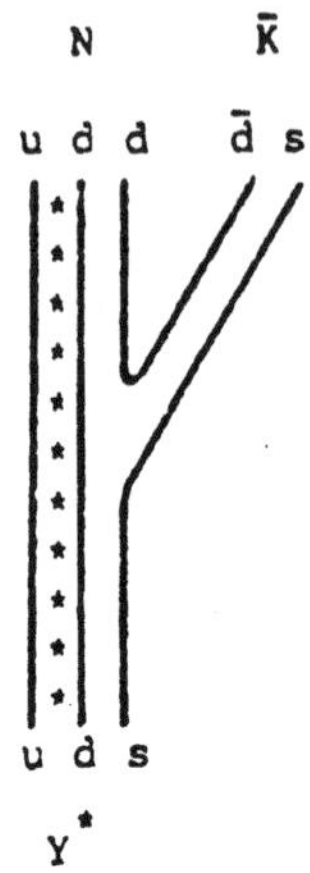

Fig. 16.: A graphic demonstration of the decoupling of $S=-1$ ρ excitations from the $\bar{K}N$ channel.

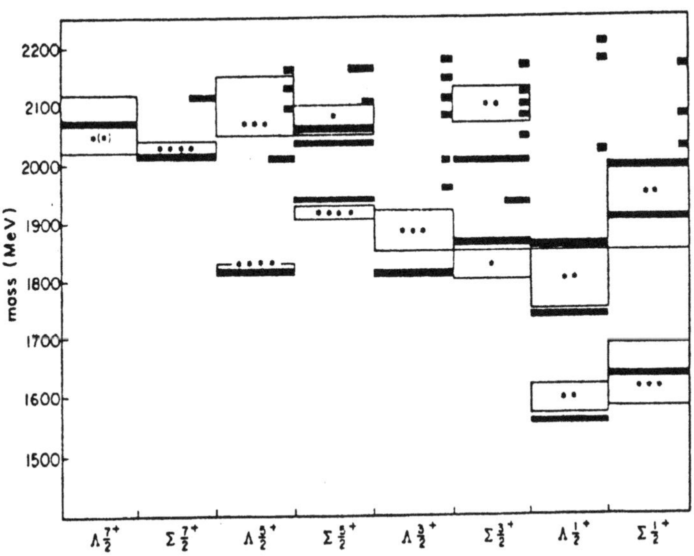

Fig. 17.: As in Fig.15, but for the $S=-1$ positive parity baryons, and with short bars representing states which are predicted to decouple from $\bar{K}N$.

remarkable results are obtained, including:

1) baryon spectroscopy emerges automatically: e.g., the Δ-N splitting and the P-wave multiplet centre-of-gravity are reproduced within a few percent, confirming the excepted relationship between Δ-N and ρ-π and between, e.g., $N^*(1675)5/2^-$ - Δ and $A_2(1320)$ - ρ.

2) spin orbit forces are small automatically, thereby rationalizing the success of the usual non-relativistic model discussed above.

3) some old puzzles (like the $\Delta\frac{3}{2}^- - \Delta\frac{1}{2}^-$ splitting) are solved.

There is thus very strong evidence now that essentially all of the low-lying states of $q\bar{q}$ and qqq can be understood in terms of the QCD based quark model.

References for Chapter 3

1. For a recent discussion of the "QCD-improved quark model" see, e.g., N. Isgur in Testing the Standard Model, AIP Conf. Proc., Vol. 81, editors C. Heusch and W.T. Kirk (American Inst. of Physics, New York, 1982), p.1.
2. The semirelativistic treatment of $q\bar{q}$ states can be found in S. Godfrey and N. Isgur, to appear in Phys. Rev. D. This paper has an extensive list of further references on mesons in the quark model.
3. The model of baryons discussed here was developed in collaboration with Gabriel Karl. For a pedagogical review of this model, see N. Isgur in The New Aspect of Subnuclear Physics, edited by A. Zichichi (Plenum, New York, 1980), p. 107 and references therein.
4. The relativized quark model for baryons can be found in S. Capstick and N. Isgur, University of Toronto preprint.

4. MULTIQUARK SYSTEMS

A. General considerations

We should not be misled into believing that we understand hadron dynamics in QCD on the basis of our being able to understand mesons and baryons. The $q\bar{q}$ and qqq systems are very special cases since for these states the lowest adiabatic potential (corresponding to the gluon ground state in the presence of fixed quark sources) is very well isolated from excited glue adiabatic potentials. In this sense the whole of the ordinary quark model spectrum is analogous to the vibrational and rotational bands of a molecule moving in the lowest adiabatic surface of the electronic ground state of the molecule. We have yet to find experimentally the "hybrid" mesons and baryons which are the analogs of the vibrational and rotational molecular bands built on excited electronic (versus gluonic) states.

While the hybrid states depend on the properties of the gluonic fields, their main characteristics should be quite similar to those of ordinary mesons and baryons. In contrast, we should expect multiquark states to behave in entirely new ways [1]: multiquark states characteristically have many low-lying adiabatic surfaces which cross. Figs.18 and 19 illustrate this characteristic behaviour in the case of the $qq\bar{q}\bar{q}$ system. Fig.18 illustrates very clearly the fact that the state of a $qq\bar{q}\bar{q}$ system depends not only on the positions of the quarks and antiquarks, but also on the state of the flux which flows between them. This is also true for simpler systems like $q\bar{q}$ (where the state of the glue distinguishes the ordinary mesons of the quark model from hybrid mesons), but in multiquark systems we cannot expect to be able to segregate the dynamical states into sectors with an approximately fixed

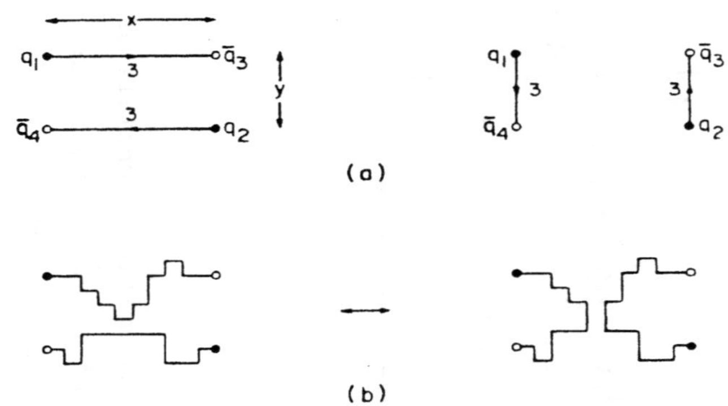

Fig. 18.: (a) two low-lying $qq\bar{q}\bar{q}$ configurations,
(b) topological mixing between these two configurations

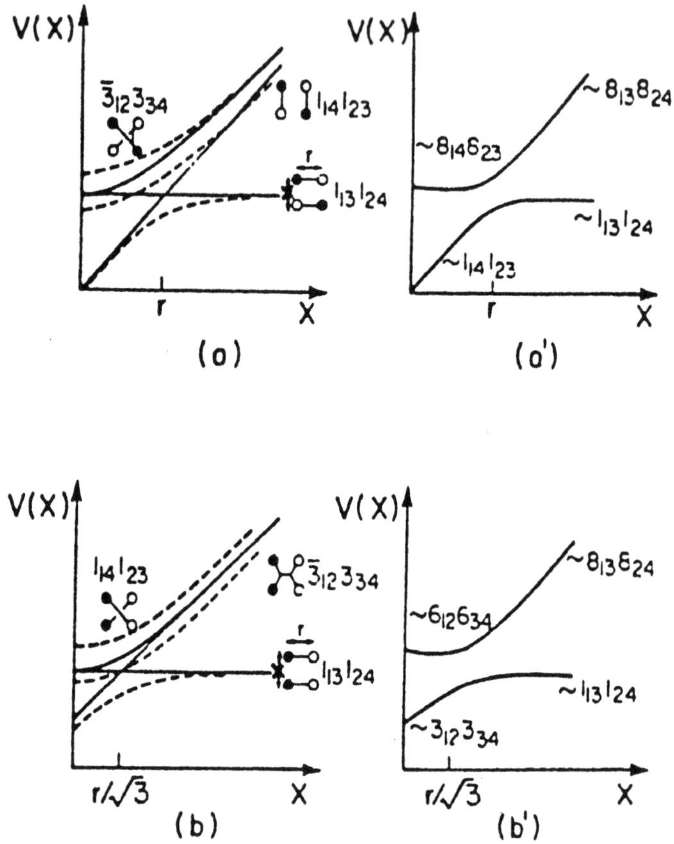

Fig. 19.: The adiabatic potentials of the flux tube model and of the $\vec{F}_i \cdot \vec{F}_j$ potential model for two $qq\bar{q}\bar{q}$ geometries: (a) and (b) show the flux tube potentials before (solid curves) and after (shown schematically by dashed curves) topological mixing, while (a') and (b') show the related $\vec{F}_i \cdot \vec{F}_j$ potentials.

(i.e., adiabatically evolving) glue state. Fig.18(b) shows why this is true: the two flux configurations of Fig.18(a) will be dynamically mixed and this mixing <u>will always be strong</u> in those regions of the lowest adiabatic surface of the $qq\bar{q}\bar{q}$ system where unmixed adiabatic surfaces of a given topology would have crossed. This effect is illustrated more explicitly in Figs.19(a) and 19(b), which show slices through the three lowest adiabatic potentials of the $qq\bar{q}\bar{q}$ system for two different rectangular arrangements of the quarks and antiquarks. They are shown as a function of the length x of one of the sides of the rectangle, with the other side kept at fixed length r. (We will return to Figs. 19(a') and 19(b') when we discuss possible models for this complex behaviour). We note that the lowest adiabatic surface in this prototypical multiquark system has completely different flux tube topologies dominant in different regions of $qq\bar{q}\bar{q}$ configuration space. In a dynamical situation, we cannot expect to be able to describe this system in terms of only a quark wavefunction: for a given $qq\bar{q}\bar{q}$ configuration we will in general have a quantum state that is a strong coherent superposition of the possible states of the glue.

The physics of this complex situation is, as indicated in Fig.18, controlled by flux tube topological mixing. Since this is an effect that becomes important for $g<1$ where strong coupling perturbation theory is not useful, we cannot provide a very quantitative description of this process. Nevertheless, there are some qualitative features of this mixing which we believe we can safely expect and which are sufficient to form a basis for discussing most of its effects. Consider the example of $qq\bar{q}\bar{q}$ depicted in Fig.18(a) once again. We would like to know the amplitude for conversion between these two topologies as a function of x and y. To estimate this function we

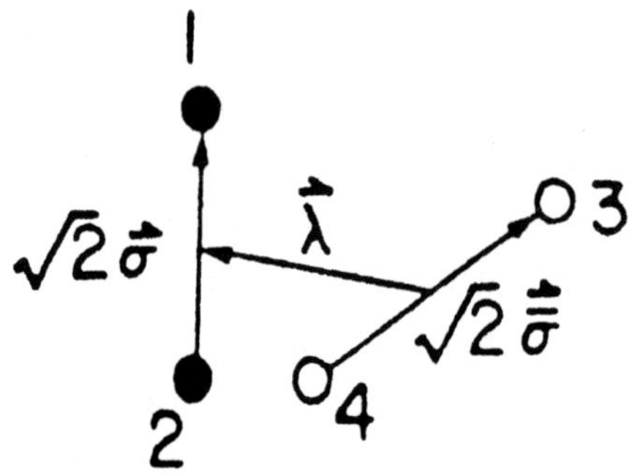

Fig. 20.: The relative coordinates of the system $q_1q_2\bar{q}_3\bar{q}_4$

first imagine that both x and y are much greater than λ_0 so that a lattice spacing $a\sim\lambda_0$ where $g\sim1$ can be used to describe the system. In this region topological mixing is not especially weak, but we can expect that a lowest order treatment will give a qualitatively sensible picture of the physics. Such lowest order mixing can occur when two (discrete) strings find themselves, in the tail of their vibrational wavefunctions, in a situation where they each have an element occupying a link on opposite sides of an elementary plaquette (see Fig.18(b)). They can then mix with an amplitude of order $1/ag^2\sim\lambda_0$. The amplitude that their wavefunctions will allow them to mix is, however, damped like a Gaussian in the required displacements of the flux tubes from equilibrium so that, even after summation over all plaquettes, the mixing will be very small if x and y are both much larger than the transverse size $b^{-1/2}$ of the string wavefunctions. Conversely, if there is good overlap between all four relevant string wavefunctions, then this mixing amplitude will be of order $1/\lambda_0$ and will be strong.

There is an immediate application of these ideas

to the question of residual forces between hadrons. If y corresponds to typical hadronic dimensions but $x \gg y$, then flux tube topological mixing between the ground states of the two short (vertical) flux tubes to any states of the two long (horizontal) flux tubes will be suppressed. We conclude that the residual forces between colour singlet hadrons will have no long range (power law) tail, in accord with experiment! Conversely, for distances $x \sim y \sim 1$ fermi we should expect strong topological mixing between colour singlets leading to medium range van der Waal's forces which could play a role in nuclear binding. Indeed, we will see below that such effects are important in explicit calculations, and they may indeed be responsible for the bulk of nuclear binding energies!

A full-fledged treatment of the gluon dynamics of multiquark systems along the lines imagined above would obviously be a formidable task. In the meantime much recent work on multiquark dynamics has been based on simpler models. I would say that the minimum requirements for such a model would be:

1) If it is to simplify calculations significantly, it should specify adiabatic potentials that depend only on the quark coordinates (including space, spin, and colour),

2) It should confine colour,

3) It should reduce to the known $q\bar{q}$ and qqq potentials in the appropriate limits.

4) It should produce flux-tube-like potentials between well separated quarks or clusters of quarks with, for example, the correct ratio of the string tension between $3 - \bar{3}$ clusters and $6 - \bar{6}$ clusters, and

5) It should incorporate the known short distance interquark forces from one gluon exchange.

The only model potential I know which meets these minimum requirements is the one which assumes that confinement develops smoothly from Coulomb's law so that

$$V = \sum_{i<j} V_{ij} \tag{148}$$

where

$$V_{ij}(r_{ij}) = -[-\frac{\alpha_s(r_{ij})}{r_{ij}} + \frac{3}{4}c + \frac{3}{4}br_{ij}]\vec{F}_i \cdot \vec{F}_j \tag{149}$$

where $\vec{F}_i = \vec{\lambda}_i/2$ for a quark and $\vec{\lambda}_i^c/2 = -\vec{\lambda}_i^*/2$ for an antiquark. Aside from meeting the above "minimum requirements", this potential seems to do a remarkable job of imitating the low-lying adiabatic surfaces of the full flux tube picture. This is illustrated in Figs. 19(a') and 19(b') which show the two adiabatic surfaces (since there are two distinct colour couplings in this system, it has exactly two adiabatic surfaces instead of the infinite number of QCD) of (148) corresponding to the same $qq\bar{q}\bar{q}$ configurations as Figs. 19(a) and 19(b). Unfortunately, the resulting imitation is imperfect. The $\vec{F}_i \cdot \vec{F}_j$ potential does not know about the limited transverse extension of the flux lines which form the potential and so does not cut off the analog of topological mixing fast enough to avoid a long range power law tail to hadron-hadron potentials. This flaw, however, by no means makes the $\vec{F}_i \cdot \vec{F}_j$ potential useless: it just means that one must proceed with due caution for possible spurious effects that it might introduce. In many cases in practice this is easy to do.

B. The multiquark fiasco

Multiquark physics has a somewhat unfortunate history. A confluence of dubious experimental results and dubious theoretical models in the late 1970's and early

1980's created, indeed, a multiquark fiasco. I am not competent to discuss what went wrong experimentally, but let me review the theoretical side of this fiasco in order to place it in perspective and thereby, I hope, point the way toward a better understanding of multiquark systems.

The story is basically one of throwing caution of the winds. Modelers from at least four different camps were, it seems to me, guilty:

1) bag models: The bag models posits the existence of a confining volume. Thus, by assumption, any number of quarks and antiquarks (so long as they are in an overall colour singlet), when placed in the bag, will produce a multiquark state that appears to be as legitimately confined as the $q\bar{q}$ and qqq states. On the other hand, it is obvious that the bag model should not be used in this naive way: the complicated (and nontrivial) dynamics of confinement may be approximated by a bag for states in which this property is guaranteed (like $q\bar{q}$ and qqq), but it was not intended to be a viable approximation to this dynamics in all situations. Nevertheless, some practitioners published detailed spectra of multiquarks confined to a single bag which they interpreted as having essentially the same legitimacy as bag model predictions for, say, meson spectroscopy.

2) naive potential models: The non-relativistic analog of the error made by some bag modelers was to assume that, in a multiquark system, the interquark forces would be identical to those confining forces operative in mesons and baryons. This totally unjustifiable assumption will of course also automatically produce a rich spectrum of multiquark states.

3) colour chemistry: A more sophisticated mistake was made by the adherents of the "colour chemistry" school. They recognized that a given multiquark configuration could

be connected by a variety of flux tube, or string, topologies. They went on the note that in some of the these topologies (see, for an example, Fig.5(c) all the quarks and antiquarks would be connected together in a single unit. Based on a false analogy with mesons and baryons, a rich spectrum of such states was then predicted. While closer to the truth, the fatal flaw of this school was, of course, its neglect of topological mixing.

4) $\underline{\vec{F}_i \cdot \vec{F}_j}$ potential models. The most peculiar story of this set belongs to those modelers who tried to study multiquark systems in the context of the $\vec{F}_i \cdot \vec{F}_j$ potential model. Unlike all of the other modelers, these were actually using a picture for multiquark confinement which, as explained above, should be approximately valid. Yet they too predicted rich spectra of multiquark bound states! In this case the problem was not with the model, but rather that approximations used to solve the $\vec{F}_i \cdot \vec{F}_j$ model were illegitimate. The difficulty and its partial resolution are the subject of the next Section.

To summarize: the multiquark fiasco was created by a series of unfortunate circumstances. Theorists, making a variety of errors from a number of different points of view, supported each other (often with very similar spectra) in predicted multiquark states analogous in their character to ordinary hadrons. This bad psychology/sociology was made even worse by the (temporary) experimental sighting of states. I suspect that the mirror image of this bad psychology/sociology operated on experimenters working in this field.

C. The $qq\bar{q}\bar{q}$ system: a multiquark primer

I can think of no better introduction to multiquark systems than to discuss the simplest multiquark system of two quarks and two antiquarks in the $\vec{F}_i . \vec{F}_j$ model [2]. I have already mentioned that, in early applications of the model, illegitimate approximations were made that led to erroneous predictions of a rich discrete spectrum of multiquark states. We can begin to educate ourselves in the complexity of multiquark systems by learning from these old mistakes. Recall as well in what follows that the $\vec{F}_i . \vec{F}_j$ model is a <u>simple</u> approximation to a much more complicated structure.

In fact, let us begin by considering an even simpler theoretical model which exposed most clearly the basic issues: imagine that the interquark potential were just

$$V_{ij}(r_{ij}) = \frac{1}{2} k r_{ij}^2 \; \vec{F}_i . \vec{F}_j \tag{150}$$

a purely "harmonic" force. (At first one might even imagine that this simple problem could be solved analytically, but it cannot be). With the labels and coordinate system shown in Fig.20, and taking as a colour basis the states $|\bar{3}_{12} 3_{34}>$ and $|6_{12} \bar{6}_{34}>$ (where C_{ij} means that particles i and j are coupled to the colour representation C and in each case the system is in a overall colour singlet), we have

$$|\psi> = \psi_{\bar{3}3}(\vec{\sigma}, \vec{\bar{\sigma}}, \vec{\lambda}) \, |\bar{3}_{12} 3_{34}> + \psi_{6\bar{6}}(\vec{\sigma}, \vec{\bar{\sigma}}, \vec{\lambda}) \, |6_{12} \bar{6}_{34}> \tag{151}$$

and (with "1" $\leftrightarrow$ $|\bar{3}_{12} 3_{34}>$ and "2" $\leftrightarrow$ $|6_{12} \bar{6}_{34}>$ in matrices)

$$H = \frac{1}{2m}(p_\sigma^2 + p_{\bar{\sigma}}^2 + p_\lambda^2) \begin{bmatrix} 1 & 0 \\ 0 & 1 \end{bmatrix} + \frac{1}{2}k \begin{bmatrix} 2\sigma^2 + 2\bar{\sigma}^2 + \frac{4}{3}\lambda^2 & -2\sqrt{2}\,\vec{\sigma}.\vec{\bar{\sigma}} \\ -2\sqrt{2}\,\vec{\sigma}.\vec{\bar{\sigma}} & \sigma^2 + \bar{\sigma}^2 + \frac{10}{3}\lambda^2 \end{bmatrix}. \tag{152}$$

If we now assume that mixing between the $|\bar{3}_{12} 3_{34}>$ and

$|6_{12}\bar{6}_{34}>$ sectors can be neglected (or at least treated perturbatively) then we will erroneously conclude that both the $\bar{3}3$ and $6\bar{6}$ sectors of the system are confining in all three relative coordinates, so these systems will be analogous to ordinary mesons and baryons. On this basis we would expect to find two separate infinite towers of discrete levels (in the absence of mixing), or, if mixing is treated by perturbation theory, a single mixed tower of discrete states. However, this conclusion must be qualitatively wrong since one can easily show that this Hamiltonian contains solutions corresponding to free mesons in situations where the system is split into two $q\bar{q}$ clusters in that mixture of $|\bar{3}_{12}3_{34}>$ and $|6_{12}\bar{6}_{34}>$ corresponding to having each of these mesonic clusters in colour singlets. We can thus prove that in at least one simple case the $-2\sqrt{2}\ \vec{\sigma}.\vec{\bar{\sigma}}$ terms cannot be treated perturbatively: at least an infinite number of the states of the (erroneously deduced) discrete tower of states must collapse into a continuum of free mesons. Indeed, matters are even worse: there are an infinite number of such continuous spectra, corresponding to the infinite number of possible internal excitations of the separated meson clusters.

From this perspective, one might expect that if non-trivial multiquark states exist in this system (i.e., states that are not essentially two mesons), then they are likely to correspond to either weakly bound states, or to meson-meson resonances. Certainly, it would seem, the picture of a tower of discrete states analogous to the spectrum of ordinary mesons is very inappropriate.

So far as I am aware, we still do not know the full spectrum of this harmonic multiquark model. However, a recent study [2] of its ground state using variational methods strongly suggests that it does not even have any weakly bound states or resonances; rather, it appears to be best viewed as a theory of two mesons with weak residual interactions.

It would take us too far afield to discuss in detail what happens in more realistic models of $qq\bar{q}\bar{q}$ models of $qq\bar{q}\bar{q}$ with an $\vec{F}_i.\vec{F}_j$ potential. Suffice it to say that there are strong suggestions that the physics of such meson-meson systems is roughly analogous to that of the nucleon-nucleon system. Indeed, it may even turn out that the $S^*(980)$ and $\delta(980)$ mesons are $K\bar{K}$ bound states [2,3] roughly analogous to the deuteron which we discuss next.

References for Chapter 4

1. The flux tube picture for multiquark states and its relation to the $\vec{F}_i.\vec{F}_j$ model are discussed in Ref.4 of Chapter 2.
2. The $qq\bar{q}\bar{q}$ system in the $\vec{F}_i.\vec{F}_j$ model is discussed in J. Weinstein and N. Isgur, Phys. Rev. Lett. 48 (1982) 659; Phys. Rev. D27 (1983) 588. References to the $qq\bar{q}\bar{q}$ literature can be found therein.
3. The suggestion that the S^* and δ are $qq\bar{q}\bar{q}$ states originates with R.L. Jaffe, Phys. Rev. D15 (1977) 267, 281; R.L. Jaffe and K. Johnson, Phys. Lett. 60B (1976) 201; R.L. Jaffe, Phys. Rev. D17 (1978) 1444.

5. "DERIVING" NUCLEAR PHYSICS: SIX QUARKS WITH CHROMODYNAMICS

Given the strong analogy of Section 4. A. and table 1 between atomic physics and hadron physics, it is natural to try to extend the analogy to make a connection between molecular physics and nuclear physics. This analogy is sketched in table 2 for the hydrogen molecule and the deuterium nucleus. Of course this basic analogy has some "technical" flaws, the most important being that in this case 6>>4. To see what I mean by this, note that in H_2 an adiabatic approximation (in which the electronic energy is calculated for fixed proton positions and then used as an interproton potential) works well since $m_p >> m_e$. Thus one really only needs to solve a Schrödinger problem for the spatial locations of two particles. In the case of H, no adiabatic approximation is viable, and moreover, the two spins and three colours of each quark play a crucial dynamical role.

There is one further analogy of another sort which I would like to stress. At this time it is safe to say that we understand the hydrogen molecule perfectly: enormous variational calculations have been able to reproduce its properties almost exactly. However, in some sense the very first calculation by Heitler and London in 1927 was the most satisfying since on the basis of that calculation (which only reproduced the dissociation energy and equilibrium radius to about 50 %) it was already possible to state the main physics conclusion (which had already been conjectured by Lewis in 1918): the chemical bond is a shared electron. The simplest calculation for the six quark system with chromodynamics is technically of comparable complexity to the latest calculations for H_2, but it should be compared in spirit to the situation in 1927. While such a calculation is, relative to the complexity of the system, very crude, if we are lucky we may still be able to answer some of the basic physics questions posed by the properties of nuclei.

The idea of such a calculation is an old one and I cannot possibly do justice to the history of the subject here. I would rather like to describe the particular approach which we took [1] and our results. Our calculation was very strongly based on the molecular analogy of table 2 and the belief that some of the fundamental questions we wished to answer could only be addressed if we could calculate the six quark ground state wavefunction. We accordingly set out to do this instead of to study, for example, scattering via resonating group techniques. As already implied, the calculation was extremely arduous; it would have

Table 1: Quark models in QCD versus atomic physics in QED

	QED	QCD
basic potential	$\frac{\alpha}{r}(\frac{e_1}{e})(\frac{e_2}{e})$ electric	$(-\frac{\alpha_s}{r}+\frac{3}{4}c+\frac{3}{4}br)\frac{\vec{\lambda}_1}{2}\cdot\frac{\vec{\lambda}_2}{2}$ chromoelectric
spin-dependence	for example, $\frac{8\pi\alpha}{3m_1m_2}\vec{S}_1\cdot\vec{S}_2\delta^3(\vec{r})(\frac{e_1}{e})(\frac{e_2}{e})$ from magnetic moments	for example, $\frac{-8\pi\alpha_s}{3m_1m_2}\vec{S}_1\cdot\vec{S}_2\delta^3(r)\frac{\vec{\lambda}_1}{2}\cdot\frac{\vec{\lambda}_2}{2}$ from chromomagnetic moments
typical "atom"	e^+e^-, positronium	$c\bar{c}$, charmoium; also non-abelian $q_Rq_Bq_Y$ baryonic atoms
orbital excitations	for example, the Lyman series	for example, B(1235)-π(140) in $q\bar{q}$ and N(1520)-N(940) in qqq
spin splittings	for example, the 21 cm line of atomic hydrogen	for example, ρ(770)-π(140) in $q\bar{q}$ and Δ(1235)-N(940) in qqq
v/c	<<1 in hydrogen; $\simeq$1 in uranium	<<1 in $b\bar{b}$, beautyonium; $\simeq$1 in the nucleon

Table 2: The hydrogen molecule in QED versus deuterium in QCD.

	QED	QCD
symbol	H_2	^{2}H
"atomic" structure	$H_2 \leftrightarrow H+H$	$^{2}H \leftrightarrow p+n$
basic structure	$H_2 \leftrightarrow 2e^- + 2p$	$^{2}H \leftrightarrow 3u+3d$
basic description	four body problem with Coulomb's law	six body problem with chromodynamics
"atomic" dissociation energy	4.48 eV	2.23 MeV
"atomic" separation	0.740 Å	3.9 fm
method of solution	variational	variational
"quasiparticle" description	H atoms in effective potential	nucleons in effective potential

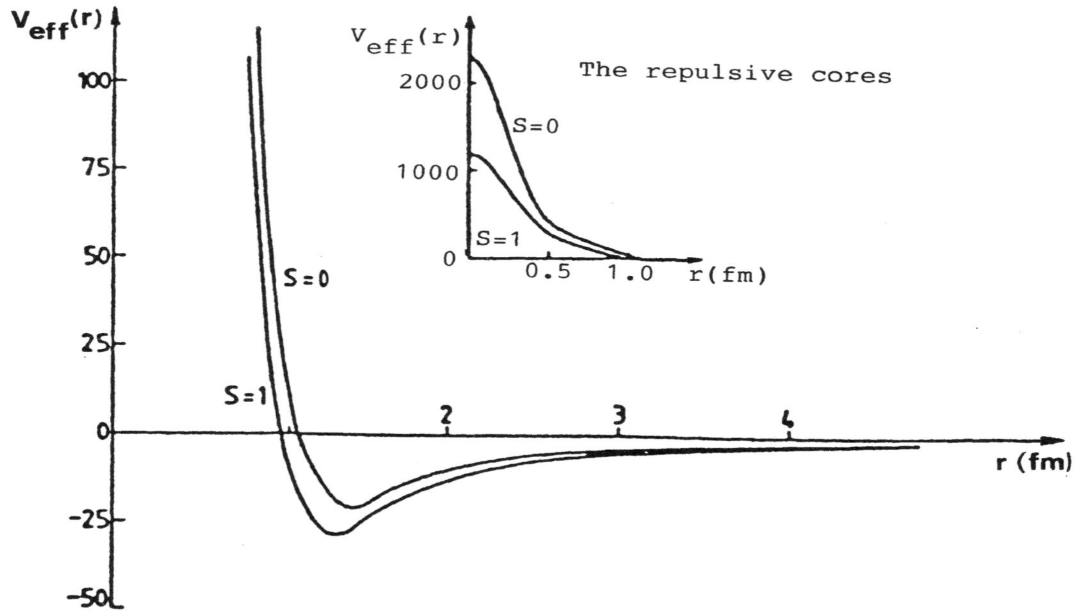

Fig. 21.: The effective nucleon-nucleon potentials in the $^3S_1$ and $^1S_0$ channels arising from residual colour forces.

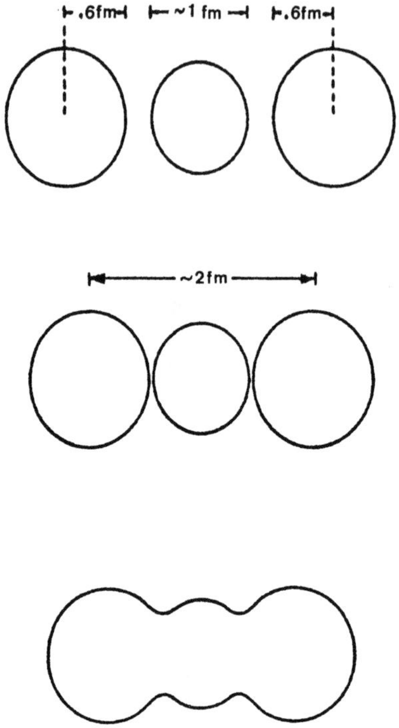

Fig. 22.: A cartoon illustrating that for $r \lesssim 2$ fm meson exchange is unlikely to be appropriate to the description of the internucleon potential.

been even more difficult if we hadn't made use of simple calculational tricks.This was to concentrate on the 3u+3d quark sector (I=0 to particle physicists, $T_3=0$ to nuclear physicists) so that we could use $S_3 \times S_3$ symmetry instead of S_6 symmetry (S_n is the permutation group on n objects). This had the disadvantage that the calculation was not automatically diagonalized in isospin. However, the penalty for this was slight: our computer had to diagonalize some larger matrices than it would have otherwise. In exchange, it was possible to do the analytic calculations required in about a year instead of at least double that.

The calculation proceeded via a cluster expansion of the five relative coordinates. Two pairs of coordinates were expanded in a three quark harmonic oscillator basis;

the fifth coordinate was described by an intercluster variational wavefunction. The cluster expansion was made in a basis space of 115 dimensions which included all possible colour, spin, flavour, and space combinations possible by allowing up to two units of intracluster spatial excitation; up to 15 parameters were used to expand the intercluster wavefunction which was allowed to be in an S, P, or D wave. Our goal was to allow the six quark wavefunction to do anything it liked. For example, our wavefunction could certainly reach any reasonable "democratic", "spherical" six quark state clustered around the origin. I emphasize that although we used a cluster expansion, we did not put clustering in by hand.

For our Hamiltonian we used the model of Section 3.C. which, as we have seen, provides a reasonably good description of baryon spectroscopy --- clearly a minimal requirement if one is to believe the results of a "two nucleon" calculation. I stress that this Hamiltonian is taken over completely from studies of baryon spectroscopy, so that our calculations for the six quark system are purely deductive with no free parameters whatsoever.

At the time we began our very lengthy calculations, this subject was in its infancy[2], with early calculations based on an adiabatic approximation seeming to show a repulsive core. In later improved calculations, however, the core disappeared. Eventually it became understood via resonating group techniques that the adiabatic approximation is invalid, and the use of such methods seemed to strongly indicate that a repulsive core emerges naturally from short distance quark dynamics. However, at least within the limited basis spaces possible for such techniques, there seemed to be no trace of nuclear binding! Our approach, while much less elegant than the resonating group methods, is considerably more powerful for studying the properties of the six quark ground states. As a measure of this, one can compare our 115 term cluster expansion with the maximum of 6 terms employed in any resonating group calculation.

Our main results can be summarized as

1) We find that the 3u+3d quark ground state wave functions with I,J = 0,1 show very strong clustering into two three quark systems with neutron and proton quantum numbers and properties. This shows that, within the context of our calculation, nuclear physics is really approximately the physics of nucleon quasiparticles. This is a nontrivial result, as evidenced by the fact that it is quantum number dependent: other six quark systems, when constrained to nuclear densities, do not exhibit such clustering.

2) Given the strong clustering it is natural to project out of the full six quark wavefunction an n-p component. This component, which nearly saturates the full wavefunction, has an amplitude as a function of the n-p separation which we can then interpret as an effective internucleon wavefunction $\psi_{eff}(r)$. This $\psi_{eff}(r)$ can in turn be used to define a bound-state effective potential $V_{eff}(r)$, via the NN Schrödinger equation, which has the following very satisfactory properties:

a) it gives a strong repulsive core for r<1 fm,
b) it provides an intermediate range binding attractive region between 1 and 1 fm, and
c) the $^3S_1$ potential is less repulsive and more binding than the $^1S_0$ potential.

These features are illustrated in Fig.21. Let me reemphasize that this potential is purely deductive: we have simply solved the six quark problem for a known Hamiltonian. Let me also immediately stress that I do not advocate this potential as a replacement for phenomenological nucleon-nucleon potentials. It is based on an approximate solution to a very rough model for the quark dynamics and so cannot be expected to be accurate. What is relevant is that this calculation may point us in the direction of eventually replacing phenomenological potentials by derived ones, much as the original H_2 calculation of 1927, while quantitatively lacking, captured the basic physics of the situation and eventually led to today's more accurate understanding of the hydrogen molecule. In this spirit, let me attempt, on the basis of our calculation, to provide a picture of the nuclear force analogous to the molecular picture that "the chemical bond is a shared electron".

The attractive region of V_{eff} is the easiest to interpret: it is actually physically analogous in origin to the molecular case. Recall that, at least for separations larger than twice the Bohr radius, there is a van der Waals attraction between two hydrogen atoms which is a weak residue of the electrical forces producing the electrically neutral ep clusters. This weak attractive potential arises because the neutral atoms have a second order perturbation in their energy $-a^2/\Delta E$ from their amplitude a to temporarily excite each other via Coulomb's law into a virtual state with energy ΔE above their ground states (in which the atoms are in P-waves coupled to L=0). Semiclassically we would say that while the atoms are neutral, it is energetically favourable for them to develop small oppositely directed electric dipole moments. Two colour neutral quark clusters can develop a weak (i.e., nuclear strength!) residual attraction by a closely related mechanism: they can temporarily excite each other (via the QCD version of Coulomb's law of table 1) into a virtual state in which the clusters are in coloured P waves coupled to L=0 and to zero net colour. Semiclassically we would say that while the atoms are colour neutral, it is energetically favourable for them to develop small chromo-electric dipole moments correlated in their orientation in both space and in colour-space.

It is much more difficult to describe the physical origin of the repulsive core. Our calculations, first of all, confirmed the conclusion which had already been reached via resonating group techniques that the repulsive core is a non-adiabatic effect. This is not surprising: unlike the molecular case where the electron motion is fast relative to the motion of the protons, in the six quark problem there are no adiabatic coordinates once the two three quark clusters begin to overlap. As shown in the earliest calculations, a repulsive core does develop if the six quark system is arbitrarily constrained to remain in two purely nucleonic clusters (a consequence of the Pauli principle and the colour hyperfine interactions). The repulsive core apparently emerges from a complete calculation because the NN configuration does not have time to evolve into more energetically favourable configurations available to it: i.e., there is a partial dynamical constraint to the two nucleonic cluster sector. One may describe this constraint in other language by noting that there are kinetic energy terms associated with the mixing angles of non-NN configurations, and these kinetic energy terms create pseudoforces (analogous to centrifugal barrier effects) which produce the NN "constraint". This very dynamical origin for the repulsive core is consistent with its known complex dependence on the spin, orbital, and total angular momenta of the six quark system; note, for example, that the $\vec{S}_i \cdot \vec{S}_j$ dependence of the core has already emerged from our calculation.

There are many weaknesses of the calculation I have just described, and before going on it seems appropriate to mention some of them. This is not the place to discuss these issues fully, however, so I will just make an annotated list of some possible criticisms:

1) non-relativistic and other approximations: Since we were mainly concerned with the properties of the six quark system relative to two separated three quark clusters, many sources of error --- including relativistic effects --- tend to cancel.

2) the spin-orbit mystery: The baryon data enforce the near absence of spin-orbit effects, and spin-orbit effects of the strength allowed by the data would not significantly effect our results.

3) the $\vec{F}_i.\vec{F}_j$ confinement potential: We have already extensively discussed the fact that such a model for confinement can only approximate the structure expected from QCD. Probably the most serious flaw of this approximation is that it leads to a spurious long-range power-law to the effective nucleon-nucleon potential. I will describe its effects on our calculations below.

4) the radial dependence of H_{sj}: In the actual calculations the anharmonic potential U was replaced by an attractive δ-function interaction. In baryon spectroscopy this replacement is known to be adequate for low-lying excitations. We have checked that, since it is averaged over clusters, it should also suffice for this problem, but this point deserves further study. We have checked other possible weaknesses of our calculation and believe that it has no fundamental problems, even though our model for the Hamiltonian _is_ very crude. The reason for this is that, if we are right, the basic features of nuclear physics are not very subtle consequences of QCD: they will emerge from any model which confines colour in a sensible way and which recognizes some of the basic phenomenological constraints from hadron spectroscopy (like the existence of important interquark spin-spin interactions).

Before finally applying our results to an actual nuclear system, we paused to ask ourselves if there were any regions in which our calculation would fail to take into account the dominant physics of the six quark system. We decided that there was one: when the system has segregated into two three quark clusters which are far apart, our infinitely stretchable flux tubes should be breaking to produce

Table 3: Some properties of the six quark ground states

property	theoretical value	experimental value
E_d (MeV)	$-2.9^{+0.8}_{-0.3}$	-2.23
$(r_d^2)^{1/2}$ (fm)	2.2±0.5	1.95
Q_d (mb)	+2.1±0.5	+2.86
μ_d *(nuclear μ_N)	+0.859±0.003	+0.857
$E(^1S_0)$ (MeV)	$-0.4^{+0.4}_{-0.1}$	unbound

*μ_d is calculated assuming that the departure from $\mu_p+\mu_n$ is due only to the calculated 3.6 % D-wave mixing.

meson exchange. We further argued --- on both the naive grounds represented by the cartoon of Fig.22 and on more rigorous grounds --- that only the region beyond about 2 fm should be affected. In this region only pion exchange is significant, so to proceed from the results of our calculation to a real nucleus we considered a "hybrid" model in which $V_{NN} = \tilde{V}^{NN}_{eff}+\tilde{V}^{ope}$. Here $\tilde{V}^{NN}_{eff}$ is our derived NN potential cut off at some internuclear separation $r_c \sim O(2fm)$ and $\tilde{V}^{ope}$

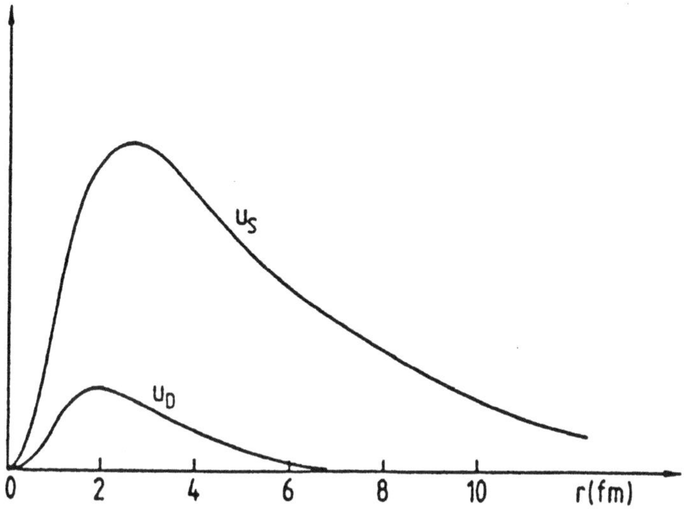

Fig. 23.: The deuteron wavefunctions u_S and u_D shown with their correct relative normalization.

is one pion exchange "cut in" at r_c. Our results for the properties of the simplest nucleus, deuterium, were not very sensitive to r_c in the range 2 to 3 fermis: table 3 compares our "hybrid" calculation with the data. The theory errors shown in the table reflect both sensitivity to r_c and also estimates of the errors arising in the various quantities shown from our basis truncation and our limited variational wavefunctions. The calculated deuterium wavefunctions are shown in Fig.23. The results are better than they deserve to be, given the quality of the model.

References for Chapter 5

1. K. Maltman and N. Isgur, Phys. Rev. Lett. 50 (1983) 1827; Phys. Rev. D29 (1984) 952.
2. A partial history and a rather complete set of references to related work may be found in the latter of Refs.1.

6. CONCLUSIONS

It will be clear from these lectures that the derivation of nuclear physics from the quark model, and ultimately from QCD, is a subject that is still in its infancy. At this time it is even unclear whether the route I have suggested in these lectures will prove to be the best one. Nevertheless, I believe that the picture we have been able to sketch based on this approach is attractive and suggestive of the possibility of further progress.

Among the most important conclusions we have drawn for nuclear physics are these:

1) Nuclear physics is multiquark physics and, as such, involves gluon dynamics in a much less trivial way than mesons and baryons.

2) Nuclear physics is approximately, but far from exactly, multi-nucleon physics. Explicit quark and colour effects exist, but they are subtle effects which will arise from configurations in which the nucleonic quark clusters overlap.

3) It is quite likely that the importance of meson exchange in nuclear physics has been overestimated. For example, it now seems very unlikely that ω exchange is responsible for the repulsive core of the NN interaction.

However, perhaps the most important conclusion that one can draw from this approach would be that it is now appropriate to view the relationship of nuclear physics to hadron physics and QCD as being roughly analogous to that of condensed matter or molecular physics to atomic physics and QED.

Acta Physica Austriaca, Suppl. XXVII, 267–448 (1985)

RESULTS FROM EXPERIMENTS AT THE C.E.R.N. ANTIPROTON-PROTON COLLIDER[+]

by

J. STRAUSS
Institut für Hochenergiephysik, Wien
and
CERN

ABSTRACT

After a presentation of the "expected" results from experiments at the CERN $\bar{p}p$ collider (soft hadrons, large-p_T jets, intermediate vector bosons and heavy quarks) the various exotic, unexpected results are discussed. In particular, the analysis procedures are presented in some detail in order to enable the reader to appreciate the accomplishments and to get some understanding of the problems. This leads to the discussion of the motivation for the present upgrading program and the presentation of a possible future large hadron collider in the LEP tunnel.

[+]Lectures given at the XXIV. Internationale Universitätswochen für Kernphysik, Schladming, Austria, February 20-March 1, 1985.

TABLE OF CONTENTS

1. INTRODUCTION

The physics of elementary particles has undergone a remarkable developement during the past decade. A host of experimental results made accessible by a new generation of particle accelerators, one of them being the CERN proton-antiproton collider, and the accompanying convergence of theoretical ideas have brought to the subject a new coherence. Our current view has been shaped by the identification of quarks and leptons as the fundamental constituents of matter and by the gauge theory synthesis of the fundamental interactions. Expositions of the current paradigm can be found in recent text-books [1].

The 100 GeV energy scale at the CERN collider

Analyses of experiments probing the energy scale of $\lesssim$ 10 GeV have yielded results which are consistent with the celebrated standard model of QCD plus $SU(2)_L \times U(1)$ electroweak theory; comprehensive reviews can be found e.g. in ref.2. The importance of the energy scale of 100 GeV comes from the mass of the bosons which mediate electroweak processes. The discovery of the intermediate vector bosons was the premium for first arriving at a centre-of-mass energy of 100 GeV for pointlike scattering with sub-nanobarn sensitivity.

Figs. 1 and 2 illustrate how pointlike processes between partons at $\sqrt{s} \simeq 100$ GeV are induced by collisions between protons and antiprotons at the CERN $\bar{p}p$ collider running at $\sqrt{s} = 540$ GeV. A hard collision initiated by the interaction of partons of type i and j is described by the diagram shown

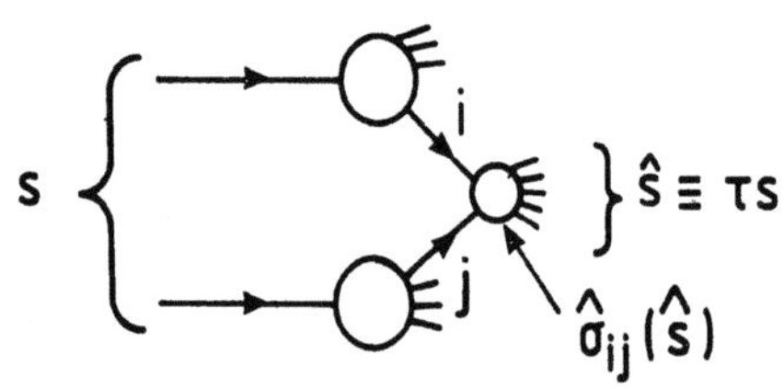

Fig. 1

in fig.1. The cross section may be written as

$$\frac{d\sigma_{ij}}{d\tau} = \int\int f_i(x_i,\hat{s}) f_j(x_j,\hat{s}) \frac{\hat{s}}{s} \tilde{\sigma}_{ij}(\hat{s}) \delta(\tau - x_i x_j) dx_i dx_j \equiv$$

$$\equiv \tau \frac{dL_{ij}}{d\tau} (s,\tau) \hat{\sigma}_{ij}(\hat{s}) \quad ,$$

where f_i are parton distributions. The differential luminosity $\tau dL_{ij}/d\tau$ represents the number of parton-parton collisions with scaled c.m. energies in the interval $(\tau,\tau+d\tau)$ per hadron-hadron collision. The cross sections for hard sub-processes behave as

$$\hat{\sigma}(\hat{s}) = \frac{c}{\hat{s}} \quad ,$$

with $c \sim 10^{-2}$ for a strong process and $\sim 10^{-4}$ for an electroweak process. A measure of event rates is therefore provided by the pseudo-cross section

$$\frac{d"\sigma"_{ij}}{d\tau} = \frac{\tau}{\hat{s}} \frac{dL_{ij}}{d\tau} \quad .$$

As an example, pseudo-cross-sections for glue-glue and $\bar{u}$-u scattering using $\bar{p}p$ collisions at $\sqrt{s}$ = 540 GeV are shown in fig.2 [3]. With standard couplings for two generations of

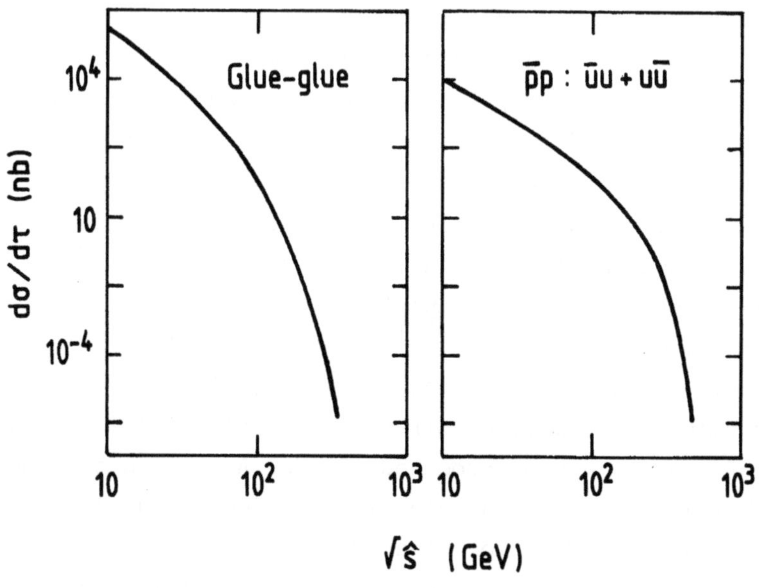

Fig. 2.: Pseudo-cross-section (see text) for Glue-Glue and $\bar{u}$-u initiated processes in $\bar{p}p$ collisions at $\sqrt{s}$ = 540 GeV.

quarks and leptons the integrated cross section for production of charged intermediate vector bosons becomes

$$\sigma_{W^+} = \frac{G_F\pi\sqrt{2}}{3} [\tau\frac{dL_{u\bar{d}}}{d\tau} \cos^2\theta_c + \tau\frac{dL_{u\bar{s}}}{d\tau} \sin^2\theta_c]$$

$$\approx 3.2 \text{ nb } \tau \frac{dL_{u\bar{d}}}{d\tau} \quad ,$$

and likewise for the neutral intermediate vector boson

$$\sigma_{Z^o} \approx 2.5 \text{ nb } \tau\frac{dL_{u\bar{u}}}{d\tau} \quad ,$$

i.e. several nanobarn at the CERN collider. A simple order-of-magnitude estimate for σ_{W^+} and σ_{Z^o} can be obtained from fig.2 by using $c = (\alpha/\pi)^2 = 10^{-5}$ and the fact that the photon propagator $1/\hat{s}$ is cancelled at the position of the resonance $\hat{s} \approx m^2_{W,Z}$. Fig.2 also implies that two-jet production at the two-jet invariant mass $m_{ij} \approx m_{W,Z}$, in a mass bin $\Delta m_{jj} \approx 20$ GeV, is a factor of $\sim$100 more abundant than production of intermediate vector bosons. Hence the identification of intermediate vector bosons from their hadronic decays is difficult and the experiments have chosen to use the leptonic decay modes $W^{\pm}\to\ell^{\pm}\nu$ and $Z^o\to\ell^+\ell^-$ in the first place. For comparison, at LEP with $\tau dL/d\tau \equiv 1$ and $L = 10^{32}$ $cm^{-2}s^{-1}$ about 10^5 Z^o's will be produced per day when running at $\sqrt{s} \approx m_Z$. All decay modes other than $Z^o \to\nu\bar{\nu}$ will be directly detectable. A single $W^{\pm}$ can, of course, not be produced at LEP. At the $\bar{p}p$ collider with $L = 10^{29} cm^{-2}s^{-1}$ and selecting the decay mode $Z^o\to\ell^+\ell^-$ with a branching fraction of about 3 %, about one Z^o per day is detected. This scenario has been well anticipated long before the collider came into operation [4,5]; it was, in fact, the driving force behind the construction of the collider.

The aim of these lectures

After the pioneering work on electron cooling at Novosibirsk [6] and on stochastic cooling at the CERN ISR [7] both cooling techniques have been tested in the CERN ICE ring in view of application in a scheme for collecting antiprotons. The results [8] confirmed the feasibility of fast precooling as needed in the proposed antiproton accumulator [9]. The CERN SPS was then converted to a $\bar{p}p$ collider as suggested in ref.4. The first collisions between protons and antiprotons at $\sqrt{s}$ = 540 GeV in the SPS accelerator operating as a storage ring were observed in the early summer of 1981 [10] not even three years after approval of the project. In spring 1983 the design performance of the collider (at the smaller than anticipated antiproton accumulation rate of $5 \cdot 10^9$ $\bar{p}$'s/hour) was surpassed. An average luminosity of $L_{av} = 10^{29} cm^{-2} s^{-1}$ was achieved in runs which were typically 15 hours long. The average performance over 6 weeks was about 3 nb^{-1} per day, and was limited mainly by the antiproton accumulation rate. Peak performances of about 20 nb^{-1} per day were reached in 1984. The total luminosity accumulated so far is about 600 nb^{-1}.

More details concerning the accelerators and the two experimental areas with the six detectors [11]have already been presented at various occasions, and have been described in previous Schladming School proceedings [12]. There is no need to repeat them here. The main purpose of these lectures is to describe in some detail the experimental procedures and to discuss and understand the results.

Most of the results presented in chapters 2 to 6 were quite well anticipated already at the time when the collider enterprise was started. This remark is valid in particular for production of (i) soft hadrons, (ii) large-p_T jets and (iii) intermediate vector bosons. As a consequence, the detectors were properly designed and the analysis procedures followed by the experimental groups were adequate and satisfactory. The detection of heavy flavours, especially of the as yet undiscovered Top, was always considered - and indeed turns out to be - a more difficult task; the reasons are (i) the absence of a simple and unambiguous signal and (ii) various background processes which may simulate Top.

In the systematic search for the intermediate vector bosons various new, unexpected phenomena were found. Events with mono-jets associated with large missing energy were found by the UA1 group (section 5.2). If confirmed by further investigations, these events may constitute the first experimental signature for supersymmetry. The UA2 group found in their 1983 data a couple of charged IVBs produced in association with very hard jets (section 4.4); no additional events of this type were, however, reported from the 1984 data. A discussion of the radiative decay modes of the Z (UA1 and UA2) is included in section 4.2. A few peculiar events which contain equal sign dimuons but no jet activity are presented at the end of chapter 6.

The possibly new phenomena constitute part of the motivation for the recent improvement program, since both the accelerators and the detectors have already been stretched to the limits of their present performances. A future outlook is given in chapter 7. The accelerators are hoped to be able to deliver an average luminosity of at least $L_{av} = 10^{30}cm^{-2}s^{-1}$ in 1986/1987. The detectors are being adapted for a better measurement of the various exotica and for a precision measurement of electroweak parameters to test radiative corrections in electroweak theory. The latter measurement can be viewed as the electroweak analog of the determination of (g-2) in QED. Finally, the Large Hadron Collider (LHC) option is presented. The LHC is motivated by the opening-up of the

1 TeV energy scale, a tool needed to understand the electroweak symmetry-breaking mechanism. The LHC represents a major new device and not only a modification of an existing machine.

A summary is presented in chapter 8.

2. SOFT HADRON PRODUCTION

Almost the entire cross section for the reaction $\bar{p}p \rightarrow$ anything is characterized by the production of soft hadrons. The two initial state particles collide only gently, because of the large volume of space they occupy. The result of the collision is emission of many hadrons of small mass and small transverse momentum with respect to the collision axis. In this chapter we first discuss the measurement of the total and elastic $\bar{p}p$ cross section at the collider. We then discuss the particle density in the central region, the total particle multiplicity, correlations between particles, particle composition and transverse momentum. For completeness we also present results from early and hitherto unsuccessful searches for Centauro events, free quarks and magnetic monopoles. For recent reviews concerned with soft hadron physics see ref. 13.

2.1 Elastic and total cross section

The elastic and the total cross sections at $\sqrt{s}$=540 GeV were measured by the experiments UA4 and UA1 [14]. A schematic layout for an experiment to measure the elastic cross section is shown in fig. 3. The scattered proton and antiproton are detected in telescopes which are located about 40 meters up- and downstream of the beam intersection point. Each telescope consists of two modules above and two mo-

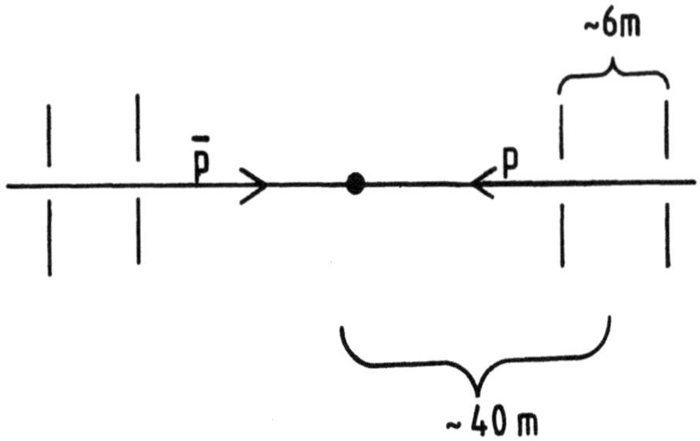

Fig. 3.: Lay-out for an elastic scattering experiment.

dules below the SPS beam tube. Since the elastic scattering is observed in the centre-of-mass, both the incoming and the outgoing $\bar{p}$ and p are collinear. The small deviation from collinearity of the incoming $\bar{p}$ and p is minimized by operating the SPS in the high-β mode which means large transverse beam size ($\sigma \approx 1.4$ mm), but small beam divergence ($\sigma \approx 0.014$ mrad) in the intersection region of interest. The measured scattering angle and the known centre-of-mass energy of the collision fully describe the elastic process. The background is kept at a small level by requiring that no other track is seen in any of the modules. The collinearity in the vertical and horizontal planes as measured by the experiment UA4 is shown in fig.4. The widths of the distributions are understood in terms of detector resolution and beam divergence. The elastic differential cross section $d\sigma_{el}/dt$ is shown in fig.5. It has been fitted, in various intervals of t, to exponential forms $d\sigma_{el}/dt \sim e^{-b|t|}$. The results of the

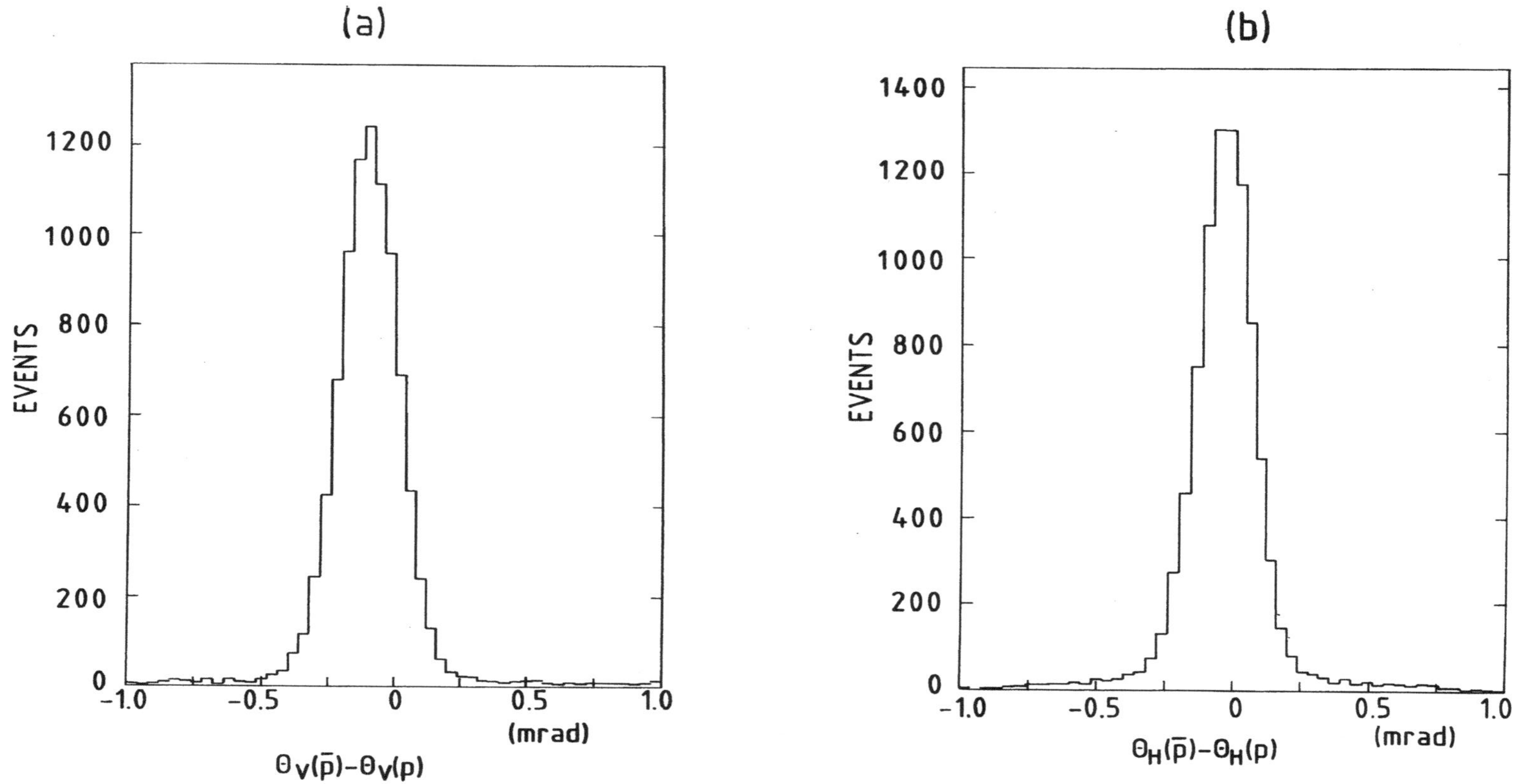

Fig. 4.: Collinearity in (a) vertical and (b) horizontal plane (UA4).

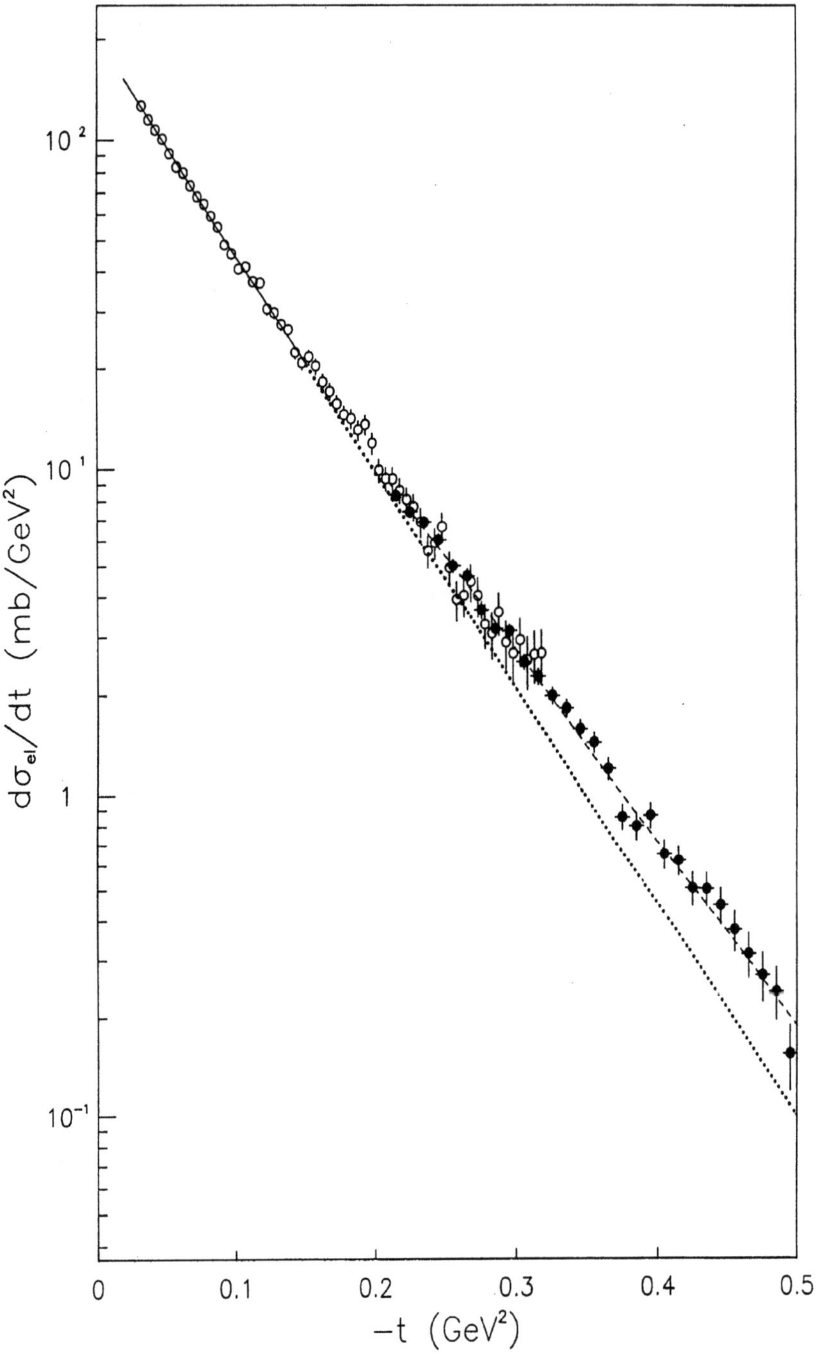

Fig. 5.: Elastic differential cross section (UA4).

fits are presented in table 1. Note the change of slope by about 2 GeV^{-2} at $-t \approx 0.15$. When comparing to ISR data, $\sqrt{s}$ = 56 GeV, the forward slope ($-t < 0.1$ GeV^2) has increased by about 2 GeV^{-2} [15].

Table 1: Elastic slopes

Collaboration	t interval [GeV^2]	b [GeV^{-2}]	Comments
UA4	$0.05<-t<0.19$	17.2 ± 1	early result
	$0.03<-t<0.15$	15.2 ± 0.2	final result, corrected for Coulomb interference, using the true $\sqrt{s}$ = 546 GeV
	$0.21<-t<0.50$	13.7 ± 0.3	early result
		13.4 ± 0.3	final result
UA1	$0.04<-t<0.18$	17.1 ± 1	
	$0.21<-t<0.45$	$13.7\pm0.2\pm0.2$	

The total cross section σ_{tot} is deduced from $d\sigma_{el}/dt$ by using the optical theorem

$$\sigma_{tot}^2 = \frac{1}{1+\rho^2} \, d\sigma_{el}/dt \Big|_{t=0}$$

where ρ denotes the ratio of the real to the imaginary part of the forward elastic scattering amplitude. From dispersion relation fits [16] this ratio is known to be small: $\rho = 0.15$. Since the directly measured quantity in the experiment is the rate and not the cross section, the UA4 group has rewritten the optical theorem to express σ_{tot} as a ratio of elastic rate over total rate:

$$\sigma_{tot} = \frac{1}{1+\rho^2} \cdot \frac{1}{\text{total rate}} \cdot \frac{d(\text{elastic rate})}{dt}\bigg|_{t=0} \cdot$$

The total rate was measured with three central telescopes up- and downstream of the beam-crossing point which were sensitive to particles in the pseudo-rapidity regions $2.5 < |\eta| < 5.6$. The final UA4 results are:

$$\sigma_{tot} = 61.9 \pm 1.5 \text{ mb}$$

$$\sigma_{el}/_{tot} = 0.215 \pm 0.005 \text{ .}$$

At the ISR, $\sqrt{s} = 52.8$ GeV, the total cross section has been measured to be $\sigma_{tot} = 42.7 \pm 0.2$ mb [17].
The ratio σ_{el}/σ_{tot} is thought to converge to a constant value if the energy approaches the asymptotic region. At the ISR it has been measured to be $\sigma_{el}/\sigma_{tot} = 0.175 \pm 0.005$ [17].

A different technique to measure σ_{tot} was used by the UA1 group. The cross section σ is related to the counting rate N by the luminosity L : $N = \sigma \cdot L$. With the SPS running in a medium-β configuration the beam profiles were measured using wire scanners situated in the intersection point diagonally across from UA1 in the SPS and then extrapolated to the UA1 beam-crossing point. The intensities of the antiproton and proton bunches were measured with directional couplers [18]. The UA1 results are:

$$\sigma_{tot} = 67.6 \pm 5.9 \pm 2.7 \text{ mb}$$

$$\sigma_{el}/\sigma_{tot} = 0.209 \pm 0.018 \pm 0.008 \text{ .}$$

The quoted second error arises from a 8 % uncertainty in the integrated luminosity. The UA1 measurement is inherently less precise than the luminosity independent measurement by UA4.

Theoretical (=impractical) comments on the elastic and total cross section are presented in ref.19.

2.2 Particle density

The density of charged particles in the central region has been measured by the experiments UA5 and UA1 [20]. The characteristic feature of the UA5 streamer chamber detector is visual recording of produced charged particles over almost the complete 4π solid angle. About 10000 events were photographed with three stereoscopic cameras. An example of an event is shown in fig.6. The photographed events were carefully scanned, measured and analysed. Various corrections had to be applied:

(i) clear cases of e^+e^- pairs from photon conversions were removed;

(ii) a correction for acceptance losses (about 25 %) was applied;

(iii) effects of secondary interactions, photon conversions and strange particle decays were corrected;

(iv) scanning and measuring losses were taken into account;

(v) very low multiplicity inelastic events were systematically lost by the trigger and had to be corrected;

(vi) the background was estimated to be small; no correction was applied.

The obtained charged particle density per event $(1/\sigma)dN/d\eta$ as a function of the pseudorapidity η is shown in fig.7.

A similar measurement was performed by the UA1 group using track information from the UA1 central detector, a 6m long × 2.4m diameter drift chamber with image read-

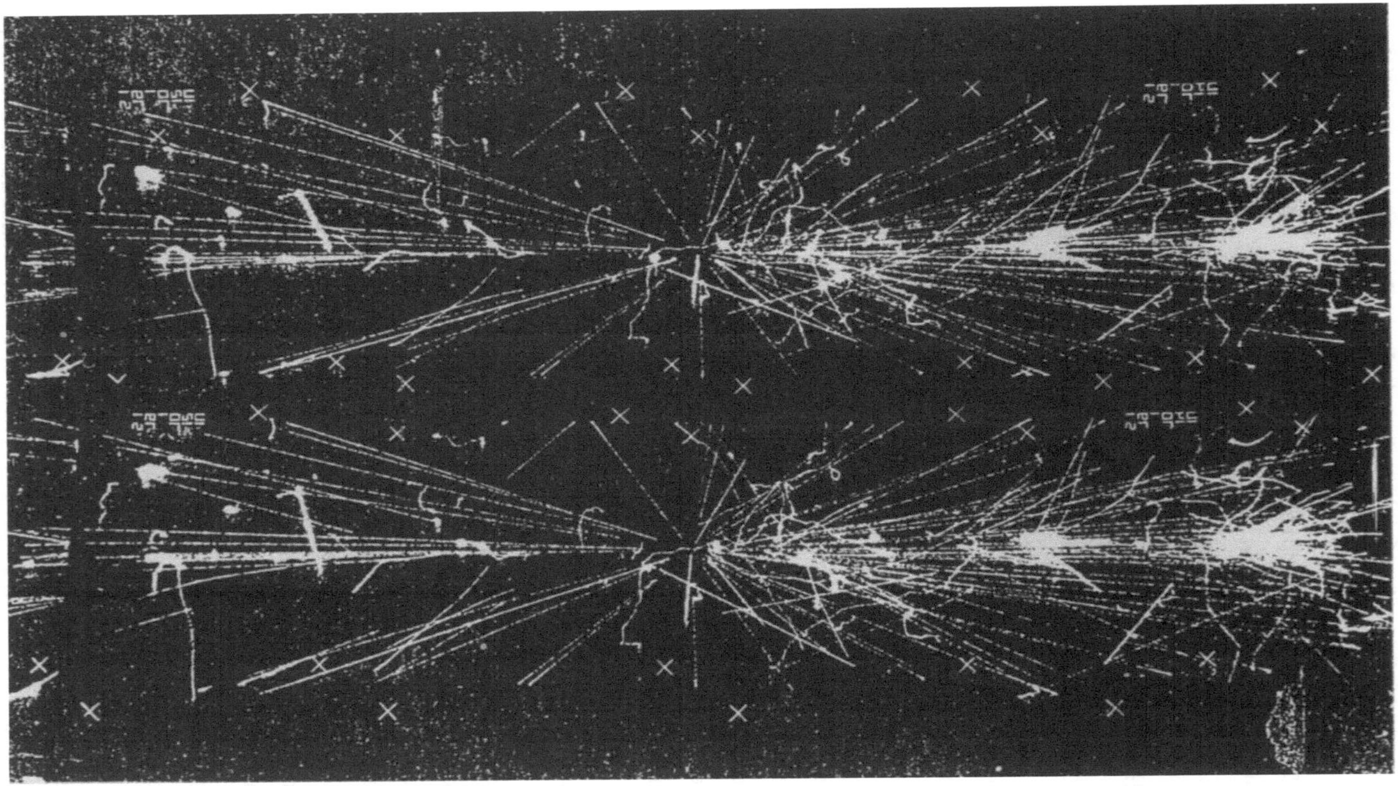

Fig. 6.: A typical UA5 event.

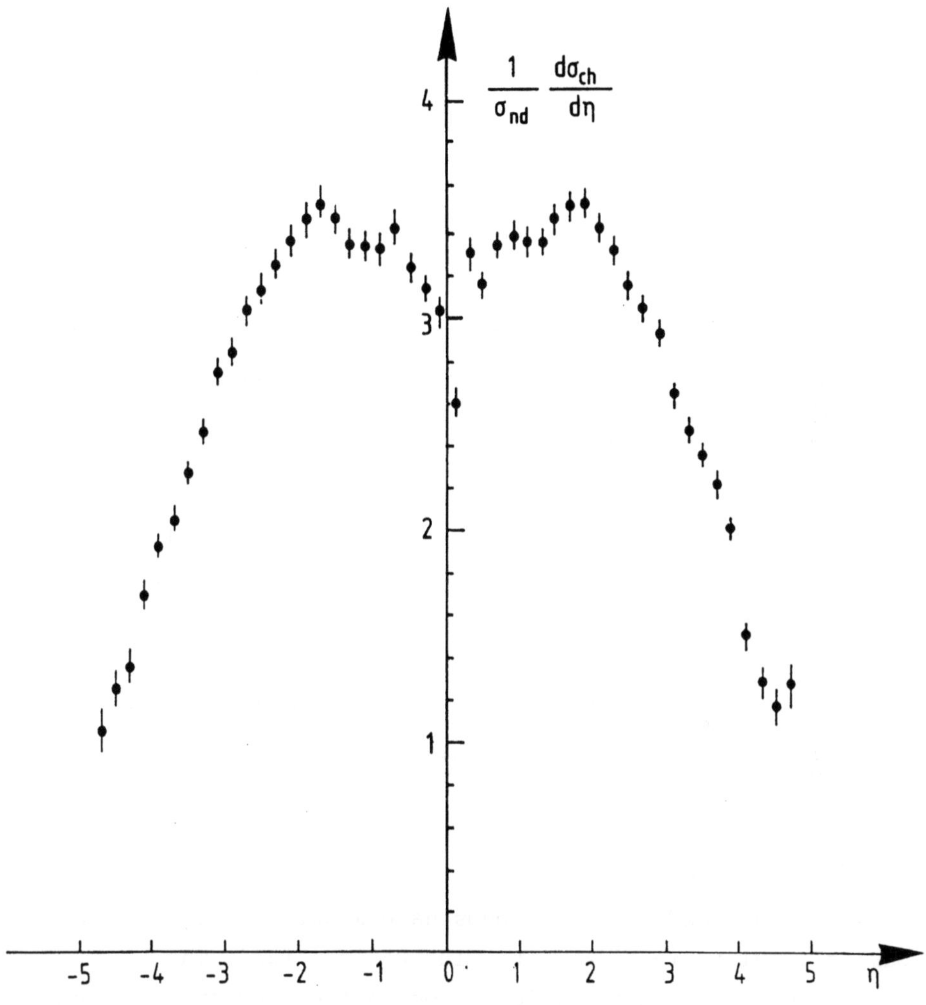

Fig. 7.: Rapidity distribution (UA5).

out on 6200 wires. The results on the charged particle density at $\eta = 0$ as given by the UA5 and UA1 groups are summarized in table 2. The observed value of the central rapidity density is larger than obtained from a log s extrapolation from lower energies [21] which yields 2.6 particles.

Table 2: Central rapidity density

Collaboration	$(1/\sigma)\ dN/d\eta\|_{\eta=0}$	Comments
UA5	3.0±0.1	non-diffractive events
UA1	(3.9±0.3)	(requiring $\geq$ 1 track within $\|\eta\|<1.3$; uncorrected for K^0 and Λ^0 decays)
	3.3±0.2	non-diffractive events

2.3 Multiplicity

The UA5 and the UA1 groups have measured the charged particle multiplicity and its fluctuations [22]. The experimental procedure is described in the previous section. The UA1 measurements of the multiplicity distributions in the central region are shown in fig.8.
The curve represents a model [23] which gives a good fit to ISR data and therefore shows how well scaling is valid in the KNO variable $n/\langle n\rangle$ if changing the c.m. energy from 56 GeV to 540 GeV. However, the UA5 group claims to have observed some deviations from KNO scaling when considering the total charged multiplicity and excluding diffractive

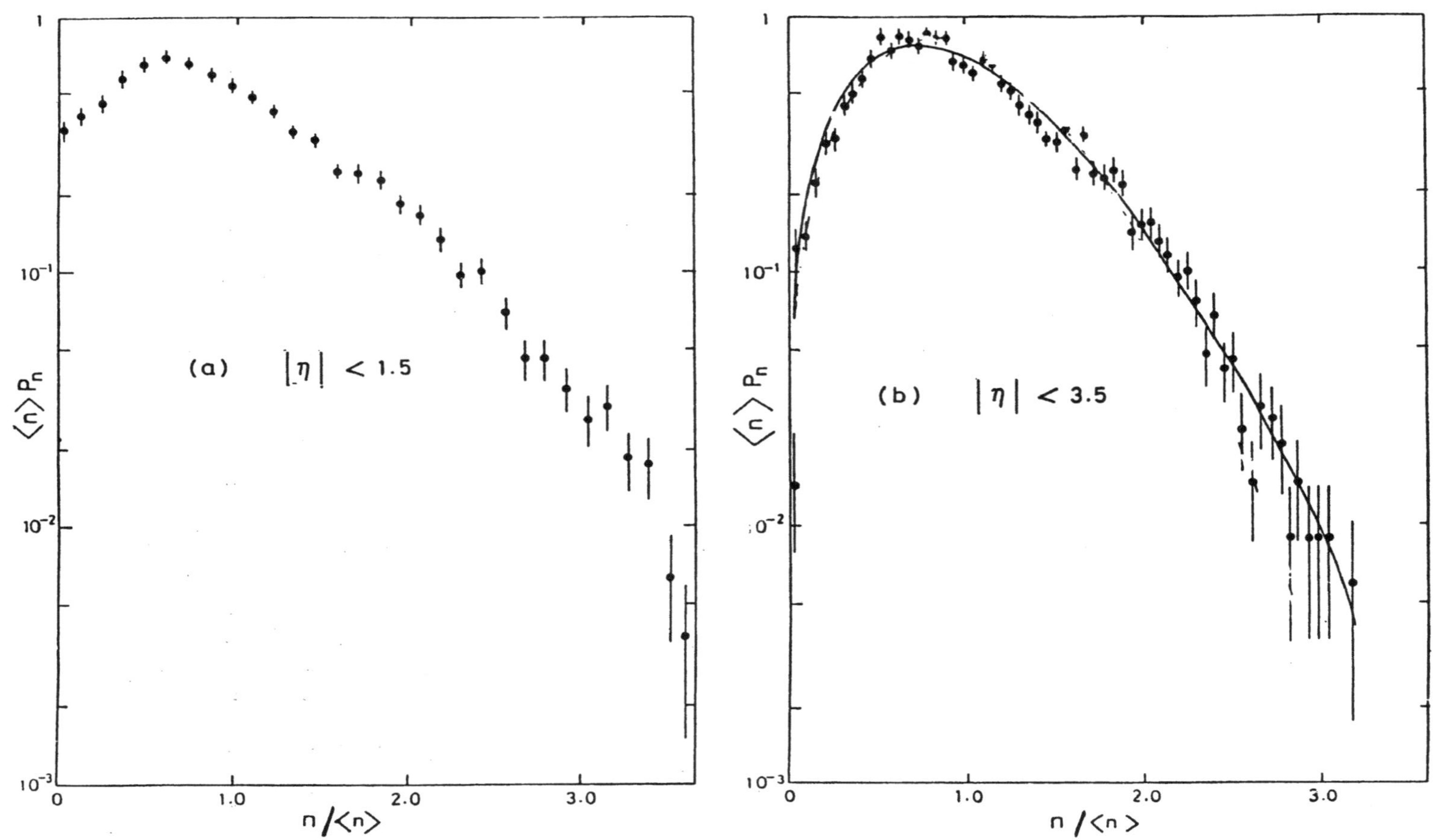

Fig. 8.: Charged-particle multiplicity (UA1).

events. The claim is that scaling is violated at the few percent level and that high multiplicity events ($z = n/\langle n\rangle > 2.0$) appear more frequently than suggested by the scaling curve. The measured multiplicities and dispersions are summarized in table 3. Comparing these results to lower energy data $\langle n\rangle$ is found to vary like $(\log s)^2$. A stronger form of energy dependence such as $s^{1/4}$ is excluded by the data.

Table 3: Charged particle multiplicity

Collaboration	η-range	$\langle n_{ch}\rangle$	$\langle n_{ch}\rangle/D$	Comments
UA5	$\lvert\eta\rvert < 4.5$	27.4±2.0	1.8±0.2	single diffraction excluded
		28.9±0.4	1.8±0.2	single diffraction included
UA1	$\lvert\eta\rvert < 1.5$	9.8±0.7	1.51±0.05	
	$\lvert\eta\rvert < 3.5$	21.1±1.5	1.84±0.07	

2.4 Correlations between soft particles

Short- and long-range correlations between soft particles have been observed by the UA5 group [24]. They reported a short range correlation length of 1 unit of rapidity and an average cluster size of about two charged particles per cluster, the same as observed at ISR energies [21]. The cluster size was found to be independent of the total multiplicity. Long-range multiplicity correlations have also been reported and interpreted as being due to the large ratio variance/mean of the multiplicity distribution.

2.5 Particle composition

The final state particles $K^{\pm}$, K^{o}, Λ^{o} and Ξ^{-} were identified by the UA5 group, while both UA5 and UA2 identified π^{o} and η^{o}; the UA2 group also presented transverse momentum spectra of charged kaons, protons and antiprotons [25].

Charged kaons were unambiguously identified by the UA5 group when selecting 3-prong decays. One-prongs could be either due to kaon or pion decays; most of the pion decays have been rejected by asking the decay angle to be greater than 15°. The rate of kaons was obtained after correcting for decay probabilities and the decay angle cut as well as for losses due to trigger inefficiency, limited acceptance, scanning and measuring.

The UA5 group identified K^{o}'s and Λ^{o}'s by visual scan, but also by an automatic search using measuring information. Photon conversions have been eliminated by demanding the opening angle to be greater than 3°. Decays of short-lived neutral kaons are coplanar (2-body decay), while decays of long-lived neutral kaons are not (3-body decay); a separation between the two samples is hence possible. The sample of coplanar decays consists of neutral kaons and lambdas. The different decay kinematics and life time probabilities allowed a K^{o}/Λ^{o} - separation despite the fact that only angles and no momenta were measured by the UA5 group. Finally, a few decays of the type $\Xi^{-} \rightarrow \Lambda^{o}\pi^{-}$ have also been observed. A summary of the results is presented in table 4.

In another analysis strange particle decays were eliminated by the UA5 group and photon conversions investigated in more detail. Photons converted in the beam pipe representing 0.05 to 2.7 radiation lengths depending on the photon angle of incidence, or in a lead glass plate, one radiation length thick, placed symetrically in the sensitive volume of one streamer chamber on either side of the beam axis [26]. Electromagnetic showers in the beam pipe were separated from hadronic showers or spurious chance associations of tracks. The cuts to separate e.m. showers exploited the fact that secondaries in electromagnetic processes tend to be produced at small angles. The shower multiplicity was required to lie in the range 2 to 10. The results obtained from conversions in the lead glass plate are based on a careful scan of 700 events of all multiplicities, which yielded 870 e.m. showers for further analysis. After all corrections the sample of conversions in the beam pipe and in the lead glass plate gave consistent results on photon multiplicity. The photon multiplicity was then compared to the charged-pion multiplicity obtained from the total charged-particle multiplicity after subtraction of the charged strange particles. The results are

$$\langle n_\gamma \rangle = 31.2 \pm 2$$

$$\langle n_{\pi^\pm} \rangle = 22.5 \pm 1 \ .$$

The excess photons were interpreted as being due to $\eta \to \gamma\gamma$ decays, using a η^o to π^o production cross section ratio of $\eta^o/\pi^o = 0.32$.

Neutral pions and etas were also observed by the UA2 group. The produced π^o's and η^o's were identified from two-photon decays. The two photons were resolved in a lead glass array located 3m away from the beam axis. For the trigger a localized energy deposition in the lead glass array in excess of 1.1 GeV was required in coincidence with the presence of charged secondaries at small angles with respect to the beam axis. Two clusters in the lead glass array of more than 0.2 GeV each were required in the analysis. The absence of tracks pointing towards the clusters within ±12cm was required. The trans-

Table 4: Particle composition

Collaboration	Result	Comments
UA2	$\eta^o/\pi^o = 0.55$	same result as at ISR, $\sqrt{s}$ = 56 GeV
UA5	$\eta^o/\pi^o = 0.33$	
UA5	K/all = 0.12	
	baryons and anti-baryons/all=0.09	assuming $\bar{p}/\bar{\Lambda} = 3$
UA2	$\Xi/\Lambda = 0.35 \pm 0.11$ (at fixed values of $m_T = \sqrt{m^2+p^2}$)	
	$K^{\pm}/\pi^{\pm} = 0.4$	∿40 % increase when compared to ISR result
	$\bar{p}/\pi^- = 1.0$	unchanged when compared to ISR

verse momentum of the two-cluster system had to be greater than $p_T(\gamma\gamma)$ = 1.5 GeV/c. Stricter criteria were applied to isolate the η signal. Either cluster had to consist of 2 cells at most, with E_{cell} < 3 GeV. No more than 3 clusters in the total lead glass array were permitted and the energy asymmetry had to be smaller than 0.4.

These requirements reduced the acceptance, but cleaned up the η signal. The obtained π^o peak in the mass distribution $M(\gamma\gamma)$ is shown in fig.9. The insert shows the η region when using strict cuts. After all corrections the ratio η^o/π^o = 0.55 was obtained, the same as measured in the ISR energy range [27].

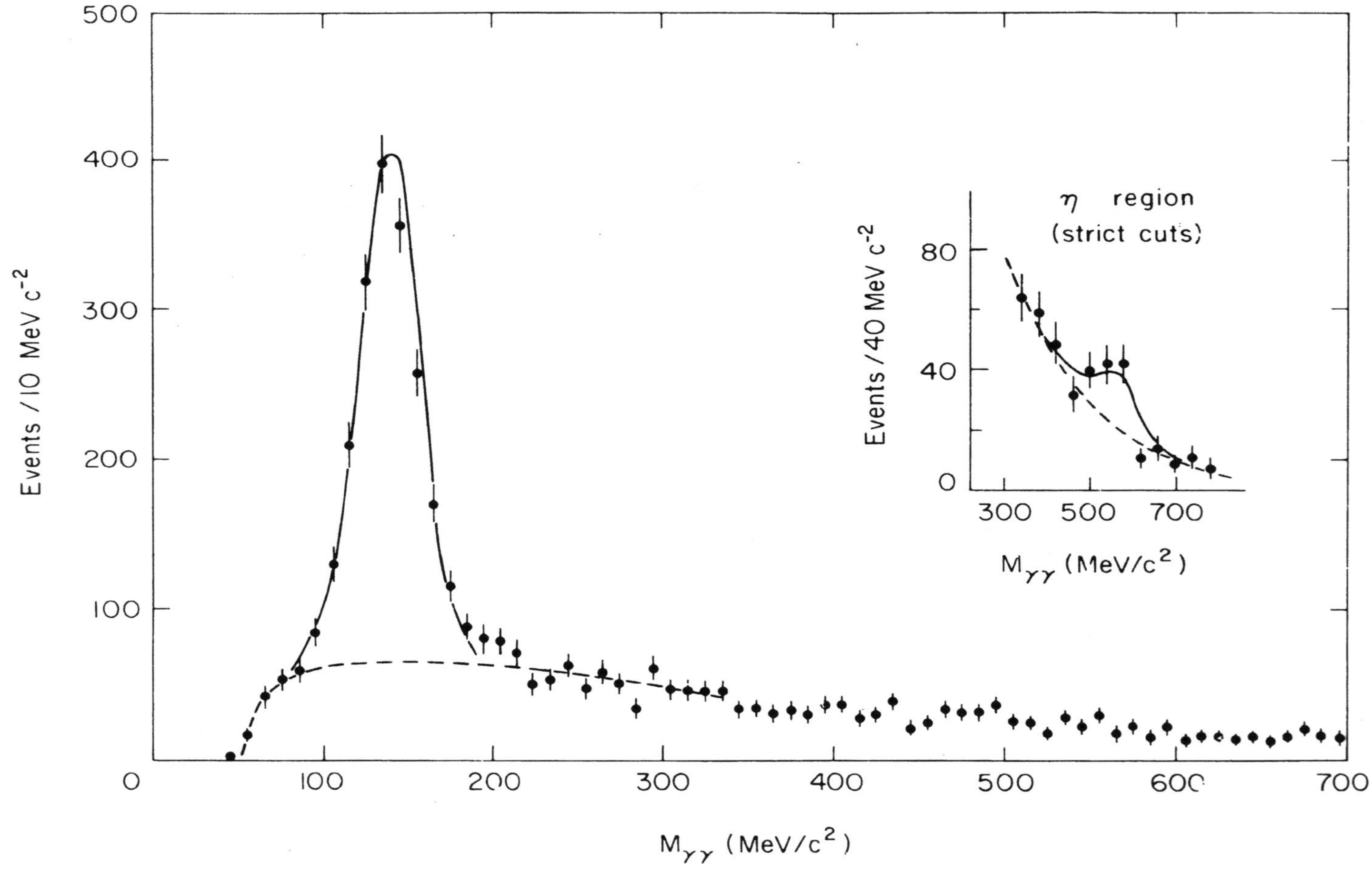

Fig. 9.: The two-photon invariant-mass distribution (UA2).

Charged kaons and baryons have been observed by the UA2 group by using the time-of-flight technique. The momenta of charged particles were measured with chambers placed before and behind a magnetic field with an accuracy

$$\Delta p/p = [(0.5)^2 + (0.5p)^2]^{1/2}\ \% ,$$

p in GeV/c. The time-of-flight resolution of the system was $\sigma = 0.49$ ns, mostly due to the time spread from the size of the antiproton and proton bunches. A good separation, defined by the probability of wrong assignment to be less than 20 %, between kaons and pions was achieved up to p = 1.1 GeV/c and between protons and pions up to p = 1.8 GeV/c. The ratios K/π and $\bar{p}/\pi^-$, at fixed values of $m_T = \sqrt{m^2+p_T^2}$, are summarized in table 4.

2.6 Inclusive transverse-momentum spectrum

The UA2 group has measured inclusive π^0 production in a range of transverse momenta between 1.5 and 4.5 GeV/c and recently up to 40 GeV/c. They have also presented data on inclusive charged-hadron production in the transverse-momentum range 0.4 to 2 GeV/c (recently up to 10 GeV/c), while the UA1 group has measured the charged-hadron spectrum up to p_T = 10 GeV/c [28].

The experimental procedure of the UA2 group to identify neutral pions and charged hadrons has been described in the previous section. The further analysis to obtain the π^0 cross section at small p_T proceeded as follows. First, the overall energy scale was readjusted by requiring the measured π^0 peak, fig.9, to agree with the π^0 peak as simulated by a Monte Carlo calculation. Then the background below the peak was subtracted. The estimated contribution of accidental clusters from another π^0 in the event was also subtracted. Corrections were applied for acceptance losses and losses due to the analysis procedure. The invariant cross section for π^o production was then obtained after unfolding the energy resolution of the calorimeters and using the integrated luminosity.

The procedure of the UA2 group to identify and measure charged hadrons has been described in the previous section. To obtain invariant cross sections corrections were applied to take into account the detector geometry, multiple scattering, decays, nuclear interactions and triggering efficiency; the integrated luminosity was used. The measured charged and neutral-pion cross sections were found to agree well and were fitted to the empirical form

$$E\frac{d^3\sigma}{dp^3} = A[2/(1+p_T)]^n .$$

The parameters A and n are presented in table 5 together with results obtained by the UA1 group and an ISR group [29].

The results of the UA1 group were obtained by track-curvature measurement in the central detector using magnetic field settings of 0.56 and 0.28 T. No particle identification was attempted. The momentum resolution at nominal field B = 0.56 T was

$$\Delta p_T/p_T^2 \approx 0.02 .$$

Data from a second field setting B = 0.28 T were used to check the momentum measurement and acceptance corrections. In the analysis strict track quality cuts were applied to reject artificially increased apparent track momenta due to reconstruction problems when dealing with overlapping tracks. The experimental p_T-resolution was unsmeared. The reduction of the cross section due to unsmearing amounted to about 10 % at p_T = 5 GeV/c and to about 40 % at

Transverse-momentum spectra

parametrized by

$$E \frac{d^3\sigma}{dp^3} = A[p_To/p_T + p_To)]^n \qquad \text{(UA1, ISR, UA2 1984)}$$

or

$$E \frac{d^3\sigma}{dp^3} = A[2/(1+p_To)]^n \qquad \text{(UA2 1982, 1983)}$$

Table 5: Parametrizations of transverse-momentum spectra

Collaboration	p_T-range [GeV/c]	A [$10^{-24}cm^2$]	p_To [GeV/c]	n
UA1 (charged particels)	0.3 - 10.0	0.47±0.01	1.3±0.2	9.14±0.9
	0.3 - 2.0	0.46±0.01	1.3	9.14±0.02
	2.0 - 10.0	0.37±0.06	1.3	8.99±0.15
UA2 1982 (charged pions and neutral pions)	0.4 - 2.0	0.00143		8.3±0.2
UA2 1984 (neutral pions)	1.5 - 41.0	$(1.66\pm.22)\cdot 10^3$	0.207±0.004	7.26±0.03
UA2 1984 (charged particles)	0.3 - 8.9	0.773±0.028	0.865±0.009	7.66±0.06
ISR ref.29 (charged particles)	0.1 - 2.0	0.28±0.01	1.3	10.7±0.03

p_T = 10 GeV/c. The invariant cross section shown in fig. 10 was obtained after corrections for geometrical acceptance and the analysis procedure were taken into account. The data of fig. 10 are compared to a fitted curve of the form

$$E \frac{d^3\sigma}{dp^3} = A[p_{TO}/(p_T+p_{TO})]^n$$

The parameters A, p_{TO} and n, using different p_T-ranges in the fit, are shown in table 5 together with results from the UA2 group and an ISR group. The average transverse momentum measured by the UA1 group is $\langle p_T \rangle$ = 0.42 GeV/c. The ISR measurement yielded $\langle p_T \rangle$ = 0.34 GeV/c. The invariant cross section as a function of p_T for charged hadrons is shown in fig. 11 for three bands of charged-particle multiplicity. A striking multiplicity dependence is observed. The p_T spectrum flattens with increasing charged multiplicity. In events with about 2.0 charged particles per unit of rapidity, as observed at the ISR, the average transverse momentum is $\langle p_T \rangle \gtrsim$ 0.35 GeV/c, consistent with the ISR result. If $n_{ch}/\Delta y$ = 10, then $\langle p_T \rangle$ = 0.48 GeV/c. An increase of $\langle p_T \rangle$ as a function of multiplicity was also observed in cosmic-ray experiments [30].

2.7 Early searches for surprises

In cosmic-ray experiments at $\sqrt{s} \gtrsim$ 1700 GeV there have been suggestions [31] of the existence of an unusual class of hadronic events, called "Centauros", which are characterized by a high charged multiplicity ($n_{ch} \sim$ 50 - 100) and a seeming absence of accompanying photons. The UA5 and the UA1 groups have investigated whether such a phenomenon or its onset could already be observed at the collider [32].

The experimental technique of the UA5 group to measure

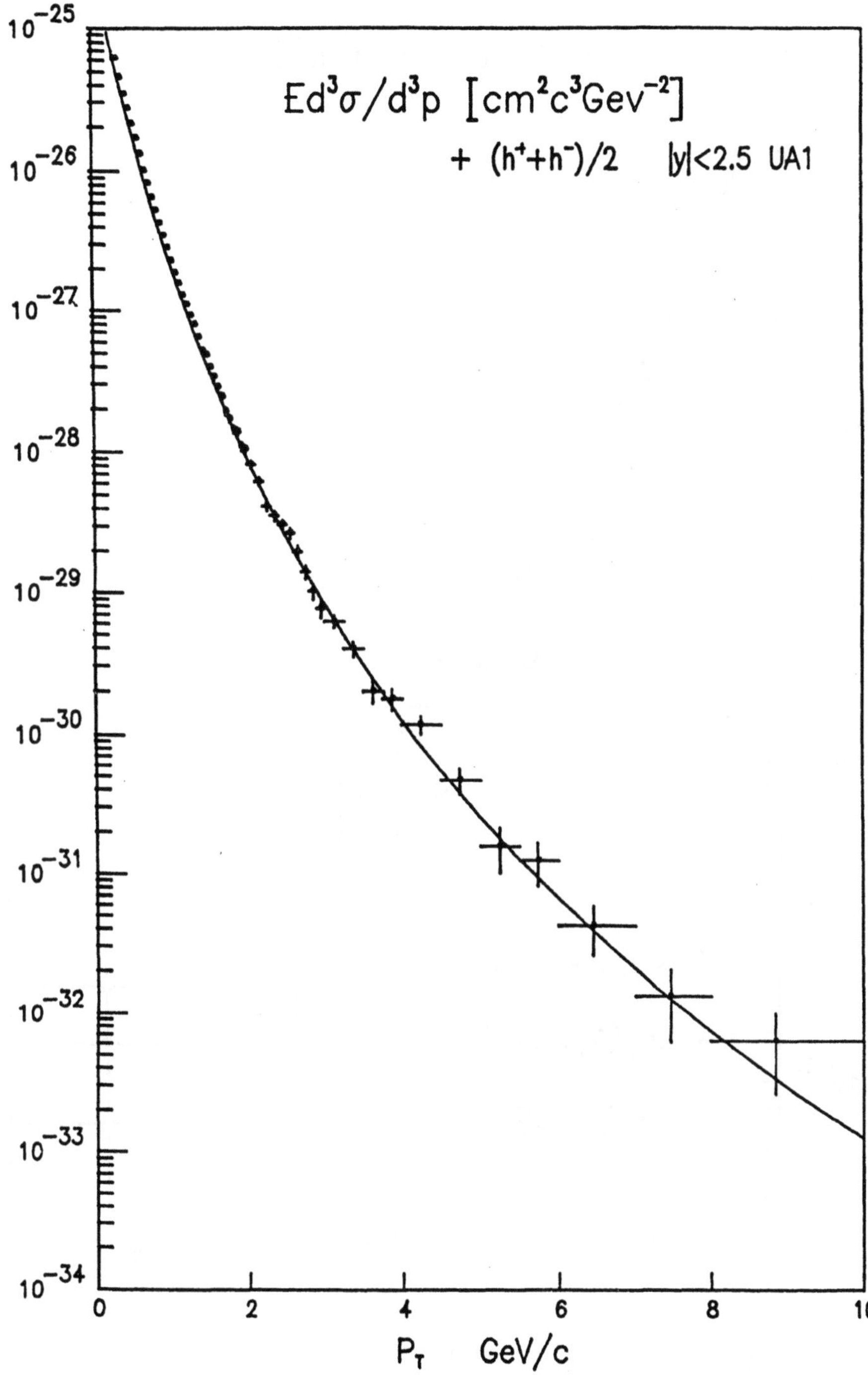

Fig. 10.: Invariant cross section as a function of p_T (UA1).

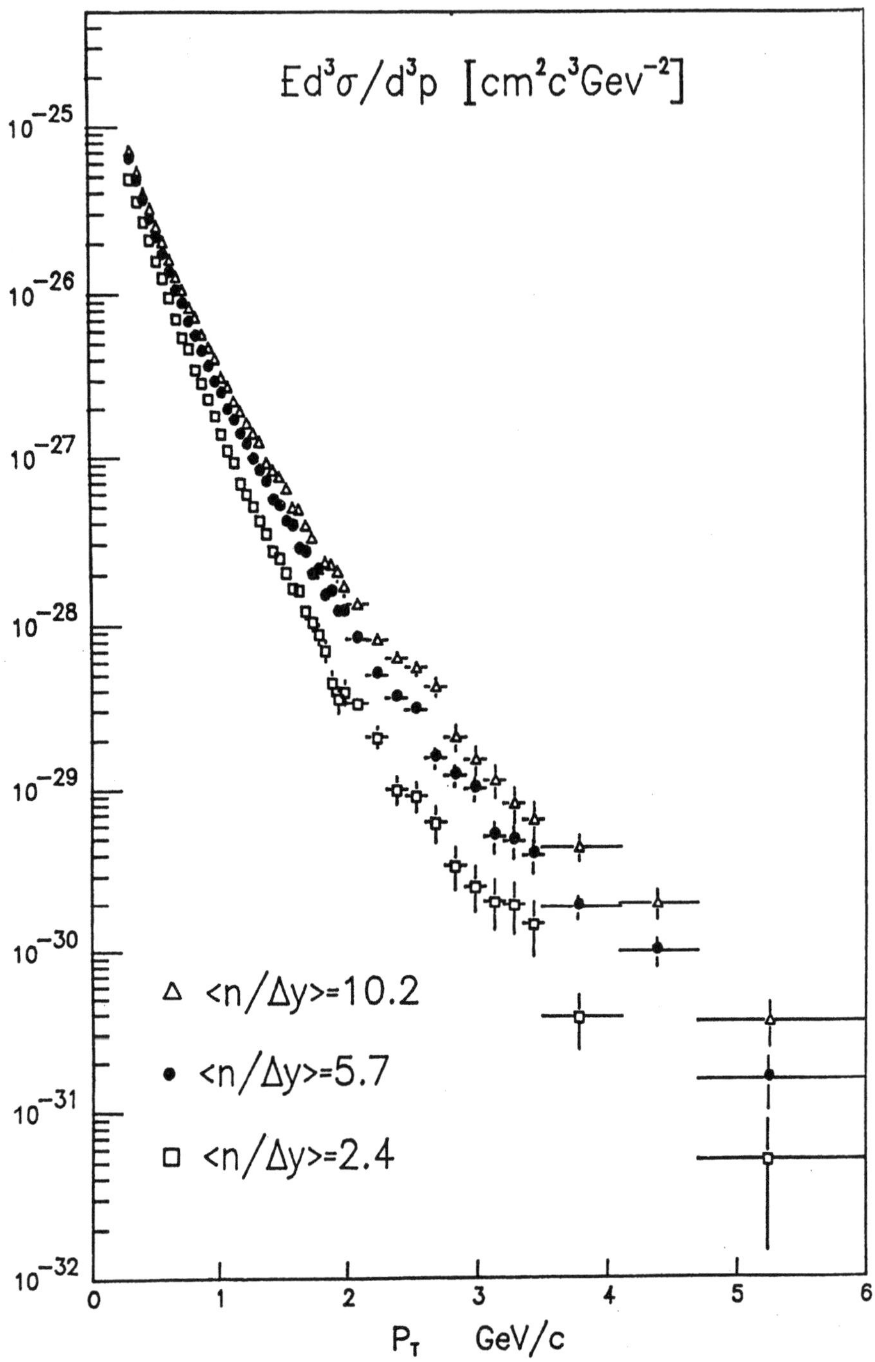

Fig. 11.: Invariant cross section as a function of p_T for three bands of charged-track multiplicity (UA1)

the charged multiplicity and the multiplicity of photons was described in section 2.5. When investigating the distribution n_{γ}^{obs} vs. n_{ch}^{obs} , fig. 12, no distinct class of events with particularly large n_{ch}^{obs} and small n_{γ}^{obs} could be seen. The few events with highest charged multiplicity and lowest photon multiplicity do not populate a separate region of the plot, but comprise the tail of the main distribution. A detailed analysis of events with low n_{γ}^{obs} and large n_{ch}^{obs} revealed no "Centauro"-candidates among about 3600 inelastic events.

The measurement of charged particles in the central detector of the UA1 experiment was described in the previous sections. Longitudinal shower profiles initiated by charged and neutral particles are measured with calorimeters. While hadrons penetrate the electromagnetic calorimeter, photons are absorbed early. No distinct class of events with many hadrons and few or no photons have been observed in a sample of about 48000 low bias (i.e. soft) events.

A search for relativistic particles with fractional electric charge was performed by the UA2 group using after momentum determination a telescope of scintillation counters to detect particles with low ionization [33]. A minimum ionizing particle of charge q = 1 (m.i.p.) initiated on average 75 and 165 photoelectrons in each phototube of the two different parts of the telescope. The trigger required an ionization in excess of 0.03 m.i.p. in coincidence with a signal from each of the two scintillator arrays up- and downstream of the intersection point. About 15000 events were collected. No free quark candidates survived the cuts applied in the analysis. The 90 % confidence-level upper limit for the ratio of quark yield to the yield of particles of unit electric charge as a function of the quark mass m_Q is shown in fig. 13. The curves labelled 2/3 and 1/3 refer to the absolute values of the quark charges. The 90 % CL upper limit deteriorates with increasing m_Q due

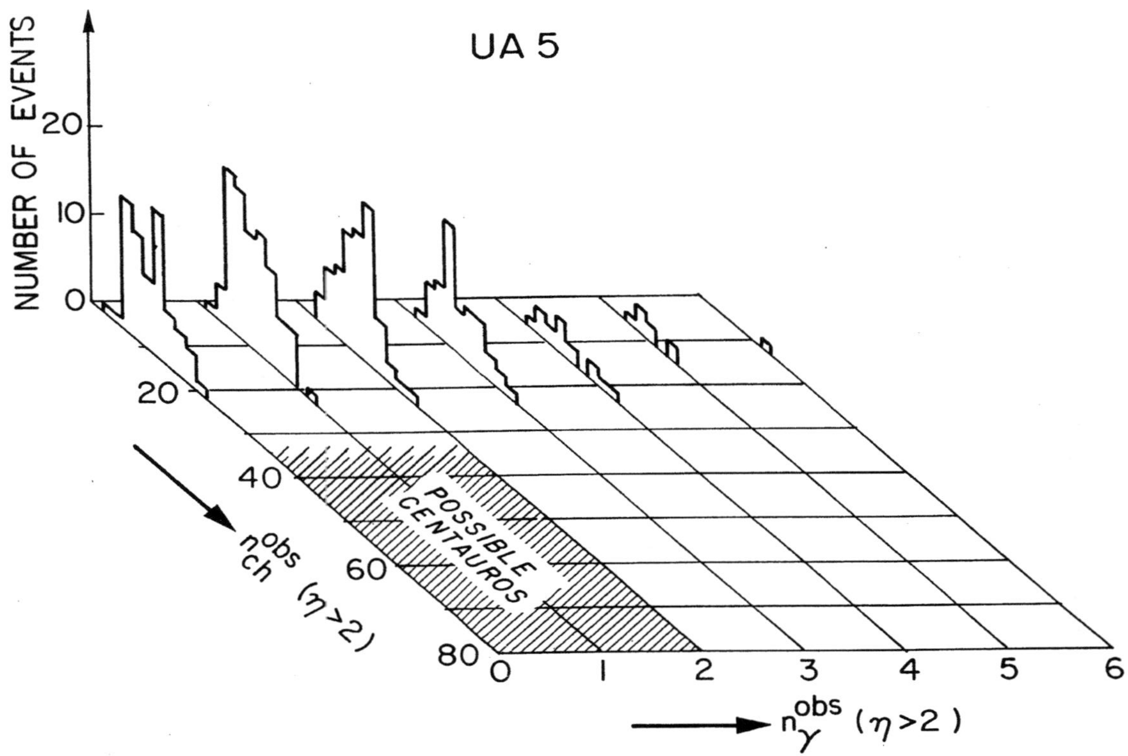

Fig. 12.: Number of events plotted against multiplicities of charged particles at pseudorapidities greater than 2 or less than -2. The distributions are drawn separately for each value of observed photon multiplicity in the same range of pseudorapidity (UA5).

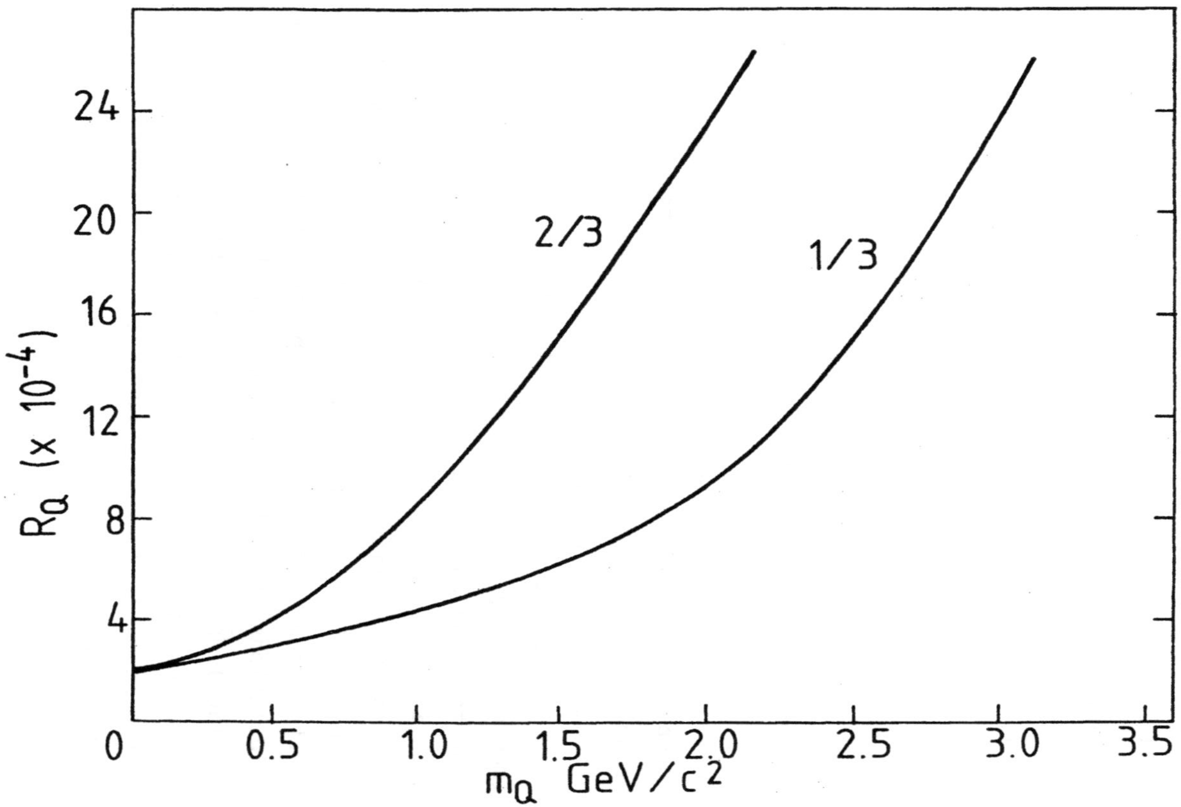

Fig. 13.: 90 % CL upper limit for production of free quarks (UA2).

to the velocity dependence of the particle ionization [34]. The thickness of the detector, about 40 gr cm^{-2}, limited this quark search to particles without strong absorption in matter. A comprehensive review of quark searches is presented in ref.35.

Magnetic monopoles, if produced at the collider, would be visible in the UA3 set-up because of their high ionization. No such monopoles have been observed so far [36].

3. HADRONIC JETS

A small part of the total cross section for proton-antiproton collisions as $\sqrt{s}$ = 540 GeV is due to hard scattering between point-like quarks and gluons (partons) as sketched in fig.1. Despite the small rate for such processes - less than one event in 10000 events - they are experimentally easy to detect by the characteristic two-jet event topology and the large total transverse energy ΣE_T which is mostly spent to build up the transverse energy of the two jets (two-jet dominance). The cross section for production of jets is computed from the flux of quarks and gluons within the hadrons and their point-like cross section. Such calculations are believed to be reliable if (i) $\sqrt{\tau} = 2E_T(\text{jet})/\sqrt{s}$ is greater than a few percent and (ii) the transverse momenta of the jets (=partons) are reasonably large, e.g. E_T(jet) > 10 GeV for $\sqrt{s}$ = 540 GeV. For smaller values of $\sqrt{\tau}$ there is theoretical (experimental) dispute, but also experimental (theoretical) uncertainty concerning the visibility of jets above the abundant soft-particle background. The large-p_T jets which stick out of the soft hadronic background serve as a tool to increase the sensitivity of experiments by about four orders of magnitude when using hadron collisions. The soft hadron processes are rejected by selecting events with jets and the fundamental processes between hadron constituents dominate the cross section despite their small rate. The events with jets themselves are clean. A modest p_T-cut of 1 or 2 GeV on tracks and calorimeter hits cuts away most of the spectator hadrons. To illustrate cleanliness we show a two-jet event with all tracks and calorimeter hits (fig.14a), and the same event after the soft hadrons have been removed (fig.14b); only tracks and calorimeter hits belonging to the two-jet structure survive the cut. The production properties of the observed jets can be studied without the need to know their fragmentation in a way very close to the ideal case in which one would directly observe the scattered partons.

In this chapter we first discuss how the two-jet topology dominates the events with large ΣE_T. We then discuss jet-production properties together with their implications concerning the properties of quarks and gluons in the proton and finally jet fragmentation.

3.1 Two-jet dominance

The UA2 group and the UA1 group observed that in events with large total transverse energy ΣE_T the two-jet topology dominates [37]. The UA2 group used a large solid-angle total-absorption hadron calorimeter subdivided into small cells. A cross section of their detector is shown in fig.15. In the centre of the apparatus a vertex detector measures particle trajectories in a region without magnetic field. The vertex detector is surrounded by a fine grain electromagnetic and hadronic calorimeter segmented into 200 cells and built in a tower structure pointing to the interaction region. The cells are segmented longitudinally into a 17 radiation-length thick electromagnetic compartment (lead scintillator) followed by two hadronic compartments

(a)

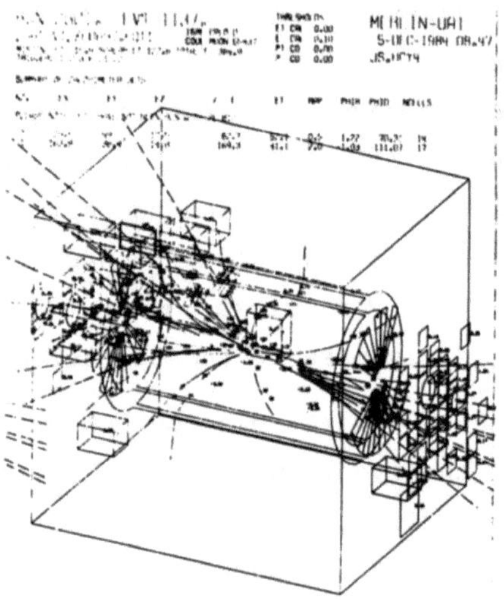

(b)

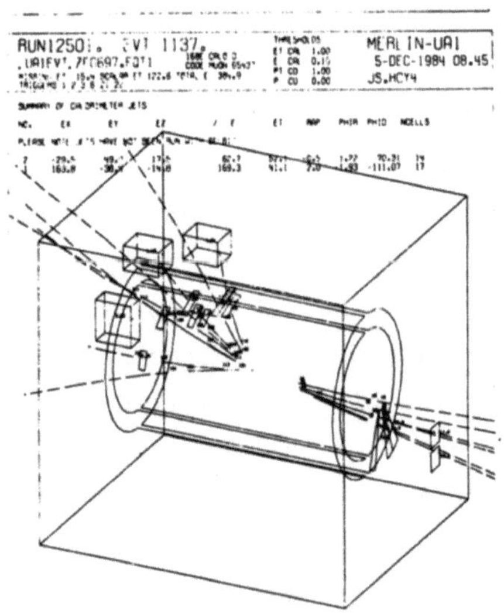

Fig. 14: A two-jet event as seen in the UA1 detector:
(a) all tracks and calorimeter hits are displayed;
(b) only tracks and calorimeter hits with p_T>1.0 GeV are shown. A striking two-jet structure emerges.

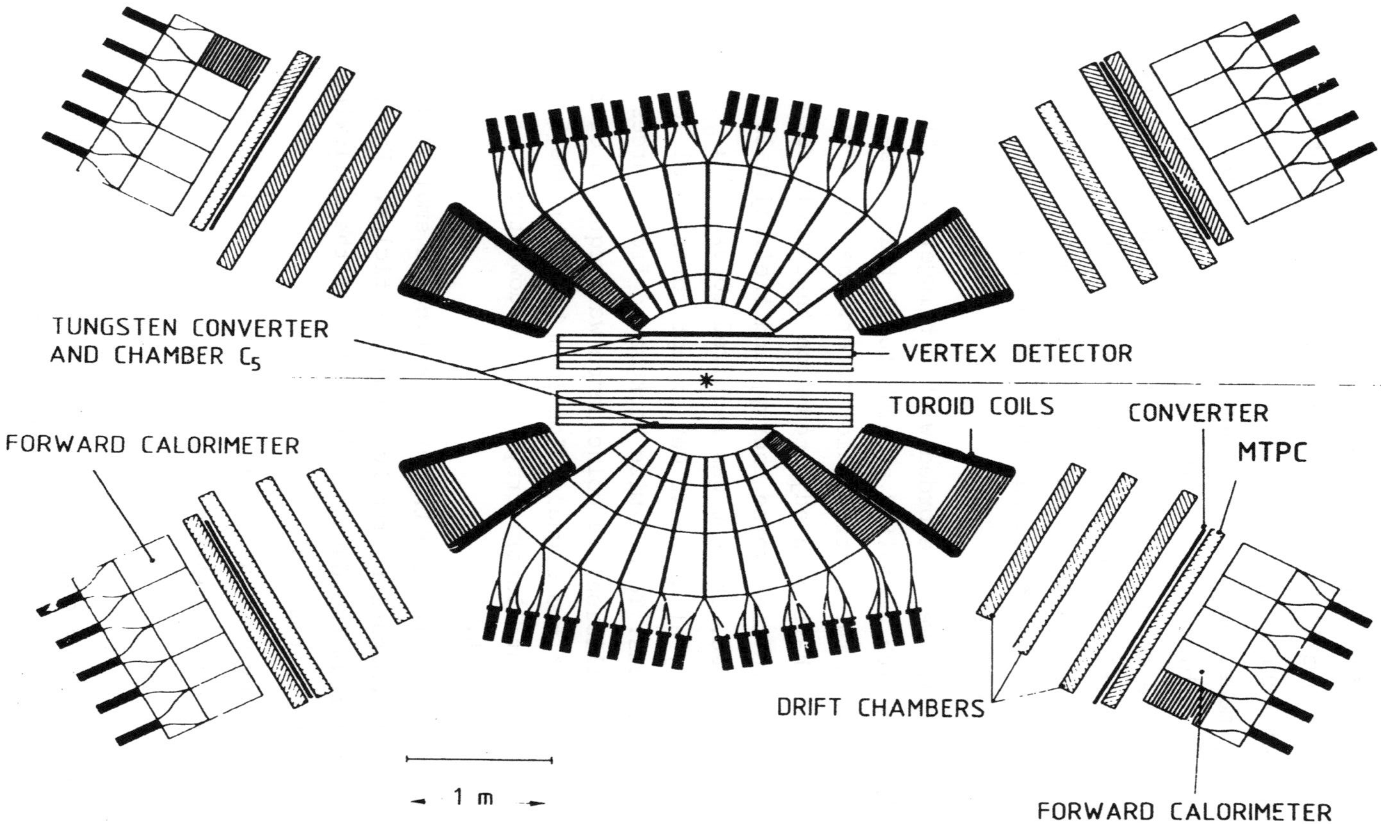

Fig. 15.: The UA2 detector; schematic cross section in the vertical plane containing the beam.

(iron scintillator) of two absorption lengths each. All calorimeter cells have been calibrated in 10 GeV electron, muon and pion beams. The initial-energy calibration has been tracked (i) with a light-flasher system, (ii) measurements of the direct current induced by a Co^{60} radioactive source and (iii) by measurements of the average energy flow into each cell for unbiased $\bar{p}p$ collisions. The systematic uncertainty in the energy calibration is less than ±1.5 % for the electromagnetic calorimeter and less than ±3.5 % for the hadronic one. The response of the calorimeter to electrons, single hadrons, and multihadrons has been measured using beams from 1 to 70 GeV.

The UA2 data were recorded with a trigger selecting $\bar{p}p$ collisions which deposited a large total transverse energy (ΣE_T) into the central calorimeter. The gains of the photomultipliers have been adjusted such that their response is proportional to transverse energy, and their signals were linearly added. The sum was required to exceed a threshold normally at about 40 GeV except for special low-threshold data runs. To minimize background from sources other than $\bar{p}p$ collisions a coincidence was required of the ΣE_T condition with two signals ("minimum bias" trigger) obtained from scintillator arrays up- and downstream of the collision point.

In the analysis the UA2 group noticed a small (< 10 %) background contamination from beam halo particles. Events were rejected from the data sample, if any of the following conditions were satisfied:

(i) if they were associated with an early signal in the small angle scintillator arrays;

(ii) if they had an abnormally large total transverse energy fraction in the hadronic compartments;

(iii) if they contained only one cluster with more than 90 % of its energy in the hadronic compartments. (Cells which share a common side and have an energy E_{cell} > 400 MeV are joined into a cluster; clusters having two or more local maxima separated by a valley deeper than 5 GeV are split).

After correcting for the different response of the electromagnetic calorimeter to energy deposited by photons and hadrons and for the losses due to the analysis procedure the ΣE_T distribution was obtained for two pseudo-rapidity intervals (fig.16). The data have been normalized using the integrated luminosity. The distributions of fig.16 show a departure from exponential when ΣE_T exceeds about 60 GeV. This value corresponds to the transverse energy where the two-jet production cross section begins to dominate over soft hadronic interactions.

For a further analysis the reconstructed energy clusters have been ranked in the order of decreasing transverse energies denoted by $E_T^1 > E_T^2 > E_T^3$ ... The clusters contain typically 3 cells for $E_T = 2$ GeV and 10 cells for $E_T = 40$ GeV.

The fractions $h_1 = E_T^1/\Sigma E_T$ and $h_2 = (E_T^1 + E_T^2)/\Sigma E_T$ describe the transverse energy accounted for in each event by the cluster (respectively the two clusters) with the largest transverse energy. The mean values of h_1 and h_2 are shown in fig.17a. Their behaviour illustrates the emergence of a dominant two-jet structure at large ΣE_T. The dependence on ΣE_T of the mean values of the ratios $r_{21} = E_T^2/E_T^1$ and $r_{32} = E_T^3/E_T^2$ are shown in fig.17b. As ΣE_T increases r_{21} approaches about 0.9 whereas r_{32} decreases below 0.1. A very large fraction of the total transverse energy of large ΣE_T events is shared on average by two clusters only.

The experimental set-up used by the UA1 group is shown in fig.18. The central detector measures particle trajetories in a magnetic dipole field of 0.56 T. The central detector is surrounded by an electromagnetic calorimeter placed inside the magnet. It consists of two

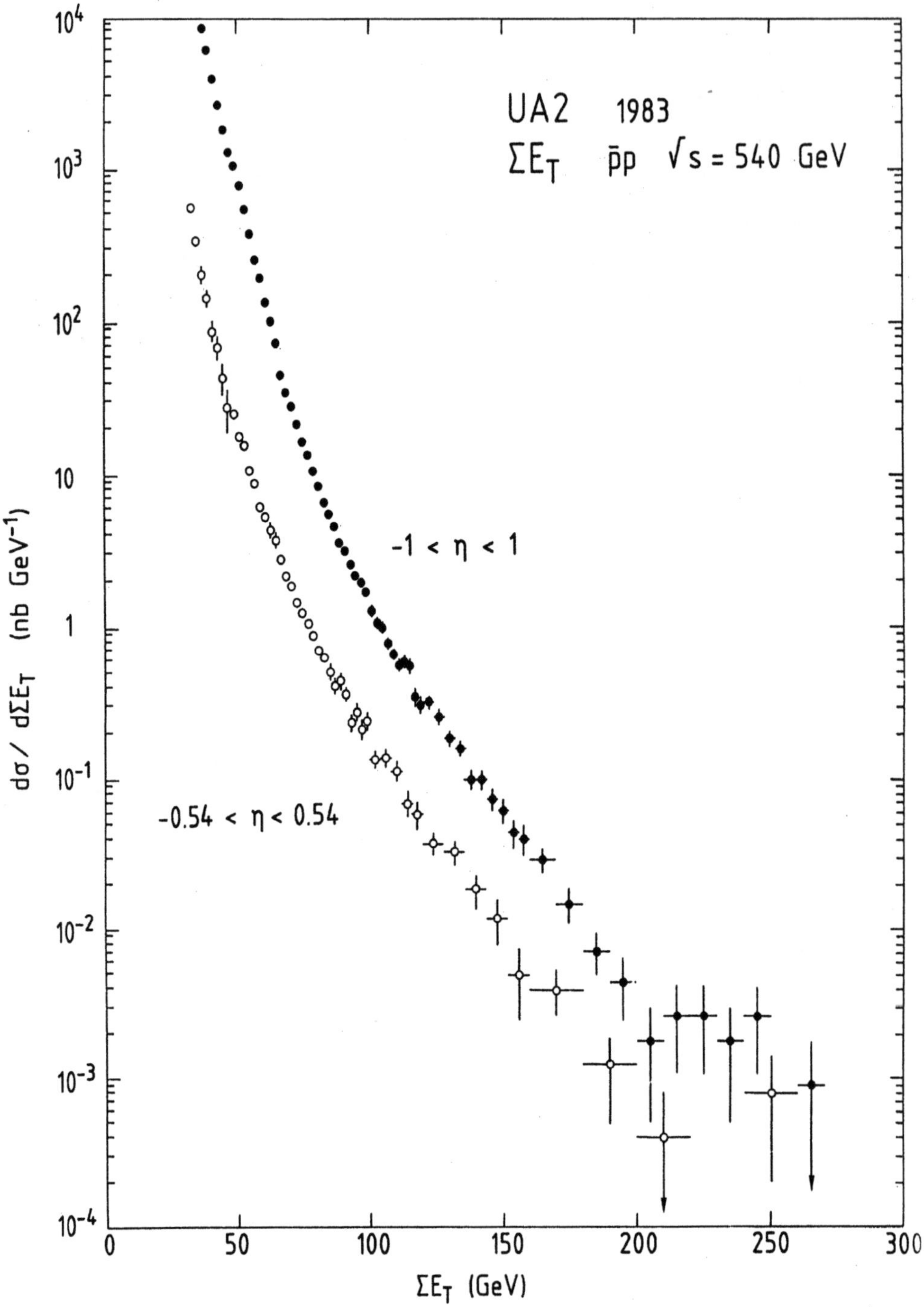

Fig. 16: ΣE_T distributions for two pseudo-rapidity intervals (UA2).

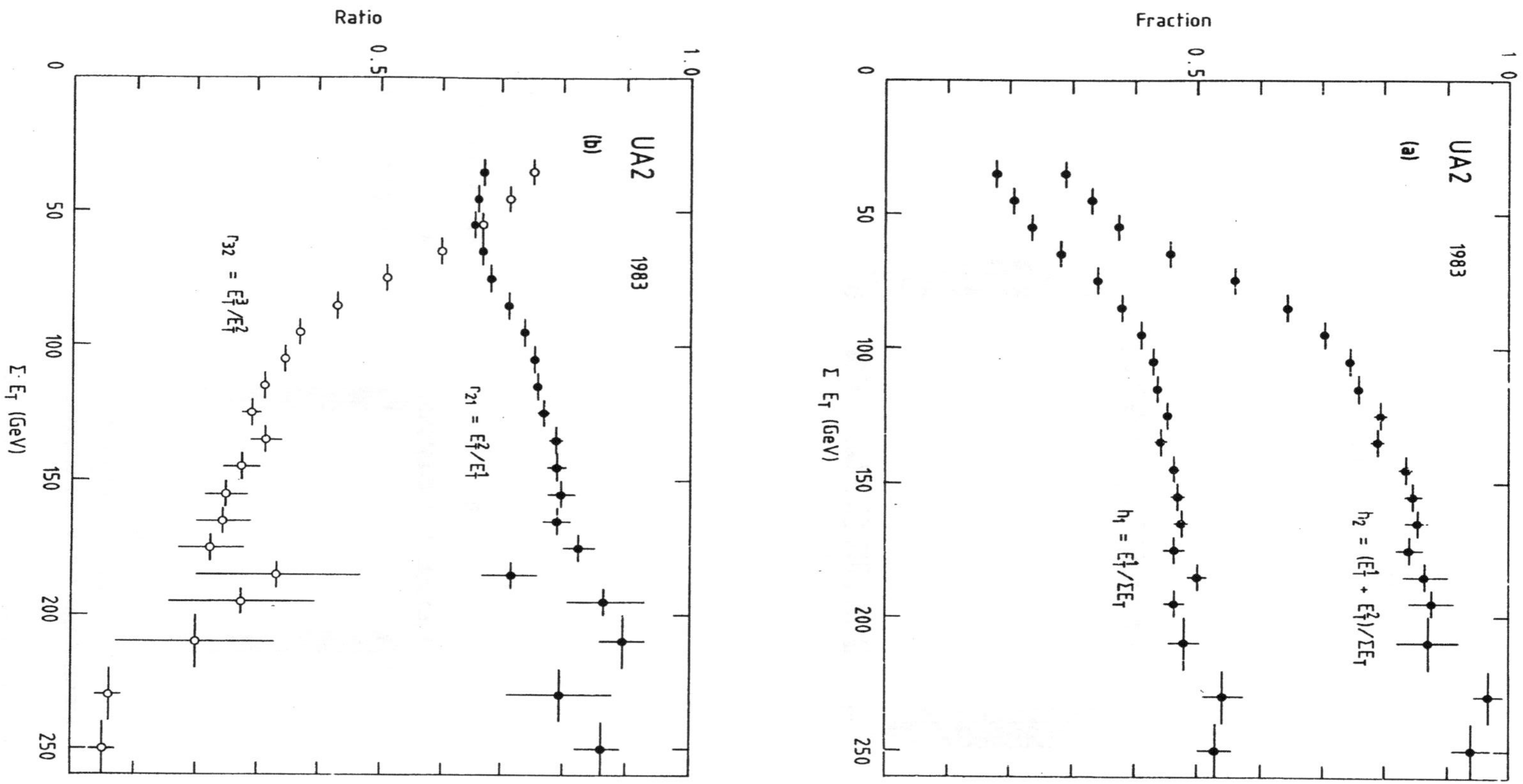

Fig. 17.: (a) The fractions h_1 and h_2 of the total transverse energy contained in the cluster and the two clusters, respectively, having the largest E_T, as function of ΣE_T; (b) The ratios $r_{21} = E_T^2/E_T^1$ and $r_{32} = E_T^3/E_T^2$ as a function of ΣE_T.

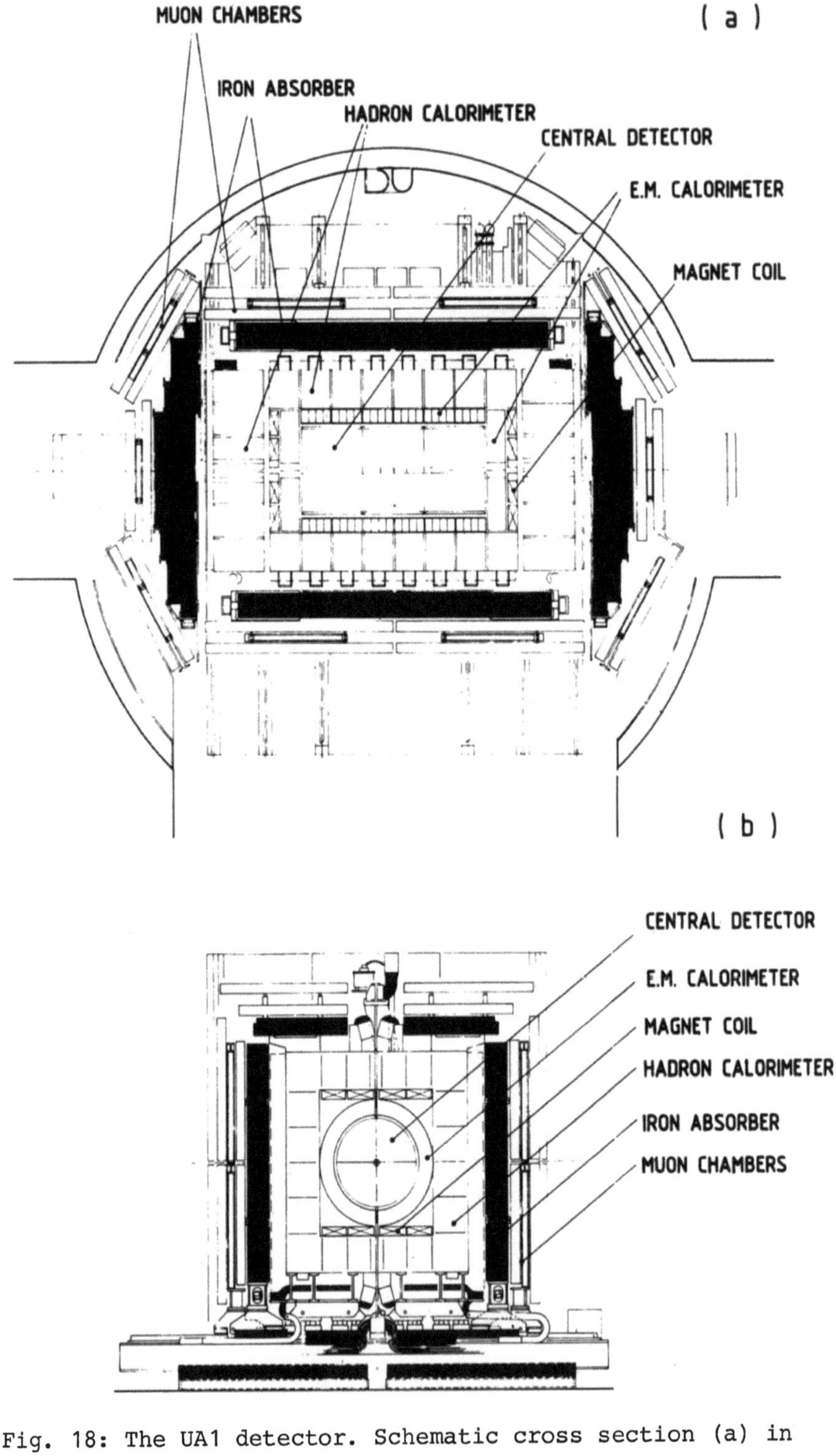

Fig. 18: The UA1 detector. Schematic cross section (a) in the horizontal plane containing the beam, (b) in the vertical plane perpendicular to the beam.

endcap detectors and two semi-cylindrical half-shells with an inner radius of 1.36 m. Each half-shell consists of 24 lead-scintillator modules, the "Gondolas". Each gondola is about 22.5 cm wide and 27 radiation lengths deep and is read out in four segments of 3, 7, 10 and 7 radiation lengths. The azimuthal position of the showers is measured by pulse height division. The endcap detectors, the "Bouchons", cover the end faces of the central detector. Each bouchon consists of 32 radial sectors and is read out in four segments which are 4, 7, 9 and 7 radiation lengths deep. The positions of showers are measured with a position detector located inside the calorimeter at a depth of 11 radiation lengths. Particles which traverse the electromagnetic calorimeters encounter the hadronic calorimeter built from scintillator plates inserted into the laminated yoke of the magnet. The hadronic calorimeter behind the gondolas (bouchons) is read out in two segments of 2.5 (3.5) absorption lengths each.

All types of UA1 calorimeter configurations have been extensively studied in test beams and their responses as a function of angle of incidence, energy and nature of the incident particle have been determined. Hadrons deposit a large fraction of their energy in the electromagnetic calorimeter. Due to nuclear effects the response to hadrons of the combined electromagnetic and hadronic calorimeter is smaller than its response to electromagnetic showers of the same energy. In the analysis an average correction factor was applied to the data.

A detailed response map of each element of the electromagnetic calorimeter has been determined and was checked before and after every run period by a fine-mesh scanning with a collimated γ-beam from a 4 Ci Co^{60} source. For a few calorimeter elements the response to the source was compared with the response to electrons in a test beam. The energy scale of the hadronic calorimeter was determined with test modules in a test beam. Cosmic-ray muons were used as an energy reference for the assembled calorimeter and Ru^{106} β-sources have been placed in reproducible positions on every scintillator at the time of setting up. The absolute calibration was adjusted by re-insertion of these sources and by measurement of the calorimeter response to cosmic-ray muons over a restricted set of calorimeter cells. The photomultiplier gains of all calorimeters were monitored by a laser system. The systematic uncertainty in the energy calibration is less than ±3 % for the electromagnetic calorimeters and less than ±5 % for the hadronic ones.

The data were recorded using an inclusive jet trigger which required a localized transverse energy deposition in excess of 15 GeV in a window corresponding to 8 gondola cells. The localized transverse-energy deposition was required to occur in coincidence with signals from a pretrigger hodoscope located up- and downstream of the collision point. This trigger was redefined in the analysis using the same calorimetric window with a threshold set at 25 GeV. Time-of-flight information from the pretrigger hodoscope was used to reject non-collision triggers ($\approx$ 5 %). The selected events were fully reconstructed in the central detector. Calorimeter depositions were corrected using detailed light attenuation maps and an average response correction factor for hadronic energy in the electromagnetic calorimeters. An energy vector was then associated to each calorimeter cell; the vector points from the interaction point to the centre of the cell. For electromagnetic cells the vector points to the energy centroid determined by pulse height measurements (gondolas) or by position detectors (bouchons). Finally, jets were defined as clusters of energy vectors in the pseudo-rapidity/azimuth (η/ϕ) space around the jet axis requiring $d = \sqrt{\Delta\eta^2 + \Delta\phi^2} < 1$ (with ϕ in radians). Lower-E_T hits (E_T <2.5 GeV) were added to the nearest jet, if their transverse momentum relative to the jet axis was less than 1 GeV and if they were not more than 45° away from the jet axis.

For further analysis an E_T threshold was applied requiring at least one jet with $E_T > 30$ GeV in the gondola region. This jet is called the trigger jet. A threshold was applied to the fraction of jet energy deposited in the electromagentic calorimeters (> 10 %) to reject signals from beam halo in the hadronic calorimeters. To avoid edge effects due to the calorimeter geometry the trigger jet had to be contained in $\eta = \pm 1.2$ and $\phi = \pm 60$ from the horizontal plane. To study possible trigger biases the UA1 group used a sample of events taken with a global (i.e. $|\eta| < 1.5$) $\Sigma E_T > 20$ GeV trigger and studied the efficiency of jet finding. The localized energy trigger and selection procedure was found to be about 80 % efficient for jets with $E_T = 30$ GeV and more than 95 % efficient for jets with $E_T > 35$ GeV.

To obtain the fraction of events with various numbers of jets the UA1 group considered all jets with $E_T > 15$ GeV reconstructed with a jet axis in $\eta = \pm 2.5$. This fraction is shown in fig.19. The two-jet topology dominates at a level of 80-85 %. The fraction of one-jet events becomes negligible at high-trigger jet E_T, whereas the fraction of three-jet events rises in the region of low E_T and levels off at about 15 %. The jet-finding algorithm and the fiducial cuts have, of course, a direct influence on the number of jets found. Note that the fractions presented in fig.18 are not corrected for algorithm effects and fiducial cuts. The occasional presence of a third jet indicated by the data suggests a gluon bremsstrahlung mechanism similar to that observed in hadronic events produced by e^+e^- collisions.

The cross section for inclusive jet production was obtained by the UA1 and UA2 groups after corrections for geometrical acceptance, experimental procedure, and detector resolution. In order to partly account for final-state

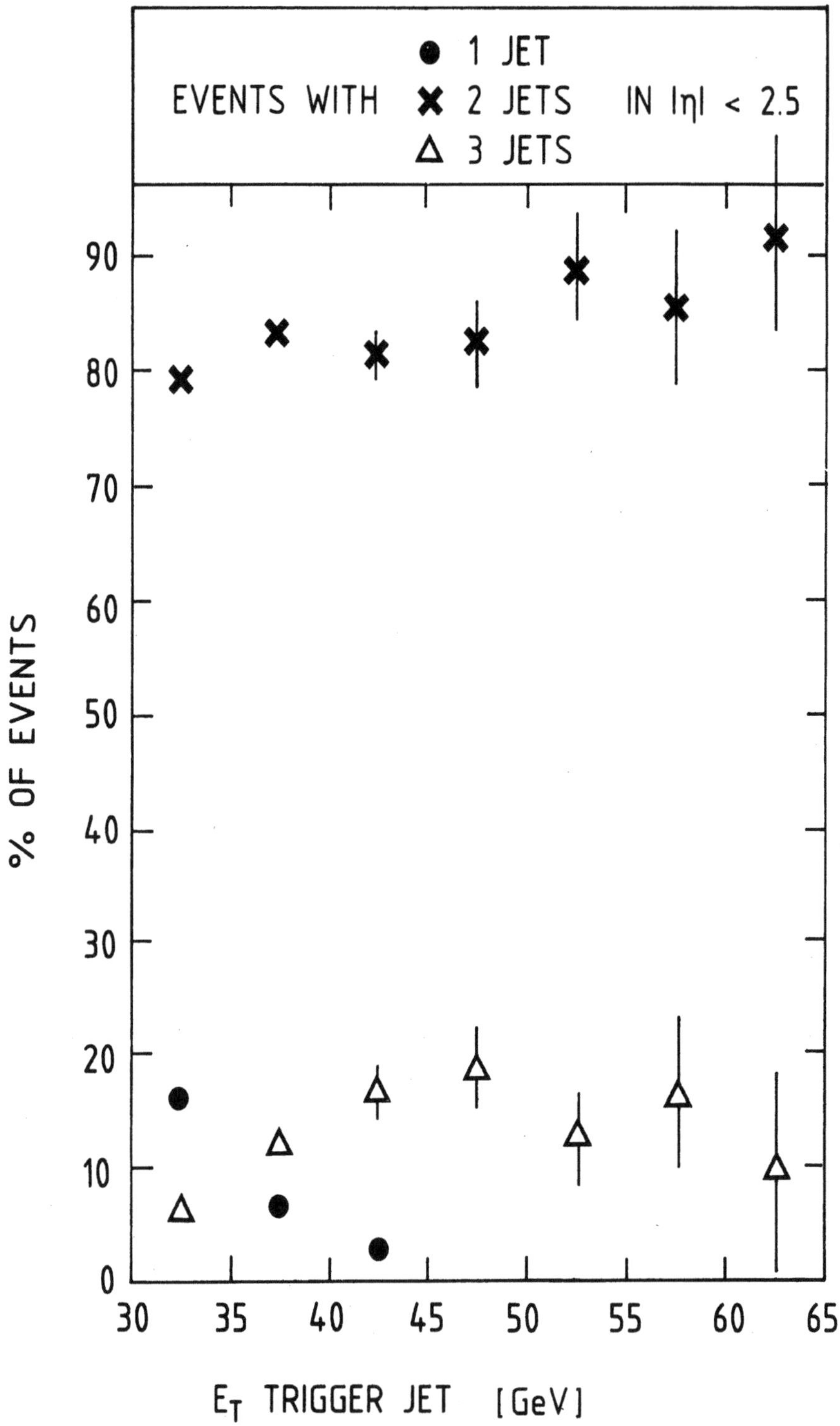

Fig. 19: Fraction of events with 1, 2,and 3 jets of E_T>15 GeV as a function of the trigger jet E_T (UA1).

gluon radiation effects the UA2 group added to the measured momentum vector of the clusters with p_T> 30 GeV/c those of all clusters having E_T> 3 GeV and separated by an angle ω from the selected cluster momentum vector such that $\cos\omega > 0.2$. The inclusion of these nearby clusters (20 % of the jets with p_T > 30 GeV/c include at least one such nearby cluster with E_T > 3 GeV) increases the observed transverse-momentum distribution for jets by about 15 % roughly independent of p_T. The UA1 group took final-state gluon radiation partly into account by tuning the jet-finding algorithm. Accounting for the integrated luminosity, the cross section for inclusive jet production $d^2\sigma/dp_T d\eta$ as presented in fig.20 was obtained. All errors shown are purely statistical. The additional systematic uncertainty in the UA2 data is ±45 %. The UA1 group reported a systematic uncertainty of a factor 1.65 which is mainly due to uncertainties in the jet finding algorithm, fluctuations in the electromagnetic/hadronic composition of the jets, and the uncertainty arising from the definition of the energy scale. This factor is independent of the jet transverse energy. The measured cross section was found to agree well with calculations based on QCD. Such calculations are themselves not more precise than to within a factor of about two [38]. The real success of QCD is the observation of the predicted increase of the yield of jets having p_T > 20 GeV at the CERN $\bar{p}p$ collider with respect to top ISR energy [39] by about four orders of magnitude [40].

3.2 Jet production properties

The jet pair, interpreted as the final state parton pair, is considered in its centre-of-mass system, where the dependence of the cross section on the scattering angle was measured and compared with QCD calculations for the elementary parton-parton subprocesses. The measurement of

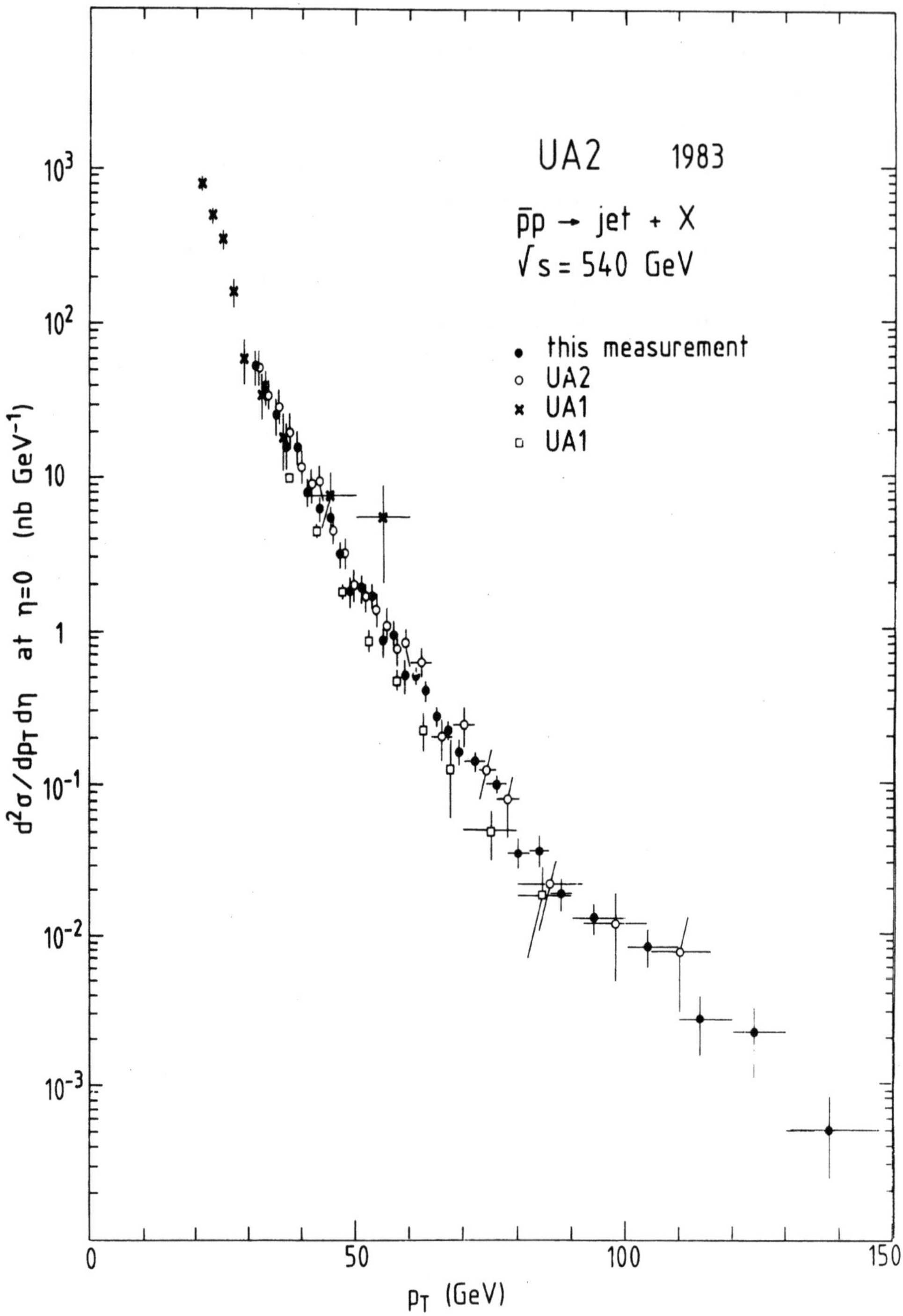

Fig. 20: Inclusive jet-production cross section (UA1) and (UA2).

the momentum of the two-jet system in the $\bar{p}p$ frame provides information on the dynamics of the initial partons. Its transverse component is due to large-angle gluon radiation by the initial-state partons, whereas the longitudinal component is directly related to the structure function of the nucleon. The latter was measured and compared with the one observed in neutrino deep-inelastic scattering at lower energy.

3.2.1 Transverse momentum of the two-jet system

The UA2 group [41] selected in their further analysis (for the initial stages of the analysis see previous section) events with two well-measured jets. The events were required to satisfy the following conditions:

(i) the invariant mass of all the clusters in the central calorimeter with an energy in excess of 3 GeV had to be larger than 45 GeV/c;

(ii) the two clusters with largest transverse energy in the event had to satisfy the conditions $E_T^1 > 10$ GeV, $E_T^2 > 10$ GeV, $E_T^1 + E_T^2 > 40$ GeV. Their momenta were required not to point to an edge cell, and their azimuthal separation had to be larger than 120°.

These criteria yielded a sample of about 3×10^4 events corresponding to an integrated luminosity of 112 nb^{-1}.

The final-state gluon radiation was partially taken into account as explained in the previous section. That is why the initial-state gluon bremsstrahlung and, to a lesser extent, the intrinsic transverse momentum of the partons in the nucleon are expected to contribute to the transverse momentum p_T^{jj} of the two-jet system. The two components p_η^{jj} (along the bisector of the two jets in the transverse plane) and p_ξ^{jj} (orthogonal to it and roughly parallel to the two-jet axis) have been studied separately, because they are associated with different resolution and systematic effects.

Figs. 21a and 21b show the normalized distributions of p_η^{jj} and p_ξ^{jj} for the two-jet system. The systematic errors due to the uncertainty on the energy scale and to the analysis procedure were estimated to be less than ±10 %. The r.m.s. of the two distributions are respectively 9.1 GeV for p_ξ^{jj} and 7.5 GeV for p_η^{jj}. Note, however, that the analysis criteria reject events with a very high value of p^{jj} and that instrumental effects contribute to the measured width of the distributions. The larger width of the p_ξ^{jj} distribution results mostly from its greater sensitivity to the energy resolution. In the case of a 40 GeV jet the resolution was measured to be 5.5 ± 1.0 GeV. The dashed lines in fig.21 show the result of a QCD calculation [42] which assumes a uniform azimuthal distribution of p_T^{jj} in the ξ-η plane. The uncertainty of the calculation results mostly from the error on the gluon structure function (gluons radiate more than quarks on the average) and is of the order of ±20 %. As an example, the dashed-dotted line of fig.20b shows the distribution expected in the case that gluons radiate like quarks. The QCD calculation assumed a fixed parton-parton invariant mass of 60 GeV. The variation of p_T^{jj} as a function of the mass of the two-jet system m_{jj} was calculated to be small, because the broadening of the p_T^{jj} distribution for a given parton type with increasing m_{jj} is compensated by a decrease of the fraction of gluons with a corresponding increase of quarks.

A similar analysis was performed by the UA1 group [43]. Two-jet events with both jet axes in a restricted fiducial region ($|\eta|<1$, $|\text{azimuth}|<45^\circ$ around the horizontal plane) have been selected. The UA1 group further required that there be no additional jet anywhere in the detector. The calorimeter hits in the event sample obtained were then grouped into A = (two-jet system) and B = (rest of event). The transverse motion of A is balanced by B

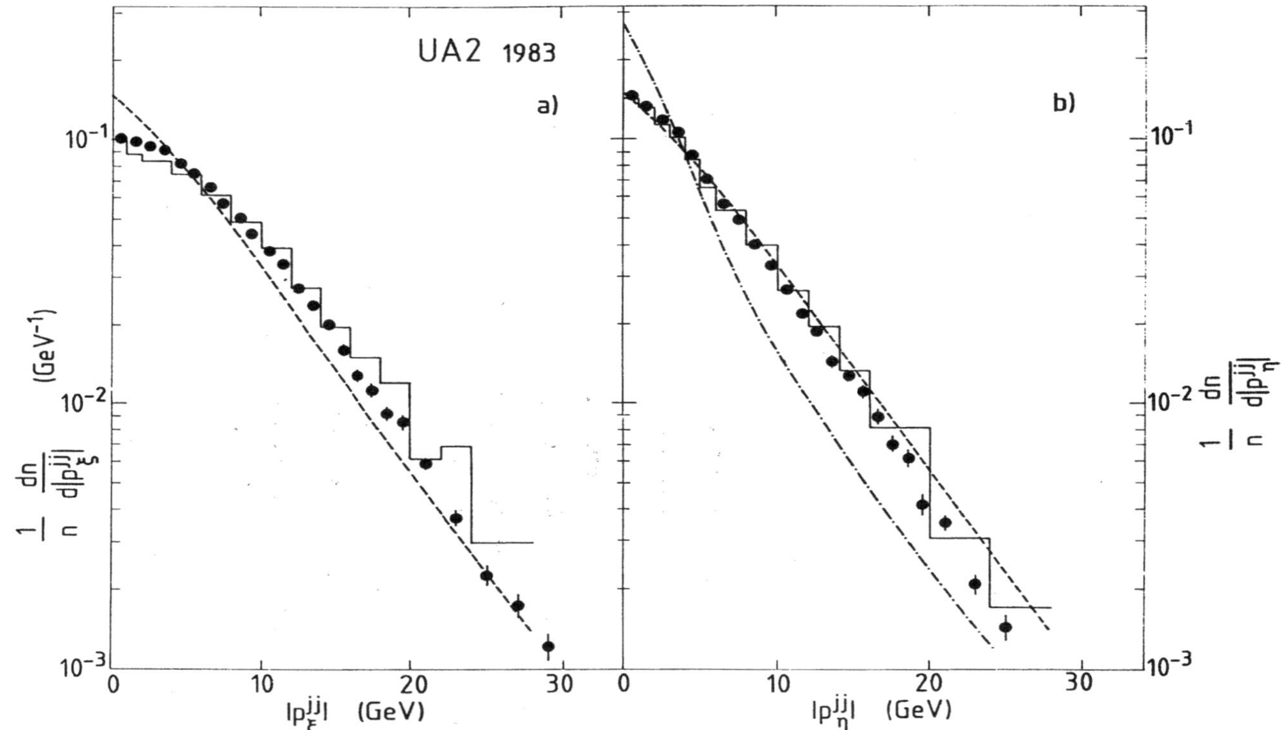

Fig. 21.: (a) Normalized distribution of $|p_\xi^{jj}|$. The dashed line is a QCD calculation, while the histogram represents the calculation after detector smearing.
(b) The same as (a) but for $|p_\xi^{jj}|$. The dotted-dashed line is the result of a QCD calculation assuming the same colour charge for gluons as for quarks.

within resolution. In events with a large p_T^{jj} as measured by A this p_T^{jj} may or may not be balanced by B depending on wether it is real or due to resolution. Thus p_T^{real} and p_T^{resol} can be distinguished. Furthermore, two independent measurements of p_T^{real} are performed, namely with A and B. In the absence of resolution effects the two-jet system and the rest of the event would both directly measure p_T^{real}. The result of the analysis is presented in table 6. The average transverse momentum of the two-jet system was found to be $\approx$ 4 GeV (parallel to the two-jet axis) and $\approx$3 GeV (transverse to the two-jet axis). The average total transverse momentum of the two-jet system is hence $\approx$ 5 GeV for events with E_T(jet1) + E_T(jet2) $\approx$ $\approx$ 70 GeV.

Table 6: Transverse momentum of the two-jet system (UA1)

$\sigma[p_\xi^{jj}]$	measured by rest of event: 3.9 GeV (resolution: σ = 2.3 GeV)
	measured by two-jet system: 4.1 GeV (resolution: σ = 7.1 GeV)
$\sigma[p_\eta^{jj}]$	measured by rest of event: 2.9 GeV (resolution: σ = 2.3 GeV)
	measured by two-jet system: 3.3 GeV (resolution: σ = 5.1 GeV)

3.2.2 Production-angle distribution and effective structure functions

After the initial procedure described in section 3.1 the UA1 group selected the events with ≥ 2 jets in the acceptance of the central calorimetry $|\eta| < 3$ [41]. While the majority of these events has a topology consistent with two balanced high-E_T jets, some 10-15 % of the events have additional jets with $E_T > 15$ GeV (see fig.19). For this analysis additional jets were ignored. A correction was applied to the energy ($\approx$ + 12 %) and momentum ($\approx$ + 7 %) of the two highest-E_T jets in order to account for final-state radiation falling outside the jet as defined by the jet-finding algorithm. In the forward direction ($|\eta| > 1.5$) it was found that the jet axis as defined by the jet-finding algorithm was systematically shifted towards the beam axis due to the misassignment of beam fragments as part of the jet. A correction was made to the jet angles to account for this effect. After all corrections the jet energy resolution is $\delta E/E \approx \pm 26$ % and the uncertainnty in jet direction $\delta\eta \approx \pm 0.05$. In the centre-of-mass of the two-jet system the differential cross-section for two-jet production may be written as

$$d^3\sigma/dx_1 dx_2 d\cos\theta = [F(x_1)/x_1][F(x_2)/x_2]\ d\sigma/d\cos\theta,$$

where $F(x_1)/x_1$ ($F(x_2)/x_2$) is a structure function representing the density of partons in the antiproton (proton) as a function of the scaled longitudinal momentum x_1 (x_2) of the partons; $d\sigma/d\cos\theta$ represents the differential cross section for a particular parton-parton subprocess as a function of the c.m.s. scattering angle and has been calculated to leading-order QCD [45]. The elastic subprocesses [gg, $gq(\bar{q})$, and $\bar{q}q$] all have a similar angular dependence [$\sim(1-\cos\theta)^{-2}$] as a consequence of vector-gluon exchange. Thus the ansatz given for the two-jet differential cross section is justified. Taking into account the relative strengths of qg and gg couplings the effective structure function F(x) becomes $F(x) = G(x) + 4/9[Q(x)+\bar{Q}(x)]$, where G and Q are the gluon and quark structure functions of the proton.

The finite angular acceptance of the apparatus and the trigger E_T-threshold requirement discriminate against events with small scattering angles and restrict the range of cosθ over which the trigger is fully efficient. Events which were close to the limits of the acceptance in cosθ were rejected by applying a fiducial cut in cosθ. This fiducial cut ($\cos\theta_{max}$) was defined for each event as the maximal value of cosθ for which both the final two jets would fall in the acceptance region $|\eta| < 2.5$ (with at least one having $|\eta| < 1.0$) and for which the mean of the transverse energies of the two-jets would exceed 35 GeV. Events with an azimuthal angle of the two-jet axis within $\pm 45^{\circ}$ of the vertial were also rejected. The final event sample consists of 2432 events. Fig.22 shows the angular distribution obtained using all events in the final event sample and accounting for the varying $\cos\theta_{max}$ values on the assumption that the angular dependence is independent of x_1 and x_2 when integrating over x_1 and x_2. In fig.22 the data are compared with theoretical forms of parton-parton elastic scattering in QCD [45] and in an Abelian scalar-gluon theory [46]. The data clearly favour the vector-gluon hypothesis, but cannot yet distinguish subproceeses, and will still not be able to do so in the near future.

The complementary approach, namely integrating over dcosθ to get F(x), leads to the distribution presented in fig.23. Factorization in x_1 and x_2 was assumed when performing this integration. The errors shown are statistical. The systematic uncertainty in the over-all normalization is

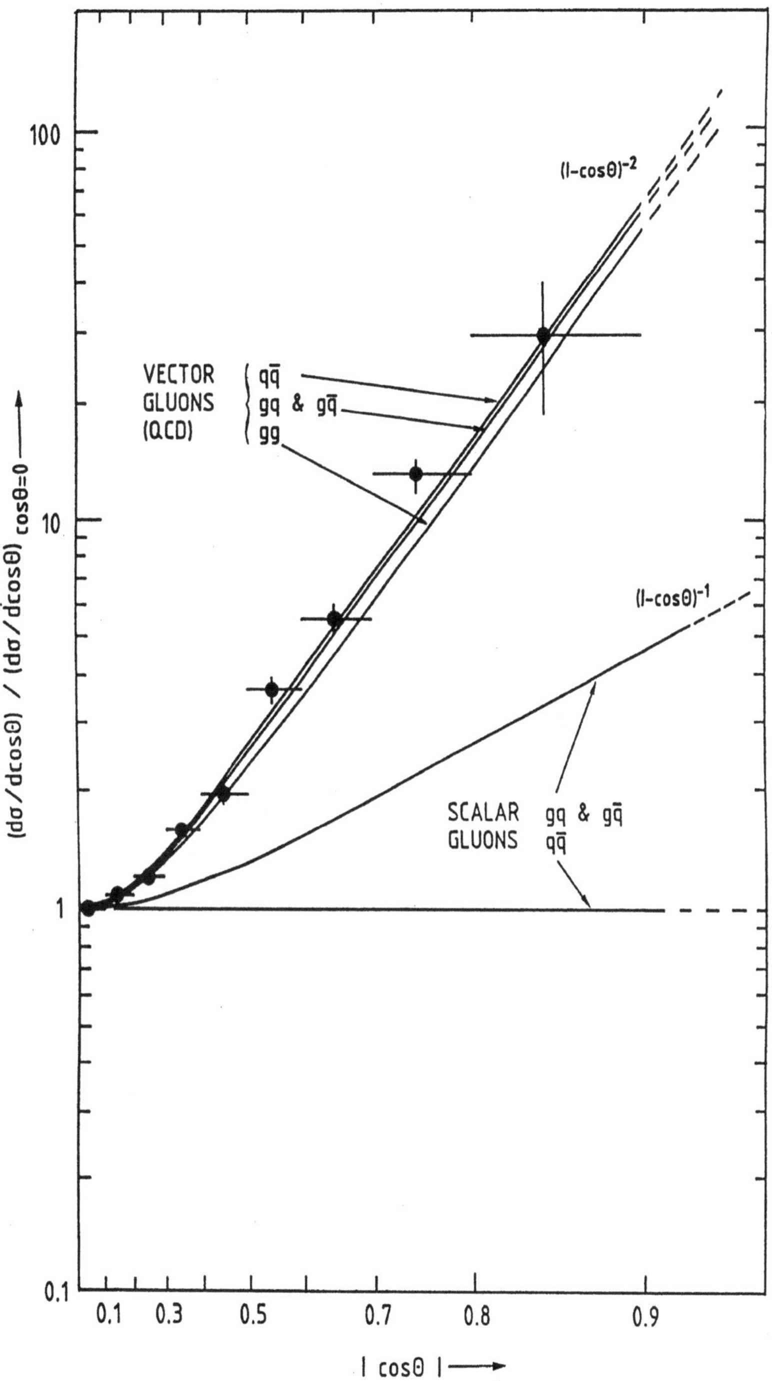

Fig. 22.: The measured angular distribution (UA1). The data are compared with theoretical forms of parton-parton elastic scattering in QCD and in an Abelian scalar gluon theory.

$\approx \pm 30$ %. Also shown in the figure is F(x) at $Q^2 = 20$ GeV^2 based on a QCD parametrization of CERN-Dortmund-Heidelberg-Saclay (CDHS) measurements [47] at low Q using a neutrino beam on an iron target and the expected effects of QCD scaling violation at the value of Q^2 appropriate for this experiment. The rapid increase in the structure function at small x ($x < 0.25$) is due to the very large flux of gluons.

A similar analysis was performed by the UA2 group [41]. Because of the smaller solid-angle coverage of the UA2 calorimeters, the cos distribution was only measured up to $\cos\theta = 0.6$. In the region of overlap the UA2 result agrees with the result of the UA1 group within the statistical and systematic errors. The UA1 and UA2 results on the effective structure function F(x) are compared in fig.24. The two results agree fairly well.

Both groups UA1 and UA2 have finally tested and verified the independence of the angular distribution of x_1 and x_2 and also the factorization of the convoluted structure function $S(x_1,x_2)$ into $S(x_1,x_2) = F(x_1).F(x_2)$.

3.3 Jet fragmentation

QCD calculations using as input lower energy structure functions predict that the jets observed at the collider are a mixture of quark and gluon jets [40]. Therefore by comparing the fragmentation of these jets with quark jets from e^+e^- annihilation one can study differences between quark and gluon jets. According to QCD inspired models [48,49] quarks and gluons are expected to fragment differently because of the larger colour charge of the gluon.

Both groups UA1 and UA2 have presented data on jet fragmentation [50]. In the earlier UA1 data only the most drastic manifestation of the QCD nature of jet production

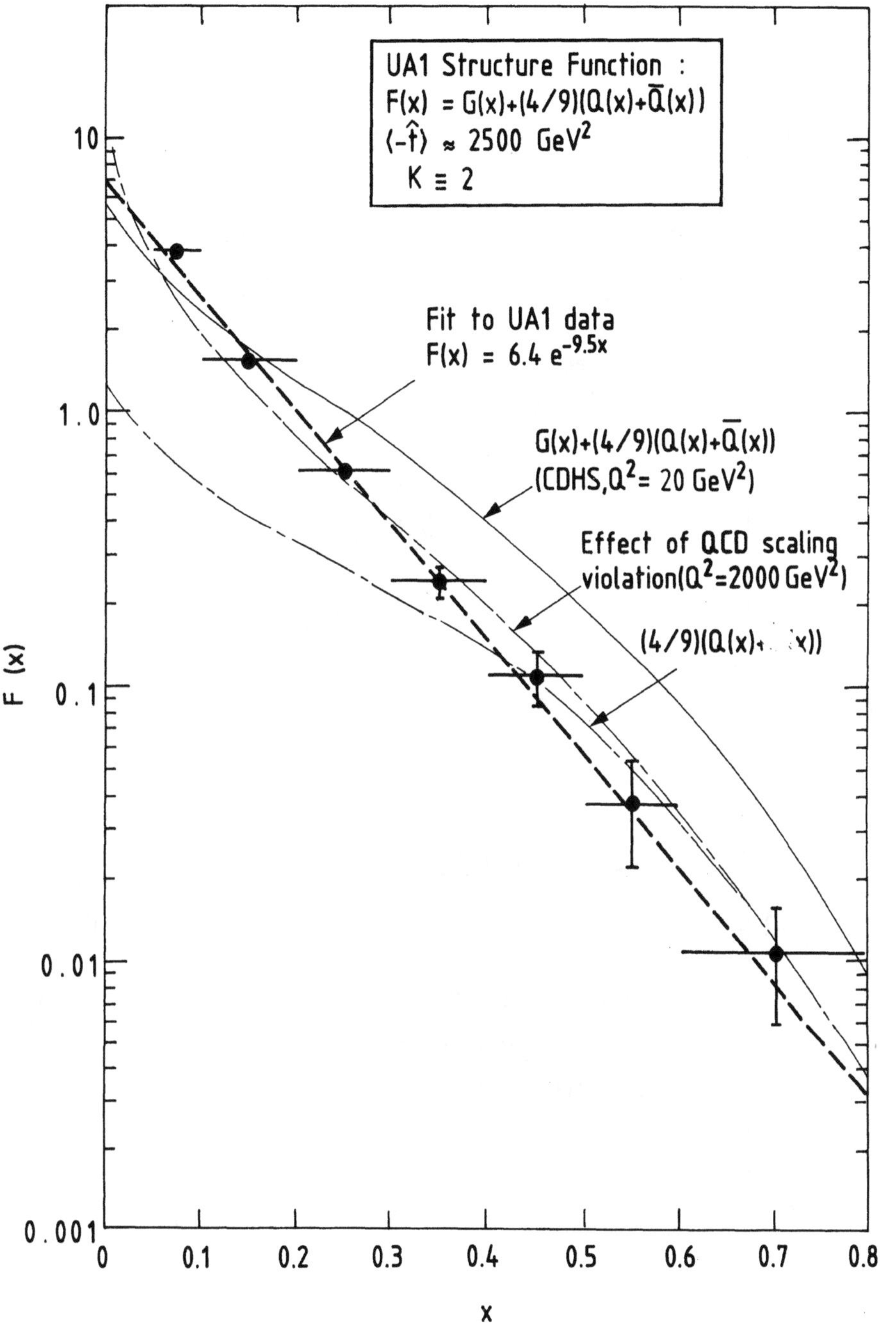

Fig. 23: Effective structure function F(x) (UA1).

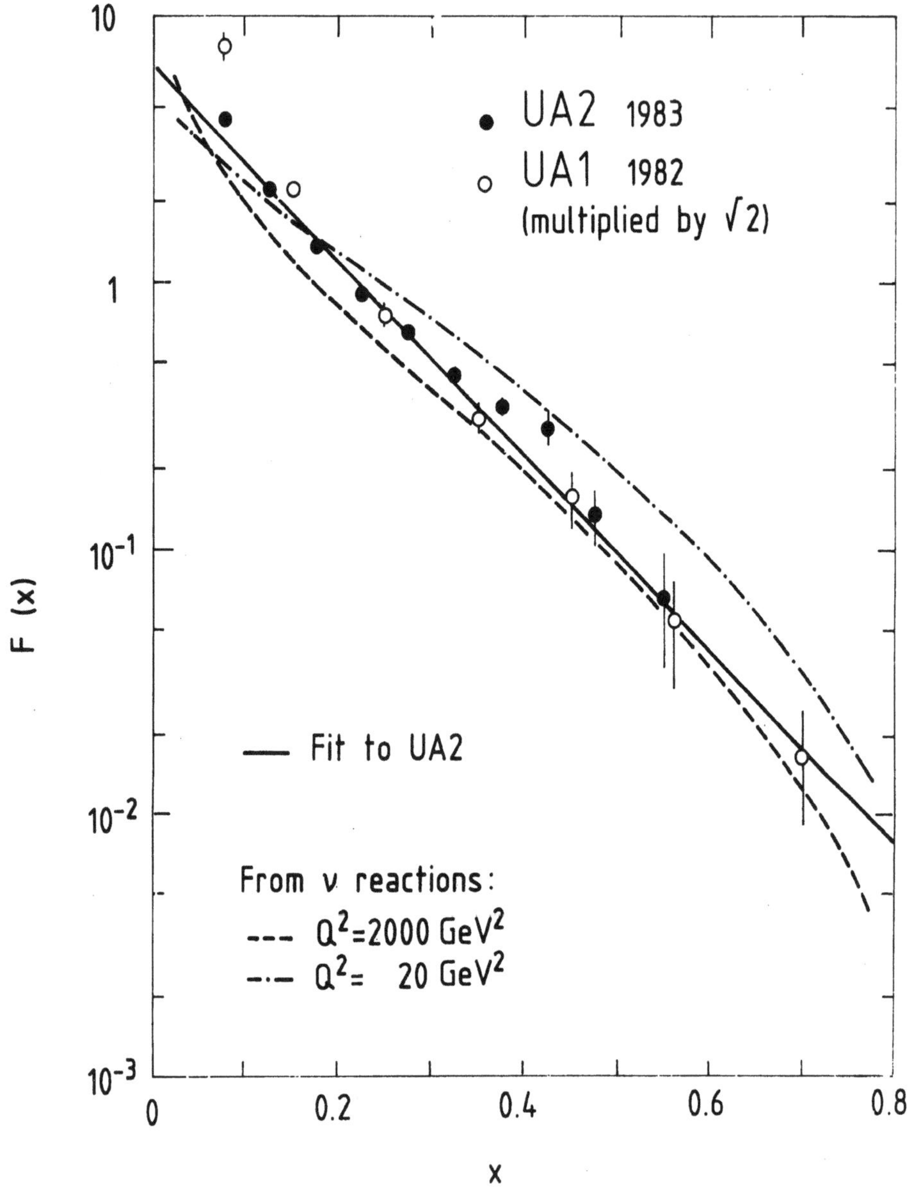

Fig. 24: Effective structure function F(x). The full line represents an exponential fit to the UA2 data. The dashed lines are computed from neutrino deep inelastic scattering taking into account scaling violation effects.

was observed, namely initial and final-state radiation of relatively hard gluons. The fragmentation of the mixture of quark and gluon jets into charged particles as measured by the UA1 group did not show a clear difference to the fragmentation of quark jets from e^+e^- annihilation. Recently the UA2 group claims to have observed an increased charged-particle multiplicity in collider jets exceeding extrapolations from lower energy e^+e^- jets; the UA1 group has reported the observation of a large yield of charm from gluon jets.

The experimental procedure followed by the UA1 group was presented in the previous sections. In the further analysis additional fiducial cuts on the jet axis had to be applied (UA1) to assure good calorimetric measurement and momentum determination for the charged particles of a given jet. The inclusive variable used for the study was chosen to be $z = p_L^{ch}$ (jet axis)/E(jet), where p_L^{ch} (jet axis) is the momentum of the charged particles projected on the jet axis). The energy of the jet E(jet) was defined as the sum of all calorimeter hits belonging to the jet. The fragmentation function is defined as $D(z) = (1/N_{jets}) \cdot (dN_{ch}/dz)$. The integral of D(z) is normalized to the mean charged multiplicity of jets.

Several corrections had to be applied to D(z). The shape of the background was obtained from a control region with particles far away from the jet axis and the number of particles uncorrelated with the jet was estimated with a Monte Carlo calculation. The loss of soft particles, because of limited acceptance, varies from 35 % for z between 0.02 and 0.03 to 5 % for z around 0.07. Charged particles with a large z value (therefore a large momentum) have a large momentum uncertainty and D(z) has been unsmeared for values above z = 0.7. The resulting distribution is shown in fig.25.

The shape and the normalization of D(z) was then com-

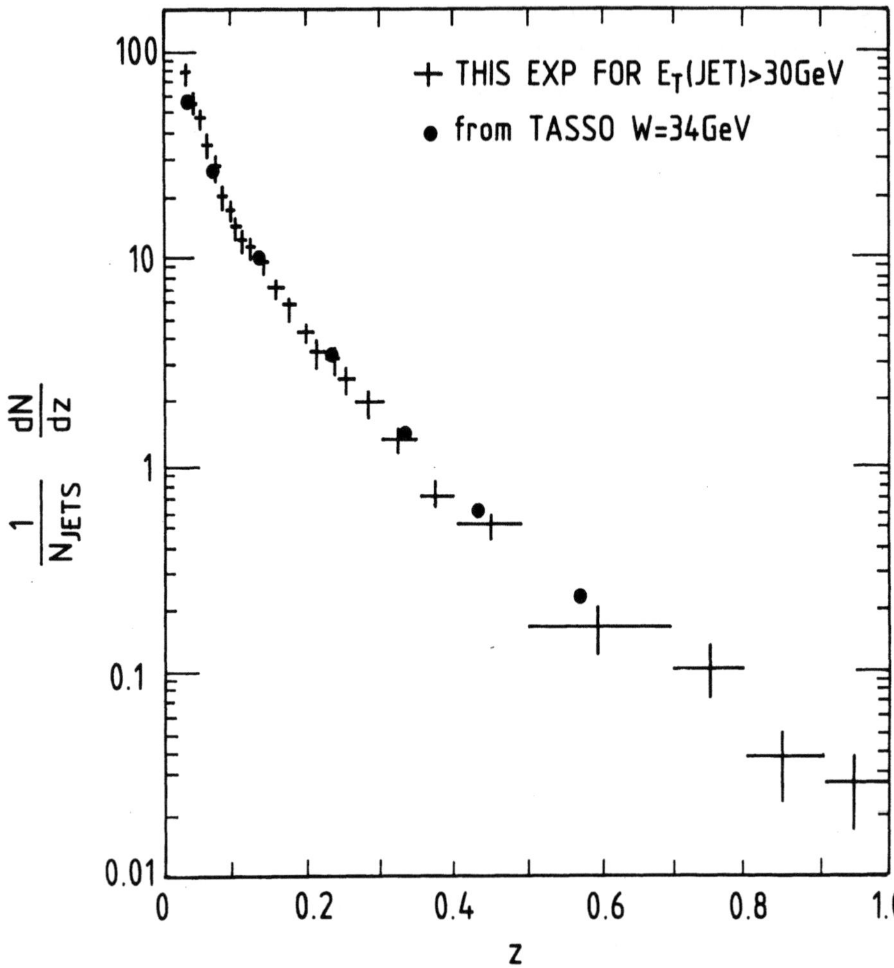

Fig. 25: Charged particle fragmentation function for E_T(jet)> >30 GeV (UA1) compared with similar results from the TASSO detector.

pared with $(1/\sigma_{tot})(d\sigma/dx_L)$ obtained by the TASSO collaboration for jet energies of 17 GeV, where $x_L = p_L^{ch}/p_{beam}$[51]. The energies of the jets are of course different. The comparison was nevertheless considered to be meaningful because scaling violation in e^+e^- annihilation are known to be small [52]. No striking difference can be observed between the two sets of data shown in fig.25. Quark dominated and gluon dominated fragmentation functions do not seem to be different from each other in the E(jet) and z ranges studied in this analysis.

Within their own data the UA1 group found D(z) to be approximately independent of E_T(jet). The mean charged multiplicity was found to vary from 6.7 ± 0.7 at E_T(jet) $\approx$ 24.3 GeV to 7.5 ± 0.7 at E_T(jet) $\approx$ 33.5 GeV and to 7.7 ± 0.8 at E_T(jet) $\approx$ 47.4 GeV.

To compute the transverse momentum with respect to the jet axis the axis given by the calorimeter was replaced by the charged jet axis, whose direction is given by the sum of the charged-particle momenta. The invariant p_T spectrum is shown in fig.26 together with the result of a fit $(1/p_T)(dN/dp_T) = A/(p_T + p_{TO})^n$ for all jets with $E_T > 30$ GeV. The p_T spectrum is well fitted by the values $p_{TO} = 4$ GeV, $n = 14.8$. The observed mean p_T value internal to the jet is $\langle p_T \rangle = 600$ MeV/c after having applied a cut $z > 0.1$ on all particles. The observed large p_T tail may be either an indication of gluon bremsstrahlung, or due to an experimental misalignment of the jet axis, or to events whose leading particles are neutrals.

The experimental procedure followed by the UA2 group was presented in the previous sections. To select a clean sample of two-jet events that were well contained in the central calorimeter the following cuts were applied:

(i) the invariant mass of the two highest transverse energy clusters, m_{jj}, exceeds 40 GeV;

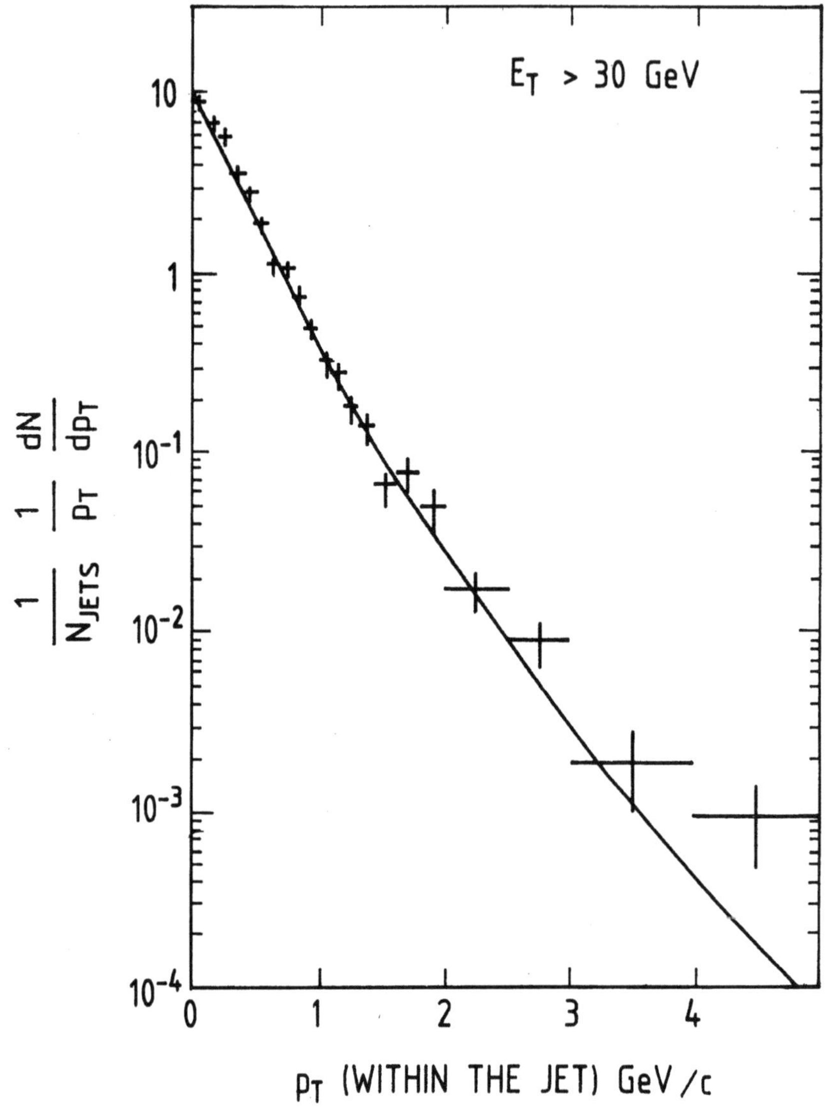

Fig. 26: p_T (within the jet) [GeV/c] (UA1).

(ii) the azimuthal separation between the two highest transverse energy clusters, $\Delta\phi^{jj}$, exceeds 150;

(iii) the transverse momentum of the two-jet system, p_T^{jj}, does not exceed 25 GeV.

These cuts retained a sample of about 3×10^4 events.

The further analysis was restricted to the transverse plane where the track reconstruction efficiency is highest and where particles not associated with a jet are expected to contribute a uniform azimuthal distribution.

There is no unique way to define the charged particle multiplicity of jets produced in $\bar{p}p$ collisions because the relative fractions of the particles coming from the jet and coming from the "underlying event" are unknown. As a lower limit on the true jet multiplicity one can define the "jet core" multiplicity as the number of charged particles in a cone of $\pm\pi/4$ above a flat level corresponding to the value at $\Delta\phi = \pi/2$. After correcting for detector inefficencies the mean jet core multiplicities $\langle n_{ch}^{jet} \rangle$ are shown as a function of m_{jj} in fig.27. The error bars shown include a 5 % statistical uncertainty in the estimate of the two-track resolution loss. The systematic error was estimated to be $\approx$ 15 %, an over-all error caused predominantly by the uncertainty in the two track resolution correction. The point-to-point systematic errors are probably much smaller.

Using the measured angular distribution of charged particles around the jet axis from e^+e^- data [53] the equivalent jet core multiplicity has been evaluated for e^+e^- data. The results from the TASSO experiment are also shown in fig.27. Neither the data from the UA2 experiment nor the TASSO data were corrected for strange particle decays. The result suggests that jets from $\bar{p}p$ collisions in the mass range $40 < m_{jj} < 80$ GeV have higher mean multiplicities than one would expect from extrapolations of

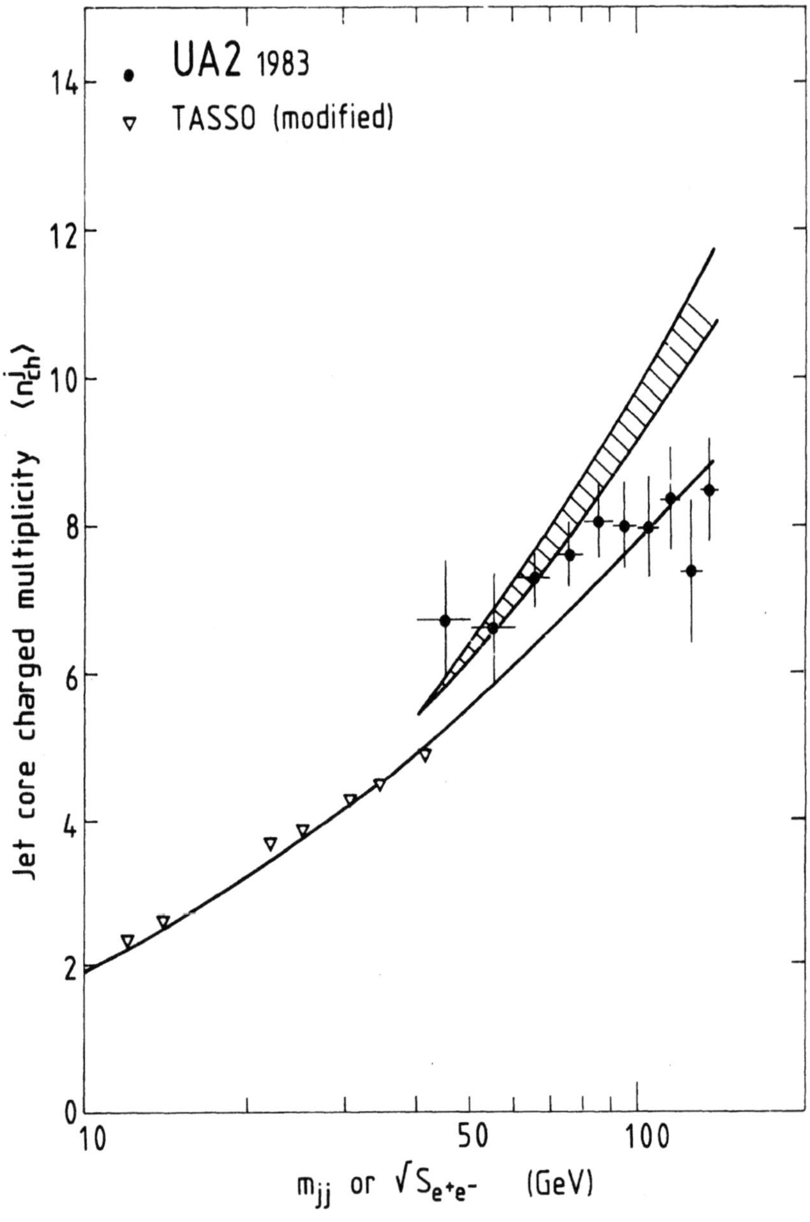

Fig. 27.: Mean charged multiplicities as a function of $\sqrt{s_{e^+e^-}}$ for e^+e^- data and of the invariant two-jet mass m_{jj} for $\bar{p}p$ data (UA2).

e^+e^- data. The solid curve in fig.27 is a calculation of the model of ref.49 for quark jets and the dashed area for gluon jets.

In the same spirit the transverse energy flow around the jet axis as measured by the UA2 group indicates the presence of a mixture of quark and gluon jets as described by ref.49.

A larger than expected fraction of $D^{*\pm}$'s in the fragments of (presumably) gluon jets has recently been reported by the UA1 group. The claim is that $N(D^{*\pm})/N(\text{jet}) = 0.65 \pm \pm 0.19 \pm 0.33$. The reported fraction of charm is large, but the measurement errors are large as well, and more data will be needed to establish a firm conclusion.

4. INTERMEDIATE VECTOR BOSONS

Experimental searches for the hypothesized [54] intermediate vector bosons (IVBs) in accelerator neutrino experiments since about 1960 have all been uniformely unsuccessful. There was, nevertheless, strong believe that forthcoming experiments would soon uncover these particles. Only few people were imaginative enough to doubt that IVBs would be found when the $\bar{p}p$ collider which is capable of producing them would run at a reasonable luminosity and when the adequately instrumented detectors UA1 and UA2 would start to operate. Most people were even more specific and expected that the IVBs would be found just where the standard model of Glashow, Weinberg and Salam [55] told us they should be, namely at $m_W \approx 82$ GeV and $m_T \approx 94$ GeV [56]. The reason for this confidence was the astonishing success of the standard model of electromagnetic and weak interactions. This model predicted the existence of a type of weak interaction, neutral currents, that had previously not been observed. The interest in the model spurred experimental searches that uncovered neutral current phenomena. Experimental and theoretical reviews of neutral currents are given e.g. in ref.2 . A theorists review of the situation of the IVBs before their discovery is given in ref.57. Experimentalists reviews of the discovery of the IVBs [58] are given e.g. in ref.59.

Measurable cross sections for production of IVBs by the $\bar{p}p$ collider were calculated [60] based on QCD and the branching fractions for the background-free leptonic decay modes $W \to \ell\nu$ and $Z^0 \to \ell^+\ell^-$ were expected to be about 8 % and 3 %, respectively [57]. Sizable leptonic branching fractions were already expected on more general grounds [61] without reference to the standard model. The experiments UA1 and UA2 were designed avoiding any compromise with respect to lepton detection capability. Since unexpected backgrounds to the leptonic decay modes did not show up in the data, i.e. nature was kind to us experimentalists, it was possible to discover the IVBs with only a few events. The collected events are surprisingly clean; a modest p_T cut of 1 or 2 GeV removes most of the soft tracks from spectator hadrons and reveals nice and clean IVB-event signatures; the cleanliness of events with IVBs is illustrated in fig.28.

An IVB is produced, if a quark and an antiquark collide at a centre-of-mass energy $\sqrt{s} \simeq m_W$ or m_Z. The process is sketched in fig.1 and the order of magnitude of the production cross section can be read off from fig.2 as explained in the introduction. The leptonic decays of the IVBs can easily be identified. A large-E_T electron (muon) together with a large missing E_T from an associated neutrino signals the process $W \to e\nu$ ($W \to \mu\nu$). Two large-E_T electrons (muons) of opposite charge and fixed invariant mass are produced by the decays $Z^0 \to e^+e^-$ ($Z^0 \to \mu^+\mu^-$). Hence the signature for $Z^0 \to \ell^+\ell^-$ is even cleaner than for

$W \to \ell\nu$; the process is, however, less frequent by approximately a factor 10 if induced by antiproton-proton collisions at $\sqrt{s}$ = 540 GeV.

In this chapter we first describe the experimental procedure. We then discuss the measured properties of the IVBs which confirm the expectations from the standard model - apart from the possible observation of exotic decay modes $Z^o \to \ell^+\ell^-\gamma$. We finally discuss the features which characterize the production of IVBs by hadrons and the jet activity in events which contain an IVB. The 1983 data yielded a couple of events where an IVB is associated with increased jet activity (UA1) or a hard jet (UA2). These data are confronted with the results of 1984.

4.1 Experimental procedure

The aim of this section is to describe how the events with IVBs (of which the event shown in fig.28 is an example) are filtered, without major losses, from a sample of many soft events and large-p_T events. Approximately 1 event in 10^9 events (1 event in 10^8 events) contains a Z^o ($W^{\pm}$) which decays leptonically. A first rejection factor of $\sim$10000 is provided for by the trigger and another factor $\sim$10 by on-line filtering. The remaining rejection factor and the understanding of the losses at the various steps of data reduction has to be accomplished by the off-line analysis. The backgrounds have to be understood in parallel to the data reduction and the detector has to be tuned for uniform response and optimal resolution.

4.1.1 $Z^o \to e^+e^-$

The UA1 apparatus was briefly described in the previous sections. A summary of the performance figures relevant to

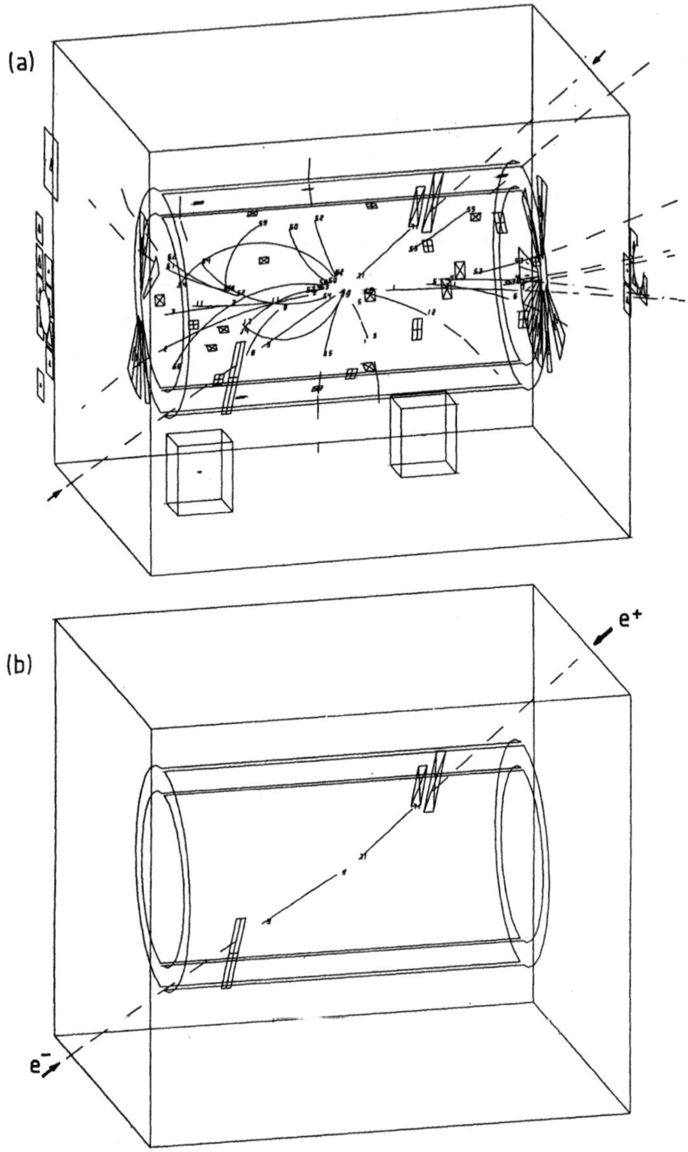

Fig. 28: An event from the $Z^o \to e^+e^-$ sample as seen in the UA1 detector: a) all tracks and calorimeter hits are displayed; b) only tracks and calorimeter hits with $p_T > 2$ GeV are shown. All spectator hadrons disappear; a clean IBV event is obtained. In the case of $W \to e\nu$ one of the two electrons is replaced by missing transverse energy.

IVB physics is given in table 7. The sample of $Z^o \rightarrow e^+e^-$ events was obtained by the following procedure. The trigger required the presence of an electromagnetic cluster of E_T > 10 GeV at angles greater than 5° from the beam axis. The data were recorded in the collider run periods at the end of 1982 and from April to June 1983 and correspond to an integrated luminosity of 136 nb^{-1}. After on-line filtering and off-line reconstruction of the recorded events about 1.5 x 10^5 events had at least one electromagnetic cluster with E_T > 15 GeV. The further selection proceeded as show in fig.29. An electron signature was defined as a localized energy deposition in two contiguous cells of the electromagnetic calorimeters with E_T > 25 GeV and a samll (or no) energy deposition (< 600 MeV) in the hadronic calorimeter cells behind the cluster. The invariant-mass distribution for two such clusters is shown in fig.29a. By requiring that at least one electromagnetic cluster be matched with a track with p_T > 7 GeV/c the distribution shown in fig.29b is obtained and fig.29c by asking that the track be isolated. Four events are left over by asking that both electromagnetic clusters be matched with isolated tracks of p_T > 7 GeV. These events (fig.29d) constitute the UA1 (1983) sample of $Z^o \rightarrow e^+e^-$ candidates.

No same charge (ee) pairs and no (eμ) pairs have been found at high invariant masses indicating that the background is probably small. Several possible background sources were nevertheless considered in detail. Ordinary large-p_T jets, both simulating electrons, were found to be neglegible (< 10^{-3} events). The possible contribution to the background from heavy flavour jets with subsequent decays into leptons was estimated from events with a large-p_T lepton accompanied by a recoil jet (< 10^{-3} events). The Drell-Yan continuum contributes a background of <10^{-2} events for masses greater than 80 GeV. Production of W W pairs was estimated to be neglegible.

Table 7: Some performance figures of the UA1 detector relevant to the intermediate vector-boson search.

What is measured?	Resolution	Comments
Momenta of charged particles	$\Delta p/p = 0.005 \times p$	For a track of 1 m length transverse to magnetic field
Energies of electrons, photons, and π^0's	$\Delta E/E = 0.15/\sqrt{E}$	±4 % (counter-to-counter variation) ±3 % (scale uncertainty)
Energies of hadrons	$\Delta E/E = 0.80/\sqrt{E}$	±5 % (scale uncertainty)
Missing E_T by global calorimetry	$\Delta[\Sigma\vec{E}_T] = 0.7\sqrt{\Sigma E_T}$	= indirect measurement of E_T of the neutrino

How are electrons identified?	By the e.m. shower profile, by E (measured by the calorimeter) $\approx$ p (measured by magnetic deflection), and by requiring that the track extrapolates to the shower centroid
How are muons identified?	By minimum ionization in calorimeters, by requiring matching of the track segments before and after the iron, and by p (measured by track curvature in the central detector) $\approx$ p (measured by displacement of track in the magnetized iron)

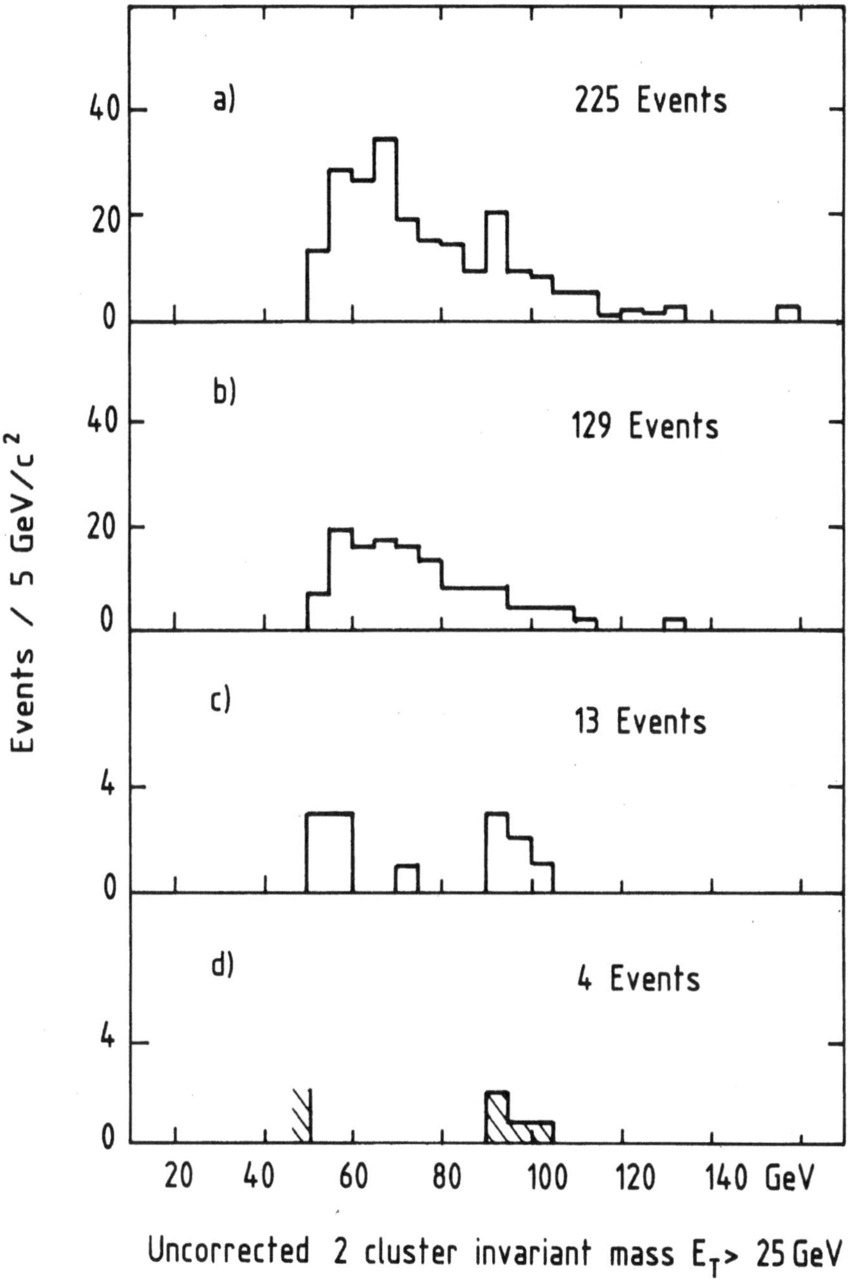

Fig. 29: Selection of the $Z^0 \to e^+e^-$ sample (UA1), see text.

A similar procedure was followed by the UA2 group. Their full data sample consists of approximately 7 x 10^5 triggers, which correspond to an integrated luminosity L = 131 nb^{-1}. An initial selection was made by rejecting all events which were identified as due to sources other than $\bar{p}p$ collisions (< 10 % of the entire data sample, mainly beam-gas background and cosmic rays). In the surviving events a search was made for configurations consistent with the presence of electrons among the collision products. An electron was defined from the observation of:

(i) a track measured in the wire chambers;

(ii) a large signal detected in the pre-shower counter (C5 in the central detector or the MTPCs in the two forward regions, see fig.15);

(iii) an energy deposition in the calorimeters with small lateral size and limited penetration into the hadronic compartments (i.e. e.m. shower profile);

and from the quality of matching in space among these three properties.

The further selection to isolate the $Z^o \rightarrow e^+e^-$ sample from the UA2 data proceeded as follows. The total electromagnetic transverse energy was required to exceed 30 GeV and a pair of energy clusters had to have an invariant mass in excess of 50 GeV/c. The energy clusters in the central calorimeter were obtained by joining all electromagnetic cells which share a common side and contain at least 0.5 GeV. A contribution of cells having at least one side in common with a cluster cell was also added. The forward calorimeter clusters consist of at most two adjacent cells having the same azimuth (here the cell is far from the interaction point and much larger than the lateral extension of an electromagnetic shower and the dead region between cells at different azimuths is too large to allow clustering across it). The energy deposited in the corresponding hadronic compartments was added to the cluster energy. The remaining data sample contains 7427 events. These events were then fully reconstructed.

At this stage the event sample is dominated by two-jet events. However, while the reconstructed cluster energy measures correctly the energy of jets produced in the central region, it is in general a gross underestimate of that of forward jets, for which the calorimeter thickness is only 88 % of an absorption length. As a consequence the sample contains many more events having both clusters in the central calorimeter than events with at least one cluster in the forward regions, because the jet momentum distribution falls off steeply with increasing jet transverse momentum (see fig.20). In order to select event with similar characteristics in the central and forward regions and to enhance the electron signal the UA2 group required that both clusters have a small lateral size in the electromagnetic compartment of the calorimeter and a limited energy leakage in the hadronic compartment.

For clusters in the central calorimeter cluster sizes R_θ, R_ϕ were calculated from the cluster centroid and the values of the angles θ and ϕ at the cell centres weighted by their energy depositions. The conditions R_θ, $R_\phi < 0.5$ cell sizes were required. In the forward calorimeters the sum of the energies deposited in the cells adjacent to the cluster cell was required not to exceed 3 GeV.

The condition that the showers have only a small energy leakage in the hadronic compartments of the calorimeters was applied by requiring that the ratio $H = E_{had}/E$ does not exceed a value H_o, equal to 0.02 for the forward calorimeters, and $0.023 + 0.034 \times \ln E$, where E is in GeV, for the central one.

The cuts applied at this stage are very loose and are satisfied by more than 95 % of isolated electrons between 10 and 80 GeV as verified experimentally using test beam data. They reduced the event sample to 24 events, whose invariant mass is shown in fig.30a. There are 12 events with both clusters in the central region, 8 events with one cluster in the central and the other in the forward regions, and 4 events with both clusters in the forward regions. The sample with both clusters in the central region was reduced by a factor ∿430 by the cuts on cluster size and hadronic leakage.

In the following step a series of additional criteria was defined for electron identification. The UA2 group used their measurements of the response of various parts of the detector to isolated electrons from which approximate cut efficiencies η were evaluated. The cuts are described in table 8 for both the central and forward region. The efficiencies resulting from the simultaneous application of all selection criteria are at least as large as the products of the individual efficiencies η, namely 67 % for the central region and 83 % for the forward regions. The rejection power of these cuts was estimated by applying them to the original event sample. This procedure also provided an estimate of the background contribution from two-jet events.

Fig.30b also shows the mass distribution for the events of fig.30a with at least one cluster satisfying all of the electron identification criteria. There are eight events in this plot which cluster around a mass value of ∿90 GeV/c.

An upper limit on the background contribution to the eight events in fig.30b was inferred from fig.30a under the assumption that background events have the same mass distribution in both samples. By comparing the event population above and below 80 GeV/c an upper limit (90 % CL) of 0.32 background events to the signal in fig.30b was found. A more stringent background estimate was obtained by starting from a much richer sample by simply releasing cut 1

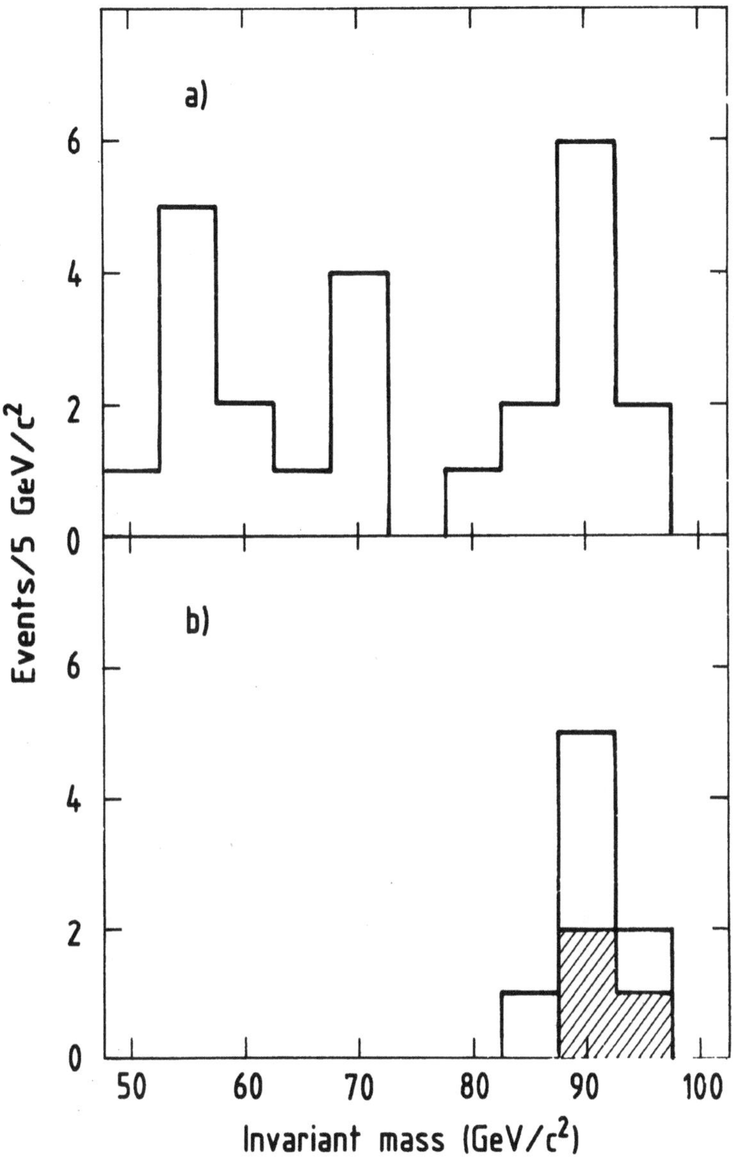

Fig. 30: Invariant-mass distribution (a) of the 24 pairs which pass cut 1 of table 8, (b) of the 8 of these 24 pairs for which all cuts of table 8 are satisfied by at least one electron. The three events in which both electrons pass all cuts of table 8 are cross-hatched.

Table 8.

Requirements	Central region: Description	η(%)	Forward regions: Description	η(%)
1. Presence of a calorimeter cluster (preselection)	Electromagnetic cluster size $R_{\theta,\phi} < 0.5$ Hadronic leakage: $H < .023 + .034 \ln E_{cl}$	98	Cluster size ≤ 2 cells Energy in adjacent cells $E_{adj} < 3$ GeV Hadronic leakage $H < 0.02$	98
2. Presence of a track	Reconstructed in the vertex detector in both transverse and longitudinal projections	85	Reconstructed a) in the vertex detector in transverse projection with $N \geq 1$ signal in the two inner chambers b) in the forward drift chambers. Azimuth difference between a) and b) $\Delta\phi(a,b)$ less than 40 mrad	92
3. Track/cluster match	Track impact to cluster centroid distance $\Delta < 1$ cell size. Compare the energy distribution observed in the 3x3 cells centered on the impact cell to that expected for an electron incident along the track. Light sharing between phototubes of impact cell and hadronic leakage are taken into account. Require that the probability P that the energy distribution for an electron is farther from the mean than that observed is larger than 0.01.	95	Shower position calculated from light sharing between impact cell phototubes must be consistent with the track impact point to within $\Delta x \leq 70$ mm	98
4. Presence of a preshower counter signal	Measured in C5 as a coincidence in space of charge clusters measured on the anode wires and the inner and outer cathode strips. Its charge q_5 must exceed 4 mip	a) >90	Measured in MTPC in both coordinate planes. Its charge Q(MTPC) must exceed 6 mip	95

Table 8 continued

Requirements	Central region		Forward regions	
	Description	(%)	Description	(%)
5. Track/preshower position match	Within $d \leq 7$ mm measured on the C5 surface	94	Within $\Delta x, \Delta y < 50$ mm in each direction	100
6. Momentum measurement			Momentum p measured in spectrometer and energy E measured in calorimeter must satisfy $\lvert p^{-1} - E^{-1} \rvert / \sigma(p^{-1} - E^{-1}) < 3$	99

[a] This does not include a 5 % inefficiency due to a $\Delta\phi = 18^{\circ}$ azimuthal region in which C5 was not operational because of electrical breakdown.

(hadronic leakage and cluster size) on one of the two clusters and evaluating the rejection power of cuts 2-6 (applied to reduce the sample of fig.30a to that of fig.30b) on the other cluster. In this way the UA2 group estimated a background contribution of 0.03 events to the signal of fig.30b. When applying the electron identification criteria to both clusters, only three events (shown as cross-hatched in fig.30b) survive with an estimated background of 2×10^{-4} events above a mass of 80 GeV/c. In the other five events one of the two electron candidates fails at least one of the strict selection criteria described in table 8.

A detailed evaluation of the eight events shown in fig.30b yielded the following conclusions. Five events are either identified as (three events), or compatible with (two events), e^+e^- or $e^+e^-\gamma$ configurations. For the other three events the UA2 group retains the interpretation that the cluster (which fails cut 3 of table 8 in the central calorimeter in all three cases) is in each case associated with an electron, despite the fact that its configuration is inconsistent with their best knowledge of the detailed response of the calorimeter to high-energy electrons. In one case the measured ratio of the signals from the two light guides of the impact cell is 0.71 ± 0.02 instead of 0.88 ± 0.02 as predicted from the track impact; in the other two cases additional energy ($\sim$3 GeV) is observed in neighbour cells, inconsistent with lateral and longitudinal leakages of a shower initiated by an electron of the measured energy.

4.1.2 $Z^o \rightarrow \mu^+\mu^-$

Isolated muons traverse the calorimeters and the added iron absorbers of the UA1 detector without deviations beyond those of multiple scattering. Muons furthermore do not deposite energy in excess of ionization loss in the four electromagnetic and two hadronic calorimeter segments. In order to detect the penetrating muon tracks 50 large drift chambers surround the whole detector covering an area of $\sim$500 m^2. Each chamber consists of two orthogonal layers of drift tubes with two planes per projection. The average spatial resolution is 300 μm. A second set of four drift chamber planes is placed at a distance of 60 cm from the first one. The angular resolution of the muon chamber system is a few milliradians, comparable to the multiple scattering angle of high-energy muons, about 3 mrad at 40 GeV/c.

An independent momentum determination of muons transversing the magnetized iron yoke is performed by the UA1 group using the known position of the interaction vertex and the track coordinates after the iron. In order to check this momentum determination the detector was exposed to vertical cosmic muons in the momentum range 10-50 GeV/c. Excellent agreement was found between momentum measurements in the central detector and in the magnetized iron. The precision of the momentum measurement in the iron was found to be $\Delta p/p \gtrsim 0.020$ as expected from multiple scattering. Cosmic-ray muons were also used to verify the relative alignements of the central and muon detectors.

The muon trigger consists of a dedicated set of hardware processors. The processors use the signals from the muon chambers and provide a quick decision as to wether or not an event contains a muon candidate pointing to the interaction region after having traversed 8 nuclear interaction lengths. During the 1983 data taking period a total of approximately 2.5×10^6 events have been recorded for an integrated luminosity of $L = 108\ nb^{-1}$. Of these about 1.0×10^6 events are muon triggers. Those have been passed through a fast filter program. The fast filter selected muon candidates with $p_T > 3$ GeV/c of $p > 6$ GeV/c ($\sim 7.2 \times 10^4$ events). The event sample with muon candidates satisfying p_T >5 GeV/c consists of 17326 events which were all fully reconstructed and then passed through a selection program in order to apply stricter central detector (CD) track quality criteria and (CD - muon chamber) matching requirements than were used at the filter stage. This yielded a reduced sample of 6781 events.

A dimuon selection was then made requiring two p_T > > 5 GeV/c muon tracks in the event both of them satisfying the criteria used in the inclusive muon selection. This gave 26 candidate dimuon events which were scanned on a high resolution graphics display facility. Of these events 7 were identified as cosmic rays and 6 as leakage of hadronic showers through cracks in the calorimeters leaving 13 candidates after scanning.

In addition to the dimuon selection all events with a dimuon trigger and at least one reconstructed CD track with p_T > 7 GeV/c were examined. Two extra dimuon events have been found in this way. In one event both CD tracks are very short and failed the track length requirements of the filter. In the other event one muon passed very close to the edge of the muon chamber and was seen in only one projection. Hence the final dimuon sample (both muons with p_T > 5 GeV/c) contains 15 events. Their invariant-mass distribution is shown in fig.31. The five events A-E with dimuon masses exceeding 70 GeV/c are interpreted as Z^o decays. The momentum errors of four of these events have been improved by relying on overall transverse momentum balance. For $Z^o \rightarrow \mu^+\mu^-$ events, where no neutrino is emitted, the transverse momentum of the recoil hadron debris must be equal to the transverse momentum of the $(\mu^+\mu^-)$ system. The muon momenta and the transverse energy flow have been adjusted in a fit to obtain an over-all balanced event. For event C, the topology of which is unusal, the assumption about missing energy was relaxed and the muon momenta and momentum errors have been determined without the balance con-

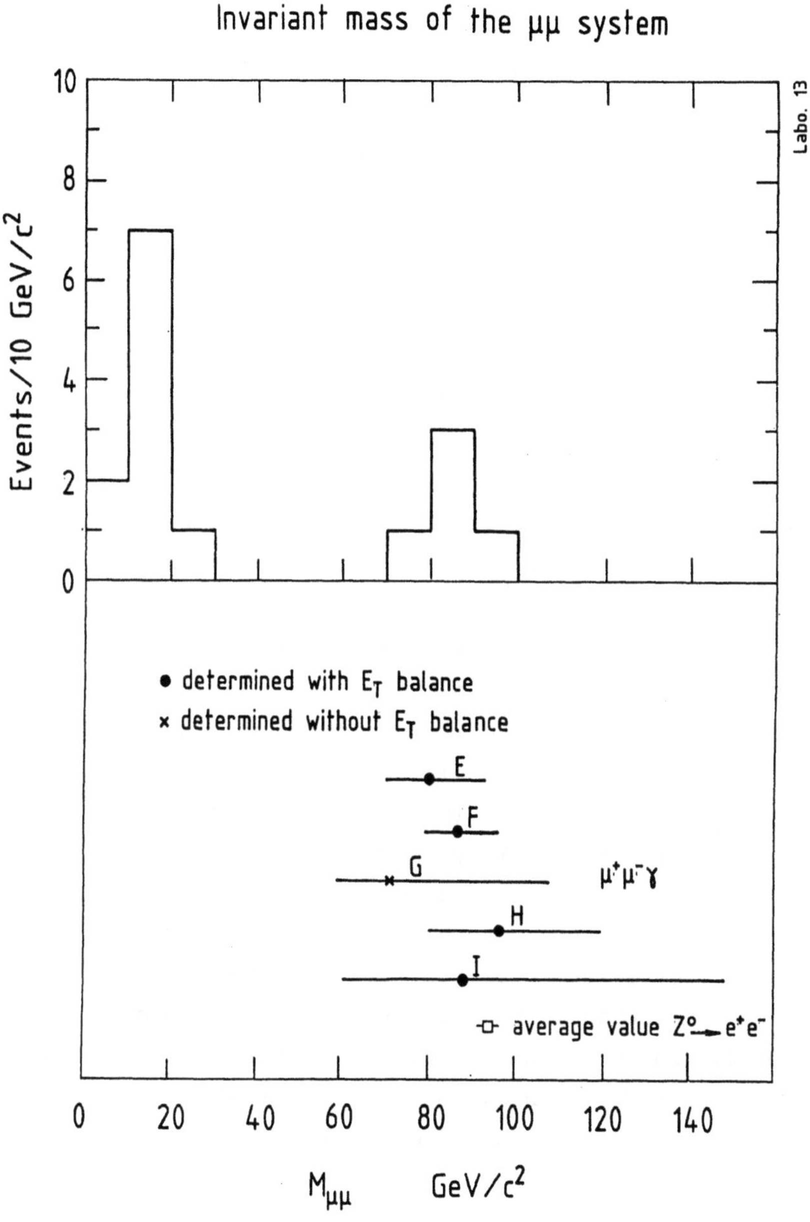

Fig. 31: Invariant-mass distribution of the $\mu\mu$ system for events where both muons have transverse momenta larger than 5 GeV/c. The bottom part of the figure shows the mass values with their errors for the 5 Z^o canditates (UA1).

straint.

The UA1 group considered a number of possible backgrounds to the dimuon sample. High-p_T charged hadrons can fake muons either by penetrating the calorimeters and additional iron shielding without interaction, or by leakage of the hadronic shower. These backgrounds have been studied in a test beam and were found to be $\leq 10^{-4}$ events per incident hadron after requiring matching between CD and muon chamber tracks. The leakage induced background is then neglegible. The dominant background process in the UA1 muon set-up is pion and kaon decay. The probability for a pion (kaon) to decay before entering the calorimeters is $\sim 0.02/p_T$ ($\sim 0.11/p_T$), where p_T is in GeV/c. A background muon of large p_T can be generated by the decay of a medium-energy kaon $K^{\pm} \rightarrow \mu^{\pm}\nu$ in the volume of the central detector, if the transverse momentum kick from the decay balances the deflection of the particle in the magnetic field. This background has been evaluated from events with a single large-p_T muon candidate by calculating the probability for decay of the other tracks (pions or kaons) in each event, folded with the probability to fake a large-p_T muon, and was found to be $< 10^{-3}$ events for $m_{\mu\mu} > 60$ GeV. The background contribution due to heavy flavour jets with semi-leptonic decays was estimated from events with a large-p_T lepton accompanied by a recoil jet ($< 10^{-3}$ events). The Drell-Yan continuum yields a background of 0.1 events for $m_{\mu\mu} > 60$ GeV. The background due to production of W^+W^- pairs was estimated to be neglegible.

4.1.3 $W^{\pm} \rightarrow e^{\pm}\nu$

Trigger, data taking and the initial steps of data reduction were the same as described when discussing the $Z^0 \rightarrow e^+e^-$ sample. The analysis presented by the UA1 group started from about 1.5×10^5 events obtained after on-line filtering and off-line reconstruction by requiring the presence of at least one electromagnetic cluster with $E_T > 15$ GeV. By requiring the presence of a matching, isolated track with $p_T > 7$ GeV/c in the central detector, the data sample was reduced by a factor of $\sim$100. Next, a maximum energy leakage of 0.6 GeV was allowed in the hadronic calorimeter cells behind the electromagnetic cluster leading to a sample of 346 events. The UA1 group then classified the events to wether there was a prominent jet activity. In 291 events there was a clearly visible jet in an azimuthal cone $|\Delta\phi| < 30^{\circ}$ opposite to the electron candidate. These events are probably contaminated by two-jet events in which one jet faked the electron signature and have been rejected. A later analysis has indeed shown that by further tightening the electron quality criteria the event sample with a jet opposite to the electron candidate is reduced by a factor of about 10. The stricter electron quality criteria are:

(i) matching between the momentum measurement and the calorimeter energy deposition within three standard deviations;

(ii) matching between the track direction measured in the central detector and the direction measured by pulse division in each of the four segments of the electromagnetic calorimeter (χ^2_G);

(iii) an overall quality parameter χ^2_R measuring the electromagnetic shape of the longitudinal shower profile and the pulse sharing between the different calorimeter segments.

In the complementary sample of 55 events there is either no jet or a jet not back-to-back with the electron candidate. This event sample is characterized by the presence of neutrino emission signalled by a large missing E_T as shown in fig.32. A detailed discussion of the missing-E_T technique is given in chapter 6. The curve in

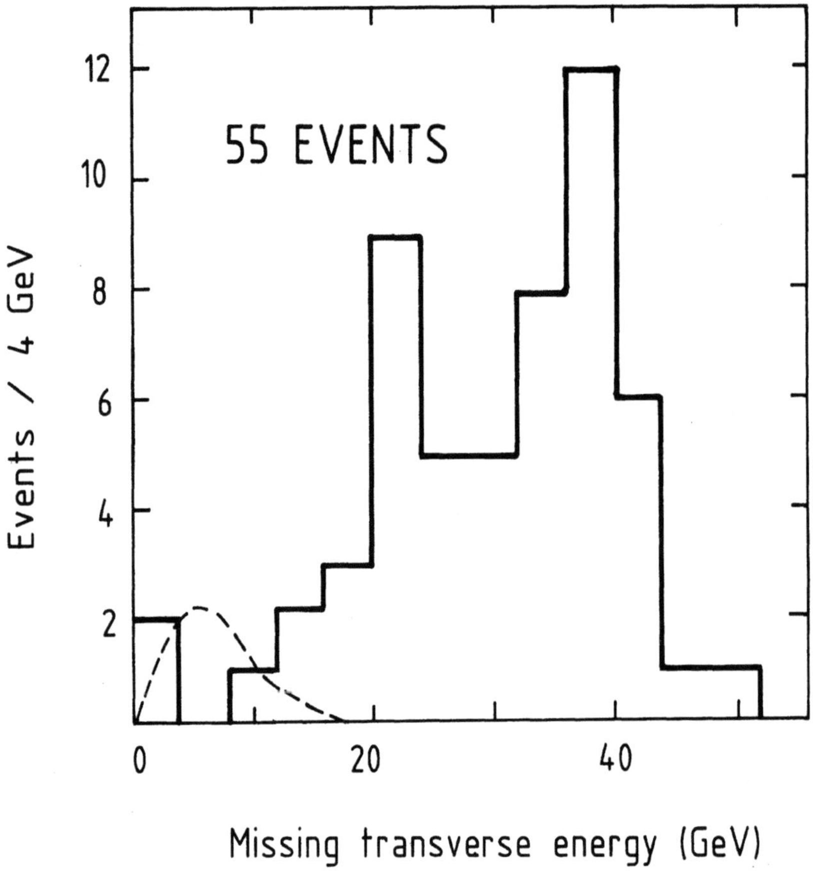

Fig. 32: The distribution of the missing transverse energy for events in which there is a single, isolated electron with E_T > 15 GeV and no coplanar jet activity (UA1).

fig.32 represents the average resolution function for no missing E_T normalized to the three lowest missing-E_T events. These have been excluded by the cut $E_T^{miss} > 15$ GeV. The electron candidates in the remaining 52 events are very clean in the calorimeters. Their longitudinal shower profiles are shown in fig.33. The leakage in the hadronic calorimeter is well below the cut of 600 MeV. the partition of energy in the four segments in the electromagnetic calorimeter follows the distribution measured for electrons in a test beam. These electron candidates are also characterized by a perfect match between the point of electron incidence as measured by the central detector track and the shower centroid in the calorimeter.Most of these electrons survive the strict electron criteria. The final analysis of the 1982/1983 data yielded a total of 68 events of the topology $W^{\pm} \rightarrow e^{\pm}\nu$. The additional events were found in the selection of events with large missing E_T. 49 events have both a well measured track in the central detector and a well measured electromagnetic cluster. The distribution of quantities in which the strict cuts are applied is shown in fig.34).

In the further analysis of the abovementioned sample of 52 events of the type $W \rightarrow e\nu$ the UA1 group retained only those 43 events in which the electron track hits the electromagnetic calorimeters more than $\pm 15^{\circ}$ away from their top and bottom edges motivated by the deterioration of the calorimeter precision when approaching the edges. These 43 events are characterized by a striking back-to-back configuration in the transverse plane of the electron and the neutrino (see fig.35) and a strong correlation between the electron transverse energy and the neutrino transverse component along the electron direction (fig.36) as expected from the two-body decay of the massive and slow $W^{\pm}$.

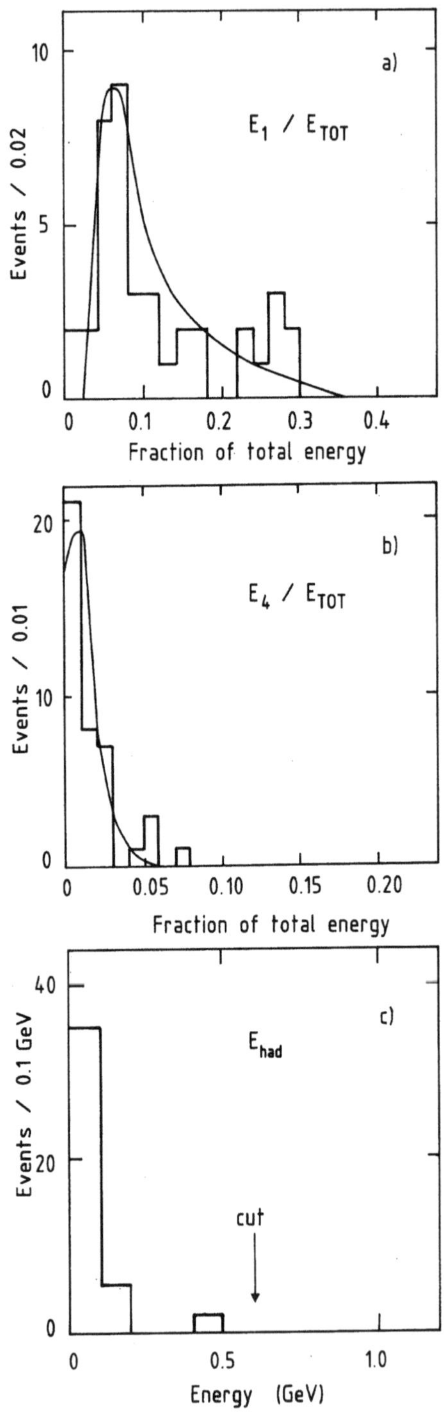

Fig. 33: Distributions showing the quality of the electron.

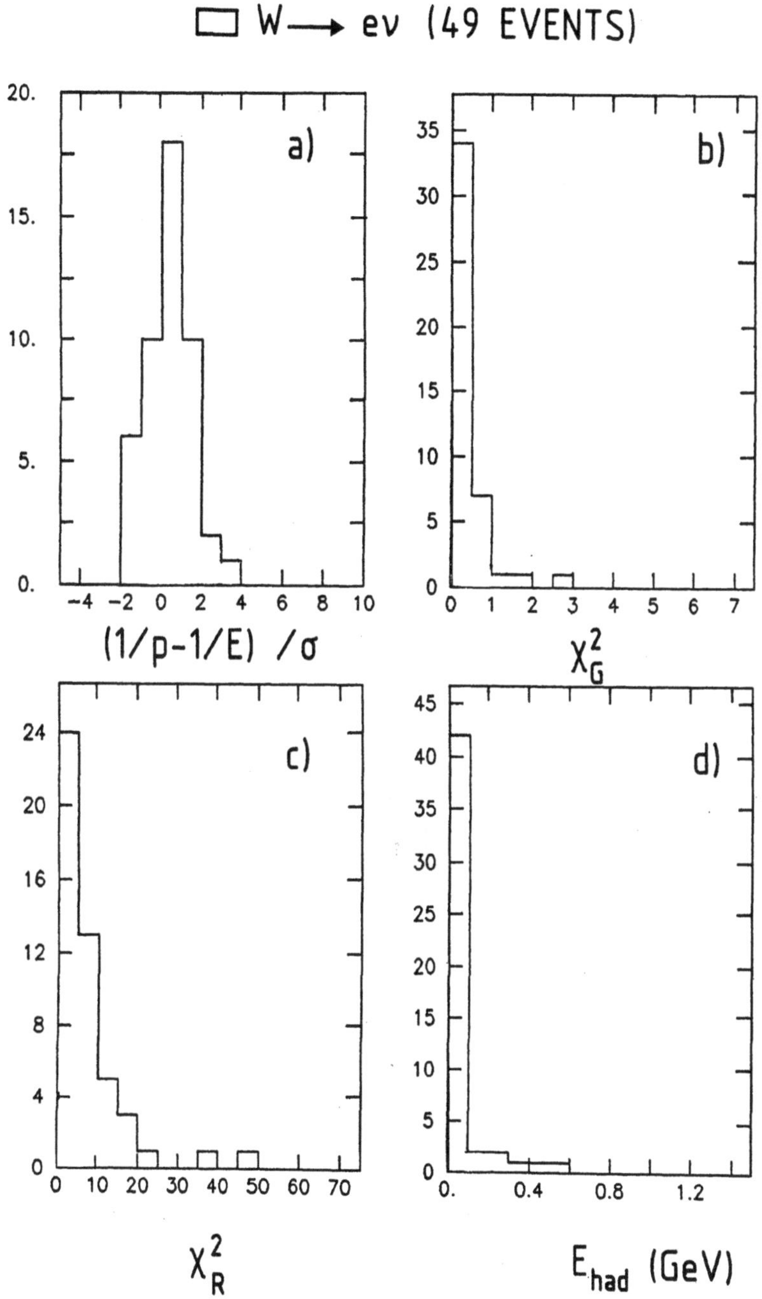

Fig. 34: Electron quality for the final-event sample (1982/1983) of electrons from $W^{\pm}\to e^{\pm}\nu$ decays. The plotted quantities are defined in the text (UA1).

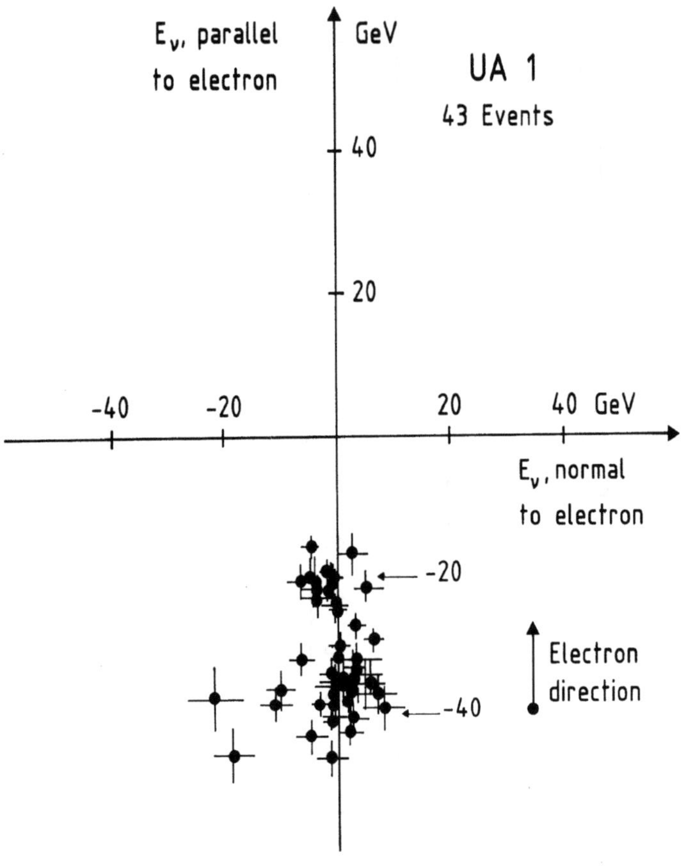

Fig. 35: Two-dimensional plot of the transverse components of the missing energy (neutrino). The events have been rotated to bring the electron direction pointing along the vertical axis. The back-to-back configuration of the electron-neutrino system is apparent (UA1).

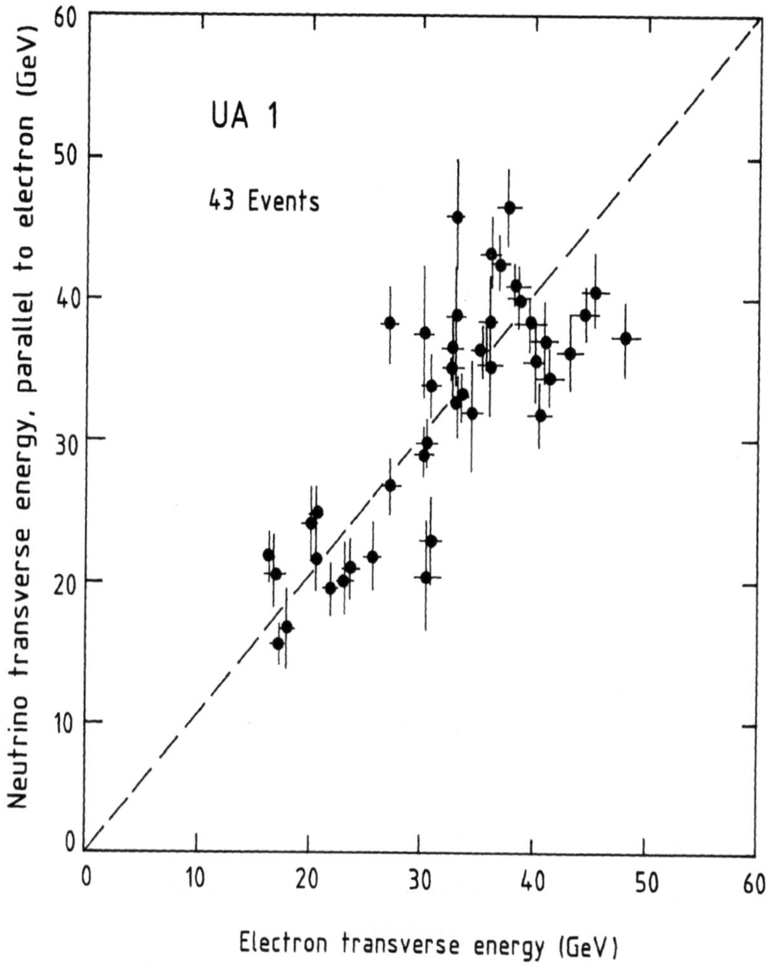

Fig. 36.: Correlation between the electron and neutrino transverse energies (UA1).

The background from misidentifying a large-p_T hadron as an electron has been studied with the help of a sample of isolated hadrons and was found to be small (<0.5 events). The background from large-p_T π^o, η^o, or γ, internally or externally converted to an e^+e^- pair with one leg missed, was estimated from a measurement of isolated electromagnetic conversions (< 0.5 events). There may be, however, a few background events from other decay modes of the W, namely

$$W^\pm \to \tau^\pm \nu_\tau \qquad (< 0.5 \text{ events})$$
$$\hookrightarrow \pi^\pm (\pi^o) \nu_\tau$$

or

$$W^\pm \to \tau^\pm \nu_\tau \qquad (< 2 \text{ events}).$$
$$\hookrightarrow e \nu_e \nu_\tau$$

These events are expected to contribute only at the low-p_T part of the spectrum and they can be eliminated in a more restricted sample.

The selection and analysis of the $W \to e\nu$ sample presented by the UA2 group is similar to their experimental procedure described when discussing their $Z^o \to e^+e^-$ sample. From about 8×10^4 triggers they obtained a sample of 225 events with electrons having $p_T > 15$ GeV/c. This sample still contains in addition to genuine electrons fake electrons from misidentified large-p_T hadrons or jets of hadrons. When the electron candidate is not genuine, but results from a misidentified large-p_T hadron or jet of hadrons, another jet of large-p_T hadrons is expected to be present in the same event at an approximately opposite azimuth. In order to examine the topology of the electron-candidate sample a search for large-p_T jets was performed. In the UA2 forward calorimeters, however, the calorimeter thickness is equivalent to only 1 absorption length and hadronic showers are not fully contained. In that case the UA2 group used the calorimeter to measure the energy of the electromagnetic showers and the magnetic spectrometers to measure charged particle momenta. The analysis of the 225 events yielded 45 events with no additional large-p_T cluster, the electron being the only large-p_T particle observed in the detector. In addition, 24 events have been found with no (or only small) jet activity in an azimuthal sector $\pm 60^{\circ}$ wide opposite to the electron direction. The p_T distribution of the electron candidates in the combined sample of 69 events is shown in fig.37a. The Jacobian peak expected for $W \to e\nu$ decay appears at $p_T \approx 40$ GeV/c and is well separated from the background. For $p_T > 25$ GeV the distribution contains 37 events with an estimated background of 1.5 ± 0.1 events. Fig.37b was obtained by applying strict electron identification criteria to the 30 electron candidates observed in the central calorimeter. A total of 26 events passed these cuts (defined in table 8), as expected from the ratio of the cut efficiencies. Further evidence that these events consist of genuine electrons is given in fig.38a-g which show the distributions of some of the variables used to define the electron identification criteria (table 8). All of these distributions are consistent with an electron behaviour.

4.1.4 $W^{\pm} \to \mu^{\pm}\nu$

Trigger, data taking and the initial steps of data reduction were the same as described when discussing the $Z^{o} \to \mu^{+}\mu^{-}$ sample. The 17326 events from the fast filter which contain a muon candidate with $p_T > 5$ GeV/c yielded 713 events containing a muon candidate with $p_T > 15$ GeV/c or $p > 30$ GeV. These events were passed through an automatic selection program which eliminated most of the

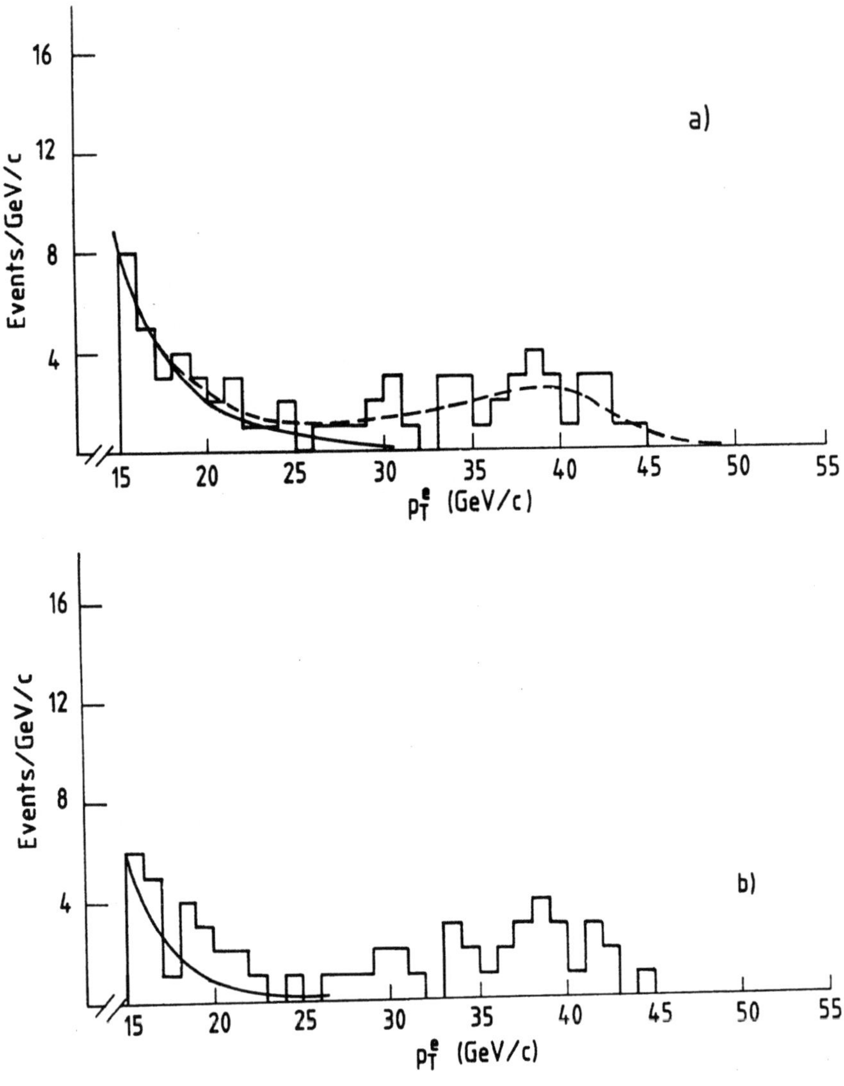

Fig. 37: (a) Transverse momentum distribution of the electron candidate in the $W^{\pm} \to e^{\pm}\nu$ event sample;
(b) strict cuts applied in the central calorimeter. Full curves: estimated background. Dashed curve: sum of all contributions (background from misidentified electrons: 1.5 events; contamination from $Z^{o} \to e^{+}e^{-}$ decay with one electron undetected: 2.5 ± 0.9 events; contribution from $W^{\pm} \to e^{\pm}\nu$) (UA2).

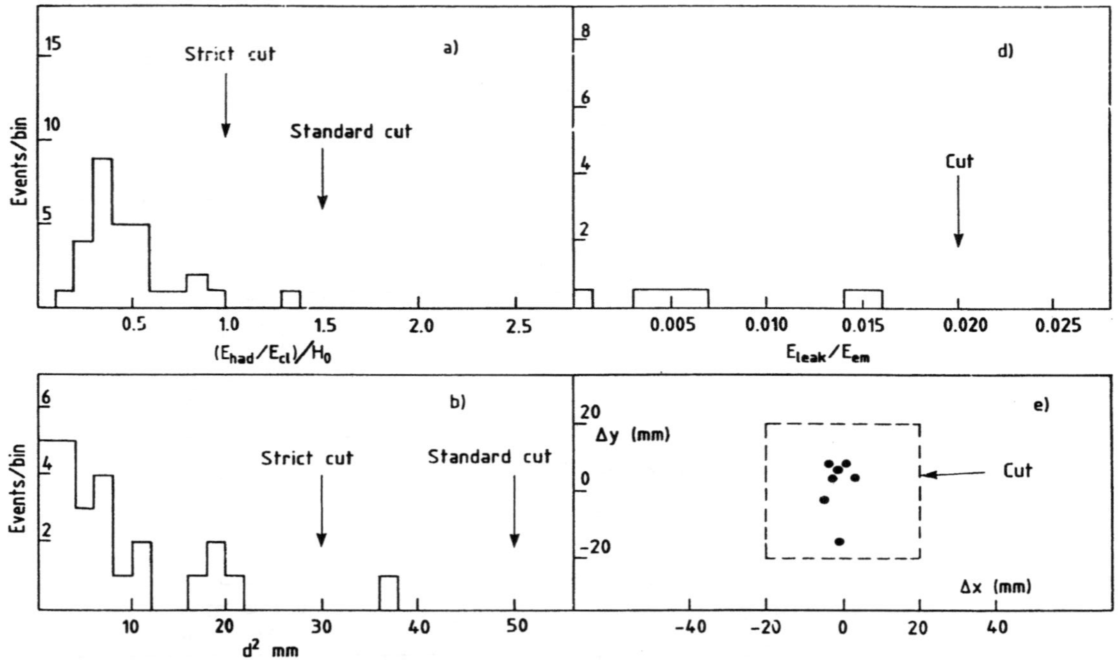

Fig. 38.: a-g: Quality parameters of the electron candidates with $p_T^e > 25$ GeV/c in the $W \to e\nu$ sample. Central detector:

a) distribution of the ratio E_{had}/E_{cl} normalised to the value of H_o used in the strict cuts (see Table 1a);

b) distribution of d^2;

d) distribution of the ratio E_{leak}/E_{cm};

e) scatter plot of the signal position in the preshower counter with respect to the track impact point;

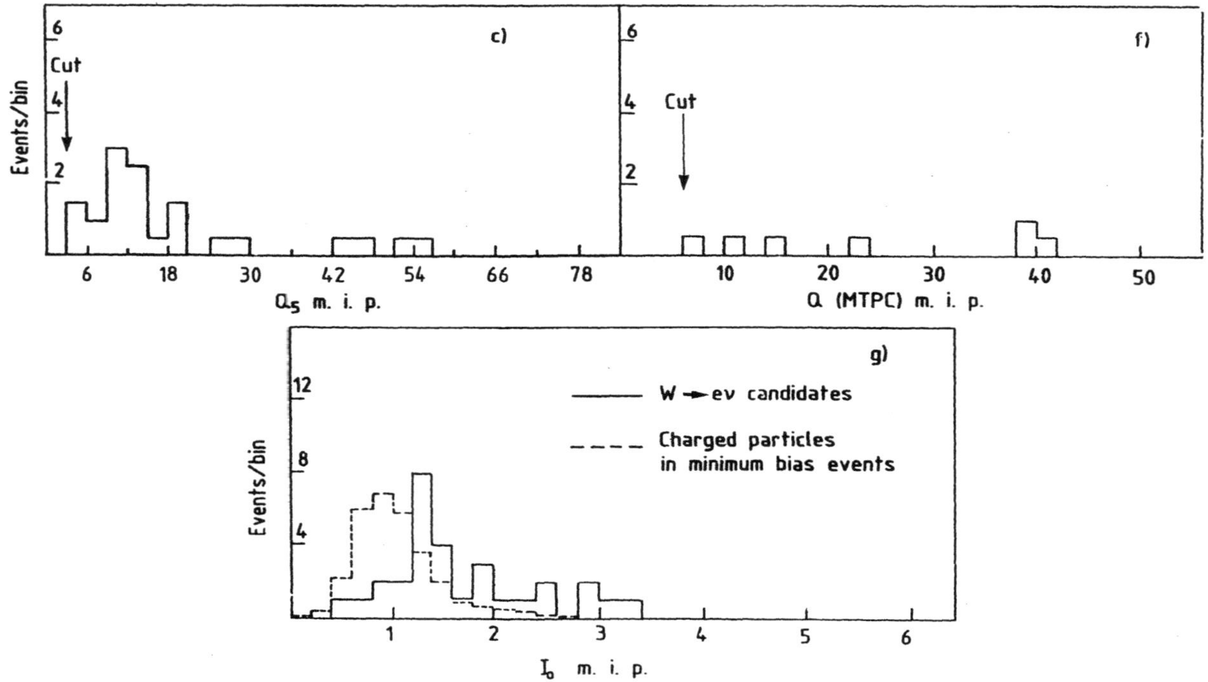

Fig. 38. ctd.: c) distribution of Q_5. Forward detectors;

f) distribution of Q(MTPC);

g) distribution of the most likely ionisation I_o associated with the electron candidate track (full histogram). The dashed histogram is the corresponding distribution for tracks in minimum bias events (UA2).

remaining background by applying strict track quality and matching cuts. Independently all 713 events were examined on an interactive high-resolution graphics display facility confirming that no W candidate events were rejected by the selection program.

The selection program imposed additional requirements on event topology in order to reject events with muons in jets or back-to-back with jets. A muon was considered to be in a jet, if the sum of the transverse momenta of the other tracks in a surrounding cone of half width $\Delta r = \sqrt{\Delta\phi^2+\Delta\eta^2} < 0.7$ (ϕ: azimuthal angle in radians, η: pseudorapidity) exceeded 3 GeV or if the jet algorithm found a calorimeter jet with $E_T > 10$ GeV and with its axis in the above cone. Events were also rejected, if the jet algorithm found a calorimeter jet with $E_T > 10$ GeV or a jet of central detector tracks with $p_T > 7.5$ GeV/c back-to-back with the muon within $\pm 30^o$ in the transverse plane (the plane perpendiular to the beams). 36 events survived these cuts and were carefully rescanned. After eliminating additional cosmics and probable $K^{\pm} \rightarrow \mu^{\pm}\nu$ decays 18 events remained. The final W sample of 14 events was obtained after the additional requirement that the neutrino transverse energy exceeded 15 GeV. (The presence of neutrino emission is signalled by an apparent transverse-energy imbalance, if the calorimeter measurement of missing transverse energy is combined with the muon momentum measurement. The accuracy of each component of the missing transverse energy is $0.4\sqrt{\Sigma E_T}$, where ΣE_T is the scalar transverse-energy sum in GeV. This determines the neutrino transverse-momentum error perpendicular to the muon p_T, whereas the error parallel to the muon p_T is dominated by the track momentum accuracy). The effects of the cuts are shown in table 9. The quality of muon identification was checked as indicated in table 7 and described when discussing the $Z^o \rightarrow \mu^+\mu^-$ sample.

The most dangerous background to the $W \to \mu\nu$ sample comes from mismeasured decays $K^{\pm} \to \mu^{\pm}\nu$, which pass the track quality and matching cuts. A mismeasurement of a $K \to \mu\nu$ decay simulates at the same time a high-momentum muon track and, in order to preserve momentum balance, a recoiling "neutrino". Most of these events are rejected by the selection procedure. The residual background as estimated by simulating $K^{\pm}$ decays in the central detector and processing them through the reconstruction and selection chain is less than one event. Releasing the E_T^{miss} cut to $E_T^{miss} >$ 10 GeV would increase this background to $\sim$3 events.

Table 9: Selection of $W \to \mu\nu$ candidates. The number of events is indicated at each stage (UA1)

events with a p_T>5 GeV/c muon selected by the fast filter program	17326
fully reconstructed events with muon p_T>15 GeV/c or p > 30 GeV/c	713
events with a good quality CD track which matches the muon chambers well	285
events remaining after rejection of cosmic rays	247
events remaining after a tight cut on the χ^2 of the CD track fit to remove decays	144
events where the muon is isolated both in the CD and in the calorimeters	53
events with no jet activity opposite the muon in the transverse plane	36
events remaining as W candidates after scanning (see text)	18
events with a neutrino transverse momentum >15 GeV/c	14

4.2 Properties of the IVBs

4.2.1 Radiative Z decays

One of the most remarkable events found in the IVB search is the UA2 $Z^o \rightarrow ee\gamma$ event shown in fig.39. Two rather similar events have been reported by the UA1 collaboration. The observed large angle between the photon and the nearby lepton in the three events (see table 10) make their explanation as being due to internal bremsstrahlung rather unlikely. The probability of observing two or more $e^+e^-\gamma$ events in the combined UA1 + UA2 sample of twelve $Z^o \rightarrow e^+e^-$ or $e^+e^-\gamma$ events is approximately 5 %.

Table 10: Properties of the $Z^o \rightarrow \ell^+\ell^-\gamma$ events (masses and energies in GeV, angles in degrees)

Measured quantity	$ee\gamma$ (UA2)	$e^+e^-\gamma$(UA1)	$\mu^+\mu^-\gamma$(UA1)
E_γ	24.4 ± 1.0	38.8 ± 1.5	28.3 ± 3
E_{ℓ_1}	11.5 ± 0.7	9.0 ± 1.0	42.2^{+44}_{-14}
E_{ℓ_2}	69.9 ± 1.8	61.0 ± 1.2	50.6^{+6}_{-5}
angle(ℓ,γ)	31 ± 1	14.4 ± 4.0	7.9 ± 0.1
$m_{12\gamma}$	90.6 ± 1.9	98.7 ± 5.0	88.4^{+46}_{-15}
m_{12}	50.4 ± 1.7	42.7 ± 2.4	70.9^{+37}_{-12}
$m_{1\gamma}$	9.1 ± 0.3	4.6 ± 1.0	5.0 ± 0.4
$m_{2\gamma}$	74.7 ± 1.8	88.8 ± 2.5	52.4^{+27}_{-9}

External bremsstrahlung is even less probable. Consequently there were many attempts to interprete these three events as manifestations of new, exotic decay modes of the Z^o, but the various proposed models have difficulties in explaining

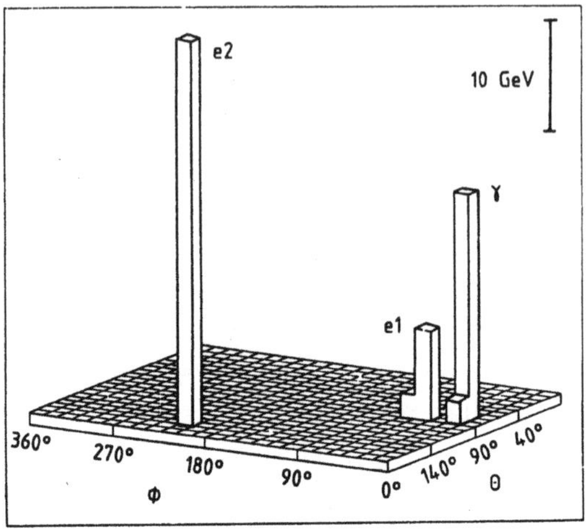

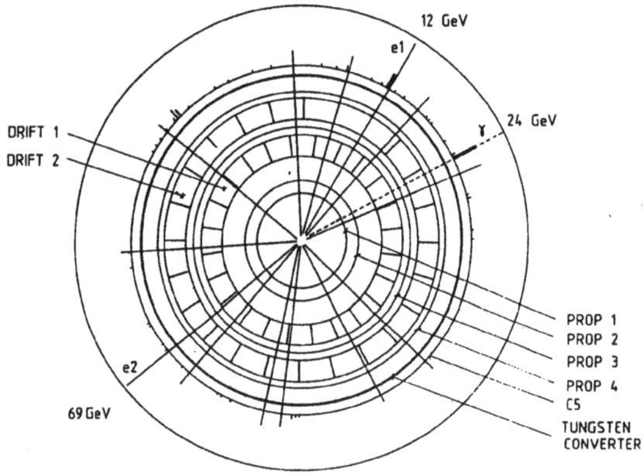

Fig. 39: (a) The cell transverse-energy distribution of the $Z^0 \to ee\gamma$ event (UA2) in the (θ, ϕ) plane. Electron e1 and photon γ are observed in the central region, electron e2 is in the forward region.

(b) The transverse view of the event. Signals from proportional and drift chambers are indicated as dots. Electron e1 and the photon γ are observed in the central region and associated with C5 signals (indicated as heavy lines proportional to pulse height). Electron e2 is observed in the forward region (not covered by C5).

why the photon prefers the direction of one of the two leptons and why the $(\ell_1\gamma)$, $(\ell_2\gamma)$, and $(\ell^+\ell^-)$ masses have a different value in each of the three events. Therefore the interpretation in terms of new physics seems at present not more likely than the bremsstrahlung hypothesis. Furthermore the twice as large Z^0 sample from the 1984 data did not yield any additional radiative decay candidate.

4.2.2 Masses and widths of IVBs

A first estimate of the mass of the W is obtained from the electron E_T spectrum (fig.40). The weakness of this determination of the W mass comes from the smearing of the Jacobian peak due to the transverse momentum of the W. Taking QCD predictions for $p_T^{(W)}$ [62] the UA1 group obtained $m_W = 80.5 \pm 0.5$ GeV.

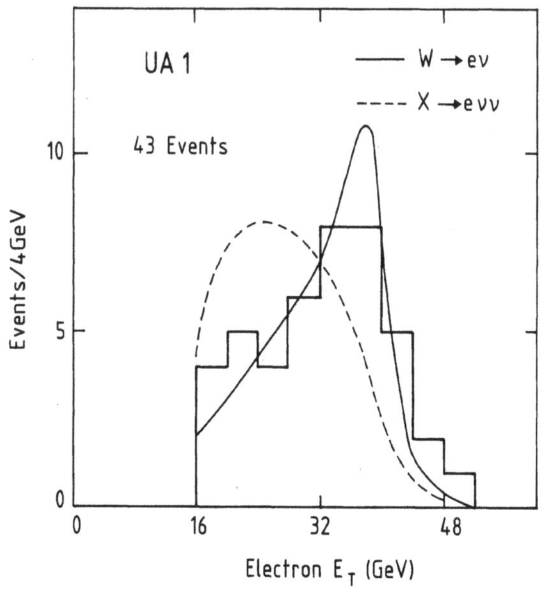

Fig. 40: Transverse momentum distribution of electrons of the restricted $W^\pm \to e^\pm\nu$ sample (UA1). The two curves show results of a fit to the hypotheses $W \to e\nu$ and $X \to e\nu\nu$.

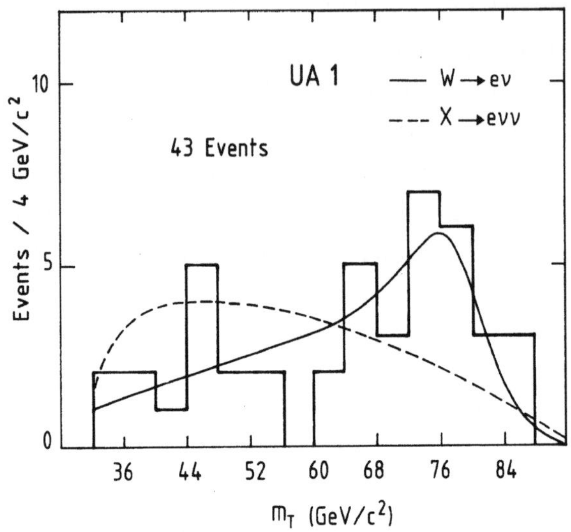

Fig. 41: Transverse-mass distribution of the restricted $W^{\pm} \rightarrow e^{\pm}\nu$ sample (UA1)

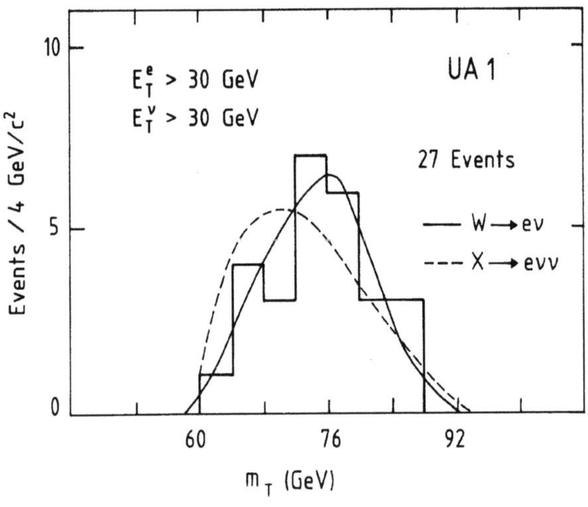

Fig. 42: Enhanced transverse-mass distribution of the restricted $W \rightarrow e\nu$ sample (UA1).

The transverse-mass variable, $m_T^2 = 2p_T^{(e)} p_T^{(\nu)} \cdot (1-\cos\theta)$, is to first order independent of $p_T^{(W)}$. Fitting the experimental distribution (fig.41) the UA1 group obtained $m_W = 80.3^{+0.4}_{-1.3}$ GeV. It should be noted that the lower part of the distribution in m_W may be slightly affected by $W \rightarrow \tau\nu_\tau$ decays and other backgrounds. This is why an enhanced transverse-mass distribution was defined using the cuts $p_T^{(e)}$, $p_T^{(\nu)} > 30$ GeV/c. The resultant distribution (fig.42) shows a narrow peak at approximately 76 GeV. Model-dependent corrections now contribute only to the difference between the average-mass value and the fitted m_W value, $m_W = 80.9 \pm 1.5$ GeV. This value is the least affected by systematic effects, even if it gives the largest statistical error. The uncertainty in the absolute energy scale, as mentioned in table 7(±3 %), remains, of course, the most stringent limitation in this measurement. In addition, an upper limit to the width of the W can be derived from the enhanced transverse-mass distribution, namely $\Gamma_W < 7$ GeV (90 % CL).

The UA2 group extracted a value of the W mass from their 37 $W \rightarrow e\nu$ candidates with $p_T > 25$ GeV/c by comparing the two-dimensional distribution $d^2n/dp_T^e d\theta_e$, where θ_e is the measured electron polar angle, to that expected from $W \rightarrow e\nu$ decay. A Monte Carlo program was used to generate the distribution $d^2n/dp_T^e d\theta_e$ for different values of the W mass. The W longitudinal-momentum distribution was obtained using the quark structure functions of the proton with scaling violations, as given in ref.63. A $p_T^{(W)}$ distribution was generated using the shape given in ref.62, where the average value $\langle p_T^{(W)} \rangle$ was allowed to vary in order to take into account uncertainties of QCD predictions. The decay angular distribution was described by the standard V-A coupling and the detector response was taken into account. A fixed value of the W width, $\Gamma_W = 2.7$ GeV/c, was used.

The best fit to the experimental distribution $d^2n/dp_T^e d\theta_e$ gave $m_W = 83.1 \pm 1.9$ GeV, where the error includes an uncertainty of ± 1 GeV resulting from the effect of varying $\langle p_T^{(W)} \rangle$ between 4 and 10 GeV/c in the fit. This error was added in quadrature to the statistical error. In addition to the quoted error the systematic uncertainty in the calorimeter energy calibration ($\pm$ 1.5 %) and the error resulting from the cell-to-cell uncertainty ($\approx \pm$ 0.4 %) were taken into account. The final values quoted by the UA2 group are:

$$m_W = 83.1 \pm 1.9 \text{ (stat.)} \pm 1.3 \text{ (syst.) GeV.}$$

A similar fit was performed by the UA1 group using their $W \to \mu\nu$ event sample. The resulting mass is $m_W = 81^{+6}_{-7}$ GeV.

The average dielectron mass after corrections using the four events of the UA1 $Z^0 \to e^+e^-(\gamma)$ sample is $m_Z = 95.6 \pm 1.4$ (stat.) $\pm$ 3.0 (syst.) GeV. The corresponding value of the UA2 group using the four events for which the energy of both electrons (and that of the photon in the $ee\gamma$ event) is unambiguously determined is $m_Z = 92.7 \pm 1.7$ (stat.) $\pm$ 1.4 (syst.) GeV. The 90 % confidence-level upper limits on the Z width Γ_Z are $\Gamma_Z < 8.5$ GeV (UA1) and $\Gamma_Z < 6.5$ GeV (UA2). Averaging the observed mass of the five $Z^0 \to \mu^+\mu^-$ (γ) the UA1 group finds $m_Z = 85.8 {}^{+7.0}_{-5.4}$ consistent with the value measured for $Z^0 \to e^+e^-$. The average value from all 9 Z^0 events found in the UA1 data is $m_Z = 93.9 \pm 2.9$ GeV, where the error includes the systematic uncertainty. The measured values and the standard model parameters

$$\sin^2\theta_W = \left[\frac{38.65 \text{ GeV}}{m_W}\right]^2$$

Table 11: Measured and predicted electroweak-interaction parameters

	UA1	UA2	Standard model with $\sin^2\theta_W$ = = 0.220 ± 0.006
m_Z [GeV]	95.6 ± 1.4 ± 2.9	92.4 ± 1.1 ± 1.4	93.3 ± 0.9
electron and muon data combined	93.9 ± 2.9		
Γ_Z [GeV], (data: 90 % CL upper lim.)	<8.5	<6.5	2.8
m_W [GeV]	80.9 ± 1.5 ± 2.4	81.2 ± 0.8 ± 1.5	82.4 ± 1.1
Γ_W [GeV], (data: 90 % CL upper lim.)	<7	-	2.8
$m_Z - m_W$ [GeV]	14.7 ± 2.1 ± 0.4	11.2 ± 1.3 ± 0.2	10.9 ± 0.2
$\sin^2\theta_W = [38.65\ \text{GeV}/m_W]^2$	0.226±0.008±0.014	0.227±0.004±0.009	0.220 ± 0.006
$\rho = m_W^2/m_Z^2\cos^2\theta_W$	0.925 ± 0.05	0.998 ± 0.0[illegible]	1

and

$$\phi = m_W^2/m_Z^2\cos^2\theta_W$$

are summarized in table 11 and compared with standard model values [56].

4.2.3 Number of light neutrino species

The measurement of the total width of the Z^o yields information about the number N_ν of light neutrino species via the relation Γ_Z (measured) $= \Gamma_Z^{total} + (N_\nu - 3) \times \Delta\Gamma_Z$. Using $\Gamma_Z = 2.8$ GeV and $\Delta\Gamma_Z = 0.178$ GeV [56] one obtains:

UA1: $\Gamma_Z < 8.5$ GeV $\rightarrow N_\nu < 34$ (90 % CL)
UA2: $\Gamma_Z < 6.5$ GeV $\rightarrow N_\nu < 25$ (90 % CL).

More stringent limits can be derived from a measurement of the ratio R of cross section times leptonic branching fraction for W and Z:

$$R = \frac{\sigma_Z \cdot B(Z \rightarrow \ell^+ \ell)}{\sigma_W \cdot B(W \rightarrow \ell\nu)} = \frac{\sigma_Z}{\sigma_W} \cdot \frac{\Gamma(Z \rightarrow \ell^+ \ell^-)}{\Gamma(W \rightarrow \ell\nu)} \cdot \frac{\Gamma_W}{\Gamma_Z} .$$

Using the calculated values for the ratio of cross sections [60], the ratio of the leptonic width to the total width of the W, and the measured value of R (table 12) the two experiments obtain the upper limits (90 % CL)

$N_\nu < 18$ (UA1) and <17 (UA2, 1984 data included)

4.2.4 V-A structure of weak interactions

The decisive test to prove that the observed heavy particles are indeed the hypothesized mediators of weak interactions is the verification of their V-A nature. At the collider weak interactions act as longitudinal polarizers of $W^{\pm}$ particles, since quarks (antiquarks) are provided by the proton (antiproton) beam. The weak decay acts as a polarization analyzer. A strong backward-forward asymmetry is therefore expected; electrons (positrons) prefer to be emitted in the direction of the proton (antiproton). Fig.43 shows the angular distribution for electron (positron) emission with respect to the proton (antiproton) direction in the W centre-of-mass for the UA1 event sample. Only events with no ambiguity in the reconstruction of the centre-of-mass of the W can be used. The over-all information of the event yields momentum and energy conservation bounds to resolve this ambiguity in 70 % of the cases. The data have been corrected for the "unambiguous selection" and for geometric acceptance. The obtained distribution is found to agree with the shape $(1 + \cos\theta^*)^2$ which is expected from V-A theory. More generally it has been shown [64] that for a particle of spin J we expect

$$\langle\cos\theta^*\rangle = \langle\lambda\rangle\langle\mu\rangle/J(J+1)$$

where λ and μ are the global helicities of the production system ($u\bar{d}$) and the decay system ($e\nu$), respectively. For V±A we then have $\langle\lambda\rangle = \langle\mu\rangle = \pm 1$, $J = 1$, leading to the maximal value $\langle\cos\theta^*\rangle = 0.5$. For $J = 0$ one obviously expects $\langle\cos\theta^*\rangle = 0$ and for any other spin value $J \geq 2$, $\langle\cos\theta^*\rangle \leq 1/6$. The experimental value is $\langle\cos\theta^*\rangle = 0.4 \pm 0.1$ and thus supports both the $J = 1$ assignement and maximal helicity at production and decay. The ambiguity of the

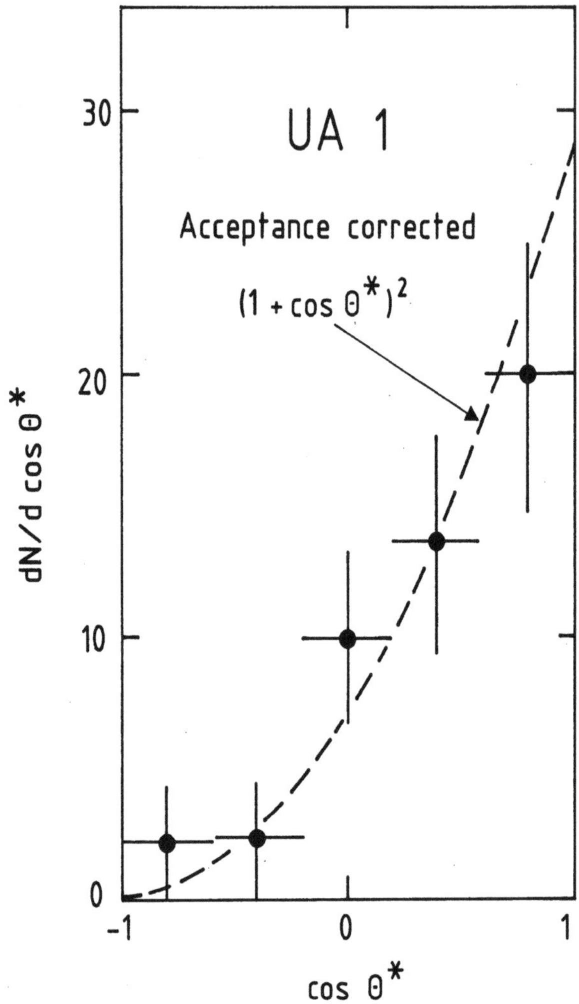

Fig. 43: The distribution of the electron emission angle θ^* in the rest frame of the W after corrections. Only those events have been used, in which the electron charge is determined and the kinematic ambiguity in the reconstruction of the W rest frame has been resolved (UA1).

sign $\langle\lambda\rangle = \langle\mu\rangle = \pm 1$ cannot, of course, be resolved without a polarization measurement.

The decay asymmetry in the UA1 $W \to \mu\nu$ sample is $\langle\cos\theta^*\rangle = 0.3 \pm 0.2$. This measurement is less significant than in the UA1 $W \to e\nu$ sample since the ambiguity for the longitudinal movement of the W could be resolved only in a few cases. This is due to the large momentum errors, and the limited acceptance in pseudorapidity ($|\eta| < 1.3$) for the muons.

4.3 Production of IVBs at the collider

4.3.1 Cross sections

A summary of the numbers of observed IVB events and the associated efficiencies of the experimental procedures described in section 4.1 is given in table 12. The corresponding cross sections for the leptonic decay modes (σ.BR) are obtained by normalizing to the integrated luminosities and are also summarized in table 12. The quoted standard-model values for (σ.BR) are taken from ref.60. The measured and the calculated values agree well within the uncertainty due to higher-order QCD terms.

No event with p_T^e or p_T^ν in excess of the expected distribution for $W \to e\nu$ events has been observed. This fact can be used to set a limit on the possible existence of very massive W-like objects decaying into electron-neutrino pairs. The UA1 group quotes (σ.BR)$<$ 0.030 nb (90 % CL) corresponding to $m_{W_2} > 170$ GeV, if standard couplings and quark distributions are used to evaluate the cross sections.

Table 12: Summary of the IVB samples, the measured cross sections for leptonic decay modes (σ.BR), and the corresponding standard-model values.

	Integrated luminosity (nb^{-1})	Observed number of events	Efficiency	(σ.BR) Experiment (nb)	Standard model ref.60 (nb)
Results from 1982 and 1983 data ($\sqrt{s}$ = 540 GeV):					
UA1 $Z^0 \to e^+e^-$	136 ± 20	4	0.68	0.043±0.022±0.009	$0.042 {+0.013 \atop -0.006}$
UA2 $Z^0 \to e^+e^-$	131 ± 26	8 (at least one $e^\pm$ satisfies strict cuts	0.57·η (e-identific.)	0.110±0.040±0.020	
UA1 $Z^0 \to \mu^+\mu^-$	108 ± 16	4 (both tracks longer than 40 cm)	0.37	0.100±0.050±0.015	
UA1 $Z^0 \to e^+e^-$ and $Z^0 \to \mu^+\mu^-$ combined				0.058±0.021±0.009	
UA1 $W^\pm \to e^\pm\nu$	136 ± 20	47 (p_T^e>20 GeV) backgr. 2±2 events	0.65	0.530±0.080±0.090	$0.360 {+0.110 \atop -0.050}$
UA1 $W^\pm \to \mu^\pm\nu$	108 ± 16	14 (backgr.<1 event	0.20 ± 0.035	0.650±0.170±0.150	
UA2 $W^\pm \to e^\pm\nu$	131 ± 26	32.1 ± 6 (p_T^e >25 GeV, backgr.sub-tracted).	0.460± 0.030	0.530±0.100±0.100	

Table 12: continued

	Integrated luminosity (nb^{-1})	Observed number of events	Efficiency	(σ.BR) Experiment (nb)	Standard model ref.60 (nb)
Results from 1984 data ($\sqrt{s}$ = 630 GeV):					
UA2 $Z^0 \to e^+e^-$	310	8	0.46	0.056±0.020±0.009	$0.051 {+0.016 \atop -0.010}$
UA2 $W^{\pm} \to e^{\pm}\nu$	310	87 (p_T^e> >25 GeV) backgr. 15±1 events	0.43	0.540±0.070±0.090	$0.460 {+0.140 \atop -0.060}$

4.3.2 Transverse momenta of IVBs

As an example the experimental distribution of p_T^W (UA1) is shown in fig.44. The distribution was obtained by adding neutrino and electron momenta. The measured average value is $\langle p_T^W \rangle = 6.3$ GeV/c. Those five events, which have a clearly visible jet (shown in black in the figure) have also the highest values of p_T^W. Transverse-momentum balance is almost exactly restored, if the vector momentum of the jet is added. The experimental distribution is in good agreement with expectations from QCD represented by the various lines in the figure for production of a massive state via the Drell-Yan quark-antiquark annihilation [62]. The small fraction (10 %) of events with a jet is explained as hard-gluon bremsstrahlung in the initial state [65]. Similar results have been reported by the UA1 group when analyzing the $W \to \mu\nu$ event sample and by the UA2 group.

Some evidence for increased jet activity in the UA1 $Z^o \to \ell^+\ell^-$ sample (1983 data) is discussed in section 4.4 together with the three remarkable UA2 events (1983 data) where a W may have been produced together with a hard jet; no additional events with these unexpected features were, however, found in the 1984 data.

4.3.3 Longitudinal motion of IVBs and proton structure functions

The longitudinal momentum of the electron-neutrino system has been measured by the UA1 group. The twofold ambiguity arising from the unmeasured longitudinal component of the neutrino momentum was resolved in 70 % of the cases by momentum and energy conservation bounds. Most of the remaining events have solutions that are quite close. The distribution of the fractional beam energy x carried by the W particle is shown in fig.45. A similar distribution was measured for the Z^0.

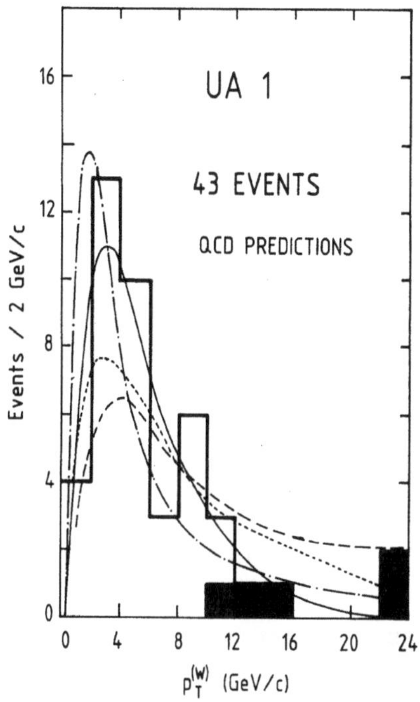

Fig. 44

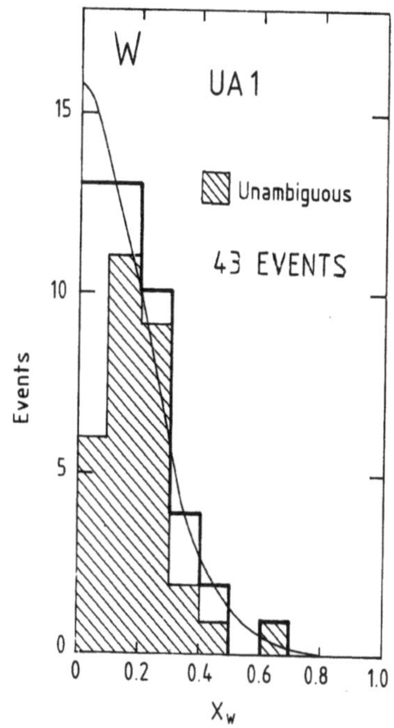

Fig. 45

Fig. 44: The transverse momentum distribution of the W (UA1 W→eν). The highest p_T^W events (shown in black) have a visible jet. The various lines represent QCD calculations [62].

Fig. 45: The fractional beam energy x_W carried by the W (UA1, W→eν sample). The curve was computed assuming the W has been produced by $\bar{q}q$ fusion [60]. For events, where the ambiguity due to the unmeasured longitudinal motion of the neutrino was not resolved, the more probable solution with the smaller x_W was plotted.

Using the relations $x_W = x_p - x_{\bar{p}}$ and $x_p . x_{\bar{p}} = m_W^2/s$ the UA1 group has further determined the parton distributions in the proton and antiproton. Contributions of the u quark and d quark have been separated by looking at the charges of the produced W particles, since $(u\bar{d}) \rightarrow W^+$ and $(\bar{u}d) \rightarrow W^-$. The resulting distributions using events where the electron charge is well determined are shown in fig. 46. The full lines represent the result of a calculation assuming $\bar{q}q$ fusion [60]. The measured distributions agree well with the expected structure functions.

4.4 Anomalous jet activity in some IVB events?

Two events in the UA1 (1982-83) $Z^0 \rightarrow \mu^+\mu^-$ event sample show large jet activity, figs. 47 and 48. In both events the jet activity is shared by four jets (table 13), one of them forward and the other three in the central rapidity region $|\eta| < 1.2$. The multijet masses alone are in the IVB mass range. The masses of the muons and the central jets exceed 120 GeV. Similar high masses were found in one UA1 (1982-83) $Z^0 \rightarrow e^+e^-$ event that also shows large jet activity. These events contain more jet activity than expected from QCD. However, no additional events with a Z^0 and many central jets were found in the 1984 data.

A preliminary analysis [66] of the transverse momentum and angular distributions of all jets produced in association with a Z^0 or $W^\pm$ indicates their QCD-initial-state bremsstrahlung nature. The jets are emitted primarily along the beam directions and have a rapidly falling transverse-energy distribution.

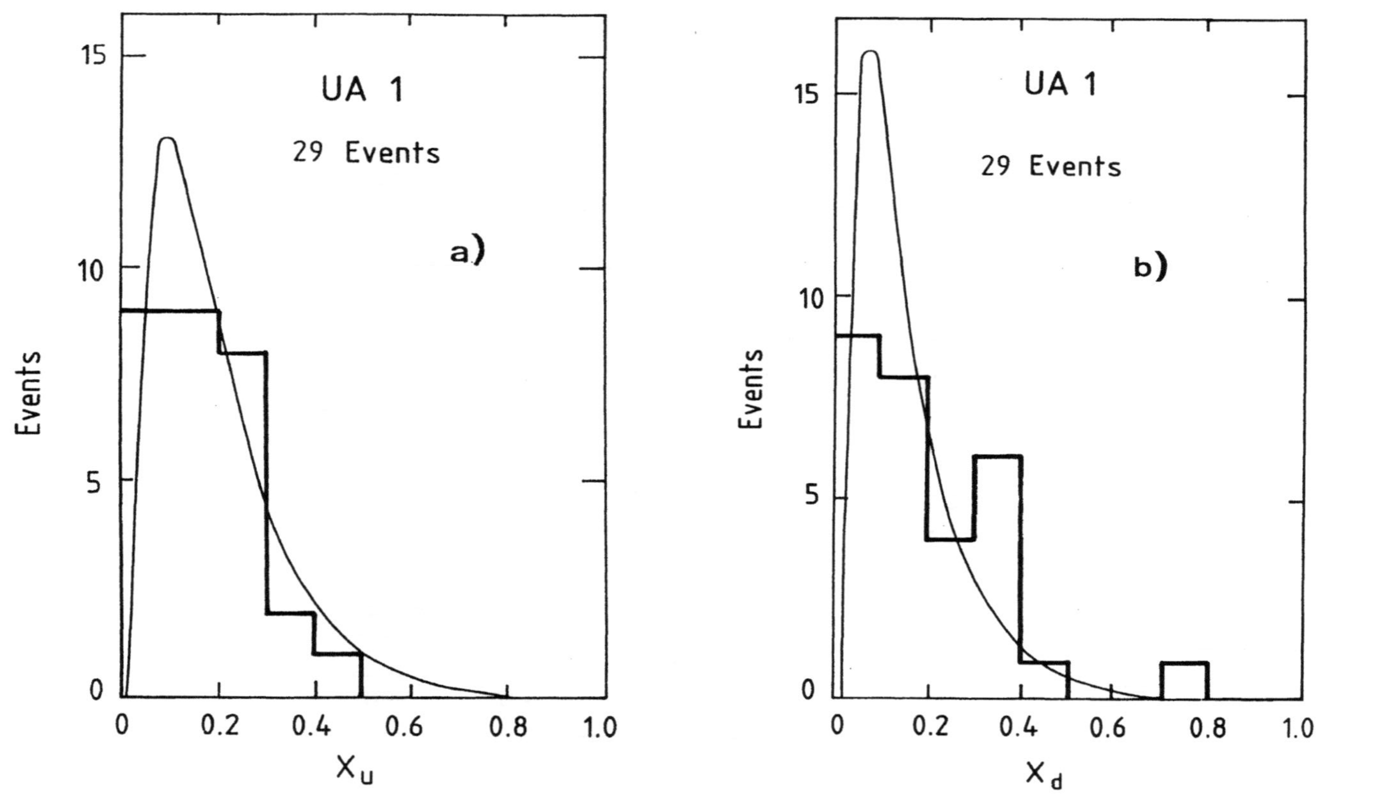

Fig. 46.: The x distributions of the u and d quark as measured by using the $W \to e\nu$ sample (UA1).

Table 13: Jet parameters of the $Z^o \to \mu^+\mu^-$ events A and B, see figs. 47 and 48 (UA1).

Run event	Jet No.	E_T (GeV)	η	ϕ (deg)
A 6219 947	1	22.8 ± 3.3	0.1	177
	2	13.0 ± 2.2	-0.7	- 7
	3	11.5 ± 4.4	-1.1	-162
	4	16.9 ± 3.0	-2.3	32
B 6600 222	1	11.1 ± 4.1	0.6	6
	2	20.6 ± 3.5	2.1	-115
	3	10.1 ± 3.8	-0.5	-156
	4	>7.1 ± 0.1[a)]	1.2	- 88

a) Only the charged component of the jet is given (CD jet), since the jet points towards an insensitive region of the calorimeter.

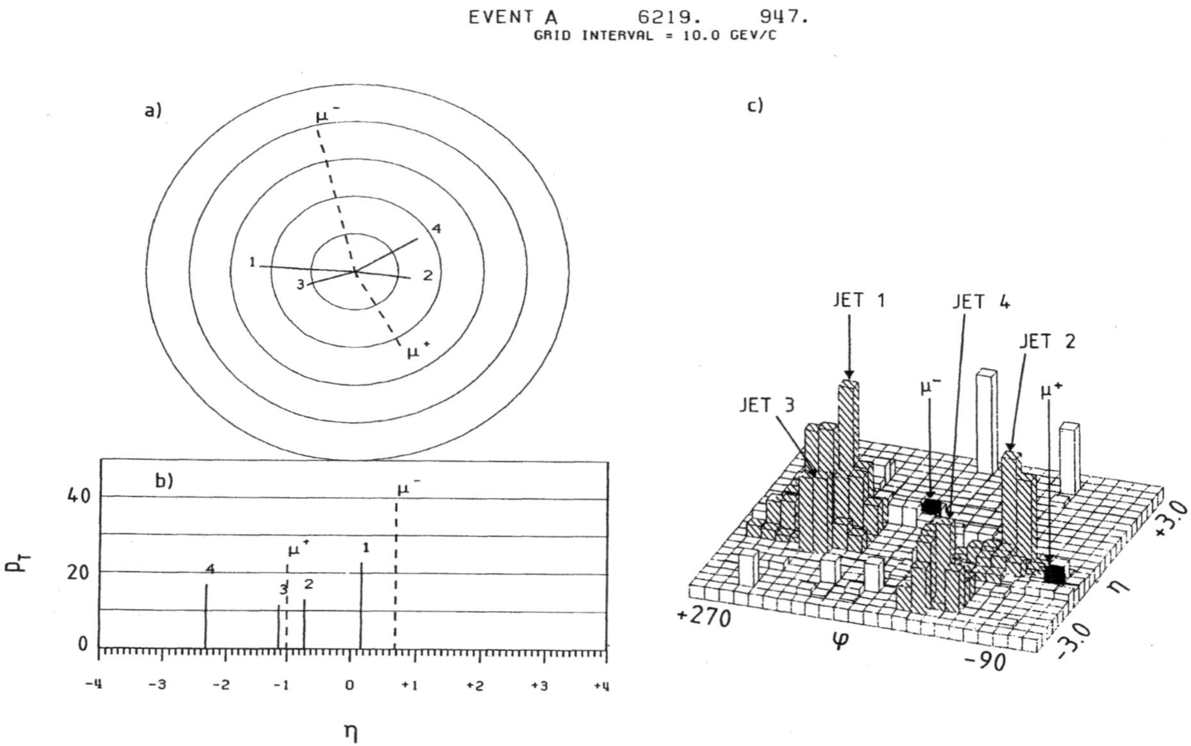

Fig. 47.: Configuration of the UA1 $Z^o \to \mu^+\mu^-$ event A with jets:
(a) in the plane transverse to the beam axis;
(b) in pseudorapidity η;
(c) "Lego"-plot of the energy deposition in the calorimeters as a function of η and azimuth ϕ.

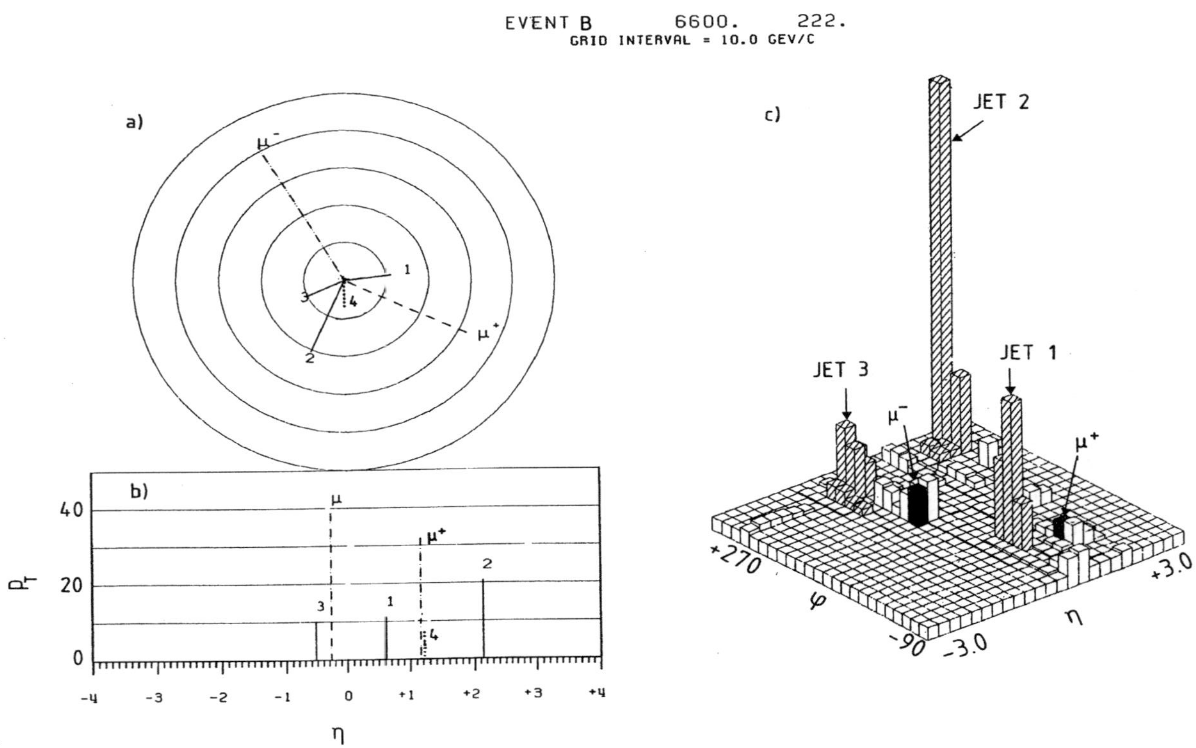

Fig. 48.: Same as for fig.47, but for the UA1 $Z^o \to \mu^+\mu^-$ event B. The CD jet (No.4) is indicated with a dotted line; it does not appear on the Lego plot of the calorimeter cells.

The UA2 group has reported [67] three events (1983 data) with a large-p_T electron and large missing energy, associated with a jet, or a system of jets, having large transverse energy. The electron and the missing transverse energy may be interpreted as due to the decay $W \to e\nu$. The jet activity $E_T(J)$ for the event sample defined by $p_T(e) > 15$ GeV and $E_T^{miss} \equiv p_T(Je) > 25$ GeV is shown in fig.49. Four outstanding events appear above $E_T(J) = 35$ GeV. The transverse-momentum configuration between electron, neutrino, and jet(s) in these events is shown in fig.50. While in three events (A-C) the $(e\nu)$-pair has a large azimuthal opening ($\Delta\phi > 120^o$), event D contains a narrow $e\nu$ pair ($\Delta\phi \approx 17^o$) and is hence unlikely to be produced by a $W \to e\nu$ decay. The background contamination in the kinematical region which contains the events A-C has been estimated to be < 0.02 events (90 % CL).

In the three events A-C the transverse mass of the $e\nu$ pair ranges between 56 GeV and 82 GeV suggesting an interpretation in terms of a $W \to e\nu$ decay. It is indeed possible to adjust the unmeasured value of $p_L(\nu)$ to obtain $m(e\nu) = m(W)$ and to describe the events in terms of associated W-jet(s) production. There are in general two solutions to this problem associated with different values of $p_L(WJ) \equiv x_F(WJ)\sqrt{s/2}$ and of $m(WJ)$. Choosing the solution giving the smaller value of $x_F(WJ)$ the three events populate the region $160 \leq m(WJ) \leq 180$ GeV, which might suggest an interpretation in terms of a heavy object decaying into a W boson and a system J of other particles. The 1984 data did, however, not yield any additional event of the described topology.

The UA1 experiment is also sensitive to the type of events reported by UA2, but the UA1 group did not confirm

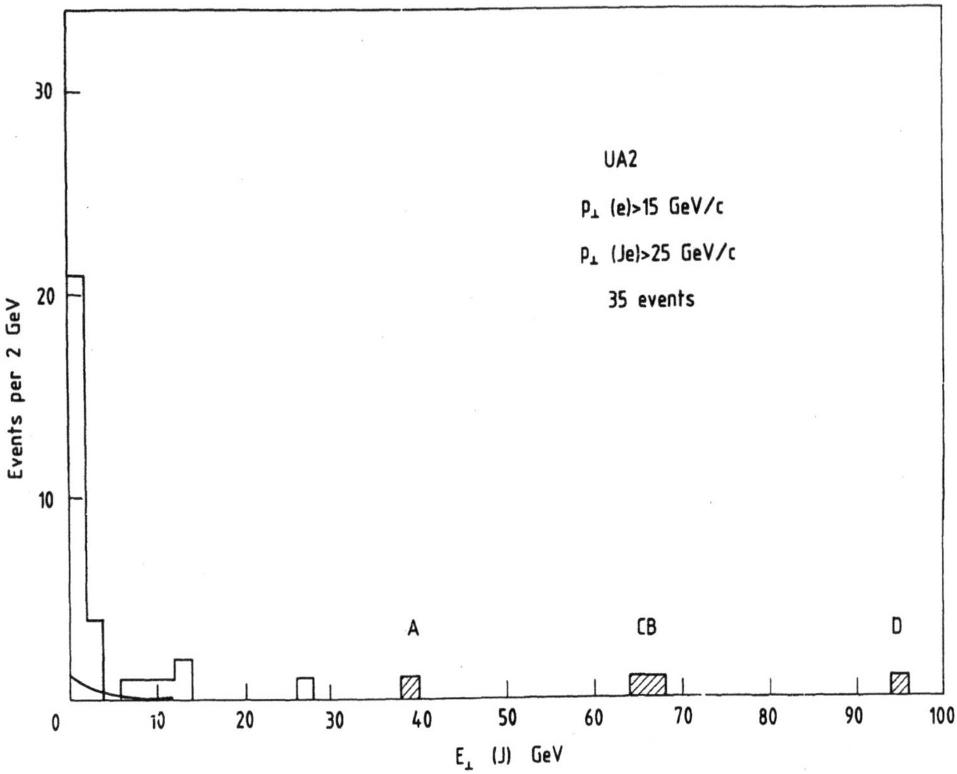

Fig. 49: Transverse-energy distribution of the system of jets in the sample of 35 events with a large p_T electron and large missing energy (UA2).

the UA2 result. The UA1 group found one event (1984 data) with $W \to \mu\nu$, $p_T(W) \simeq 40$ GeV, where the transverse momentum of the W is balanced by a jet of $p_T \approx 40$ GeV. However, with the improved total luminosity ($\gtrsim 400\ nb^{-1}$) such an event on the tail of the $p_T(W)$ distribution [62] is not unlikely to occur.

In conclusion the possibly anomalous jet activity in some IVB events as reported by UA1 and UA2 (1983 data) is not confirmed by the 1984 data.

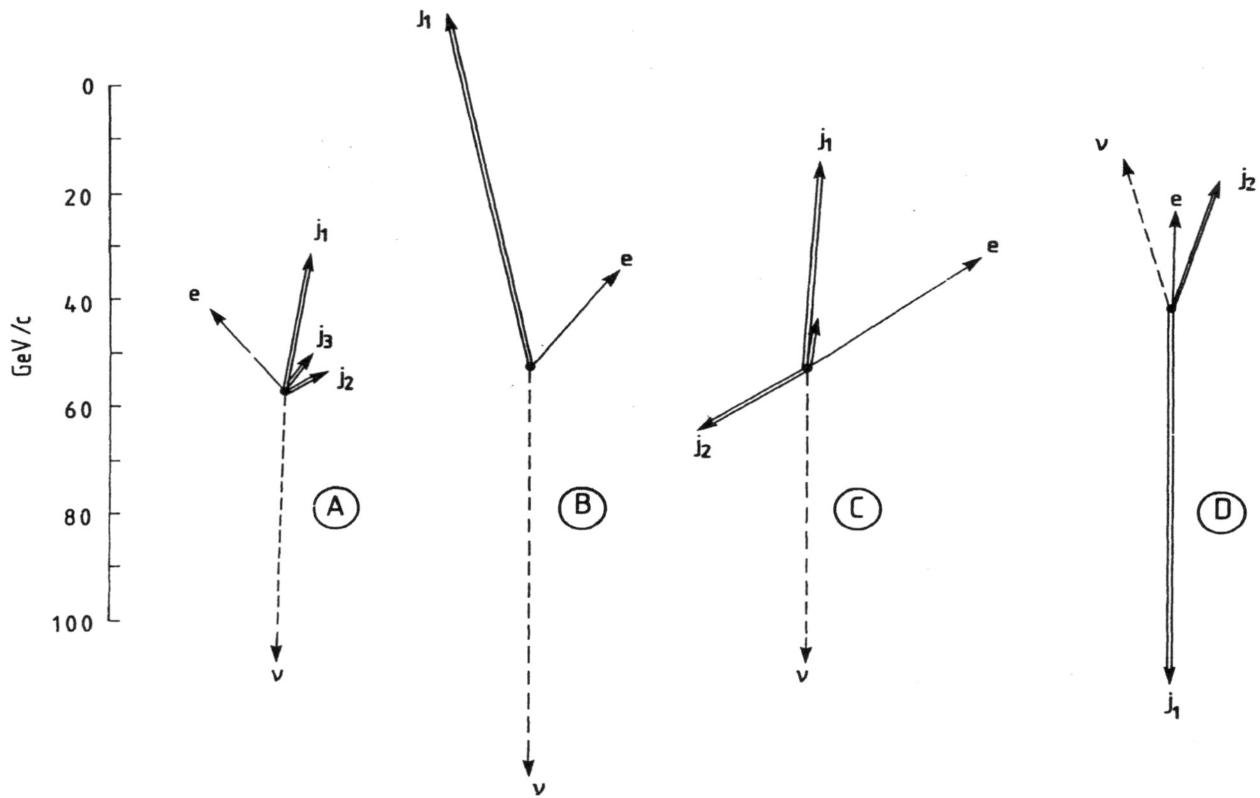

Fig. 50.: Transverse-momentum configuration of the four events having $p_T(Je) > 25$ GeV, $E_T(J) > 30$ GeV. Their relative orientation is arbitrary (UA2).

5. EVENTS WITH A JET AND LARGE MISSING TRANSVERSE ENERGY

The measurement of missing transverse energy plays a crucial role in the analysis presented in this chapter and a few general comments on the missing-E_T technique may be in order.

In the absence of non-interacting particles (e.g. neutrinos) and in the case of an ideal detector a perfect energy and momentum balance would be observed. This can, of course, not be achieved in practice, but the transverse components can be fairly well measured already with the present UA1 detector. The longitudinal component of momentum balance cannot be measured due to energetic particles (typically 100 GeV) escaping down the beam pipe. Fig.51 shows how well the vertical component of transverse energy (E_y) is observed to balance in soft-hadron events. The missing-E_T resolution for each transverse component may be parametrized as $\sigma = 0.43\sqrt{\Sigma E_T}$, where ΣE_T is the scalar sum of transverse energies observed in the entire calorimeter. For events, which contain high transverse momentum jets, the parametrization of the missing-E_T resolution also holds. This is shown in fig.52: the missing transverse energy observed for a sample of two-jet events is shown together with a Monte Carlo calculation of the expected distribution based on the parametrization of the resolution for each transverse component. For a typical event with ΣE_T = 80 GeV the missing transverse-energy resolution is about 6 GeV. An illustration of the use of missing-E_T is the observation of missing transverse energy in a sample of $W \rightarrow e\nu$ decays, fig.32 of section 4.1.3. A sharp Jacobian peak is seen at about one half of the W mass. Proposals of UA1 and UA2 to improve the missing-E_T resolution by better calorimetry are discussed in chapter 7.

A copious source of events with a jet associated with

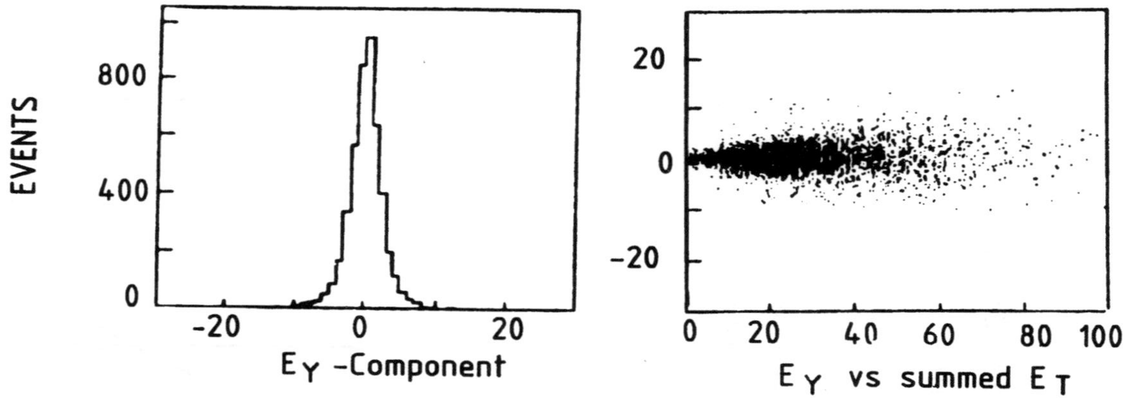

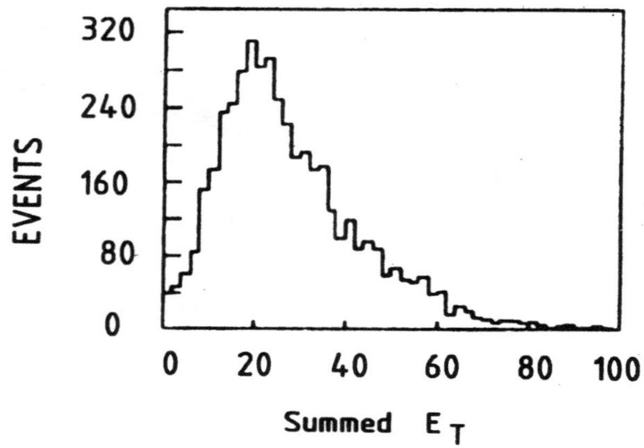

Fig. 51.: Scatter plot of the vertical component of missing transverse energy versus the total transverse energy observed in all calorimeter cells (UA1).

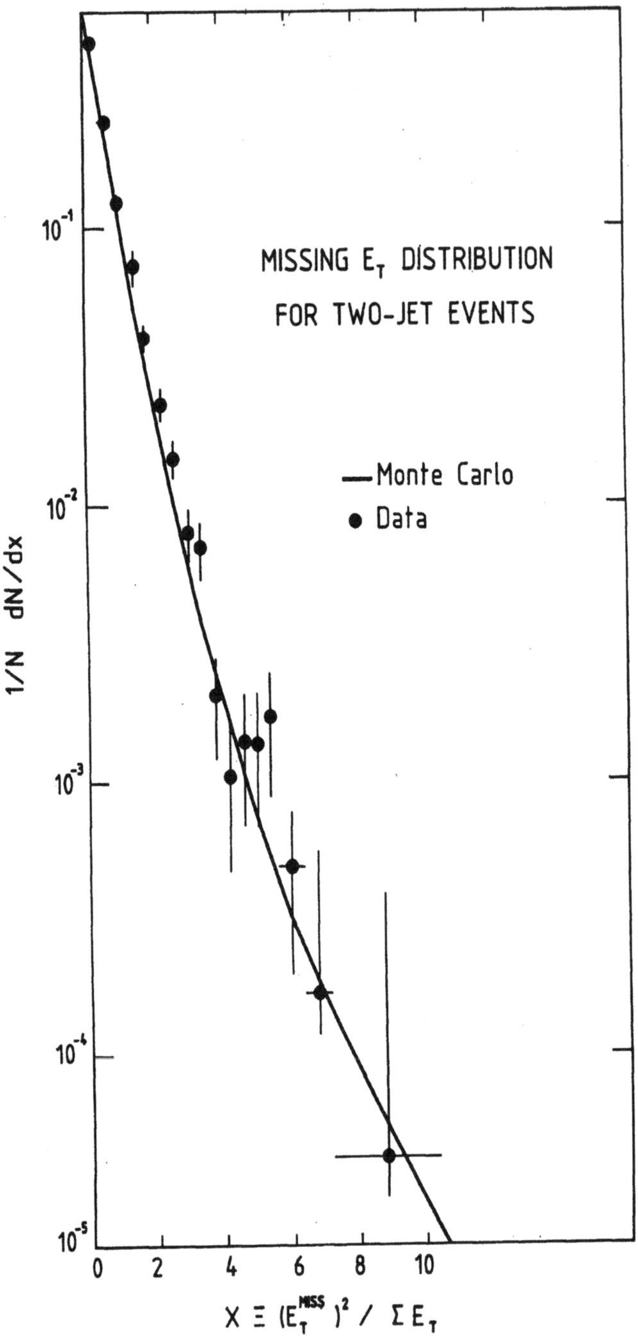

Fig. 52: Transverse-energy balance for a sample of two-jet events (UA1). To convert the horizontal scale to number of standard deviations (n), use the relationship $n^2 = 2x$.

missing-E_T is the decay $W \to \tau\nu$, $\tau \to$jet. Almost half of the τ's decay into one charged π (10 % without neutrals and 38 % with) and a neutrino. 17 % of the τ's decay into three charged pions, neutrals, and a neutrino. Due to the presence of the ν_T in the decay fagments of the τ the Jacobian peak in E_T^{jet} and E_T^{miss} is shifted down to $\approx$25 GeV. Such events have indeed been observed by the UA1 group and the measured rate supports the hypothesis that they are due to $W \to \tau\nu$, $\tau \to$jet decays (section 5.1). In a couple of events the jet energy is higher ($E_T^{jet} \geq 40$ GeV) than expected from $W \to \tau\nu$ or the charged multiplicity ($n_{jet} > 3$) is inconsistent with the τ decay hypothesis (section 5.2). These events, if confirmed by further experimentation, may be due to the tail of supersymmetric processes, where E_T^{miss} signals the production of photino(s), and the single jet is the remainder of the multijet structure surviving the cuts applied in the analysis. A substantial improvement of the measurement of missing E_T will be needed, paired with a better measurement of events with many jets - in general rather soft jets, $E_T^{jet} \approx 10$ GeV - to be able to judge better the supersymmetry hypothesis. The possibility to discover Supersymmetry is one of the motivations for the proposed improvement of the UA1 calorimeter.

5.1 $W \to \tau\nu$, $\tau \to$jet

The signature for $W \to \tau\nu$, $\tau \to$jet is one or three high-p_T tracks of hadronic type and some missing transverse energy. A preliminary analysis of such events has been presented by the UA1 Collaboration [68].

The data have been selected from the total statistics of the 1982 and 1983 $p\bar{p}$-runs and processed through an off-line filter as described at the beginning of section 4.1.1. By asking that the missing E_T be greater than 15 GeV the UA1 group obtained approximately $3 \cdot 10^4$ events, which have

been subjected to further selection criteria:
i) A validation of the missing transverse energy by asking $E_T^{miss} > 2\sigma$, where $\sigma = 0.7\sqrt{\Sigma E_T}$. Because of the reduced efficiency of the calorimetry in the region $\pm 15^o$ around the vertical axis, events have been excluded, if they have a jet axis in this region, or if the sum of the p_T of all the central detector tracks in this region exceeds 15 % of the total transverse energy.
ii) A validation of the trigger jet, defined as the calorimeter jet with the highest transverse energy in the pseudorapidity range $|\eta| < 1.5$. A matching is required between the calorimeter jet and the central-detector jet.
iii) No jet back-to-back with the calorimeter jet; the backward region is defined by the angle $\phi^{back} = 180^o \pm 30^o$ around the jet axis. It is required that there be no jet inside this backward region, with E_T(calorimeter jet) > >7 GeV, or p_T(central-detector jet) > 5 GeV, for a jet rapidity $|\eta^{jet}| < 2.5$. 550 events are selected at this stage.
iv) Events with activity back-to-back to the trigger jet, but with no jet reconstructed by the jet algorithm are rejected. The total sum of p_T of the tracks included in the backward region is required to be less than 4 GeV and the sum of the transverse energy in the calorimeter cell in the backward region is required to be less than 6 GeV.
v) The trigger jet is required to have a leading track with a minimum p_T of 4 GeV/c.

Finally those events were removed which have been already recognized as $W \to e\nu$ decays. The remaining events were scanned on an interactive high-resolution graphics display. 16 validated events were obtained, which are subdivided into three topological categories:

- Five events with a single high-p_T charged track, which is called the 'one-branch sample'. Two of them have an associated jet with E_T > 10 GeV not back-to-back with the trigger jet.

- Four events with two or three charged tracks called the 'three-branch sample'.
- Seven events with more than three charged tracks or $E_T^{jet} \geq 40$ GeV called the 'multi-branch sample'.

An analysis of the single high-p_T charged tracks from the one-branch sample has shown that they are hadrons. (Tracks with $p_T > 10$ GeV are recognized as hadrons, if the deposition in the hadronic calorimeter cell behind two adjacent e.m. cells is at least 500 MeV or 1 % of the total energy, or if the energy deposition in the four e.m. calorimeter segments is not consistent with the profile of an e.m. shower. If the charged track has a $p_T < 10$ GeV, it is difficult to recognize the shape of the shower development. On the other hand a charged hadron can be identified by a mismatch in space and/or energy when comparing the central detector and the calorimeter information.)

The scatter plot E_T^{jet} versus E_T^{miss} for the one-branch and the three-branch events is shown in fig.53. The plot is populated as expected from $W \to \tau\nu, \tau \to$ jet decays. Furthermore the mass of the system $\pi^{\pm}(n\pi^{o})$ or $\pi^{\pm}\pi^{\pm}\pi^{\mp}(n\pi^{o})$ was found to be consistent with the τ hypothesis.

The main source of background is due to standard bi-jets where one or both jets are mismeasured because of fluctuations in the calorimeter response. The calculated rejection factors (see also fig.52) relevant to this analysis are $\gtrsim 10$ for jets with $E_T < 25$ GeV and $\gtrsim 5$ for jets with $E_T > 25$ GeV.

The multiplicity of the trigger jet has so far not been constrained. To define the multiplicity all tracks with $p_T > 0.5$ GeV in a cone $\Delta R < 0.5$ around the jet axis are counted. The measured multiplicity for jets is plotted in fig.54. It shows that for one-charged track jets an additional reduction factor of 10 for jets with $15 < E_T <$

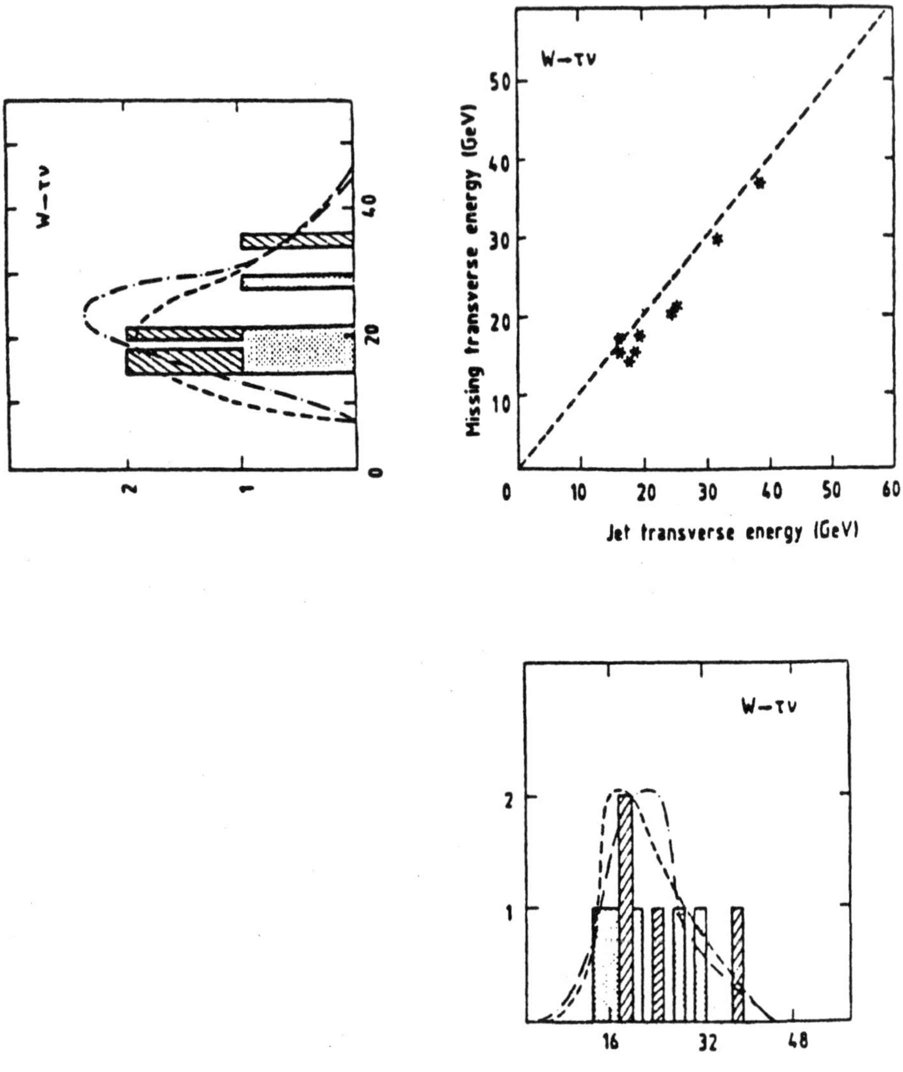

Fig. 53: Missing transverse energy versus jet transverse energy of the one-branch and three-branch $W \to \tau\nu$ event candidates (UA1). The projected distributions for the one-branch and the three-branch (shaded) samples are compared to theoretical curves for the one-branch (dashed curve) and the three-branch τ's (dashed-dotted curve). The jet includes one or three charged pions and possible additional neutral pions.

< 25 GeV and a factor of 20 for jets with E_T > 25 GeV is obtained. For a multiplicity of two or three the reduction factor is only two. So the overall rejection factor against background from bijets is $\approx$100 for the one-branch sample and $\approx$20 for the three-branch sample when using the central detector information.

The background due to jet fluctuations is 0.5 ± 0.1 (stat.) for the five one-branch events sample. For the four events in the three-branch sample the background is 2 ± 1 (stat.).

The overall detection efficiency for the τ has been estimated to be 15 % yielding

$$R = (W\to\tau\nu)/(W\to e\nu) = 0.96 \pm 0.45 \text{ (stat.)}$$

for the one-branch sample.

To conclude, a τ signal has been evidenced for the first time in the channel $W\to\tau\nu$ ($\tau\to$hadronic modes); the branching ratio with respect to the channel $W\to e\nu$ is found to be consistent with the universality of the weak current. The background coming from QCD bijets is supressed by using central-detector information on jet multiplicity.

The corresponding analysis of the 1984 data is in progress. A trigger asking for energy imbalance (>17 GeV) in coincidence with a low-E_T(>15 GeV) jet was used during the 1984 data taking in order to improve the detection efficiency for the process $W\to\tau\nu$ and other events of similar topology. A first analysis of a limited data sample (corresponding to 130 nb^{-1}) has yielded two one-branch and three-branch events when requiring a strict validation of missing transverse energy defined by $E_T^{miss} > 4\sigma$. Two of these events are shown in fig.55.

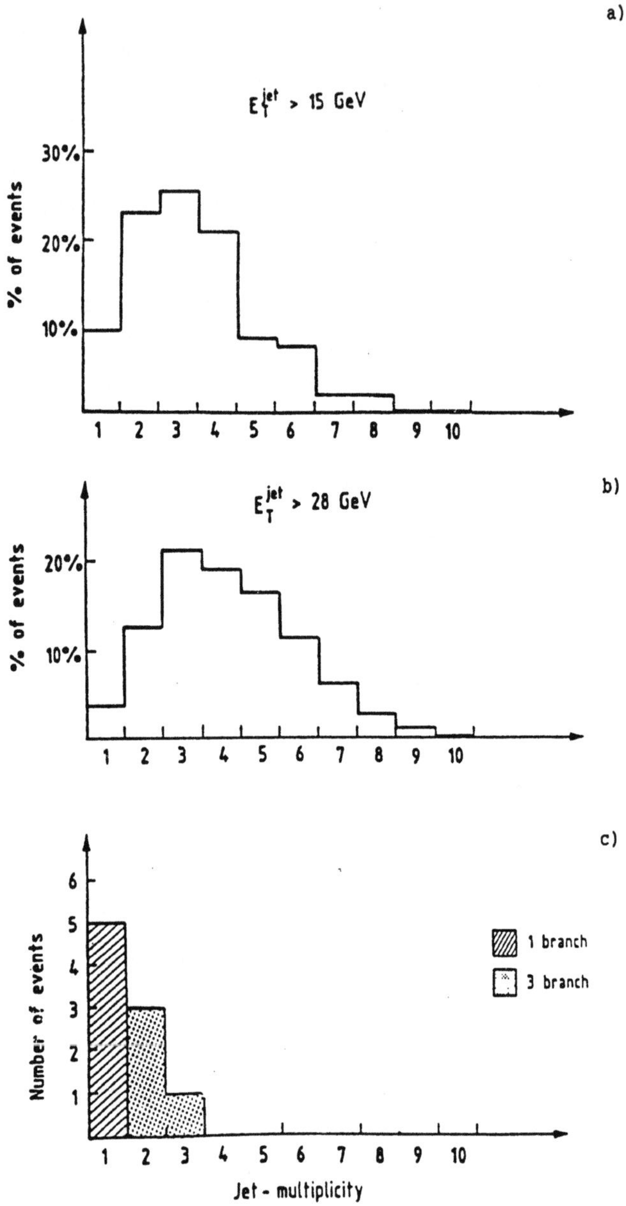

Fig. 54.: Multiplicity distribution. The multiplicity is defined in a cone of $\Delta R = 0.5$ and counting only tracks with $p_T > 0.5$ GeV in this cone.

(a) Multiplicity distribution for jets which have $15 < E_T < 25$ GeV and satisfy the electron trigger.

(b) As (a), but for jets which have $E_T > 25$ GeV.

(c) Multiplicity of the one- and three-branch events.

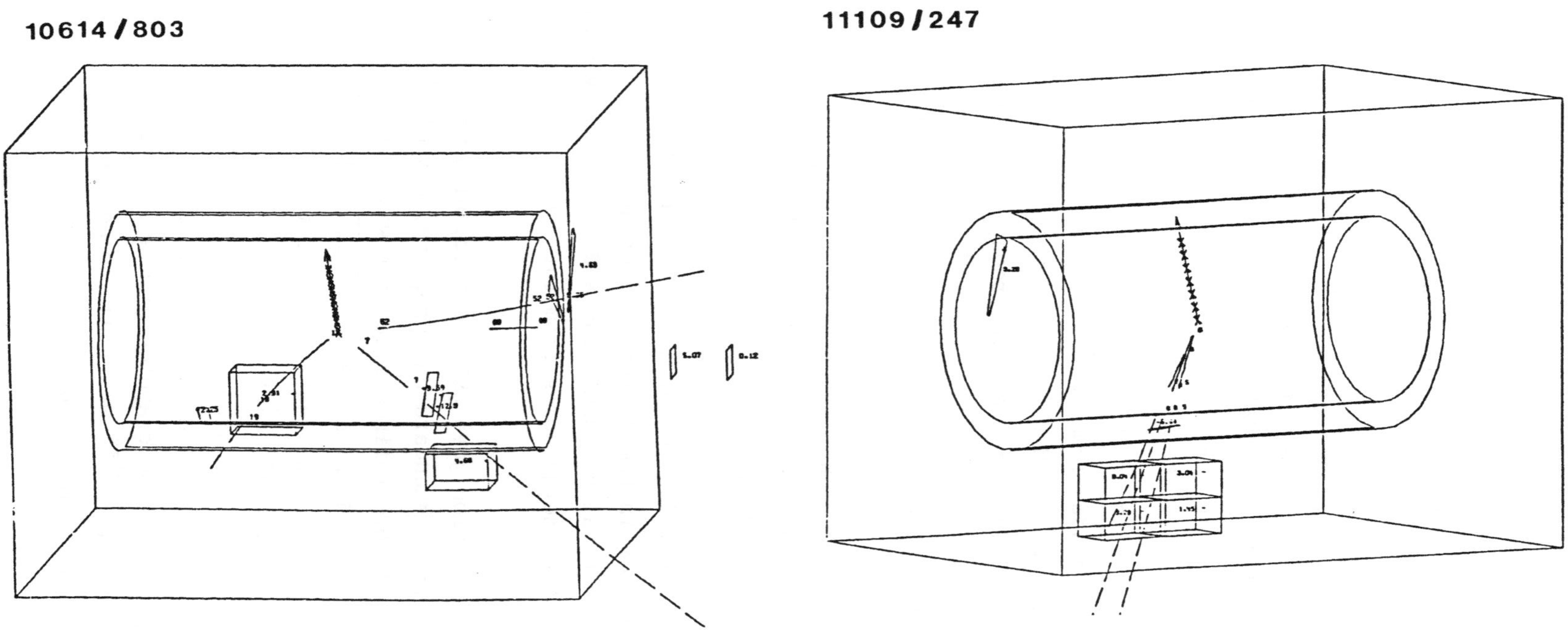

Fig. 55.: Visual display of (a) a one-branch and (b) a three-branch event selected from the 1984 data with the requirement of having a validation of the missing-E_T bigger than 4σ. A 1 GeV cut is applied on the transverse energies of the calorimeter cells and on the transverse momenta of the charged tracks. (UA1).

5.2 Events with missing-E_T and a jet which cannot be due to τ-decay

The UA1 group has found a couple of interesting events in the 1983 data sample when selecting events with significant (4σ validated) missing-E_T [69]. These events will be discussed first. Then a few remarks on similar events from the 1984 data sample will be added.

In the data taking of 1983 no dedicated trigger for E_T^{miss} was in use. An additional signature was therefore necessary in order to record the event, namely either (i) a jet of $E_T > 25$ GeV, (ii) an electromagnetic cluster of $E_T > 10$ GeV, (iii) a muon of $p_T > 5$ GeV, or (iv) a scalar transverse energy $\Sigma|E_T| > 60$ GeV in the region $|\Delta\eta| < 1.5$.

The initial selection of events started from a sample of $1.5 \cdot 10^6$ calorimeter triggers. A first selection consisted of the following requirements: (1) $|E_T^{miss}| >$ 15 GeV; (2) total energy smaller than 700 GeV to remove multiple interactions; (3) technical cuts to remove reconstruction errors in the forward electromagnetic calorimetry; (4) removal of cosmic ray and beam halo events, in which most of the energy comes from the outer calorimeter segments.

The first selection gave 29962 events,which were fully reconstructed by standard UA1 programs, and E_T^{miss} was recalculated. A more refined selection was then performed: (1) $|E_T^{miss}| > 4\sigma$ with $\sigma = 0.7\cdot\sqrt{\Sigma|E_T|}$; (2) E_T^{miss} must not point to within ±20 degrees of the vertical. This cut is necessary because of the reduced efficiency of calorimetry in that region. After this second selection 1157 events were left. The events were than scanned on an interactive high-resolution graphics display. In most of the cases the value of E_T^{miss} was found to be faked by (1) cosmic rays; (2) beam halo particles; or (3) reconstruction problems. A total of 77 events have been declared genuine. They are plotted in fig.56a. There are 50 identified $W \to e\nu$ events which are removed from the sample leaving 27 additional events (fig.56b). The new events are topologically separated into:

(i) 3 events with more than two jets.

(ii) 5 events with two jets.

(iii) 2 events with a single, isolated neutral electromagnetic cluster.

(iv) 17 events with a single jet.

Seven of the eight events with two jets or more than two jets are very near to the 4σ cut. They are most likely compatible with background from QCD and heavy flavour decays. One three-jet event (the one with the largest E_T^{miss} in fig.56b) does not seem to be explicable in terms of background from jet fluctuations. The jets have E_T's of about 55 GeV, 20 GeV,and 15 GeV, and the event has a missing transverse energy of about 55 GeV.

Two events have a single electromagnetic cluster as expected for a $W \to e\nu$ decay, but without associated hard track. One of the two events (event G in fig.56b) is nevertheless most likely to be a $W \to e\nu$ event because the shower points to a small insensitive gap between the chambers. For the other event (even H) such an interpretation has only a sub-percent probability because the shower points to an active region of the central detector.

As apparent from fig.56b there are several events with a single isolated jet which show spectacular missing energies. In order to determine the background from two-jet fluctuations, the cut on E_T^{miss} has been relaxed from 4σ to 2σ and a cut has been applied on $\Delta\phi$, the angle in the transverse plane that the rest of the event (jet not included) makes with respect to the jet. Events A-F, which pass the $\Delta\phi$ cut, are plotted in fig.57 together with the background from jet fluctuations. It is evident

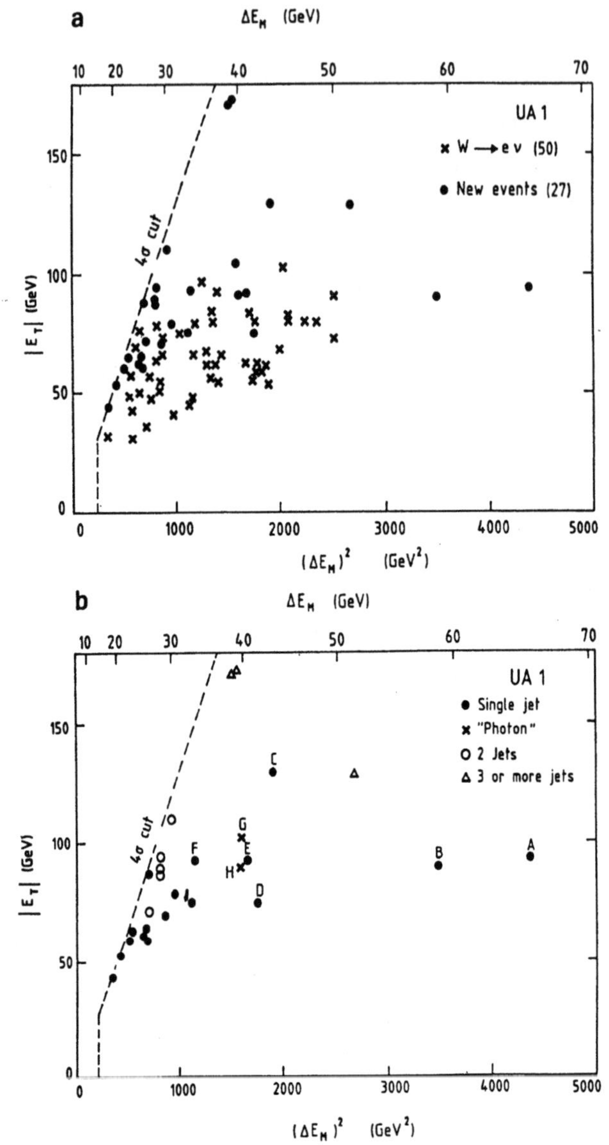

Fig. 56: Scatter plot of missing transverse energy squared versus total scalar transverse energy for the events which have missing E_T in excess of 4 standard deviations. Events identified as $W \to e\nu$ decays are labeled with a cross and the other events are labeled with a circle.

(b) Same as (a), but with $W \to e\nu$ events removed. The event topology is indicated as single jet (solid circle), photon(s) (cross), two-jet event (open circle) and three-or more-jet event (triangle)

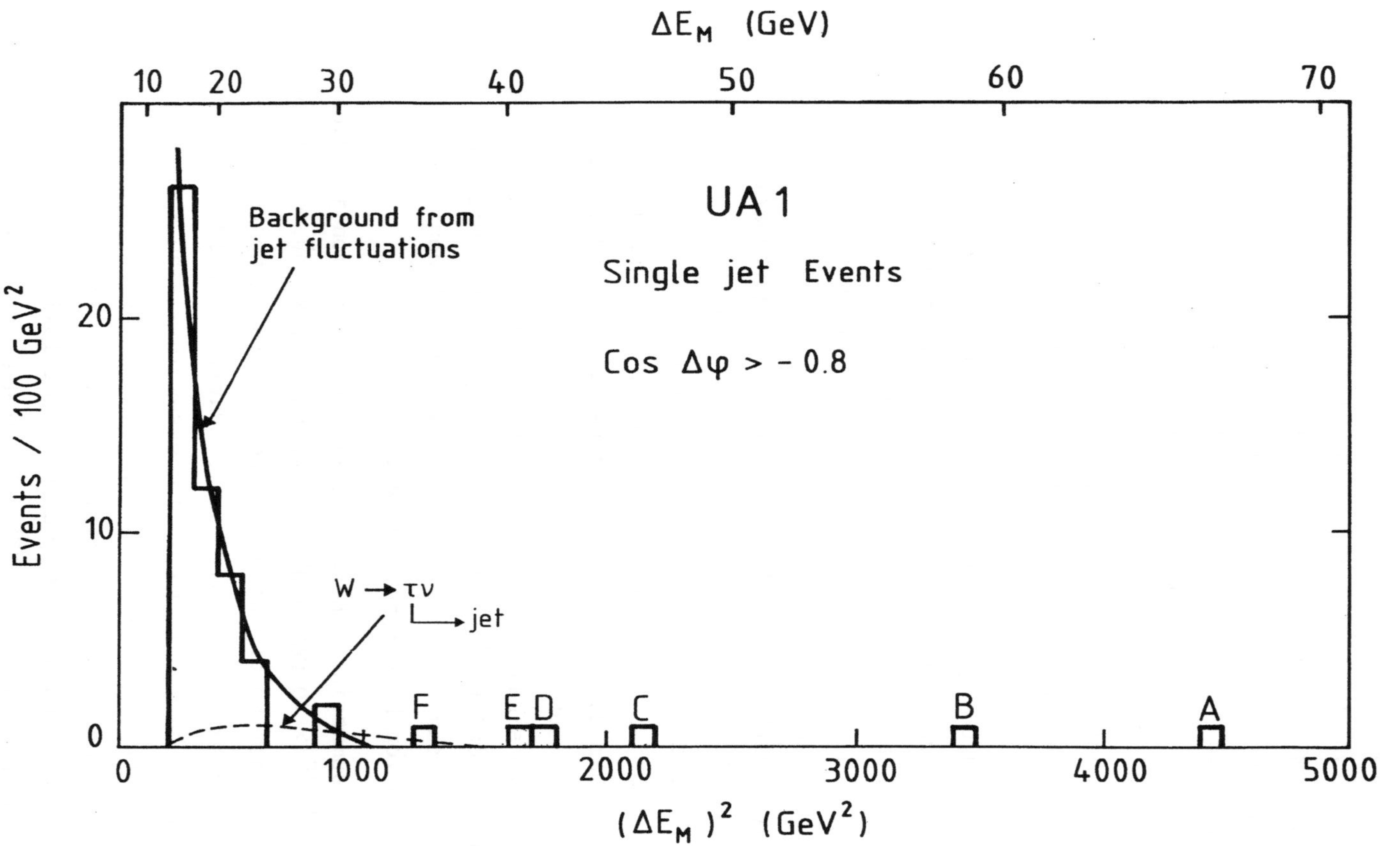

Fig. 57.: Distribution of missing E_T squared for events passing the 2σ cut on missing E_T and the cut $\cos\Delta\phi < 0.8$ (UA1).

that this background does not explain the monojet events A-F. Several other possible background sources have been considered
1) $W\to\tau\nu$. Two events are expected with a missing-E_T distribution shown in fig.57. Only event F, by nature of its lower jet energy and three-prong multiplicity, is consistent with this process. This event is also contained in the three-branch sample of $W\to\tau\nu$ candidates discussed in section 5.1.
2) $W\to L\nu$, where L is a new sequential heavy lepton. The background for any heavy lepton is smaller than that for the τ due to the phase space and decay kinematics.
3) gZ^o. A gluon jet could be produced in association with a Z^o, which could then decay into $\nu\bar{\nu}$ (expected branching ratio is nearly 20 %). From the E_T distribution of jets produced in association with a $W^{\pm}$ a contribution of $\gtrsim$ 1 event has been estimated.
4) Heavy quark decays. There could be a heavy quark pair with one of them decaying to a leading neutrino. Less than 0.1 events are expected from this process. A direct search for lepton-jet events with electron or muon and jet E_T greater than 30 GeV gives no events. The method of monitoring the neutrino spectrum by the electron and muon distributions has, however, recently been questioned [70]. The pseudoscalar $b\bar{c}$ meson B_c has a $B_c\to\tau\nu$ decay mode, but does not decay directly to $e\nu$ or $\mu\nu$. Event F could be accounted for by a $B_c\to\tau\nu$ decay.

To summarize, one of the five UA1 1983 monojet events with E_T^{miss} > 40 GeV could be due to gZ^o, $Z^o\to\nu\bar{\nu}$. The other events cannot be explained by the background sources considered so far. As a consequence, various authors have tried to explain them in terms of processes involving supersymmetric particles[71]. However, the supersymmetry scenario will only then be truly assessed when a full analysis of more data including multijet events of lowish ($\gtrsim$10 GeV) jet E_T's and smaller E_T^{miss} will be available. This will probably only be possible with the upgraded calorimeter. A first analysis of the events with 4σ validated missing transverse energy from the 1984 data seems to confirm the 1983 results on the existence of monojets.

6. SEARCH FOR HEAVY FLAVOURS

This chapter deals with the heavy quarks c, b and t. The c- and b-quarks have been discovered through their hidden flavour states $c\bar{c}$ and $b\bar{b}$ before the era of the SPS collider. The t-quark may become accessible at the collider by the decay $W \rightarrow t\bar{b}$.

All kinematically accessible heavy quarks are produced by the decay of the intermediate vector bosons. Since the collected number of IVB events in the leptonic decay modes is several hundred (including the 1984 data) and since the W couples with equal strength to all weak isospin doublets, a large sample of events of the type $W \rightarrow t\bar{b}$ has been collected as well. However, only a fraction of the heavy quark events can be identified with some confidence. Until now the analysis was restricted to events where one or both quarks decay semileptonically.

A second source of heavy flavours is hadronic production of heavy quark pairs. In this case there are large uncertainties on the value of the production cross section (factor >2), and the unambiguous identification of a given heavy quark pair is experimentally even more difficult than it is in the W decay.

The experimental signature for heavy flavours is one (or two) lepton(s) from one (or two) semileptonic decay(s) associated with jets and enhanced strange particle activity from the later stages in the cascades, together with some global event characteristics. From 1985 on the UA1 group will be able to tag b and c decays by the measurement of additional decay vertices close to the interaction point using the high-spatial resolution 'microvertex detector' (see chapter 7).

A preliminary search for the decay modes $W \to c\bar{s}, Z^o \to c\bar{c}$, $b\bar{b}$ is described in section 6.1 and a search for $W \to t\bar{b}$ in section 6.2. In both cases the analysis starts by selecting one lepton in the final state. Events with two leptons are described in section 6.3. Most of the dilepton events seem to be explicable by heavy flavours or as Drell-Yan pairs. However, some events show unexpected features. In one class of events isolated unlike sign muons are found to be associated with two jets. Possible explanations are lepto-quarks, or higher-order-QCD compensation of a large-p_T Drell-Yan pair. In another class of events, like-sign dimuons, both muons isolated, are seen with no associated jet activity that would indicate heavy quark cascades.

6.1 Search for $W \to c\bar{s}$, $Z^o \to c\bar{c}$, $b\bar{b}$

Results of a preliminary search for $W \to c\bar{s}$, $Z^o \to c\bar{c}$, $b\bar{b}$ have been presented by the UA1 Collaboration [72]. This search is motivated by the comparison of the yield of jet pairs versus p_T from both weak and strong production processes. This comparison is shown in fig.58. Note that, if at least one of the jets is required to come from a heavy quark, the strong production background is reduced to a level comparable to that of W and Z decay. c- and b-jets are identifiable by the presence of a muon in the jet in the approximately 10 % of the cases in which a semi-leptonic muonic decay occurs.

The following selection criteria have been used to look for W or Z decaying to c or b with subsequent semi-

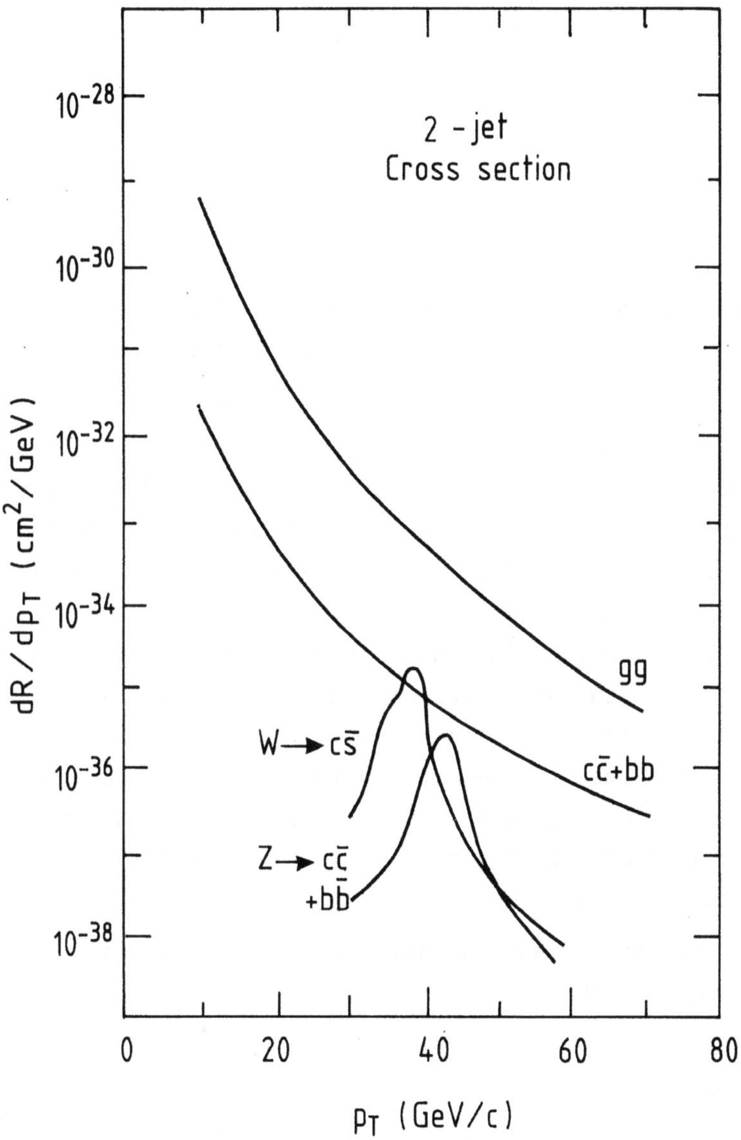

Fig. 58: $d\sigma/dp_T$ versus p_T for centrally produced jet pairs (ref.40).

leptonic decay:

- a muon with $p_T > 8$ GeV/c;
- a nearby jet ($\phi_{\mu\text{-jet}} < 40^o$) with $p_T > 6$ GeV/c and $|\eta| < 2$;
- a second jet with $p_T > 10$ GeV/c and $|\eta| < 2$;
- no additional jet with $p_T > 6$ GeV/c and $|\eta| < 2$;
- a neutrino (i.e. missing transverse energy) of $p_T > 5$ GeV/c.

Because the missing transverse energy has large measurement errors (typically 3.5 GeV/c for each component), the muon-neutrino separation in the transverse plane was allowed to be as large as 90°.

From 108 nb^{-1} of muon-triggered data from the 1983 collider run 22 events satisfied these criteria.

Fig.59a shows $M(jj\mu)$, the invariant mass of the muon-jet-jet system for these 22 events. As indicated in fig.58 this plot should contain $c\bar{c}$ and $b\bar{b}$ pairs produced by strong interactions and contributions from the decay of W and Z. The dashed line in fig.59a is a leading-log QCD calculation of the $M(jj\mu)$ distribution due to strong $c\bar{c}$ and $b\bar{b}$ production [73]. The calculation incorporates scale-violating structure functions and coupling constant and the heavy quark fragmentation function of ref.74. The resulting curve is normalized to the total number of events in the plot; this normalization is compatible, within a factor of three, with expectations for central $c\bar{c}$ and $b\bar{b}$ production cross sections[75].

The invariant mass including the neutrino, $M(jj\mu)$, is shown in fig.59b.

In both plots five to six events are observed above the strong production curve in the neighborhood of the intermediate vector boson mass. Using the measured cross sections for W and Z production, table 12, three such events are expected to be found in the proportions $Z(b\bar{b}) : Z(c\bar{c}) : W(c\bar{s}) = 3 : 1 : 2$. The dominance of $Z \to b\bar{b}$ is due predominantly to the hardness of the b-quark fragmentation

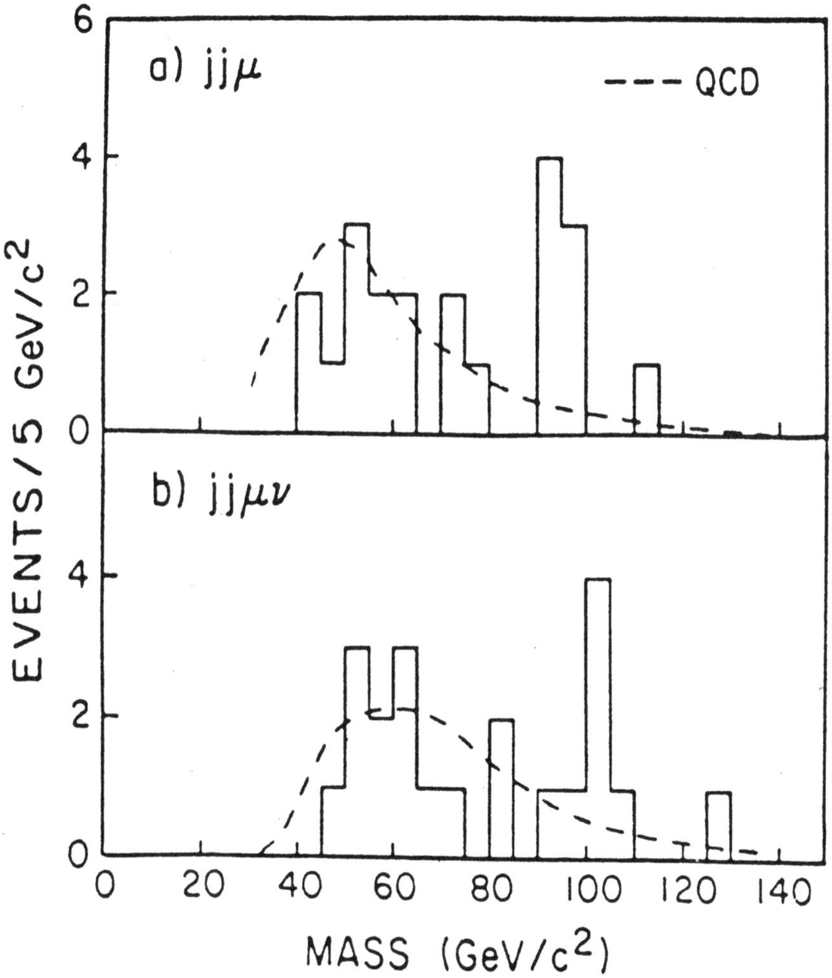

Fig. 59: (a) M(jjμ) and (b) M(jjμν) for 22 selected events together with a QCD calculation. One event is off scale at around 190 GeV/c^2 in both plots. (UA1).

($\langle z_b \rangle \gtrsim 0.8$) relative to that of the c-quark ($\langle z_c \rangle \gtrsim 0.6$)[74].

The jet energies have been corrected by approximately +20 % to account for debris which fall outside of the cone of acceptance of the jet-finding algorithm. Ultimately the observation of W/Z decay into jets should provide an experimental calibration of jet energies.

6.2 Search for $W \to t\bar{b}$

The UA1 Collaboration has recently presented results of a search for events with an isolated, large-p_T lepton (electron or muon) produced in association with two jets [76]. The six events found in the search may be considered as the first experimental evidence for production of Top via the process $W \to t\bar{b}$; $t \to \mu$ or e, jet; $\bar{b} \to$ jet. The definitive discovery of Top has, however, not yet been claimed, although the rate and the features of the events are consistent with the process $W \to t\bar{b}$.

The background to $W \to t\bar{b}$ from multijet events is more important than the background to the corresponding process $W \to c\bar{s}$. This is due to the shift of the Jacobian peak, fig.58, towards smaller transverse momentum p_T of the recoil jet. A reduction of p_T by 10 GeV/c (corresponding to a Top mass m_t = 40 GeV/c^2) translates into an one-order-of-magnitude increase of background from QCD jets. The increase of the background is somewhat compensated by the sharper experimental signature of the semileptonic t decay. Due to the expected large mass of Top the lepton from t decay is in general better isolated than the corresponding lepton from b decay.

The experimental signature for $W \to t\bar{b}$ and the subsequent semileptonic decay of t is given by:

(i) an isolated lepton,

(ii) a large-p_T recoil jet, and

(iii) a lower p_T b jet.
In the centre-of-mass frame of the (lepton, neutrino, $\bar{b}$-jet, b-jet) system θ^* is defined as the angle between the b-jet and the average beam axis. The distribution of $\cos\theta^*$ is expected to be flat for the reaction of interest. The projection of the lepton energy onto the axis perpendicular to the plane defined by the average beam axis and the recoil jet is called E_T^{out}. For the reaction of interest the average value of E_T^{out} is expected to be $\langle E_T^{out}\rangle \gtrsim m_t/4$.

The background to $W\to t\bar{b}$ is dominated by QCD two-jet events, mainly $b\bar{b}$, where one jet fakes the lepton signature and an additional jet is produced by initial-state gluon bremsstrahlung. In this case the smaller gluon jet tends to follow the beam direction, $|\cos\theta^*| \gtrsim 1$, and $\langle E_T^{out}\rangle$ is small. Possible problems arising from final-state gluon bremsstrahlung are discussed in ref.77.

An independent estimation of the absolute value and the shape of the $b\bar{b}$ cross section is obtained from events with two muons (section 6.3).

Using the 1982-1983 data the expected event rate for $W\to t\bar{b}$ is approximately one event in the electron channel and one event in the muon channel. The reduction of the signal from the produced $\gtrsim$200 events is caused by the selection of the semileptonic decay mode and by the cuts applied in the analysis.

A sample of events with isolated electrons has been selected by using the following requirements:
(i) presence of an electromagnetic cluster in the calorimeters, with E_T > 15 GeV, and a matching track in the central detector;
(ii) the electron candidate is isolated in a cone $\Delta R = 0.4$: Σp_T < 1 GeV/c (in terms of central-detector tracks), and ΣE_T < 1 GeV (in terms of calorimeter energy).

In the first step conversions have been rejected. Then all known $W \to e\nu$ events have been removed. 12 events passed the selection after applying strict quality cuts on the shower profile and on the quality of calorimeter / central detector matching. These events are subdivided into three topological categories:

1) electron + one jet (7 events)
2) electron + two jets (3 events)
3) electron + three jets (2 events).

From the point of view of isolation the (e + 1jet)-events seem to be less well isolated than the electrons from $W \to e\nu$ decays, while the (e + $\geq$ 2jet)-events are perfectly isolated. The (e + 1jet)-events seem to be remnants of two-jet events which passed all the selection cuts.

A sample of isolated muons has been selected by asking the muon candidate tracks of $p_T > 8$ GeV/c to satisfy $\Sigma p_T < 0.1 \cdot p_T^{\mu}$ and $\Sigma E_T < 0.2 \cdot E_T^{\mu}$; the summing is performed in a cone of $\Delta R < 0.4$ around the muon. In addition the distance to the closest jet was required to be $\Delta R > 1$. After the known $W \to \mu\nu$ events have been removed 12 events pass the selection. These events are subdivided into three topological categories:

1) muon + one jet (7 events)
2) muon + two jets (4 events)
3) muon + three jets (1 event).

The scatter plot of $\cos\Theta^*$ versus E_T^{out} is shown in fig.60 for electron and muon events with $\geq$ 2 jets. The distribution is roughly uniform in both variables. One of the (μ + 2jet)-events has a large value of $\cos\Theta^*$, $\cos\Theta^* = 0.93$; this event is hence likely to be a background event and was removed from the sample.

The shape of the background was obtained from π^o + 2jet events and is shown in fig.61. The absolute magnitude of the background was estimated by using the smaller sample

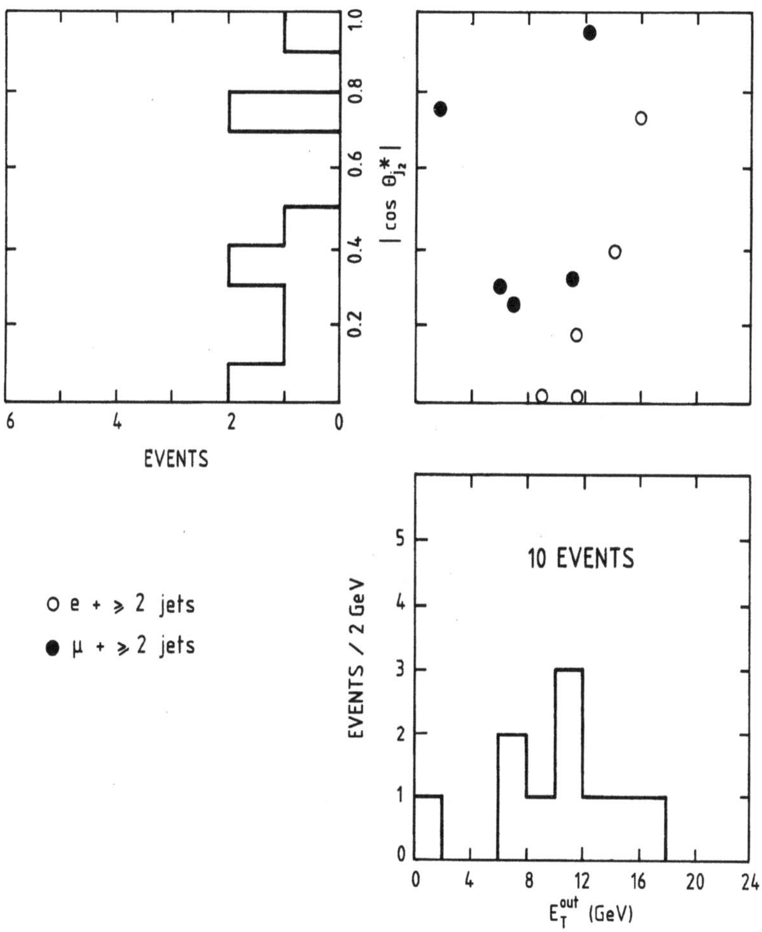

Fig. 60: Scatter plot of $\cos\theta^*$ versus E_T^{out} for the isolated electron + $\geq$ 2jet events (open circles) and the isolated muon + $\geq$ 2jet events (full circles). (UA1).

($\approx$6 nb^{-1}) of events with an isolated charged hadron + $\geq$ 2jets. The charged-hadron spectrum was then folded with the probability to fake a lepton. The resulting total background to the electron + $\geq$ 2jet sample is $\approx$0.1 events; the corresponding background to the muon + $\geq$ 2 jet sample is $\approx$0.4 events. In conclusion, the background contamination to the six isolated lepton events seems to be small and becomes even smaller, if events at large $\cos\Theta^*$ and small E_T^{out} are excluded.

The non-isolated muon + 2jet events were used in a second estimation of the background. Contrary to an electron a muon can be identified in the UA1 detector, even if it happens to be burried in the middle of a jet. 17 events with a non-isolated muon + 2jets were found (the lower E_T jet is required to be central, $|\cos\Theta^*| < 0.8$). The distribution of these events in the scatter plot $|\cos\Theta^*|$ versus $\Delta\phi$, the angle in the transverse plane between the lepton and the higher E_T jet, is shown in fig.62. The corresponding distribution for the six isolated lepton + 2jet events is shown in fig.63. The isolated and the non-isolated events are found to populate complementary regions. While all 17 non-isolated events are concentrated in the region defined by $|\cos\Theta^*| > 0.8$ or $\Delta\phi > 155^o$, five of the six isolated events are found to lie outside this region. This observation further indicates the presence of a signal.

The separation of the possible signal from the background is further elaborated in fig.64, where the 4-body mass distribution $m(\mu\nu_T jj)$ is plotted as a function of the isolation parameter ΣE_T (summed in the cone $\Delta R < 0.4$ around the muon; the average energy deposition of the muon in the calorimeter has been subtracted). The isolated muon events (shaded) show a peak at the W mass. The ΣE_T distribution of the isolated events agrees with that expected for genuinly isolated muons, the ΣE_T arising from a random contribution from the underlying event (solid curve).

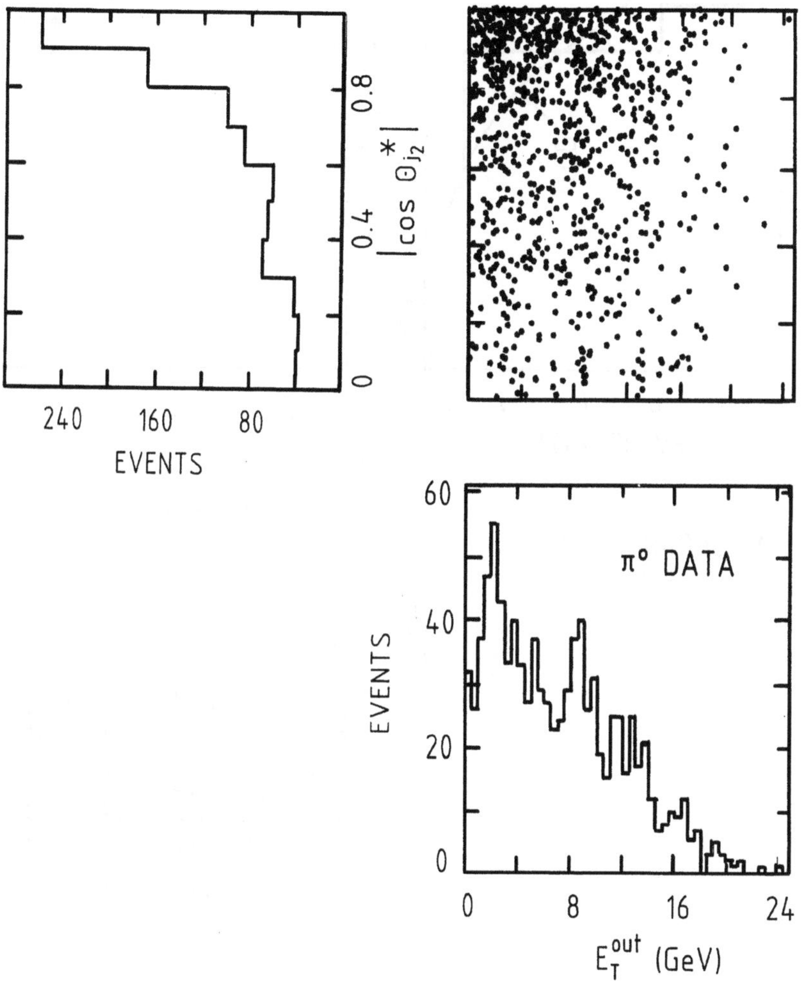

Fig. 61: As fig.60, but for $\pi^0 + \geq 2$jet events (UA1).

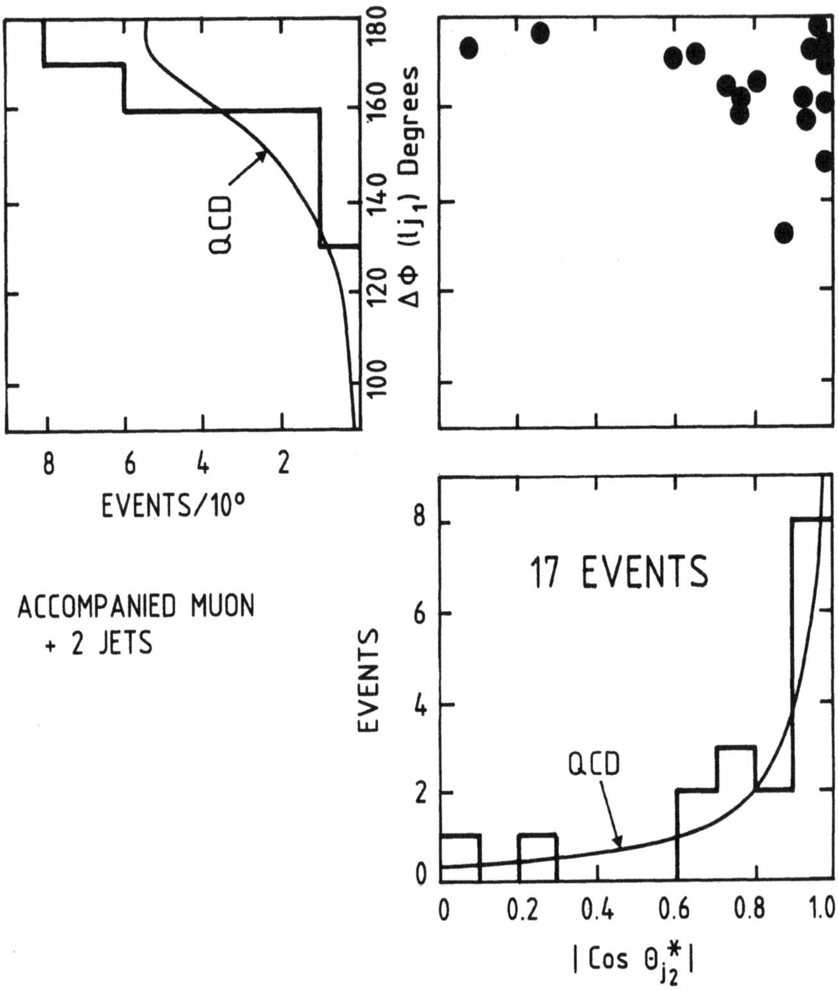

Fig. 62: Event shape for non-isolated muon + 2jet events (UA1). The full lines indicate the shape of the expected background from QCD three-jet events.

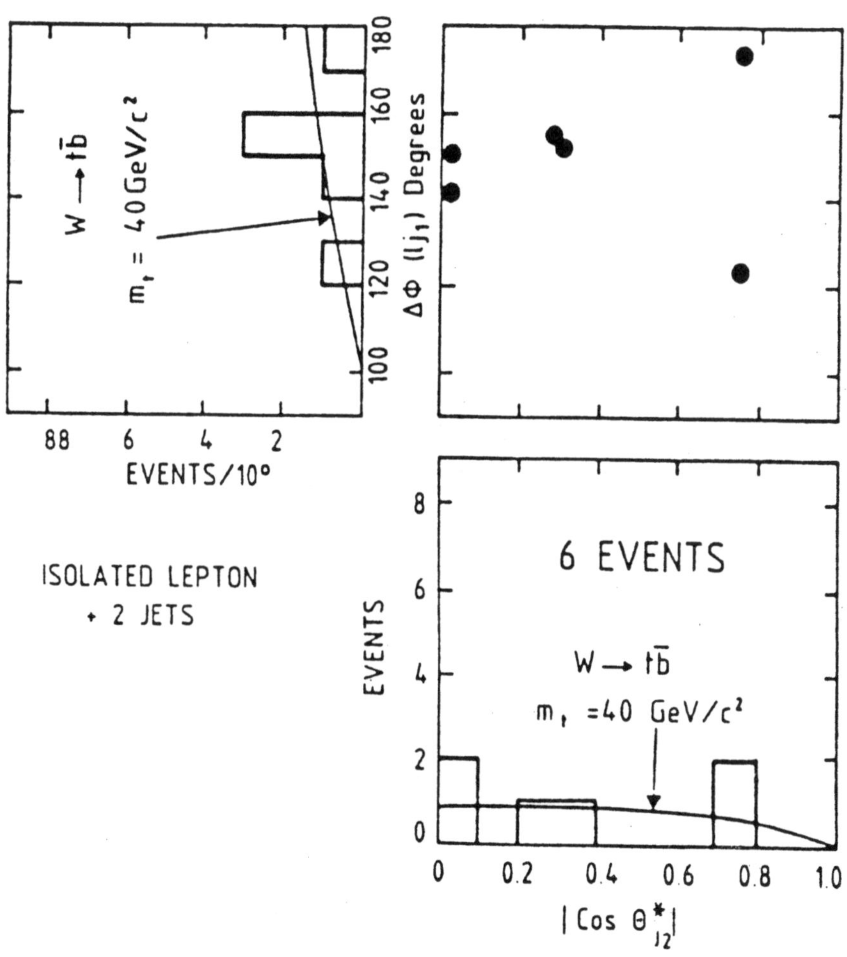

Fig. 63: Event shape for the isolated lepton + 2jet sample (UA1). The full lines show the expectation for $W \to t\bar{b}$ with a Top quark mass m_t = 40 Gev/c^2.

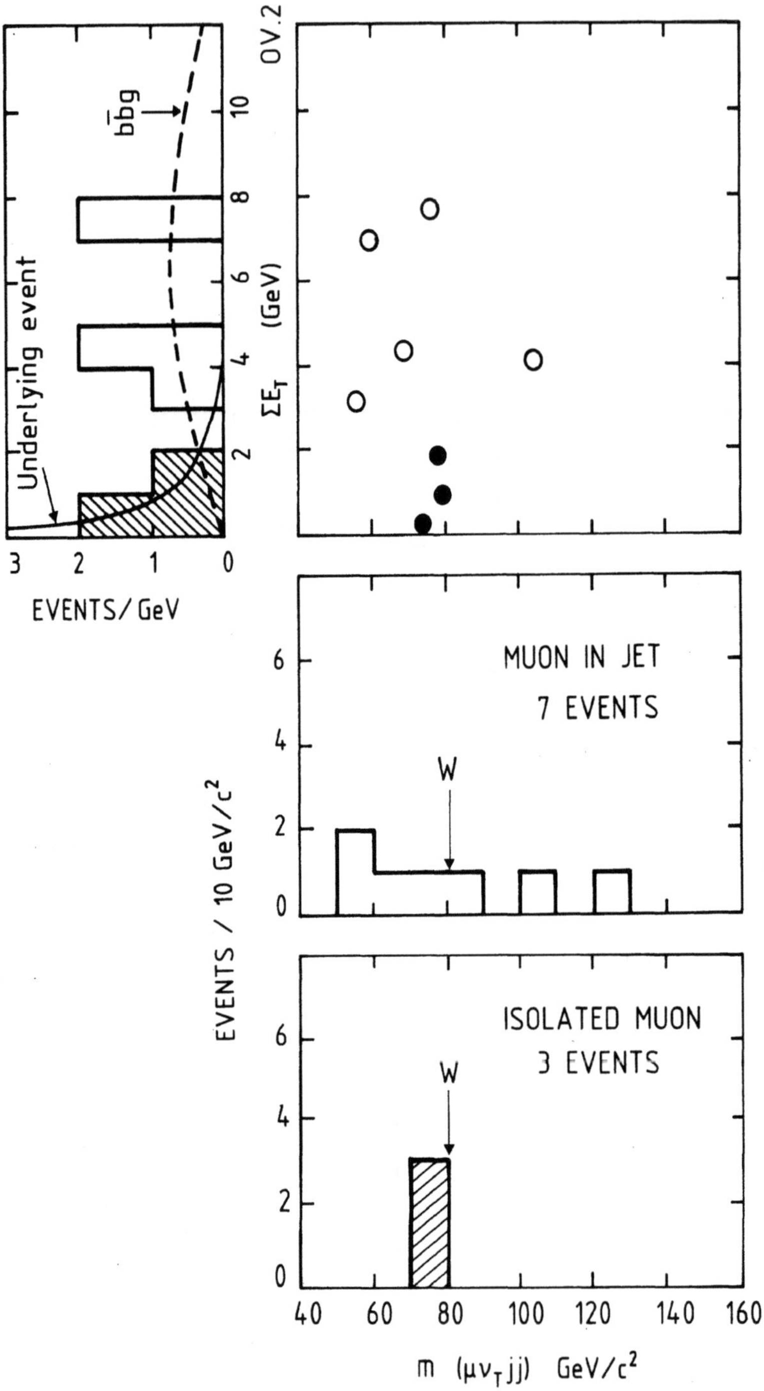

Fig. 64: Scatter plot of $m(\mu\nu_T jj)$ versus the isolation parameter ΣE_T. See text for further details (UA1).

The ΣE_T distribution for the seven non-isolated events selected by asking both jets to lie in $|\cos\theta^*| < 0.8$ agrees with an estimation of the contribution from $b\bar{b}g$ (broken curve). In the region of the isolated events there appears to be only a small contamination from $b\bar{b}g$ background. The background shows a broad distribution in $m(\mu\nu_T jj)$. The obtained absolute number of $b\bar{b}g$ background events is consistent with an estimation obtained by using dimuon events (section 6.3).

Finally, the four-body versus the three-body mass distribution for the six $W\rightarrow t\bar{b}$, $t\rightarrow b\mu\nu$ candidate events is shown in fig.65. The four-body mass peaks at the W mass. The three-body mass clusters around a value of $\gtrsim 40$ GeV/c^2. The curves show the distributions expected for $W\rightarrow t\bar{b}$. The experimental resolution has been taken into account when computing the curves. Allowance should, however, be made for a systematic error arising from uncertainties in the jet reconstruction (± 10 GeV/c^2).

In conclusion, the experimental distributions for the isolated lepton + 2jet events appear to be consistent with the $W\rightarrow t\bar{b}$ hypothesis and incompatible with c and b production. Assuming the six isolated lepton + 2jet events are indeed due to t they may be used to compute the following bounds on the Top mass m_t:
$30 \text{ GeV}/c^2 \leq m_t \leq 50 \text{ GeV}/c^2$.

The analysis of the 1984 data is in progress and should provide answers to further questions.

6.3 Dilepton events other than $Z^0\rightarrow\ell^+\ell^-$

Studies of lepton pairs below the Z^0 mass peak were presented by the UA1 and UA2 groups [78]. The starting

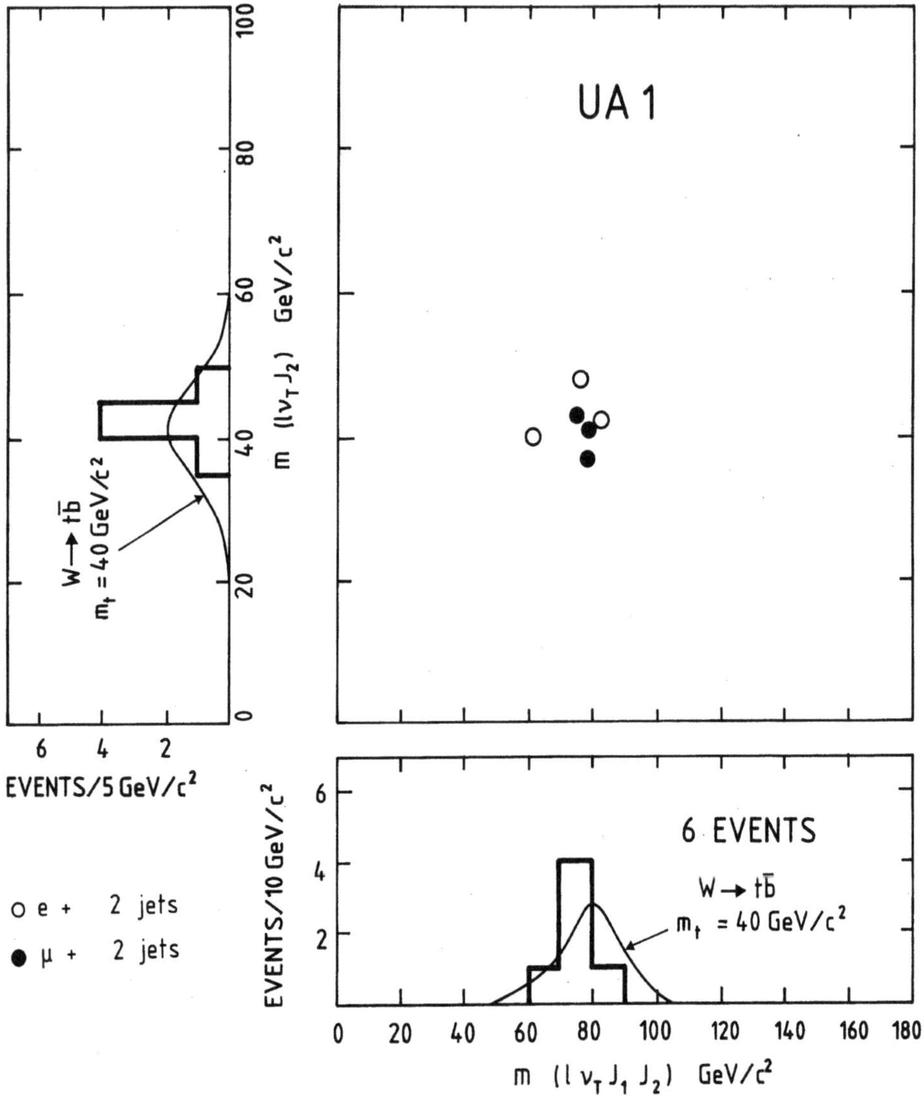

Fig. 65.: Three-body versus four-body mass distribution for the six $W \to t\bar{b}$, $t \to b\ell\nu$ candidate events. The effective mass of the lepton, lower-E_T jet, and transverse component of the neutrino is plotted against the mass of lepton, two-jet, transverse neutrino system (UA1). See text for further details.

point in the UA1 analysis was the invariant-mass distribution of muon pairs for the 1983 data shown in fig.31. Apart of the Z^o peak which contains 5 events 10 additional events were found at lower masses. The selection required that both muons have transverse momentum $p_T > 5$ GeV/c. Since the background was found to depend mostly on the sum of the transverse momenta $|p_T(\mu_1)| + |p_T(\mu_2)|$, the event selection was later extended by using the following requirements:

(i) $p_T(\mu) > 3$ GeV/c for each muon, and

(ii) $|p_T(\mu_1)| + |p_T(\mu_2)| > 10$ GeV/c.

In addition to the 5 Z^o events 21 events were found (4 like-sign, 17 unlike-sign) with dimuon masses from 2 to 22 GeV/c^2. The background in the full sample was estimated to be 1.4 events. The background is small, 0.06 events, in the sample where both muons have $p_T > 7$ GeV/c, and 0.013 events, if both muons have $p_T > 10$ GeV/c. The background for events with isolated muons is a factor $\sim$5 smaller than the quoted total background which is highest in events with muons in jets. The acceptance for dimuons in the 1983 data is $\sim$40 % at $M(\mu\mu) = 25$ GeV/c^2, and $\sim$44 % at the Z^o mass.

There are several expected sources for medium-mass dilepton events. The J/ψ and Υ resonances decay to unlike-sign lepton pairs with muon branching fractions of $\approx$ 7.5 % and $\approx$ 3 %, respectively. The Drell-Yan mechanism [79] $p\bar{p}\to\gamma^*X\to\ell^+\ell^-X$ produces unlike-sign leptons with a steeply falling dilepton mass spectrum. Strong production of heavy quark pairs [75] followed by first-generation semileptonic decays yields unlike-sign dileptons. Pairs of heavy quarks can, of course, also be produced through the decay of IVB's. While the process $W\to t\bar{b}$ gives like-sign leptons from first generation decays, the processes $Z^o\to t\bar{t}$, $b\bar{b}$, $c\bar{c}$ yield unlike-sign leptons.
However, like-sign dileptons can also originate from

strong production or Z^o decays either through second-generation decays ($b\to c\to\mu$) or $B^o\,\bar{B}^o$ mixing.

The isolation of the muons can be used to distinguish between the Drell-Yan process and heavy-flavour decays. Muons arising from the Drell-Yan process should not be embedded in significant hadronic activity, in contrast to c or b quark decays. The overall activity observed in the dimuon events covers a wide range of transverse energies. Some events have several accompanying jets in the central region with transverse energies up to 60 GeV; others are completely quiet. The muon isolation can be classified by calculating the ΣE_T around the muon, i.e. the sum of transverse energy of all calorimeter cells in a cone of a size ΔR around the muon; the average energy deposition due to the muon itself is subtracted from the sum. The sum ΣE_T in a cone $\Delta R < 0.7$ is plotted in fig.66, which shows that there is a class of events where both muons are clearly in jets. Defining an isolated muon to be one with $\Sigma E_T < 4$ GeV, 9(3) unlike- (like-)sign events with two isolated muons are found, 3(0) events with one isolated muon, and 5(1) events with neither muon isolated. The dotted curve in fig.66b is deduced from soft events in which the ΔR cone was randomly distributed in pseudorapidity and azimuth and the fluctuations in the energy loss of the muon have been folded into the distribution. It indicates the level of activity present from the underlying event. The dashed curve is a simulation for $b\bar{b}$ and $c\bar{c}$ production [80]. The heavy-flavour curve is normalized to 9.9 events and the soft-event curve to 11.1 events so that the sum of the two curves (solid line) corresponds to 21 events.

Heavy-quark decays are often accompanied by strange particles. K^o- and Λ^o-decays can be identified in the central detector as V^o's in the kinematical range $p_T > 0.3$ GeV/c

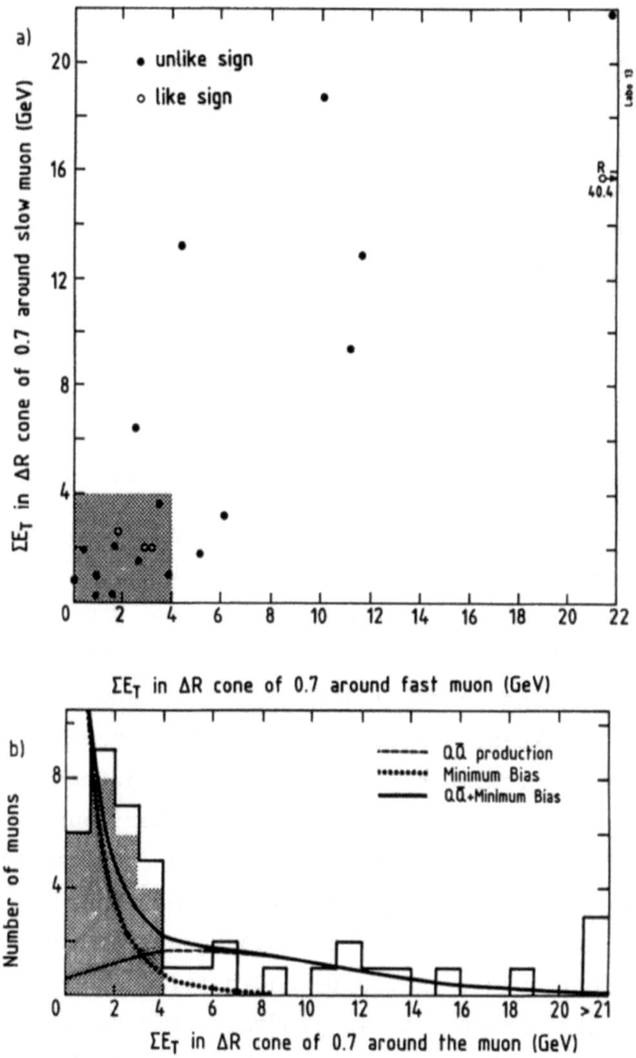

Fig. 66: The scatter plot (a) shows the transverse energy in a cone $\Delta R < 0.7$ around the faster versus the slower of the two muons. The average energy deposition of the muon itself has been subtracted from the calorimeter cells traversed by the muon. The lower plot (b) shows the transverse energy in the cone for each muon. The curves show the distributions for the expectation from $b\bar{b}/c\bar{c}$ production [80] and for a randomly oriented cone in soft events (dotted line). The sum of the two curves has been normalized to 42 muons.

and $|\eta| < 2$, if the decay length is greater than 0.5 cm. The number of V^0's produced per soft event for this kinematical range has been measured by the UA5 Collaboration [25] and was found to be 0.15 ± 0.04 (0.45 ± 0.09) for $\Lambda/\bar{\Lambda}$ (K^0). Heavy-flavour production of dimuons will also yield two neutrinos, which could give rise to missing transverse energy. However, the neutrinos will generally have low p_T because of the bias from the muon p_T cuts. Furthermore, if the neutrinos are back-to-back, their transverse momenta will tend to cancel out in the missing E_T. The observed missing energy is indeed found to be small.

Drell-Yan pairs are characterized by two unlike-sign muons that are isolated. Monte Carlo simulations have shown that the isolation for a muon from the Drell-Yan process should be $\Sigma E_T < 4.0$ GeV in $\Delta R < 0.7$. This isolation criterion is fulfilled for both muons in nine events with unlike-sign pairs. Five of these events have the two muons approximately back-to-back in azimuth with a dimuon mass from 10 to 17 GeV/c^2. The other four events have the muons in the same hemisphere, and, because of the selection cuts, with a large $p_t^{\mu\mu}$. In these events the large p_T of the muon pair is balanced by one or more jets. One of the four events is a J/ψ candidate. In two of the four events the dimuon pair is predominantly balanced by two jets; one of them (event M) is shown in fig.67. It has been speculated [81] that these two events together with another event of similar topology found at PETRA [82] may be due to a light leptoquark. The (muon-jet) invariant masses are found to be in the range of 20-25 GeV/c^2 in the three events, if the $\mu^{\pm}$-jet pairings are chosen such as to maximize the angle between the μ and the jet momenta within each pair.

Assuming that the five events with masses above 10 GeV/c^2 are from Drell-Yan pair production the total Drell-Yan cross section is found to be 0.6 ± 0.3 nb for

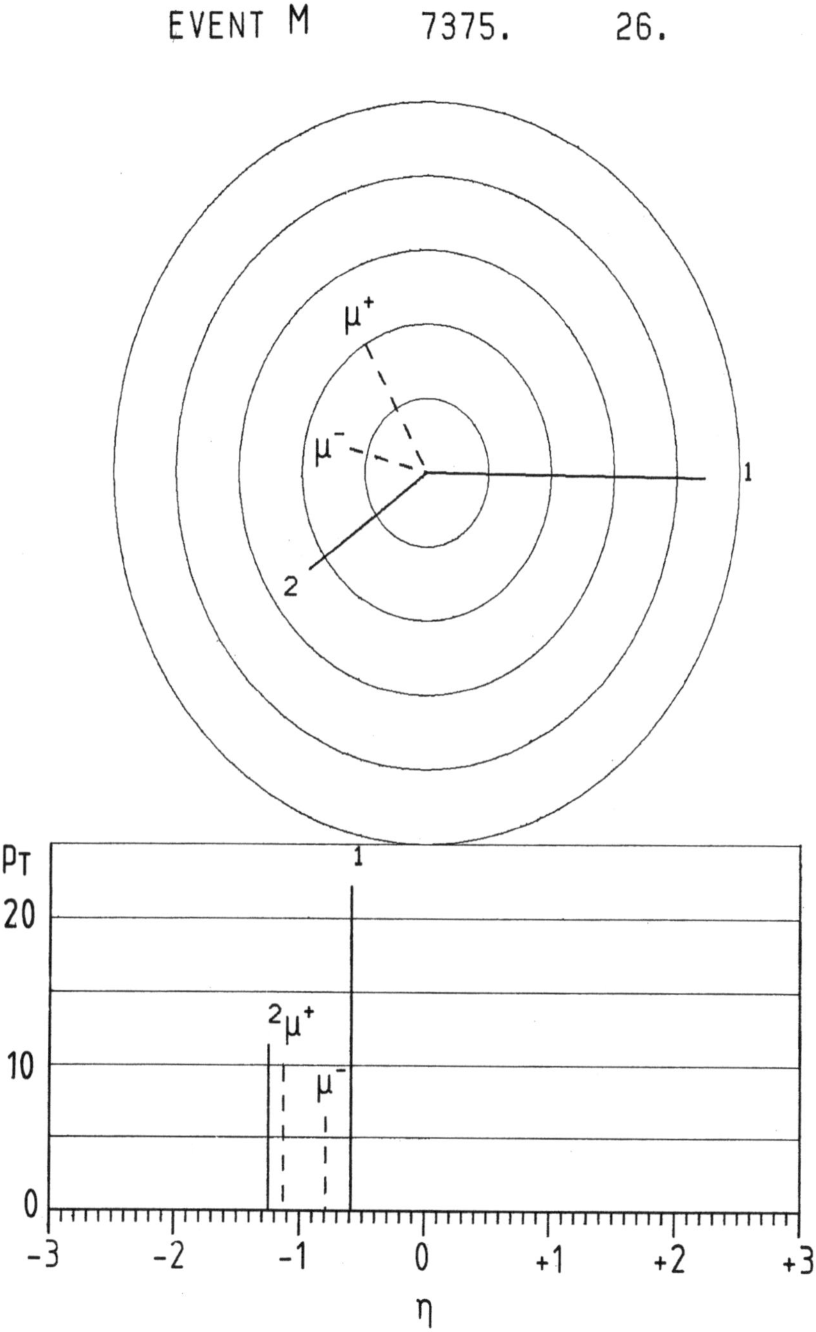

Fig. 67: Configuration of event M (7375/26) in the plane transverse to the beam axis and in pseudorapidity (η). Muons are drawn with dashed lines and jets with solid lines.(UA1).

$m_{\mu\mu} \geq 10$ GeV/c^2 in agreement with a recent calculation that yields 0.35 nb. However, the event sample may be contaminated by the Υ resonances as well as by heavy-flavour production; no correction for this has been applied to the cross section.

In another class of events the muons come from heavy-flavour decays and are expected to be accompanied by hadronic activity. If the events with at least one non-isolated ($\Sigma E_T > 4$ GeV) muon are defined as being the heavy flavour events, then there are nine such events (fig.66a). The calculated isolation distribution for $b\bar{b}/c\bar{c}$ production [80] is compared with data in fig.66b, the theoretical distribution being normalized relative to the nine non-isolated events. The two distributions are consistent. With this normalization heavy flavours are estimated to contribute about one event where both muons are isolated. As a consequence of the cuts applied in the selection $b\bar{b}$ is the dominant component, while $c\bar{c}$ contributes $\approx$10 %. The cross section for producing $b\bar{b}$ with both quarks having $p_T > 5$ GeV/c and $|\eta| < 2$ is found to be 2.2 ± 0.8 μb.

There is one like-sign event among the nine non-isolated events. Like-sign events can be produced in the cascade process from $b\bar{b}$, $t\bar{t}$, and $t\bar{b}$ final states and also from primary $b\bar{b}$ quarks with B^o-$\bar{B}^o$ mixing. The ratio of like- to unlike-sign events is consistent with the expectations of $b\bar{b}$ events without invoking B^o-$\bar{B}^o$ mixing. Many events have additional high-p_T strange particles; this favours the heavy-flavour interpretation. One like-sign event (event R, fig.68) is possibly a $Z^o \to b\bar{b}$ decay, where one muon comes from a first-generation decay and the other from a second-generation decay. The measured mass of the μ-μ-jet-jet system in this event is $\approx$110 GeV.

Three events are observed which have like-sign dimuons with both muons being isolated ($\Sigma E_T < 4$ GeV). In all these events the muon momenta are 6 standard deviations or more from infinite momentum; hence the charges are well determined. One of these events may come from a second-generation decay of a b quark, since one muon is with an identified central-detector jet in a region where the efficiency of the calorimeters is reduced; about one event with two isolated muons is expected to come from heavy flavours (see discussion of fig.66).

Two of the isolated like-sign events have no jets and the high muon p_T's make second generation $b\bar{b}$ decay unlikely. Both events have similar topology: the muons are back-to-back and a Λ is found at about 90° to the muons in the transverse plane. One event (event Q) is shown in fig.69. In this event both muons have very high transverse momenta (larger than 10 GeV/c); as a consequence the background is small ($< 3.10^{-3}$ events). Calculations have failed to reproduce events with characteristics of event Q from $b\bar{b}$ production with or without B^o-$\bar{B}^o$ mixing. The expected rate of events with both muons having $p_T >$ >9 GeV/c is about 0.2 events with unlike signs and about 0.03 events with like signs; if both muons are isolated, the expected number of events is as small as $\gtrsim 10^{-4}$ events. By invoking mixing the like-sign rate can be increased by about a factor of 2. In conclusion, event Q seems to escape any conventional interpretation. Several events of similar characteristics have also been found in the 1984 data, although none of them appears to be as spectacular as event Q.

7. FUTURE OUTLOOK

7.1 Improvement programm (1983-1986)

The detectors UA1 and UA2 were proposed about seven

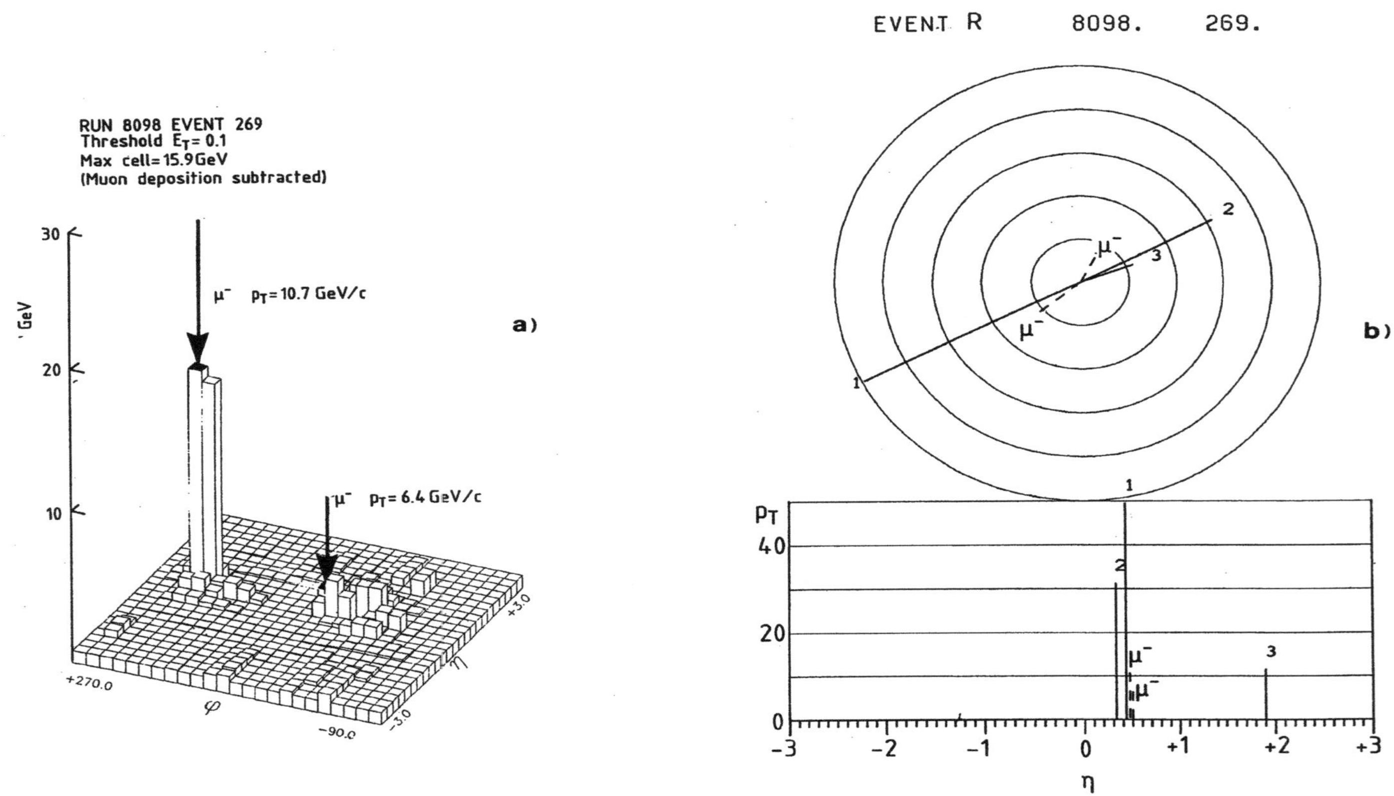

Fig. 68.: (a) The calorimeter energy flow in $\phi-\eta$ space for event R (8098/269). The arrows (length proportional to p_T) indicate the positions of the muons.
(b) Event configuration for event R as in fig.67. For this figure the muons are not included in the jet parameters. (UA1).

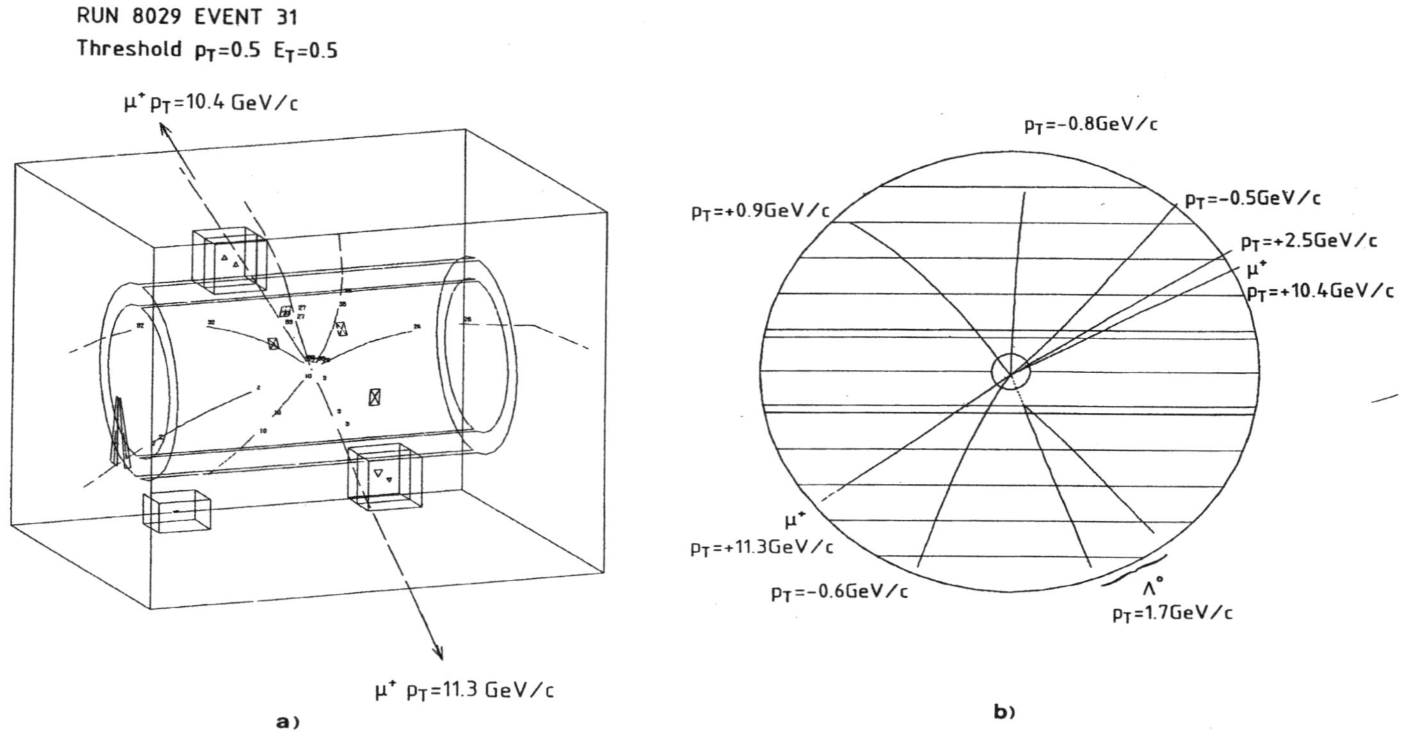

Fig. 69.: (a) Event picture for event Q (8029/31). The two muons are indicated by arrows. Only charged tracks with $p_T > 0.5$ GeV/c and calorimeter cells with $E_T > 0.5$ GeV are shown.

(b) Transverse view of the central detector for event Q showing all charged tracks with $p_T > 0.5$ GeV/c and the π^- track from the Λ^o, which is only 0.2 GeV/c. The sign of p_T indicates the particles charge. (UA1).

years ago. At that time the proposals were motivated by three well-defined physics goals:

(1) the discovery of the charged intermediate vector bosons $W^{\pm}$ in the electron and muon decay modes and the measurement of the charge asymmetry in the decay;

(2) the discovery of the neutral intermediate vector boson Z^{o} in the electron and muon decay channels;

(3) the observation of jets and a detailed comparison with QCD calculations.

These goals have all been fulfilled by 1983. In addition and perhaps more surprising has been the observation of the extreme cleanliness of the events, where a modest p_T threshold cut can unravel the interesting phenomena from the spectator background.

Although the detectors have performed according to specifications, the picture of high-energy physics and the significance of high-energy hadron-hadron collider experiments have further evolved:

(i) The experiments UA1 and UA2 have verified the predictions of the standard SU(2) x U(1) theory within the accuracy achievable with the present detectors.

(ii) the remarkable accuracy of the predictions of SU(2) x U(1) suggests a close connection between the massive vector bosons and the massless photon. Still, the symmetry must be broken in some way. The only known way of doing this at present is with the aid of Higgs scalars. The standard model introduces a doublet of Higgs fields leading to one "physical" neutral spinless Higgs particle. The mass of such a particle cannot be predicted, although it must be larger than about 8 GeV/c^2. Alternatively, the Higgs scalars could be composite objects, generated as a part of a whole new particle spectroscopy, the so-called extended technicolour. Some of these particles may already be produced at collider energies.

(iii) Searches for the t-quark at e^+e^- colliders so far

have been fruitless. The charged IVB decay should be a copious source for this new flavour (provided $m_t \leq 60$ GeV/c^2), but an unambiguous identification of the t-quark from W decays has not yet been achieved. The t-particle should be produced with and decay into the b-particle, for which a long life time, $\tau \approx 1.5 \times 10^{-12}$ sec, has been measured [83], which permits the observation of its decay with high precision detectors.

(iv) New particles could be produced at the collider: new generations of quarks and leptons, new W's, Z's, supersymmetric particles, etc. An important measurement in this search for new particles is the precise verification of the Z^o width, since almost all kinematically allowed particle pairs are produced with a detectable rate by the decaying Z^o. A broader width is also expected for composite Z^o models, since charged, coloured constituents may open additional decay channels. The W decay offers a possibility of detecting heavy sequential leptons or supersymmetric partners of quarks, leptons, and IVBs provided they are not too heavy. New non-interacting particles can be detected like neutrinos by missing-E_T; improvements should allow to measure missing-E_T in the UA1 detector reliably down to 5 GeV.

After a long shut-down in 1985 the SPS collider will resume operation with a substantially higher luminosity than presently available. The Antiproton Collector (ACOL) which is now being constructed should allow for a total integrated luminosity of nearly 10 pb^{-1} by the end of 1989. In order to make the best use of the increased luminosity in terms of physics results, the performances of the present UA1 and UA2 detectors have to be improved.

The improvement program for the UA1 detector consists of four parts:

(a) improved muon detectors,

(b) new data-acquisition system with distributed intelligence for fast decision-making procedures,

(c) microvertex detector for tagging b-particle decays, and

(d) improved calorimetry for better missing-E_T and jet-energy measurement.

The first three items are well advanced and will already be operational in the collider period of 1985:

(a) Improved muon detection: additional iron absorbers (60 cm on the sides and 40 cm on the roof, see fig. 18) have been installed. The resulting trigger-rate reduction allows to trigger the detector on large-angle single muons up to a luminosity of 10^{30} $cm^{-2}s^{-1}$. The side iron walls can be magnetized. The magnetization further reduces the trigger rate and, after instrumentation with a detector having a spatial resolution of $\gtrsim$0.5 mm, provides an independent momentum measurement with a momentum resolution $\approx$40 %. This is in particular useful in regions, where because of the unfavourable dipole field direction the momentum measurement with the central detector is poor. It also helps to identify high-energy muons.

(b) New data acquisition system: the present parallel readout and event collection will be replaced by a multiprocessor system based on the VME bus. This is to improve the reliability of the system by complete autotest and monitor facilities made available by the microprocessor system, and to allow a modular implementation of existing and future data filtering and data-sampling tasks.

(c) Microvertex detector: a reasonable efficiency of the measurements of flight paths of b-particles demands a resolution of better than $\tau \gtrsim 10^{-12}$ sec corresponding to $300\beta\gamma$ microns, where $\beta\gamma = p/m$. Independent of the properties of the localization detectors, the first effect one has to contend with is the multiple scattering in the vacuum chamber. The geometry of the vacuum chamber, in particular its diameter, influences the precision of the vertex reconstruction. The UA1 group has chosen a thin Beryllium tube of 5 cm diameter. The localization detector is a pressurized drift chamber with a spatial resolution of $\leq$ 50 microns.

As many as eight distinct decay vertices may be produced by the decay $W \to t\bar{b}$. In about half of the events at least two decay vertices will be detected, providing an unique experimental tag for the top quark.

The standard model predicts a fermion-Higgs coupling proportional to the fermion mass. The coupling of the Higgs to the $W^{\pm}$ and Z^{o} are particularly strong, offering two ways of looking for the Higgs particle: (i) bremsstrahlung emission $W^{\pm} \to H^{o} + f + f'$, where (ff') are possible decay channels of the W, either $q\bar{q}$ pairs (jets) or lepton pairs, and (ii) associated production of a $W^{\pm}$ particle and of a Higgs mediated by a virtual boson propagator: $p\bar{p} \to W^{\pm}H^{o}+X$. In the latter case the intermediate vector boson in the final state can be used to trigger the event. Both methods require in addition to the presence of a virtual or real W a positive identification of the decay of the H^{o}. The Higgs is expected to decay most abundantly into a pair of the most massive fermions kinematically accessible. For $m_H \geq$ 15 GeV the branching fraction into $b\bar{b}$ pairs is about 90 %. Above the $t\bar{t}$ threshold this last decay mode takes over rapidly. In about half of the events the microvertex detector would identify at least two decay vertices from the heavy-quark decay of the Higgs particle. The rate for associated W-H^{o} production depends on m_H and is of the order of a few 10^{-3} of the W rate [84].

<u>UA1 calorimeter upgrade</u>: the main motivations for the proposed upgrade are: (i) improve the measurement of missing transverse energy down to $\gtrsim$5 GeV, (ii) improve the measurement of jet energy for soft ($\gtrsim$ 10 GeV) jets from t-decays and possible supersymmetric decays, (iii) improve

the resolution of the e.m. calorimeter to precisely determine the W and Z masses and the Z^0 width. The main features of the proposed calorimeter are:

(a) compensation (equalization of the responses to electromagnetic and hadronic showers) by use of Uranium as converter material,

(b) position detection of the showers with a precision of $\leq$ 1 cm,

(c) absolute energy determination (in the electromagnetic calorimeter) of better than 1 % paired with a good resolution $\Delta E/E \approx 0.13/\sqrt{E}$, and

(d) complete solid-angle coverage and shower containment.

The proposed calorimeter will improve the simultaneous measurement of electrons, muons, neutrinos, and other non-interacting particles, if they exist, photons and jets.

(i) Electrons: the granularity of the calorimeter cells will be $\approx$ 13 x 13 cm^2. The electron isolation criteria can be loosened and the triggering capabilities improved. The position detector gives an improved position measurement ($\sigma \approx 0.5$ cm) for the shower and better rejection of $\pi^{\pm}/\pi^0$ overlaps.

(ii) Muons: the number of interaction lengths of absorber increases and an additional space point from the position detector provides an extra constraint.

(iii) Neutrinos and photinos: the new granularity in azimuth ϕ enables a true missing transverse-energy trigger, as compared with the present left-right imbalance. The resolution on missing transverse energy will improve by a factor $\approx$2 because of the improved hadronic resolution of the calorimeter and the improved azimuthal coverage.

(iv) Photons: the position detector will give a better localization of electromagnetic showers and will distinguish between single photons and π^0's up to $\approx$20 GeV.

(v) Jets: the compensation will give an improved energy measurement for the jets, and the better granularity will provide a better jet-direction determination, a better se-

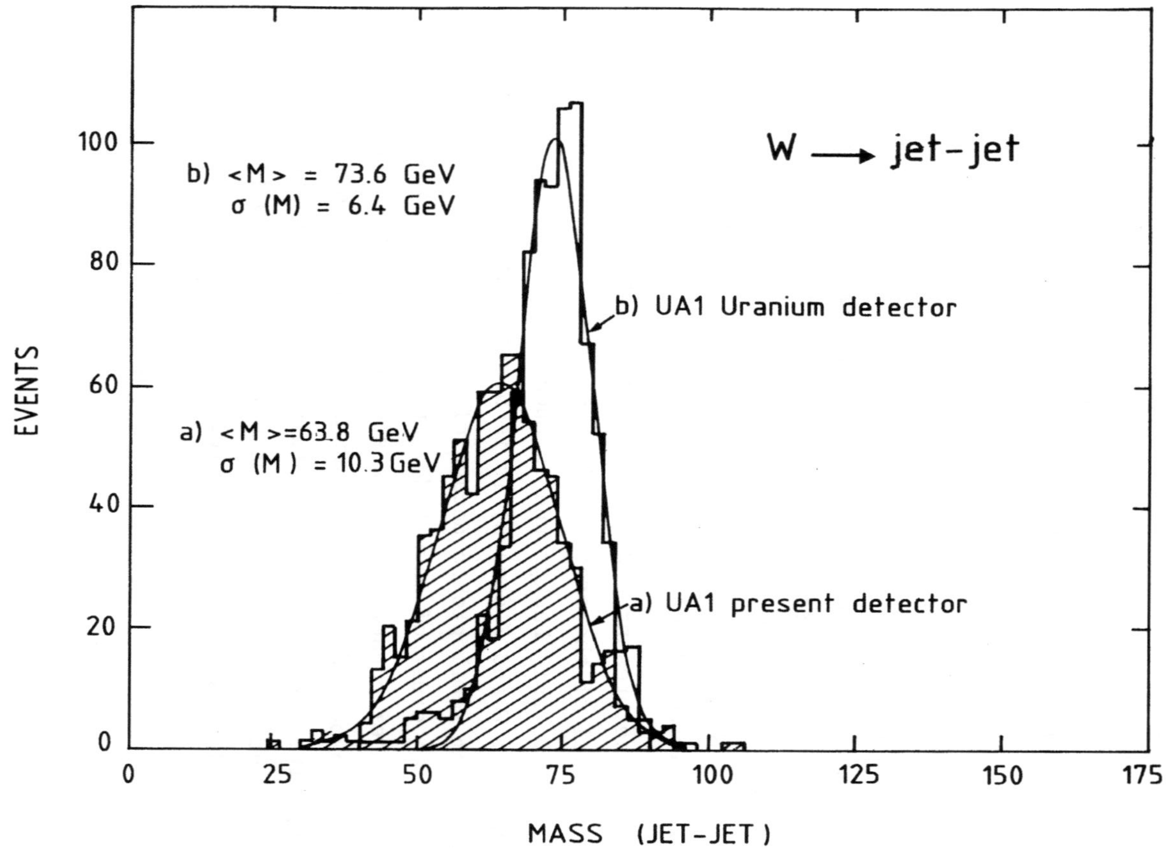

Fig. 70: Mass distribution for W decaying to jet + jet.

paration between two jets etc. The simulated mass distribution of W → jet-jet events is compared in fig.70 for the present and the new calorimeters.

The UA2 group wants to improve the missing transverse-energy measurement capability of their detector. Dead regions, such as the magnet coils of the forward UA2 spectrometers, and the lack of calorimeter coverage below 20°, preclude an accurate missing-transverse-energy measurement. For example a two-jet event having a mass of $\approx$ 50 GeV/c^2 is detected as a single jet in $\approx$2 % of the cases, the other jet escaping detection. The UA2 decided to add calorimeters down to 10° from the beam line by replacing the forward spectrometers with segmented calorimeter end caps. They will hence lose the ability to measure particle charges in this region, but gain by improving on neutrino detection.

In the present UA2 detector jets can fake electrons at the level of the order of 10^{-5} of the inclusive jet yield. While this background is much lower than the electron signal from W and Z decays, it is higher than the expected signal from semileptonic decays of t-quarks ($m_t \leq$ 40 GeV) by about an order of magnitude. In practice, because of lepton number conservation an electron in the final state is expected to be associated with another electron of opposite sign or with a neutrino, and the presence of this second lepton provides further rejection against background. However, in the lower mass range, where both leptons have small transverse momenta, the background conditions are particularly difficult and any increase of the rejection factor against jets would improve the detector performance. In the present UA2 detector the use of preshower counters (see fig.15) contributes a factor $\approx$10 to this rejection factor. An improvement over the present situation implies a simultaneous discrimination against the two major types of misidentified electrons: converted photons from π^o decays and unresolved π^o-$\pi^\pm$ pairs. The proposed modification of the UA2 vertex detector will improve the situation. Moving outwards from the intersection region the main components of the proposed central detector are:

(i) a small diameter Beryllium vacuum chamber to be installed in 1985.

(ii) a new vertex detector providing an accurate measurement of the longitudinal and transverse positions of the event vertex,

(iii) a matrix of Silicon counters with the double purpose of measuring ionization and helping in pattern recognition,

(iv) a pair of transition-radiation detectors providing an additional rejection factor of at least an order of magnitude, and

(v) a multilayer scintillating fibre detector for tracking and measuring early electromagnetic showers developing after a 1.5 radiation-length thick converter.

The key to the anticipated total integrated luminosity of nearly 10 pb^{-1} by the end of 1989 is the Antiproton Collector (ACOL) [85] which will be operational after the long 1986 shutdown of the SPS. The Antiproton Collector is a machine of large acceptance to collect the produced antiprotons and to precool them before accumulation in a second machine, the Antiproton Accumulator (AA) [9].

Presently, the AA receives typically six million antiprotons per PS pulse. With the Collector, the goal is to capture 10^8 antiprotons per pulse. The factor 16 is gained by splitting the functions of collection and accumulation between two machines. Compared to the AA the collector has four times the betatron acceptance and four times the momentum acceptance. To populate the acceptance at the best efficiency special devices are required: a pulsed target and high-magnetic-field-gradient lenses to achieve the ultimate proton focusing on the target and the initial antiproton focusing.

7.2 After LEP: A Large Hadron Collider (LHC) in the LEP tunnel?

In a recent workshop held at Lausanne and CERN a wealth of strong motivations to build an LHC in the LEP tunnel were discussed. The LHC would open up the 1 TeV energy scale. The following questions arising from our present picture of high-energy physics could then be investigated [86,87]: From where do the $W^{\pm}$ and the Z^{o} acquire their masses? Via the Higgs mechanism with elementary Higgs particles? Are Higgs masses protected by supersymmetry? Or are Higgses composite? How are flavours generated? What is the origin of CP-violation? Is there a connection between electroweak, strong, and gravitational forces? Are there new particles with masses up to ≈ 1 TeV? What are their signatures?

Although energy is of utmost importance when tackling these questions, there appears to be only a minor difference between hadron collisions at, say 18 TeV or 40 TeV in the centre of mass. The main reason can be traced both to the softening of the quark and gluon distributions and to the over-all $1/\hat{s}$ behaviour of the point-like cross sections. Interesting processes at the SPS collider ($\sqrt{s} = 600$ GeV) have typical cross sections of $\approx 10^{-34}$ cm^2 at $\hat{s} = (100\text{ GeV})^2$. They will translate into corresponding higher mass effects at $\hat{s} = (3\text{ TeV})^2$ (at $\sqrt{s} = 18$ TeV) with a cross section 10^3 times smaller, namely $\approx 10^{-37}$cm^2. It has furthermore been shown that for a large set of different phenomena the useful energy grows approximately like $\sqrt{E_{cm}}$ rather than E_{cm}. The maximum rate of events that can be studied in a practical detector sets a ceiling to the useful luminosity of the collider at about

$10^{32} - 10^{33}$ cm^{-2}s^{-1}. Phenomena at very high energies will eventually become inaccessible because of lack of events irrespective of the energy of the collider.

It has been shown that there exist no major difficulties in building an LHC with a luminosity $L \approx 10^{32} - 10^{33}$cm^{-2}s^{-1} [88] provided that high-field magnets with the degree of realiability necessary for such a large-scale device (27 km!) will become available. A proton-proton collider seems to be the most promising solution. A centre-of-mass energy of 18 TeV could be reached with superconducting magnets of 10 T.

In order to achieve this goal it would be necessary to launch a vigorous program for the development of materials and techniques necessary for the construction of such magnets. Several laboratories and institutes have expressed interest in participating in such a program. All other machine components and systems appear to be feasible with present technology [88].

Formidable detector development problems are associated with the larger luminosity and the necessity of handling many collisions at each crossing. An equivalent effort in research and development is needed in order to ensure an economical and feasible solution to a 4π universal detector capable of operating at $L = 10^{33}$ cm^{-2}s^{-1}. Some of these problems have already been investigated during the Lausanne workshop[89].

Some signatures which experiments at the LHC should be able to detect are:

(i) $t\bar{t}$: useful for hunting Higgs bosons and technipions.

(ii) Multiple $W^{\pm}$, Z^{0}: Also useful for Higgs and technicolour searches. In view of the low rates for many Higgs production mechanisms one should not rely solely on leptonic decay modes of the W and Z, but should also have some efficiency for picking them out via their decays into hadronic jets.

(iii) Missing E_T: This is very important for supersymmetry searches. Complete calorimeter coverage and containment is essential.

(iv) jet-jet mass bumps: Good dijet mass resolution is clearly important for the $t\bar{t}$, $W^{\pm}$, and Z^{o} searches previously mentioned, as well as for such signals as excited quark $\rightarrow q + g$ decay.

(v) γ-jet mass bumps: Are also potentially useful in the search for excited quarks.

(vi) leptons, e, $\mu,\tau,\nu,\ldots$: A high capability to detect these is a sine qua non for many of the other particle searches.

Based on various analyses [86] and on the likely cleanliness of the above signatures the following tentative conclusions about the difficulty of observing different species of new particles at the LHC can be formulated:

(a) Supersymmetry: relatively easy

(b) Technicolour: possible

(c) Conventional Higgs boson: difficult.

Despite the enthusiasm that has been generated it should not be forgotten that the LHC is an expensive new device which would call for an investment of a similar order of magnitude as the LEP project. More constraints on machine and detectors will be known after results from the upgraded SPS collider and from the Tevatron will become available.

8. SUMMARY

The cross section for production of Intermediate Vector Bosons at the Collider represents only a small part of the total cross section. Approximately 10^{10} $p\bar{p}$

interactions have to be investigated to find 10 decays of the type $p\bar{p} \to Z^o + X$, $Z^o \to \ell^+ \ell^-$ (table 14). The discovery of the $W^{\pm}$ and the Z^o was nevertheless an easy job, because the UA1 and UA2 detectors provided enough rejection power against soft-hadron events and events with large-p_T jets. The events which contain Intermediate Vector Bosons appear to be clean and their features are not disturbed by the presence of soft spectator hadrons. It could have been different. Centauro events or a dramatic broadening of the hadronic transverse momentum distributions, as indicated by cosmic ray data, would probably have necessitated a different experimental approach to the Intermediate Vector Bosons.

The performance and the reliability of the SPS Collider

Table 14: Hiearchy of cross sections.

Process	Cross section (in cm^2) at $\sqrt{s} \gtrsim 600$ GeV	Total rate normalized to 10^{10} soft events
$p\bar{p} \to$ soft hadrons	6.10^{-26}	10^{10}
$p\bar{p} \to$ jet + jet +X	$\leq 10^{-30}$	$\leq 2.10^5$
$p\bar{p} \to W^{\pm} + X$		
$W \to \ell\nu$	$0.5 \cdot 10^{-33}$	100
$p\bar{p} \to Z^o + X$		
$Z^o \to \ell^+\ell^-$	$0.05 \cdot 10^{-33}$	10

and the detectors UA1 and UA2 is probably best illustrated by the number of IVB's found so far. Preliminary numbers are shown in table 15.

Table 15: Number of leptonic IVB decays found in the Collider periods 1982/1983/1984.

	UA1	UA2
$Z^o \to e^+e^-$	18	16
$Z^o \to \mu^+\mu^-$	10	-
$W^{\pm} \to e^{\pm}\nu$	172	119
$W^{\pm} \to \mu^{\pm}\nu$	47	-

One decay of the type $W^{\pm} \to e^{\pm}\nu$ in the UA1 detector corresponds to approximately $2nb^{-1} = 2.10^{33}cm^{-2}$. The integrated luminosity delivered in the collider periods 1982/1983/1984 amounts to about 600 nb^{-1}, of which 2/3 was used by the experiments. The future plans are to reach an integrated luminosity of about 10 pb^{-1} (= $10^{37}cm^2$) by 1989 with improved machine and detectors.

By 1984,the following leptonic decay modes of IVB's have been seen:

$W^{\pm} \to e^{\pm}\nu_e$

$W^{\pm} \to \mu^{\pm}\nu_{\mu}$

$W^{\pm} \to \tau^{\pm}\nu_{\tau}$ $(\tau \to jet+\nu)$

$Z^o \to e^+e^-$

$Z^o \rightarrow \mu^+\mu^-$

$Z^o \rightarrow \tau^+\tau^-$ (both or one $\tau \rightarrow e\nu$ or $\mu\nu$).

First indications of hadronic IVB decays have also been reported. The hadronic decays are particularly tantalizing, since the mode W→tb is a source of the as yet undiscovered Top. Besides possible candidates of the expected decays, a few events of the dilepton topology have been found which seem to escape conventional interpretations.

In the search for events with large missing E_T a few peculiar events have been found characterized by the production of one jet associated with large missing transverse energy. Such events may be due to the tail of supersymmetric processes, where E_T^{miss} signals the production of photino(s) and only one jet survived the cuts applied in the analysis. A better assessment of the Supersymmetry scenario will be possible with the upgraded UA1 calorimeter, which will provide jet identification and measurement even at low ($\lesssim$ 10 GeV) E_T^{jet} and a reliable measurement of small ($\gtrsim$ 5 GeV) missing transverse energy.

The success of the SPS collider project stimulates discussions concerning the use of hadron colliders for the exploration of the mass scale of 1 TeV. The natural step to achieve this goal may be a Large Hadron Collider in the LEP tunnel.

ACKNOWLEDGEMENTS

I would like to thank Prof. H. Mitter for the invitation to the Schladming School, and R. Baier, P. Darriulat, K. Eggert, Th. Müller, and C. Rubbia for help and suggestions concerning the understanding and the presentation of the material covered in these lectures. The impressive amount and quality of the described results represents the work of the whole CERN Collider community.

REFERENCES

1. L.B. Okun, Leptons and Quarks (North-Holland, Amsterdam, 1981). D.H. Perkins, Introduction to High-Energy Physics, 2nd ed. (Addison-Wesley, Reading, Massachusetts, 1982).
I.J.R. Aitchison and A.J.G. Hey, Gauge Theories in Particle Physics (Hilger, Bristol, 1982).
E. Leader and E. Predazzi, An Introduction to Gauge Theories and the New Physics (Cambridge, 1982).
C. Quigg, Gauge Theories of the Strong, Weak and Electromagnetic Interactions (Benjamin/Cummings, Reading, Massachusetts, 1983).
F. Halzen and A.D. Martin, Quarks and Leptons (Wiley, New York, 1984).
2. J.E. Kim et al., Rev. Mod. Phys. 53 (1981) 211.
J. Panman, Inv. talk XI-th Int. Conf. on Neutrino Physics and Astrophysics, Dortmund, 1984, report CERN-EP/84-150 (1984).
3. C.H. Llewellyn Smith, Proc. ECFA-CERN Workshop, Lausanne and Geneva 1984, report CERN 84-10(1984) (ed. M. Jacob), p.25. The figure is based on a fit due to D. Duke and J. Owens, Phys. Rev. D30 (1984) 49, and Erratum. See also: A. Ali et al., same Proceedings, p.469. E. Eichten et al., Rev. Mod. Phys. 56 (1984) 579.
4. C. Rubbia et al., Proc. Int. Neutrino Conf., Aachen, 1976 (ed. H. Faissner et al.)(Vieweg, Braunschweig, 1977) p. 683.
5. C. Rubbia, Proc. EPS Int. Conf. on High-Energy Physics, Geneva 1977 (CERN, 1977) p.791. R. Horgan and M. Jacob, Proc. CERN School Malente 1980, report CERN 81-04 (1981) p.65.
6. G.I. Budker et al., The 1966 Proc. Int. Symp. Electron and Proton Storage Rings, Saclay; Atomnaya Energiya 22 (1967) 346.
7. S. Van der Meer, Stochastic Damping of Betatron Oscillations in the ISR, CERN/ISR-PO/72-31 (1972); Stochastic

Cooling and the Accumulation of Antiprotons, Nobel lecture, report CERN PS 84-32 AA (1984). W. Schnell, Proc. First Course Int. School Part. Accel. Erice 1976, and CERN 77-13 (1977).

8. G. Carron et al., Phys. Lett. 77B (1978) 353.
9. Design Study of a $\bar{p}p$ Colliding Beam Facility, CERN/PS/AA 78-03 (1978).
10. The Staff of the CERN $\bar{p}p$ Project, Phys. Lett. 107B (1981) 306.
11. UA1 proposal,CERN/SPSC/78-6/P92(1978), and Add. 1-5.
UA2 proposal, CERN/SPSC/78-8/P93(1978), and Add. 1-3.
UA3 proposal, CERN/SPSC/78-15/P96 (1978).
UA4 proposal, CERN/SPSC/78-105/P114 (1978).
UA5 proposal, CERN/SPSC/78-70/P108 (1978).
UA6 proposal, CERN/SPSC/80-63/P148 (1980).
12. M. Spiro, Acta Physica Austriaca Suppl. XXIV (1982) 125.
H. Wahl, Acta Physica Austriaca Suppl. XXV (1983) 399.
13. R.N. Cahn, Proc. 10th SLAC Summer Institute (ed. A. Mosher), SLAC Report 259 (1982) 1. M. Jacob, report CERN TH-3693 (1983).
14. UA4 Collaboration, R. Battiston et al., Phys. Lett. 117B (1982) 126; 115B (1982) 333; and 127B (1983) 472; M. Bozzo et al., 147B (1984) 385; 147B (1984) 392; and Phys.Lett. 155B (1985) 197. UA1 Collaboration, G. Arnison et al., Phys.Lett. 121B (1983)77,and 128B(1983)336.
15. U. Amaldi et al., Phys. Lett. 66B (1977) 390. N. Amos et al., Phys. Lett. 120B (1983) 460; and 128B (1983) 343. L. Baksay et al., Nucl. Phys. B141 (1978) 1.
16. M.M. Block and R.N. Cahn, Phys. Lett. 120B (1983) 224.
17. U. Amaldi and K.R. Schubert, Nucl. Phys. B166 (1980) 301.
18. R. Bossert et al., Report SPS/ABM/RB/80-12 (1980), presented at the XIth Int.Conf. on High-Energy Accelerators, CERN, Geneva 1980.
19. A. Martin, Proc. IVth Topical Workshop on Proton-Antiproton Collider Physics, Berne, 1984, report CERN 84-9 (1984), (eds. H. Hänni and J. Schacher) p. 308.

20. UA5 Collaboration, K. Alpgård et al., Phys. Lett. 107B (1981) 310. UA1 Collaboration, G. Arnison et al., Phys. Lett. 107B (1981) 320; and 123B (1983) 108.
21. W. Thomé et al., Nucl. Phys. B129 (1977) 365.
22. UA5 Collaboration, K. Alpgård et al., Phys. Lett. 107B (1981) 315; and 121B (1983) 209; and G.J. Alner et al., Phys. Lett. 138B (1984) 304. UA1 Collaboration, G. Arnison et al., Phys. Lett. 107B (1981) 320; and 123B (1983)108.
23. E.N. de Groot, Phys. Lett. 57B (1975) 159.
24. UA5 Collaboration, K. Alpgård et al., Phys. Lett. 123B (1983) 361.
25. UA5 Collaboration, K. Alpgård et al., Phys. Lett. 115B (1982) 65; 115B (1982) 71;and 121B (1983) 209; and G.J. Alner et al., CERN-EP/84-111 (1984), submitted to Phys. Lett.B. UA2 Collaboration, M. Banner et al., Phys. Lett. 115B (1982) 59; and 122B (1983) 322.
26. UA5 Collaboration, Phys. Scr. 23 (1981) 642.
27. F.W. Büsser et al., Phys. Lett. 55B (1975) 232. K. Eggert et al., Nucl. Phys. B98 (1975) 49. C. Kourkoumelis et al., Phys. Lett. 84B (1979) 277.
28. UA2 Collaboration, M. Banner et al., Phys. Lett. 115B (1982) 59; 122B (1983) 322; and report DPhPE 84-09 (1984), submitted to Z. Phys.C. UA1 Collaboration, G. Arnison et al., Phys. Lett. 118B (1982) 167.
29. B. Alper et al., Nucl. Phys. 100 (1975) 237.
30. C.M.G. Lattes et al., Phys. Rep. C65 (1980) 151.
31. see ref.30. R.W. Ellsworth et al., Phys. Rev. D23 (1981) 771.
32. UA5 Collaboration, K. Alpgård et al., Phys. Lett. 115B (1982) 71. UA1 Collaboration, G. Arnison et al., Phys. Lett. 122B (1983) 189.
33. UA2 Collaboration, M. Banner et al., Phys. Lett. 121B (1983) 187.
34. H.A. Bethe, Z. Phys. 76 (1938) 293. F. Bloch, Ann. Phys. (Leipzig) 16 (1933) 285.
35. L.W. Jones, Rev. Mod. Phys. 49 (1977) 717.

36. B. Aubert et al., Phys. Lett. 120B (1983) 465.
37. UA2 Collaboration, M. Banner et al., Phys. Lett. 118B (1982) 203; P. Bagnaia et al., Z. Phys. C20 (1983) 117; and Phys. Lett. 138B (1984) 430. UA1 Collaboration, G. Arnison et al., Phys. Lett. 118B (1982) 173; 123B (1983) 115; and 132B (1983) 214.
38. Z. Kunszt and E. Pietarinen, Phys. Lett. 132B (1983) 453.
39. T. Åkesson et al., Phys. Lett. 118B (1982) 185; and 118B (1982) 193. A.L.S. Angelis et al., Phys. Lett. 126B (1983) 132. A review of earlier results is given in P. Darriulat, Ann. Rev. Nucl. Part. Sci. 30 (1980) 159.
40. R. Horgan and M. Jacob, Nucl. Phys. B179 (1981) 441. For a recent review see R. Baier, these Proceedings.
41. UA2 Collaboration, P. Bagnaia et al., Phys. Lett. 114B (1984) 283.
42. M. Greco, Back-to-Back Jets as a Test of the Three-Gluon Coupling, LNF preprint 84/21 (1984), submitted to Z. Phys. C.
43. UA1 Collaboration, Hadronic Jets and Intermediate Vector Bosons at the CERN $\bar{p}p$ Collider, preprint CERN-EP/83-196 (1983), presented by J. Strauss at "Hadron Structure 83", Int. Conf. on Elementary Particle Physics, Bratislava 1983. M. Greco, QCD p_T Effects in W/Z and Jet Production, Proc. IVth Topical Workshop on $\bar{p}p$ Collider Physics, Berne, 1984, report CERN 84-09 (1984) (eds. H. Hänni and J. Schacher) p.202.
44. UA1 Collaboration, G. Arnison et al., Phys. Lett. 136B (1984) 294.
45. B. Combridge et al., Phys. Lett. 70B (1977) 234. J.C. Collins and D.E. Soper, Phys. Rev. D16 (1977) 2219.
46. D. Drijard et al., Phys. Lett. 121B (1983) 433. P. Söding and G. Wolf, Ann. Rev. Nucl. Part. Sci. 31 (1981) 231.
47. CDHS Collaboration, H. Abramowicz et al., Z. Phys. C12 (1982) 289; C13 (1982) 199; C17 (1983) 283.

CHARM Collaboration, F. Bergsma et al., Phys. Lett. 123B (1983) 269.

48. G. Sterman and S. Weinberg, Phys. Rev. Lett. 39 (1977) 1436. K. Shizuya and S.-H.H. Tye, Phys. Rev. Lett. 41 (1978) 787. M.B. Einhorn and B.G. Weeks, Nucl. Phys. B146 (1978) 445.
49. G. Marchesini and B.R. Webber, Nucl. Phys. B238 (1984) 1. B.R. Webber, Nucl. Phys. B238 (1984) 492.
50. UA1 Collaboration, G. Arnison et al., Phys. Lett. 132B (1983) 223; (1983) 214; and 147B (1984) 222.
51. TASSO Collaboration, private communication.
52. TASSO Collaboration, R. Brandelik et al., Phys. Lett. 114B (1982) 65.
53. TASSO Collaboration, M. Althoff et al., Z. Phys. C22 (1984) 307.
54. E. Fermi, Nuovo Cimento 11 (1934) 1; Z. Phys. 88 (1934) 161; and La Ricerca Scientifica 4 (1933) 491. H. Yukawa, Proc. Math. Soc. Jpn. 17 (1935) 48. O. Klein, On the Theory of Charged Fields, in New Theories in Physics, Conf. Int. Union Phys. and Polish Intellectual Co-operation Committee, Warsaw, 1938, reprinted in Proc. 1983 JINR-CERN School of Physics, Tábor, Czechoslovakia (Dubna, 1984).
55. S.L. Glashow, Nucl. Phys. 22 (1961) 579. S. Weinberg, Phys. Rev. Lett. 19 (1967) 1264; A. Salam, in Elementary Particle Theory (ed. N. Svartholm) (Almqvist and Wiksell, Stockholm, 1968) p. 367.
56. W.J. Marciano, Electroweak Interaction Parameters, Proc. IVth Topical Workshop on $\bar{p}p$ Collider Physics, Berne, 1984, report CERN 84-09 (1984) (eds. H. Hänni and J. Schacher) p. 165.
57. J. Ellis et al., Ann. Rev. Nucl. Part. Sci. 32 (1982) 443, and references therein.
58. UA1 Collaboration, G. Arnison et al., Phys. Lett. 122B (1983) 103; 126B (1983) 398; 129B (1983) 273; 134B (1984) 469; 147B (1984) 241. UA2 Collaboration, M. Banner et al., Phys. Lett. 122B (1983) 476; P. Bagnaia et al., Phys. Lett. 129B (1983) 130; Z. Phys. C24 (1984) 1; Preliminary

Results on W and Z Production at $\sqrt{s}$ = 630 GeV, presented by L. Mapelli on V-th Topical Workshop on $\bar{p}p$ Collider Physics, St.Vincent,1985,report CERN-EP/85-84(1985).

59. C. Rubbia, Physics Results of the UA1 Collaboration at the CERN Proton-Antiproton Collider, IXth Hawaii Topical Conf. Part. Phys., 1983, report CERN-EP/84-55 (1984). E. Radermacher, The Experimental Discovery of the Intermediate Vector Bosons, preprint CERN-EP/84-41 (1984), submitted to Prog. Part. Nucl. Phys. L. Di Lella, Hard Hadron Collisions at Very High Energy, Proc. 1983 JINR-CERN School of Physics, Tâbor, Czechoslovakia (Dubna, 1984) Vol.2, p.82.
60. F.E. Paige and S.D. Protopopescu, report BNL 29777 (1981). B. Humpert and W.L. van Neerven, Phys. Lett. 93B (1980) 456. G. Altarelli et al., preprint CERN-TH 4015 (1984).
61. R. Bertlmann and H. Pietschmann, Phys. Rev. D15 (1977) 683.
62. F. Halzen and D.M. Scott, Phys. Lett. 78B (1978) 318. P. Aurenche and R. Kinnunen, Phys. Lett. 135B (1984) 493. A. Nakamura et al., Z. Phys. C21 (1984) 243. P. Chiapetta and M. Greco, Nucl. Phys. B221 (1983) 269; Phys. Lett. 135B (1984) 190. G. Altarelli et al., Nucl. Phys. B246 (1984) 12.
63. M. Glück et al., Z. Phys. C13 (1982) 119.
64. M. Jacob, Nuovo Cimento 9 (1958) 826. J. Werle, Relativistic Theory of Reactions (Polish Scientific Publishers, Warsaw, 1966).
65. P. Aurenche and J. Lindfors, Nucl. Phys. B185 (1981) 274.
66. UA1 Collaboration, W and Z Production at the CERN SPS Collider, presented by S. Geer at Int. Symp. on Physics of $p\bar{p}$ collisions, Tsukuba (Japan), 1985, report CERN-EP/85-63 (1985).
67. UA2 Collaboration, P. Bagnaia et al., Phys. Lett. 139B (1984) 105.

68. UA1 Collaboration, Experimental Evidence for the $W \to \tau\nu$ Decay, presented by A. Savoy-Navarro at the V-th Topical Workshop on $p\bar{p}$ Collider Physics, St.Vincent, Italy, 1985, report CERN-EP/85-64 (1985). $W^{\pm}$ and Z^{o} Production in the UA1 Experiment, presented by V. Vuillemin at the Conference on Collider Physics at Ultra-High Energies, Aspen,Colorado,1985, report CERN-EP/85-29 (1985). G. Arnison et al.,Experimental Evidence for the $W \to \tau\nu$ Decay, in preparation.
69. UA1 Collaboration, G. Arnison et al., Phys. Lett. 139B (1984) 115. J. Rohlf, Proc. XXII-nd Int. Conf. High-Energy Physics, Leipzig, 1984 (eds. A. Meyer and E. Wieczorek) Vol.II, p.12. Observation of Events with Large Missing Transverse Energy, presented by M. Mohammadi at the Conference on Collider Physics at Ultra-High Energies, Aspen, Colorado, 1985,report CERN-EP/85-52 (1985).
70. R. Odorico, $\bar{b}bX$ and $W \to \tau\nu$ Backgrounds for the UA1 Monojet Signal, report IFUB 85/1 (1985).
71. see e.g. R.M. Barnett et al., Phys. Rev. Lett. 54 (1985) 1983. W. Buchmüller, these Proceedings.
72. UA1 Collaboration, Search for $W \to c\bar{s}$, $Z^{o} \to c\bar{c}$, $b\bar{b}$ at the CERN $p\bar{p}$ Collider, presented by J. Ransdell at the DPF Annual Meeting, Santa Fe, 1984.
73. J. Lindfors and D.P. Roy, Z. Phys. C24 (1984) 271.
74. C. Peterson et al., Phys. Rev. D27 (1983) 105.
75. R. Horgan and M. Jacob, Nucl. Phys. B238 (1984) 221. F. Halzen and D.M. Scott, Phys. Lett. 129B (1983) 341. R. Kinnunen,Z. Phys. C25 (1984) 167.
76. UA1 Collaboration, G. Arnison et al., Phys. Lett. 147B (1984) 439.
77. A. Grosso and R. Odorico, QCD Background under the UA1 $W \to t\bar{b}$ Signal, report IFUB 84/18 (1984).
78. UA1 Collaboration, G. Arnison et al., Phys. Lett. 155B (1985) 442. See also presentations of N. Ellis and H.G. Moser at the V-th Topical Workshop on $p\bar{p}$ Collider Physics,

St.Vincent, 1985, report CERN-EP/85-71. UA2 Collaboration, P. Bagnaia et al., Phys. Lett. 154B (1985) 338.

79. S.D. Drell and T. Yan, Phys. Rev. Lett. 25 (1970) 316; and Ann. Phys. 66 (1971) 578. For a recent review see G. Altarelli, Phys. Reports 81 (1982) 1. For experimental reviews see Proc. Drell-Yan Workshop, Fermilab, Batavia, 1982 (ed. E.L. Berger); and I.R. Kenyon, Rep. Progr.Phys. 45 (1982) 1261.
80. A. Ali et al., Eurojet Program, 1984, to be published.
81. B. Schrempp and F. Schrempp, Phys. Lett. 153B (1985) 101.
82. CELLO Collaboration, H. Behrend et al., Phys. Lett. 141B (1984) 145.
83. For a review see e.g. R. Klanner, Proc. XXIInd Int. Conf. High-Energy Physics, Leipzig, 1984 (eds. A. Meyer and E. Wieczorek) Vol.II, p.201.
84. S.L. Glashow et al., Phys. Rev. D18 (1978) 1724.
85. E.J.N. Wilson (ed.), CERN 83-10 (1983). B. Autin, Proc. CAS (eds. P. Bryant and S. Newman), report CERN 84-15(1984) p.525.
86. C.H. Llewellyn Smith et al., Proc. ECFA-CERN Workshop on Large Hadron Collider in the LEP Tunnel (ed. M. Jacob), report CERN 84-10 (1984) p.543 and p.27. R.D. Peccei, same Proceedings, p.329. H. Fritzsch, same Proceedings, p.354. J. Ellis et al., same Proceedings, p.393. R. Barbieri, same Proceedings, p.457. A. Ali et al., same Proceedings, p.469. B. Anderson, same Proceedings, p.531.
87. C. Rubbia, same Proceedings, p.363.
88. A. Asner et al., same Proceedings, p.49.
89. T. Åkesson et al., same Proceedings, p. 165. A. Ali et al., same Proceedings, p. 209. I. Corbett et al., same Proceedings, p.243. A. Barone et al., same Proceedings,

p.265. M. Haguenauer et al., same Proceedings, p.301. R. Batley et al., same Proceedings, p.367. D. Linglin et al., same Proceedings, p.387.

Acta Physica Austriaca, Suppl. XXVII, 449–515 (1985)

QCD IN HIGH-ENERGY PROTON-PROTON AND PROTON-ANTIPROTON COLLISIONS+

by

R. BAIER
Fakultät für Physik
Universität Bielefeld
D-4800 Bielefeld, FRG

INTRODUCTION

The experimental and theoretical investigation of nucleon-nucleon collisions at high energies allows to explore the structure of the nucleon by large momentum transfer (deep-inelastic) processes. In these lectures the structure of the nucleon from momentum scales $Q > 1$ GeV/c (<0.1 fm) to $Q \sim 100$ GeV/c ($\lesssim 10^{-16}$cm) is discussed. In the first lecture the basic concepts of the parton model and of perturbative quantum chromodynamics (QCD) are introduced, and applied to deep inelastic lepton-nucleon scattering. The following lectures cover large transverse momentum, p_T, hadronic processes, massive dilepton production and production of prompt real photons at large p_T. The present status of the theoretical understanding of these processes is summarized. Because of the

+) Lectures given at the XXIV. Internationale Universitätswochen für Kernphysik, Schladming, Austria, February 20-March 1, 1985.

enormous amount of work performed in this field of research during the past years completeness is impossible to pretend. Nevertheless, together with the lectures by J. Strauss [1] on the recent CERN Sp$\bar{p}$S collider results I hope that a comprehensive overview is gained.

FIRST LECTURE

Structureless projectiles, like charged leptons or neutrinos, are the ideal tools for exploring structures of target particles as nuclei or nucleons in inelastic large momentum transfer collisions [2],

$$\text{lepton + target} \rightarrow \text{lepton + anything,}$$

where the incoming (outgoing) lepton energy $E(E')$ and the scattering angle θ are measured in the laboratory system. Analysing these processes the convenient variables are the energy transfer $\nu = E - E'$ and the momentum transfer squared $q^2 = -Q^2 = -4EE'\sin^2\theta/2$. The cross section is expressed in terms of dimensionless structure functions $F_i(\nu,Q^2) = F_i(x = Q^2/2M_{target}\nu,Q^2)$, e.g. i = 1,2 for $e+p \rightarrow e+X$. When increasing Q beyond the intrinsic scale of the target, Λ_{target}, it is the characteristic feature of these collisions that elastic scattering at $x \simeq 1$ vanishes by an inverse power of Q/Λ_{target} and a quasi-elastic peak centered near $x \sim 1/N$ becomes approximately independent of Q. N counts the number of "point-like" constituents of the target. Consequently the functions F_i scale as $F_i(x,Q^2) \simeq F_i(x)$ for $\Lambda_{constituent} >> Q >> \Lambda_{target}$ and fixed x. A beautiful example of this behaviour is shown in Fig.1 for electron-nucleus scattering ($\Lambda_{Nucleon} \simeq 1$ GeV $> Q > \Lambda_{Nucleus}$) making evident that the nucleus consists of nucleons behaving as almost noninteracting (= point-like) particles when probed in the deep inelastic domain. Indeed the quasi-elastic peak is to be understood as due to incoherent and elastic scattering on the protons,

$e+p \to e+p$.

Inelastic experiments on nucleons [2-4], which started at SLAC in 1967, give analogous evidence for point-like sub-structures within the nucleons when $Q > \Lambda_{\text{Nucleon}}$ (Fig.2). Consequently R.P. Feynman [5] proposed to regard the nucleon N as a collection of free point-like constituents, partons, which interact incoherently with the virtual photon in $e+N \to e+$anything. This interaction is best described in the infinite momentum frame, in which the nucleon is very fast moving. Treating the massless partons collinear with the nucleon and on-shell with momenta p_i, and denoting their momentum fraction with respect to the nucleon momentum P by x_i, $p_i = x_i P$, one derives in the impulse approximation

$$
\begin{aligned}
F_2^{eN}(x,Q^2) &= \sum_i \int_o^1 dx_i f_i(x_i) F_2^{e,\text{parton } i}(x/x_i, Q^2) \\
&= \sum_i \int_o^1 dx_i f_i(x_i) e_i^2 \delta(1-x/x_i) = \sum_i e_i^2 \, x \, f_i(x) = F_2^{eN}(x) \, .
\end{aligned}
\tag{1}
$$

The nucleon structure function $F_2^{eN}(x,Q^2)$ scales, when the photon-nucleon interaction is reduced to incoherent photon-parton interactions, γ + parton i $\to$ parton i. e_i is the charge of the struck parton, and $f_i(x_i)$ is the number distribution of this parton within the nucleon N.

Subsequent experimental results, including those from neutrino scattering, allowed to determine the properties of the partons coupled to the electro-weak currents: they carry spin 1/2 and the quantum numbers of the quarks (up, down, strange,..), as they are known from hadron spectroscopy. Their distributions $f_i(x) = u(x), d(x), s(x), ..$ are measured and available. Fig.3 shows an example of testing the quantum number assignment, namely

$$F_2^{ep} + F_2^{en} = 5/9\ x\ (u(x) + \bar{u}(x) + d(x) + \bar{d}(x)) + \dots$$
$$= 5/18\ (F_2^{\nu p} + F_2^{\nu n})\ . \qquad (2)$$

From the momentum sum rule

$$\sum_i \int_0^1 dx\ x\ f_i(x) = 1\ , \qquad (3)$$

one expects $9/5 \int_0^1 dx(F_2^{ep} + F_2^{en}) \simeq 1$; instead experimentally a value of 0.45 is deduced. The missing momentum fraction has to be carried by neutral partons with no weak interactions. In the framework of QCD it is natural to identify these as the gluons.

The parton picture of the nucleon can be extended and applied to various hadronic processes in kinematic domains with large momenta and/or masses. Since the basic assumption is that no intrinsic scale exists at the sub-structure level, simple dimensional scaling laws are predicted:

(i) the cross section for single inclusive hadrons, produced with large p_T [6] , should behave as

$$E\,d^3\sigma/d^3p(a + b \rightarrow c + X) \simeq f(x_T = 2p_T/\sqrt{s}, \theta)/p_T^n\ , \qquad (4)$$

with n = 4. Fig.4 indicates clearly two kinematically well separated regimes in $pp \rightarrow \pi^o x$, one with low p_T pions-$\langle p_T \rangle \simeq 300$ MeV/c and independent of the incident energy, and one with large p_T pions with energy-dependent cross section. The power n resulting from this dependence is shown in Fig.5, $4 < n < 9$ [7];

(ii) a similar behaviour should hold for the production of a hadronic jet, defined as a bunch of hadrons with limited momentum transverse to a jet axis interpreted as the struck parton direction, as well as for single prompt photons at large p_T. The scaling properties for jet production are summarized in Fig.6, with $n = 5.1 \pm 0.3$ for FNAL-$Sp\bar{p}S$ energies [8], whereas for photons $n = 6.6 \pm 0.3$ [9,10] for FNAL-ISR; (iii) for the production of massive lepton (Drell-Yan) pairs, $a+b \to \ell^+\ell^- + X$ the prediction is

$$Q^3 d\sigma/dQ = F(\tau), \quad \tau = Q^2/s \tag{5}$$

for a large invariant mass Q of the pair. A recent compilation of data from FNAL-ISR and a successful comparison with equ. (5) can be found in Fig. 7-15 of [11].

Indeed, this short summary tells us that the parton free-field model works as a first approximation, and that it is experimentally established. Further information about strong interactions can be obtained by studying in more detail the deviations from the scaling behaviour discussed above. Therefore the model has to be extended to allow for interactions between partons. QCD is the most encouraging theory.

QCD is a SU(3) colour gauge theory which describes the interactions between quarks (colour triplets for each flavour) and colour octet gluons. In the following we concentrate on the perturbative treatment of QCD in terms of its dimensionless coupling constant g. The theory is then defined by its Feynman rules [12] completed by the renormalization program [13]. The physics underlying renormalization is nicely explained e.g. by K.G. Wilson [14]. A renormalizable field theory is to be considered as an infinite set of theories, one for each fixed momentum (renormalization) scale μ. The dynamical equations of the theory are the same for all μ, but there are different parameters $g(\mu)$ for each theory. Changing the scale μ, $g(\mu)$ has to be adjusted correctly by the renormalization group equation,

$$\mu \frac{\partial g(\mu)}{\partial \mu} = \beta(g) \; . \tag{6}$$

The function β can be calculated from the relation between $g(\mu)$ and the bare coupling constant g_o. For QCD with N_f quark flavours and massless quarks the relation is given in the one-loop approximation by

$$g_o^2 = g^2(\mu)\ \{\ 1- 2b\ \frac{g^2(\mu)}{(4\pi)^2}\ [1/\varepsilon + \ln 4\pi - \gamma - \ln\mu^2] + \ldots\}, \tag{7}$$

$b = 11/2 - N_f/3$.

The ultraviolet divergencies are regularized by performing the calculations in $n = 4 - 2\varepsilon$, $\varepsilon > 0$, space-time dimensions [13]. $g(\mu)$ is not unique, because one is free to include finite parts in equ.(7); here it is the term $\ln 4\pi - \gamma \simeq 1.954$. This choice specifies a renormalization scheme (RS) and equ.(7) holds for the $\overline{MS}$ scheme [15]. Because of its attractive features most of the QCD calculations are performed in this scheme, and it will be used throughout the lectures. However, the evaluation of physical quantities, when calculated at finite order perturbation theory, does depend on the RS prescription. For illustration let us consider e^+e^- annihilation. The total cross section to order g^4 is

$$R(Q) = \sigma(e^+e^- \to \text{hadrons})/\sigma(e^+e^- \to \mu^+\mu^-)$$

$$= 3\ \sum_i e_i^2 \{1 + a(\mu) + r_1(Q^2/\mu^2)\, a^2(\mu)\}, \tag{8}$$

with $a(\mu) \equiv g^2(\mu)/4\pi^2$. Like $a(\mu)$ the function r_1 is RS dependent: for $\overline{MS}$ it is [16]

$r_1^{\overline{MS}} = 1.986 - 0.115\ N_f - b/2\ \ln\ (Q^2/\mu^2)$.

Dropping e.g. the constant piece $\ln 4\pi - \gamma$ in equ.(7), which defines the MS scheme [15], r_1 is changed to

$$r_1^{MS} = r_1^{\overline{MS}} + \frac{b}{2}(\ln 4\pi - \gamma), \tag{9}$$

and the coupling constants are related by

$$a_{MS}(\mu) = a_{\overline{MS}}(\mu) - a^2_{\overline{MS}}(\mu)\frac{b}{2}(\ln 4\pi - \gamma) + O(a^3_{\overline{MS}}) \ , \qquad (10)$$

when μ is kept fixed [17]. Definite quantitative values for R(Q) can only be obtained after specifying the (arbitrary) RS and μ. Different numerical values will result; the difference is of $O(a^3)$ in our example, equ. (8). In [18], where the RS ambiguity is discussed in great detail, the following numerical example is given: for $\mu = Q$, $N_f = 3$, at $a_{\overline{MS}} = 0.048 \hat{=} a_{MS} = 0.039$ (equ. (10)), $R_{\overline{MS}} = 2(1+0.0517)$ and $R_{MS} = 2(1+0.0483)$.

Differentiating equ. (7) with respect to μ, the β-function in leading order of g is deduced. Including the second order term it is

$$\mu\frac{\partial a(\mu)}{\partial\mu} = \hat{\beta}(a) = -ba^2(1+ca)+O(a^3) \ ,$$

$$c = (153 - 19N_f)/(66 - 4N_f) \ . \qquad (11)$$

At this order $\hat{\beta}(a)$ is unique, independent of the choice of RS. The solution of this first order differential equation (6,11) has to be supplemented by one boundary condition. Following the standard convention the solution of equ. (11) is written as,

$$b \ln \mu/\Lambda = -\int_a^\infty \frac{dx}{x^2(1+cx)} + c \ln(b/2c)$$

$$= \frac{1}{a} + c \ln \left[\frac{ba}{2(1+ca)}\right], \qquad (12)$$

neglecting $O(a^3)$ terms. Expanding this equation in powers of $1/\ln(\mu/\Lambda)$ one approximates the coupling constant by

$$1/a = b \ln(\mu/\Lambda) + c \ln(\ln(\mu^2/\Lambda^2)) - \frac{c^2}{b} \frac{1}{\ln(\mu/\Lambda)} + \ldots \quad (13)$$

Using the first two terms a plot of $\alpha_s(\mu = Q)$ for different values of Λ is shown in Fig.7. The important property of QCD, called asymptotic freedom [12], is clearly seen: for large values of μ the coupling constant vanishes as

$$a(\mu) = \alpha_s(\mu)/\pi \xrightarrow[\mu\gg\Lambda]{} \frac{1}{b\ln(\mu/\Lambda)} \quad . \quad (14)$$

This property motivates the perturbative QCD calculations for deep inelastic phenomena, whenever $\mu \sim Q$ is large, and it serves as the theoretical justification of the parton model. The parameter Λ appearing in equs.(12-14) is a free parameter of QCD. It has to be determined by comparing QCD predictions with data of deep inelastic processes. However, because of the RS dependence of these predictions, the deduced actual value of Λ is evidently RS-dependent. Fig.7 illustrates furthermore the difficult task of extracting a precise value of Λ: at large values of $\mu \sim Q$ the dependence of α_s on Λ is very weak; at small values of the scale it is usually hard to justify the approximation of the perturbative expansion by a few terms. Contrary to the parton model QCD contains the intrinsic scale Λ and deviations from the parton-scaling predictions are to be expected; because of equ.(13) they are of logarithmic nature.

We first consider deep inelastic leptoproduction in QCD [19], but we restrict ourselves to the discussion of the $O(\alpha_s)$ calculation following [20]. In addition to the basic $\gamma q \to q$ interaction the contributions of the real gluon emission $\gamma q \to qG$ and of the virtual emission in $\gamma q \to q$ have to be evaluated. In this calculation one encounters infrared and collinear (mass) singularities. Treating the partons massless and on-shell it is again convenient to regulate these divergencies by dimensional regularization [21], with $\varepsilon<0$. The structure function for $eq_i \to eX$ is then given by

$$F_2^{eqi}(z,Q^2)/e_i^2 = \delta(1-z) + \alpha_s H_{qq}(z,\mu^2/Q^2,\varepsilon) + \frac{\alpha_s}{2\pi} f_q(z) \qquad (15)$$

$$\text{with } \alpha_s H_{qq} = -\frac{1}{\varepsilon}\frac{\alpha_s}{2\pi} P_{qq}(z) \left(\frac{4\pi\mu^2}{Q^2}\right)^{\varepsilon} \frac{\Gamma(1-\varepsilon)}{\Gamma(1-2\varepsilon)}, \; z = x/x_i \; . \qquad (16)$$

P_{qq} is the Altarelli-Parisi splitting function [22] for the transition $q\to q$,

$$P_{qq}(z) = \frac{4}{3}\left[\frac{1+z^2}{(1-z)_+} + \frac{3}{2}\delta(1-z)\right]. \qquad (17)$$

The function $f_q(z)$ [20] does not depend on ε, but on the process under consideration and on the regularization scheme. Adding the contribution from the subprocess $\gamma G \to q\bar{q}$ one is able to define Q^2-dependent quark distributions [20] by

$$q_i(x,Q^2) = q_i(x)+\frac{\alpha_s}{2\pi}\int_x^1 \frac{dy}{y}\Big\{\Big[\left(-\frac{1}{\varepsilon}\frac{(4\pi)^{\varepsilon}\Gamma(1-\varepsilon)}{\Gamma(1-2\varepsilon)} + \ln Q^2/\mu^2\right)P_{qq}(x/y)$$

$$+ f_q(x/y)\Big]q_i(y) + \Big[\Big(-\frac{1}{\varepsilon}\frac{(4\pi)^{\varepsilon}\Gamma(1-\varepsilon)}{\Gamma(1-2\varepsilon)} +$$

$$+ \ln Q^2/\mu^2\Big) P_{qG}(x/y) + f_G(x/y)\Big]G(y)\Big\}. \qquad (18)$$

The functions $q_i(x)$ and $G(x)$ denote the bare distributions ($\hat{=} f_i(x)$, equ.(1)) for the quark i and the gluon, respectively, which cannot be calculated in the framework of perturbative QCD. The advantage of this definition (18) is that the relation

$$F_2(x,Q^2) = \sum_i e_i^2 \, x \, q_i(x,Q^2) \tag{19}$$

keeps the form of the parton model, equ.(1), with no extra corrections proportional to α_s. Differentiating equ.(18) with respect to Q^2 then replacing $q_i(y)(G(y))$ by $q_i(y,Q^2)\cdot(G(y,Q^2))$ and evaluating α_s at $\alpha_s(\mu = Q)$ the renormalization group improved evolution equation in leading logarithmic order is derived [22] with $\alpha_s = \alpha_s^{LO} = \pi/(b\ln(Q/\Lambda))$, equ.(14),

$$Q^2\frac{\partial q_i(x,Q^2)}{\partial Q^2} = \frac{\alpha_s(Q^2)}{2\pi}\int_x^1 \frac{dy}{y}[P_{qq}(\tfrac{x}{y})q_i(y,Q^2)+P_{qG}(\tfrac{x}{y})G(y,Q^2)]. \tag{20}$$

By this procedure one goes beyond the lowest order $\alpha_s(\mu)$-result, equ.(18), since all the terms of order $[\alpha_s(\mu)\ln(Q^2/\mu^2)]^n, n\geq 1$, get summed. Correspondingly the gluon distribution $G(x,Q^2)$ satisfies,

$$Q^2\frac{\partial G(x,Q^2)}{\partial Q^2} = \frac{\alpha_s(Q^2)}{2\pi}\int_x^1 \frac{dy}{y}[P_{Gq}(x/y)\sum_i q_i(y,Q^2)+P_{GG}(x/y)G(y,Q^2)]. \tag{21}$$

The pattern of scaling violations emerging from these equs.(20,21) is qualitatively the following: as Q^2 increases $F_2(x,Q^2)$ increases mainly at small x, because the gluon in the nucleon generates additional $q\bar{q}$ pairs, and it decreases at large x, because the emitted gluon carries away momentum of the quark interacting with the virtual photon. This is in accord with the measured Q^2-dependence shown in Fig.3. More important, detailed quantitative tests of these equs.(20,21) including the terms of order α_s^2 have been performed [23] in order to determine Λ.

It is concluded [23] that all deep inelastic data are consistent with a value of $\Lambda_{\overline{MS}}$ = 250 ± 150 MeV. An example of fitting data with extreme values, $\Lambda_{\overline{MS}}$ = 124 MeV and $\Lambda_{\overline{MS}}$ = 404 MeV, is shown in Fig.8 [24]. The higher order QCD corrections appear to be small and existing data are not yet precise enough to permit an unambigous measurement of Λ. This may also be seen in Fig.9, in which the difference of structure functions, $F_2(x,Q^2) - 2xF_1(x,Q^2)$, measured in neutrino scattering [25] is compared with the QCD prediction which is directly proportional to $\alpha_s(Q^2)$. In leptoproduction the gluon distribution $G(x,Q^2)$ is not directly measured, but it is constrained by the momentum sum rule equ. (3). Concerning the shape of $G(x,Q^2)$ there exists a correlation with the parameter Λ and the distribution of sea quarks [24]. Starting from a broad(=hard) gluon distribution at some reference $Q = Q_o$ a larger value of Λ is preferred by the data than for a soft $G(x,Q_o^2)$. The larger Λ the stronger is the evolution of $G(x,Q^2)$ and the difference diminishes for $Q^2 > Q_o^2$. With increasing Q^2 antiquarks at x are generated by gluons $G(x \leq y \leq 1, Q^2)$ according to equ.(20). Since $\bar{q}(x,Q^2)$ is small at large x and therefore difficult to measure precisely, existing data do not allow to distinguish between a hard and a soft gluon distribution. Input distributions $G(x,Q_o^2 = 4\ \text{GeV}^2)$ as determined by a recent analysis [26] based on leading logarithm (LO) accuracy, equs. (20,21), are shown in Fig.10. The corresponding Λ is Λ_{LO} = 400 MeV (200 MeV) for the hard (soft) gluon distribution. The functions $G(x,Q_o^2)$ determined by the CDHS [4] and by the EMC [27] Collaborations do agree with the soft one in Fig.10.

SECOND LECTURE

Fig.4 shows that secondary hadrons at large p_T are observed in a small fraction of hadron-hadron collisions $a+b \rightarrow$hadrons with large p_T + X. In the framework of the parton model it is natural to interpret these events as the result of parton-parton scattering. Already twelve years ago J.D. Bjorken put forward the question: "Can we measure the parton-parton cross section" [28] and thereby study the interactions between quarks and gluons. The idealized picture is that partons within the colliding hadrons a and b, respectively, undergo dominantly two-body scattering at large angles (= large momentum transfer) giving rise to two opposite partons which fragment into hadrons. Since partons are confined within the size of the hadron, it is expected that the produced hadrons are confined to the direction of the scattered partons: large p_T hadrons are produced as hadronic jets. Like for deep inelastic scattering, equ.(1), the impulse approximation should be applicable: the production cross section for hadronic hard scattering events factorizes in terms of parton distributions $f_i^{a,b}(x)$ and the parton cross section. The master formula is

$$d\sigma(a+b\rightarrow \text{hadronic jets} +X) = \sum_{i,j} \iint dx_a dx_b f_i^a(x_a) f_j^b(x_b) \cdot$$

$$\cdot\ d\hat{\sigma}\ (i + j \rightarrow \text{partons}). \qquad (22)$$

Spectacular two-jet structures as predicted on the basis of equ.(22) are observed at the Sp$\bar{p}$S collider since 1982 [29]. A clean example is shown in Fig.11, which displays the angular distribution of the transverse energy E_T of the final state hadrons. Almost one-half of the incident energy is distributed on two localized hadronic jets produced near 90° in opposite hemispheres, $\Delta\phi \simeq 180^{o}$, with respect to the $p\bar{p}$-beam direction. The advantage of high incident energy is clearly demonstrated by this figure. Jets with $p_T \simeq E_T \sim 100$ GeV are kinematically possible and well separated in phase space. At lower energies this structure is not as pronounced. Fig.12 shows the density of large p_T final states averaged over a large number of events as observed at ISR energy, $\sqrt{s} \simeq 63$ GeV [6]. It is important to note that the observation of jets at $Sp\bar{p}S$ energy supports equ.(22) concerning the energy dependence: at fixed p_T the rates are increasing with energy according to the (approximate) scaling behaviour of the jet cross section (equ.(4) and Fig.6). These exciting results are reviewed recently including further aspects and details in [6,30-34].

Equ.(22) is quite simple for the two-jet cross section:

$$\frac{d\sigma(a+b\to jet_1+jet_2+X)}{dx_a dx_b d\cos\theta^*} = \sum_{i,j} f_i^a(x_a) f_j^b(x_b) \frac{d\hat{\sigma}(i+j\to c+d)}{d\cos\theta^*}, \qquad (23)$$

where θ^* is the scattering angle in the parton CMS. Two-body kinematics determine the parton variables uniquely, once the momenta of the outgoing jets ($\hat{=}$partons c and d) are measured. In the CMS of the hadrons a and b, which is a useful infinite momentum frame, $y_{1,2}$ denote the rapidities of the jets with transverse momenta, $|\vec{p}_{T_1}| = |\vec{p}_{T_2}| = p_T$. The corresponding kinematical relations are

$$0 \le x_a = p_T(e^{y_1}+e^{y_2})/\sqrt{s} \le 1, \quad 0 \le x_b = p_T(e^{-y_1}+e^{-y_2})/\sqrt{s} \le 1, \qquad (24)$$

$$\hat{s} = x_a x_b s, \ \hat{t} = -x_a \sqrt{s}\, p_T\, e^{-y_1} = -\hat{s}/2(1-\cos\theta^*), \ \hat{u} = -\hat{s}-\hat{t},$$

and

$$\cos\theta^* = (1-e^{y_2-y_1})/(1+e^{y_2-y_1}),$$

which lead to

$$\frac{d\sigma(a+b\to jet_1+jet_2+X)}{dy_1\, dy_2\, dp_T} = \frac{4p_T}{s} \frac{d\sigma(a+b\to jet_1+jet_2+X)}{dx_a\, dx_b\, d\cos\theta^*} . \tag{25}$$

This cross section is straightforward to calculate using the quark and gluon distributions $f_i = q_i$, G determined from lepton scattering, and using $d\hat{\sigma}/d\cos\theta^*$ for all the eight two-body quark-gluon interactions [35], for which the corresponding Feynman diagrams are shown in Fig.13. For the dominant quark flavours in the proton (antiproton), $q_i = u,d,s,\bar{u},\bar{d},\bar{s}$, and the gluon G, equ.(23) is a linear combination of 49 terms. Instead it is more transparent to follow an analysis by the UA1 and UA2 Collaborations [36], which is inspired by a previous one performed for $\pi^o - \pi^o$ pairs at ISR [37]. It starts with the observation (see [32]) that all the eight two-body $d\hat{\sigma}/d\cos\theta^*$ have to a very good approximation an universal θ^*-dependence and that the dominant ones are related by

$$\frac{d\hat{\sigma}(GG\to GG)}{d\cos\theta^*} : \frac{d\hat{\sigma}(qG\to qG)}{d\cos\theta^*} : \frac{d\hat{\sigma}(qq\to qq)}{d\cos\theta^*} = 1:(4/9) : (4/9)^2 . \tag{26}$$

The colour factor $4/9 = C_F/N_c$ reflects the relative strength of the qqG and GGG couplings, respectively. The angular dependence is dominated by gluon, spin-one, exchange (Fig.13),

$$\frac{d\hat{\sigma}}{d\cos\theta^*} \simeq \frac{\alpha_s^2}{\hat{s}} \frac{1}{\sin^4(\theta^*/2)} , \tag{27}$$

i.e. Rutherford's formula in strong interaction physics. Fig.14 illustrates the agreement of equ.(27) with the measured θ^*-dependence [36]. Scalar gluon exchange is not compatible with this set of data (Fig.14). In the approximation under discussion equ.(23) factorizes

$$\frac{d\sigma(p\bar{p}\to jet_1+jet_2+X)}{dx_a dx_b d\cos\theta^*} \simeq F(x_a)F(x_b) \frac{d\hat{\sigma}(GG\to GG)}{d\cos\theta^*} \tag{28}$$

in terms of an effective structure function for the proton

$$F(x) = G(x) + 4/9\,[u(x) + d(x) + sea(x)], \tag{29}$$

which can be extracted from the two-jet data [36]. The result is shown in Fig.15 together with the fit to the UA2 data, which is in fair agreement with $F(x,Q^2)$ as determined from neutrino scattering [4] ($\langle Q^2\rangle = 20\ GeV^2$) and evolved according to equs.(20,21) to $Q^2 = 2000\ GeV^2$, the typical scale for the collider experiment. Although the scale Q is not necessarily the same over the p_T-range of Fig.15, one may indeed conclude that the two-jet data favour evolving structure functions, $q_i(x)$, $G(x) \to q_i(x,Q^2)$, $G(x,Q^2)$ in equ.(23) as required by QCD.

Integration of equ.(25) with respect to y_2, the rapidity of the second jet, yields the inclusive single jet cross section,

$$E\,\frac{d^3\sigma}{d^3p}\,(a+b\to jet+X) = \frac{1}{\pi} \sum_{i,j} \int_{x_a^{min}}^{1} dx_a \frac{x_a f_i^a(x_a)\, x_b f_j^b(x_b)}{(x_a - p_T\, e^y/\sqrt{s})} .$$

$$\frac{d\hat{\sigma}(i+j\to c+d)}{d\hat{t}} \quad . \tag{30}$$

It obeys scaling (equ.(4)) with n = 4. Experimentally n = 5.1 ± 0.3 [8] (Fig.6), which is consistent with replacing $f_i(x) \to f_i(x,Q^2)$ in equ.(30): Fig.16 shows a comparison of single-jet data from ISR and $Sp\bar{p}S$ with typical QCD calculations in this leading order [38,39]. The predictions are impressive, since especially the increase in rate by five orders of magnitude from ISR to $Sp\bar{p}S$ energies is correctly described. However, one has to keep in mind that large theoretical uncertainties are involved. Especially at large $p_T \simeq 100$ GeV/c predictions may even differ by a factor ten. In LO (equ.(30)) approximation the actual values of Q^2 in $f_i(x,Q^2)$ and of μ^2 in $\alpha_s(\mu^2)$ are not constrained except $Q^2 \simeq \mu^2 \simeq O(p_T^2)$. Different parametrizations for $f_i(x,Q^2)$ reflecting mainly the uncertainty in $G(x,Q^2)$ discussed in the first lecture give in general different results for $Ed^3\sigma/d^3p$. A recent detailed investigation of all these LO ambiguities can be found in [40], from which Fig.17 is taken for illustration. The curves are evaluated in LO, fixing $Q^2 = \mu^2 = 2\hat{s}\hat{t}\hat{u}/(\hat{s}^2+\hat{t}^2+\hat{u}^2)$ $\simeq 4p_T^2/3$ at y = 0, for different sets $f_i(x,Q^2)$ compatible with deep inelastic lepton scattering. In Fig.17 the freedom in choosing the scales Q and μ is rephrased by varying $\Lambda = 0.1$ to $\Lambda' = 0.7$. With the help of equ.(14) one can expand

$$\alpha_s(\mu/\Lambda') = \alpha_s(\mu/\Lambda) + \frac{b}{\pi}\,\alpha_s^2(\mu/\Lambda)\ln(\Lambda'/\Lambda)+\ldots \tag{31}$$

This equation tells us that the uncertainty of the LO calculation, here of $O(\alpha_s^2)$, is of the order of the α_s^3 terms, which are not yet completely calculated for the jet cross section. Therefore at present only semiquantitative QCD results are available.

Further (approximate) confirmation of the QCD-parton model comes from analysing the properties of the jet structure. One measures the fractional longitudinal momentum z of hadrons within a jet with respect to the jet momentum $\vec{p}_{jet}$, $z = \vec{p}_h \cdot \vec{p}_{jet} / \vec{p}_{jet}^{\,2}$. The distribution in z, D(z), is usually called the fragmentation function. It is an important question, if there are differences between jets, which originate from quarks or gluons. Since the parton model assumes that the jet properties are universal, independent of the production mechanism, it is interesting to compare D(z) measured in $e^+e^- \to$ jets [41] and in $p\bar{p}$ (or pp) $\to$jets + X [42,43]: indeed the jet properties are very much the same. Since e^+e^- annihilation is dominated by quark jets, whereas in $p\bar{p} \to$ jets + X the gluon jets are more frequently produced at Sp$\bar{p}$S [39], the present conclusion is that quarks and gluons have very similar fragmentation (hadronization) properties, namely $D_q^{hadrons}(z) \simeq D_G^{hadrons}(z)$ within the experimental errors [41-43].

Beyond the leading order in perturbative QCD the definition of a jet is faced by the problem of infrared and collinear singularities due to the transitions, e.g. $q \to q + G$ or $G \to G + G$ as already mentioned in the previous lecture. To cure the problem angular jets are proposed [44] with production cross sections remaining finite when they are calculated perturbatively. As a consequence the gluon jet is expected to be broader than a quark jet [45]: for small opening angle δ, $\delta \ll 1$, the relation for energetic jets is

$$\delta_{Gluon} \simeq [\delta_{quark}]^{4/9} \quad , \qquad (32)$$

where the power is the same colour factor which appears in equ.(26). However, this statement is contaminated by non-perturbative effects, which are responsible for the actual hadronization of partons and which are not yet understood theoretically leaving the quantitative description of jet structures a big open problem.

The splitting of partons gives rise to multijet final states. At the $Sp\bar{p}S$ three-jet events are identified well separated in phase space. The subprocesses in LO are $qq \to qqG$, $qG \to qGG$, $q\bar{q} \to GGG$ and $GG \to GGG$ [46]. For these well separated jets the production rate can then be savely calculated: it is of order α_s^3 in agreement with the observation that at $Sp\bar{p}S$ the jet sample consists of about 85 % two-jet and 15 % three-jet events [47]. Another quantitative measure of the presence of three-jet events is the p_{out}-distribution, where p_{out} is the jet momentum perpendicular to the plane defined by the jet with the highest p_T and the beam momentum. The two-jet sample is peaked around small values of p_{out}, whereas for $p_{out} > 10$ GeV/c the three-jets are well separated. Fig.18 shows clearly the presence of non-coplanar events [47] with a rate in agreement with QCD calculations [48].

Using fragmentation functions, $D_{q,G}^{hadrons}$, as input one calculates the single inclusive hadron cross section at large p_T for the processes $a + b \to H + x$. From equ.(30) and with $p_H \simeq zp_{jet}$ one obtains in LO

$$E_H \frac{d^3\sigma}{d^3p_H} (a+b\to H+X) = \sum_{c=q,G} \int_o^1 \frac{dz}{z^2} E_c \frac{d^3\sigma}{d^3p_c} (a+b\to c+X) D_c^H(z,Q^2) \quad . \quad (33)$$

Like the structure functions the distributions $D_{q,G}^H$ depend on a scale Q. Therefore one expects for the effective power n characterizing the scaling behaviour (equ.(4) and subsequent discussion) $n_H \gtrsim n_{jet} > 4$. It is obvious to remark that single-particle yields are not biased by the theoretical as well as experimental ambiguities related to precise quantitative jet definitions and jet cross sections.

Reasonable parametrizations for $q_i(x,Q^2)$, $G(x,Q^2)$ and $D^H_{q,G}(z,Q^2)$, with $Q^2 = O(p_T^2)$, indeed describe successfully the data covering FNAL, ISR and Sp$\bar{p}$S energies [33,49,50]. A very recent result [50] is shown in Fig.19. The large values of p_T for the secondary π^o's measured at Sp$\bar{p}$S, $p\bar{p} \to \pi^o x$, have to be remarked. The energy and p_T dependence of the data is correctly reproduced by equ.(33) with the distributions from [51] evaluated with $Q^2 = p_T^2$ and $\Lambda_{LO} = 310$ MeV. Since these predictions are presently based on LO calculations, the same critical remarks already given for the jet cross section have to be taken into account in judging the agreement between experiment and QCD.
A detailed quantitative test requires higher-order QCD calculations, but in addition high-precision data are necessary. The following example should support this demand: the QCD parton model predicts independent of the order of perturbation theory at large p_T

$$E \frac{d^3\sigma}{d^3p} (pp \to \pi^o x) = \frac{1}{2}[E \frac{d^3\sigma}{d^3p}(pp \to \pi^+ X) + E \frac{d^3\sigma}{d^3p}(pp \to \pi^- X)]. \quad (34)$$

This relation rests on the basic assumption of "independence of the environment" [28], i.e. the factorized structure of production cross sections (see equ.(33)). It does not follow from isospin considerations at the hadronic level, although isospin enters in relating $D_q^{\pi^\pm}$ to $D_q^{\pi^o}$, $q = u,d,\bar{u},\bar{d},\ldots$. For the pionization of the flavourless gluon one relates $D_G^{\pi^o} = D_G^{\pi^+} = D_G^{\pi^-}$. The corresponding data [52] shown in Fig.20, however, indicate that equ.(34) is missed by a factor two. Since $\pi^\pm$'s and π^o's are detected by different techniques it is too early to draw definite conclusions.

THIRD LECTURE

We consider now the reaction

$$a + b \to \ell^+\ell^- + \text{anything} , \tag{35}$$

where the invariant mass Q of the lepton pair is large. The main concern is the non-resonant continuum, which according to the parton model is produced via the elementary quark-antiquark fusion mechanism, $q + \bar{q} \to \gamma(\text{timelike}) \to \ell^+\ell^-$, i.e. the Drell- Yan mechanism [53,11]. At the level of partons the relation by crossing to deep inelastic lepton scattering and to electron-positron annihilation, $e^+e^- \to \gamma \to q\bar{q} \to$ hadrons, is evident. As we will discuss later on this mechanism is furthermore of eminent importance for intermediate boson, $W^\pm$ and Z^o, production, when the quarks are coupled to the weak current. The experimental exploration of reaction equ.(35) led also to the discovery of the heavy resonances, J/ψ and Υ: their production, however, is described by different mechanisms, e.g. gluon fusion $GG \to Q\bar{Q} \to J/\psi$ or Υ [54]. Therefore the Drell-Yan process is of additional interest as background to searches for heavy quark Q = c,b,t..bound states in hadronic collisions [55].

In the parton model the integrated cross section for dileptons of continuous mass Q is

$$\frac{d\sigma}{dQ^2}(a+b\to\ell^+\ell^-+X) = \langle\frac{1}{3}\rangle \frac{4\pi\alpha^2\tau}{3Q^4}\int_o^1 \frac{dx_a}{x_a}\frac{dx_b}{x_b}\,\delta(1-\frac{\tau}{x_ax_b})\ \cdot$$

$$\cdot\ [\sum_i e_i^2\, q_i^a(x_a)\bar{q}_i^b(x_b) + (a\leftrightarrow b)], \tag{36}$$

where $\tau = Q^2/s$, $s = (p_a+p_b)^2$ and <1/3> is the colour factor. In this approximation the scaling behaviour equ.(5) follows immediately.

We turn now to the QCD corrections to this equ.(36). At order α_s they arise from the subprocesses $q\bar{q} \to \gamma G$, virtual corrections to $q\bar{q} \to \gamma$, and $qG \to \gamma q$. Applying the same techniques as for deep inelastic lepton scattering, described during the first lecture, the resulting cross section - here for one quark flavour - can be written [20]as,

$$d\sigma/dQ^2 = \frac{4\pi\alpha^2\tau}{9Q^4}\int\frac{dx_a}{x_a}q^a(x_a)\{\int\frac{dx_b}{x_b}[(\delta(1-\hat{\tau})+\alpha_s H_{qq} + \frac{\alpha_s}{2\pi}f_q)\bar{q}^b(x_b)$$

$$+ (\alpha_s H_{qG} + \frac{\alpha_s}{2\pi} f_G)\; G^b(x_b)]\} + (q^a \leftrightarrow \bar{q}^b)$$

$$+ \frac{4\pi\alpha^2\tau}{9Q^4}\int\frac{dx_a dx_b}{x_a x_b}\{q^a(x_a)\bar{q}^b(x_b)[\frac{\alpha_s}{2\pi} f_q^{DY} - 2\frac{\alpha_s}{2\pi} f_q]$$

$$+ (q^a(x_a)+\bar{q}^a(x_a))G^b(x_b)[\frac{\alpha_s}{2\pi} f_G^{DY} - \frac{\alpha_s}{2\pi} f_G]\} + (a\leftrightarrow b). \tag{37}$$

where $\hat{\tau} = \tau/x_a x_b$. The expression in the first curly bracket is just the Q^2 dependent quark distribution $q(x,Q^2)$ defined by equ.(18). At order α_s this is an illustration of the factorization theorem [56]: the mass singular terms proportional to $1/\varepsilon$, namely H_{qq} of equ.(16) and H_{qG}, are universal, i.e. process independent; therefore they can be factored from the perturbative parton cross section and consistently be absorbed into the parton distributions (equ.(18)). However, there are ambiguities again concerned with finite terms in the definitions of $q_i(x,Q^2)$, $G(x,Q^2)$ [57]: the definition (equ.(18)) used here is specified by the non-universal convention of measuring directly $q_i(x,Q^2)$ by F_2^{eN} (see equ. (19)). This fixes the finite terms [20] appearing in the second curly bracket of equ.(37). A different possibility is to use an universal convention [58], in which the parton densities beyond leading order are defined without the finite terms f_q, f_G in equ.(18). For this process-independent convention the formula for F_2 (equ.(19)) and for Drell-Yan (equ.(37)) have to be changed correspondingly. The definition of the gluon function $G(x,Q^2)$ requires calculations at order α_s^2 for deep inelastic scattering, as well as for the process under consideration. The scale Q^2 entering equ.(37) via the functions $H \sim \ln Q^2/\mu^2$ may be chosen differently at $Q'^2 \neq Q^2$. Rewriting (equ.16))

$$-\frac{1}{\varepsilon}\,(\mu^2/Q^2)^{\varepsilon} = -\,1/\varepsilon\,(\mu^2/Q'^2)^{\varepsilon} + \ln Q^2/Q'^2 + O(\varepsilon) \tag{38}$$

this amounts to replace $q(x,Q^2)$ by $q(x,Q'^2)$ and adding a finite term to the second part of equ.(37), which is proportional to

$$2\alpha_s\, P_{qq}\, \ln Q^2/Q'^2\; q^a(x_a,\, Q'^2)\; \bar{q}^b\,(x_b,\, Q'^2) \tag{39}$$

At fixed order α_s and using equ.(18) for the evolution of $q(x,Q^2)$ with Q^2 the result for equ.(37) obviously does not change by the different procedures described above. The terms just compensate each other. However, inserting the renormalization-group improved parton distributions, obtained from equs.(20-21) - even including higher orders $O(\alpha_s^2)$, the numerical results of QCD predictions for Drell-Yan, in general for all hard processes, at finite order are actually influenced by the choices of conventions and scales. Like the dependence on the RS scheme, described before, this factorization scheme dependence [57] disappears when all orders in α_s could be taken into account; clearly physical quantities do not depend on the procedure of calculation.

The standard predictions for Drell-Yan cross sections beyond leading order are given for $\mu = Q$, factorization scale Q' = Q and for the convention equ.(19), such that all scales are set directly by the mass Q,

$$\frac{d\sigma}{dQ^2}(a+b\to\ell^+\ell^-+X) = \frac{4\pi\alpha^2\tau}{9Q^4}\int\frac{dx_a dx_b}{x_a x_b}\{\sum_i e_i^2 q_i^a(x_a,Q^2)\bar{q}_i^b(x_b,Q^2)$$

$$\cdot\ [\delta(1-\hat{\tau})+\theta(1-\hat{\tau})\frac{\alpha_s(Q^2)}{2\pi}[f_q^{DY}(\hat{\tau})-2f_q(\hat{\tau})]]$$

$$+\sum_i e_i^2 q_i^a(x_a,Q^2)G^b(x_b,Q^2)\theta(1-\hat{\tau})\frac{\alpha_s(Q^2)}{2\pi}[f_G^{DY}(\hat{\tau})-f_G(\hat{\tau})]+(a\leftrightarrow b)\}\ . \tag{40}$$

The quantitative features of the high mass data at different incident energies are successfully reproduced. An example based on a recent analysis [59] is shown in Fig.21. The magnitude of the higher order (HO) contribution, here at order α_s (equ.(40)), is usually expressed in terms of the K-factor,

$$K = (\sigma^{LO} + \sigma^{HO})/\sigma^{LO}\ . \tag{41}$$

Although the numerical values for K are convention dependent as discussed before, their magnitudes serve as indication on the applicability of perturbative expansions like equ.(40). For instance, $K \simeq 1.5$ at $\sqrt{s} = 63$ GeV and $Q \gtrsim 11$ GeV [59]. It can be seen in Fig.21 that the data at this energy are not precise enough for testing the HO contributions of equ.(40).

For the production of W and Z bosons in hadronic collisions, $a + b \to W^{\pm}, Z + X$, the basic elementary processes are: $u + \bar{d}_\theta \to W^+$, $\bar{u} + d_\theta \to W^-$ and $u+\bar{u}\to Z^o$, $d+\bar{d}\to Z^o$, respectively, where $d_\theta = d\cos\theta_c + s\sin\theta_c$ [60]. Because of the high masses of the intermediate bosons the Drell-Yan mechanism is tested at scales $Q \simeq 100$ GeV, which are one order of magnitude larger than for the continous pairs (compare Fig.21). The integrated production cross section has the same structure as equ.(40). For W^+ production in $p\bar{p} \to W^+X$ it is

$$\sigma(p\bar{p}\to W^+X) = \sqrt{2}\pi G_F \cos^2\theta_c \langle\frac{1}{3}\rangle \tau \int_\tau^1 \frac{dx}{x}[u(x,M_W^2)\,d(x',M_W^2) + d\leftrightarrow u]$$

$$+ \text{ term proportional } \sin^2\theta_c + O(\alpha_s(M_W^2)) \ , \qquad (42)$$

$\tau = M_W^2/s$ and $x' = \tau/x$; u and d are the distributions in the proton. Multiplying σ by the branching ratio into electrons, $B(W\to e\nu) = 0.089$ and $B(Z\to e^+e^-) = 0.032$, respectively, the theoretical calculations [61] for $p\bar{p}$ collisions at $\sqrt{s} = 540$ GeV give,

$$B\sigma(W^\pm) = 370^{+110}_{-60}\ \text{pb}, \quad B\sigma(Z) = 42^{+12}_{-6}\ \text{pb} \ . \qquad (43)$$

The errors are estimated by using different sets of parton distributions and scales. The corresponding experimental values quoted by the UA1 [62] and UA2 [63] Collaborations on the basis of the 1983 data are,

$$\text{UA1: } B\sigma(W^\pm) = 530 \pm 80 \pm 90 \text{ pb}, \quad B\sigma(Z) = 71 \pm 24 \pm 14 \text{ pb};$$
$$\text{UA2: } B\sigma(W\ ) = 530 \pm 100 \pm 100 \text{ pb}, \quad B\sigma(Z) = 110 \pm 40 \pm 20 \text{ pb}. \qquad (44)$$

Although the numerical agreement is presently not perfect, especially with respect to the UA2 result with the relative large number of Z's, the discovery of the W and Z bosons at $Sp\bar{p}S$ [64,65] is not only a success for the electroweak theory, but also for the QCD-improved parton model.

FOURTH LECTURE

Investigating the production of real photons at large p_T, $a+b\to\gamma+$anything, is of great interest, because of experimental and theoretical advantages: first the photon represents the complete jet with well-defined momentum, second it behaves elementary with point-like couplings to the quarks at short distances. At leading-order QCD only two subprocesses contribute, namely $q\bar{q}$ annihilation, $q\bar{q}\to\gamma G$, and Compton-like scattering, $qG\to\gamma q$. In this LO approximation the single-photon cross section is computed as the hadronic jet cross section in equ.(30), where instead the following $d\hat{\sigma}/d\hat{t}$ are inserted,

$$d\hat{\sigma}/d\hat{t}(q\bar{q}\to\gamma G) = \frac{2\pi\alpha e_q^2\alpha_s}{\hat{s}^2}\left\langle\frac{4}{9}\right\rangle(\hat{t}/\hat{u}+\hat{u}/\hat{t}) \ ,$$

$$d\hat{\sigma}/d\hat{t}(qG\to\gamma q) = \frac{2\pi\alpha e_q^2\alpha_s}{\hat{s}^2}\langle 1/6\rangle(-\hat{t}/\hat{s} - \hat{s}/\hat{t}) \ . \tag{45}$$

The main disadvantage is the small rate proportional to $\alpha\alpha_s$. When compared to the hadronic jet yield the estimate is $\gamma/\text{jet}\simeq 5.10^{-4}$ [9,10]. However, when compared to single π^o production the ratio γ/π^o is predicted to increase with p_T and even $\gamma/\pi^o\sim 1$ at large p_T. This is confirmed experimentally [10], as can be seen in Fig.22. Double photon production, $a+b\to\gamma+\gamma+X$, is dominated by the basic QED process, $q\bar{q}\to\gamma\gamma$ [66]. It is proportional to e_q^4. Experimental information on $pp\to\gamma\gamma X$ is available [67], and indeed the measured rate is compatible with fractionally charged quarks. The theoretical picture described so far becomes more involved, since photon bremsstrahlung [68], especially off quarks, has to be taken into account, e.g. $qq\to q(q\to\gamma)$ and $Gq\to G(q\to\gamma)$. In LO these processes contribute at the same order $\alpha\alpha_s^2\ln(Q^2/\Lambda^2)\simeq\alpha\alpha_s(Q^2)$

as the Born terms, equ.(45), since the fragmentation of a quark into a photon is given by

$$D_q^{\gamma}(z,Q^2) = \frac{e_q^2\alpha}{2\pi}\,\frac{1+(1-z)^2}{z}\,\ln(Q^2/\Lambda^2). \qquad (46)$$

However, at large p_T the inclusive single photon yield is mainly sensitive to the region $z \simeq 1$, therefore the bremsstrahlung contribution is only 20 % or less [68,69]. This is compatible with the experimental observation that a high-p_T photon is essentially unaccompanied by other hadrons [10].

Comparing QCD calculations in LO with data [9,10,70] for single prompt photons one finds that agreement is achieved at large p_T (Fig.23). However, like for the other hard processes discussed in the previous lectures these predictions are by no means stable against variations of scales. Fig.23 illustrates this fact: $\alpha_s = \alpha_s^{LO}(p_T^2)$, Λ_{LO} = 200 MeV, for two different scales in the structure functions $f_i(x,Q^2)$ taken from [26], $Q^2 = p_T^2/2$ (solid curve) and $Q^2 = 4p_T^2$ (dashed curve), respectively. The p_T spectrum becomes steeper with increasing Q^2, since scaling violations in f_i become more pronounced at large x ($\simeq O(x_T)$).

In order to improve the predictive power of QCD calculations the HO contributions have to be included. At order α_s^2 they are completely evaluated [71] for prompt photons. Four classes of subprocesses have to be considered: $q\bar{q} \to \gamma X$, $qG \to \gamma X$, $q_iq_j \to \gamma X$ and $GG \to \gamma X$, but the analytic form of the resulting cross section is too long to be given here. Including these HO terms the predictions indeed stabilize, which can be seen in Fig.24 using the same set of parameters as in Fig.23. Good agreement with the ISR data for $p_T \geq 5$ GeV/c is obtained. A similar good agreement is found at SPS energies ($\sqrt{s} \sim 20$ GeV) by comparison with (preliminary) data for $pp \to \gamma X$ measured by the NA24

Collaboration [72]. At the energies under consideration the corresponding K-factor, equ.(41), turns out to be rather large, $K \simeq 2$, independent of p_T, when it is calculated for the scales $\mu^2 = Q^2 = p_T^2$. Because of this large value one may even question the perturbative expansion, but this problem is postponed for a moment.

The HO corrections for double photons, originating from the subprocesses $q\bar{q} \to \gamma\gamma X$ and $qG \to \gamma\gamma X$ at order α_s, are calculated for the differential cross section $d^4\sigma/dy\,d\vec{p}_T dz$ [73]. In the CMS for $a + b \to \gamma(y,\vec{p}_T) + \gamma(\vec{p}_T{}')+X$ the invariant variable z is defined by $z= -\vec{p}_T \cdot \vec{p}_T{}'/\vec{p}_T^{\,2}$. In LO ($q\bar{q} \to \gamma\gamma$) only back-to-back photons, $\vec{p}_T{}' = -\vec{p}_T$, are predicted and consequently the above cross section is peaked at $z = 1$, $d\sigma^{LO}/dz \sim \delta(1-z)$. This peak has to be smeared by the experimental resolution in z, $z \to z \pm \Delta z/2$. Fig.25 shows the predictions for

$$E\,\frac{d^4\sigma}{d^3p\,dz} = \frac{1}{\Delta z}\int_{z-\Delta z/2}^{z+\Delta z/2} \frac{d^4\sigma}{dy\,d\vec{p}_T dz'}\,dz' \,, \tag{47}$$

as functions of z and different bin widths Δz. The example is given for $\pi^- p \to \gamma\gamma X$ at $\sqrt{s} = 23$ GeV, $y = 0$ and $p_T = 2$ GeV/c, since data are expected to come. For $z \neq 1$ this cross section (Fig.25) counts genuine three-jet events in the final state, $\gamma+\gamma+$ hadron jet. The bin around $z = 1$ is infrared sensitive, therefore it strongly depends on the value of Δz. Due to the soft gluon in $q\bar{q} \to \gamma\gamma G$ and in the virtual correction to $q\bar{q} \to \gamma\gamma$ the HO term is negative, and for $\Delta z = 0.1$ even larger in magnitude than the LO term (Fig.25). This behaviour clearly indicates a breakdown of finite order perturbation theory. In order to get reliable predictions at this order α_s the resolution Δz has to be chosen such that $8\alpha_s \ln^2(\Delta z/2)/3\pi \ll 1$; e.g. the predictions shown in Fig.25 are reliable for $\Delta z = 0.2$ but not for

$\Delta z = 0.1$. Otherwise resummation techniques, developed for infrared-sensitive quantities [74], have to be applied in order to derive results to all orders in α_s: e.g., in the double leading logarithm approximation the cross section equ. (47) at $z = 1$ would even vanish exponentially in the limit $\Delta z \to 0$.

We now return to the problem of scales μ and Q in the truncated perturbative QCD calculations, in general to the problem of RS and factorization scheme ambiguities. In order to simplify the discussion [75] we consider single-photon production for the specific non-singlet combination of processes: $(p\bar{p} + \bar{p}p - pp - \bar{p}\bar{p}) \to \gamma X$. It has the theoretical advantage of being independent of the gluon distribution; it is just determined by the valence quarks in the proton, $q_v = q - \bar{q}$. At order α_s^2 to be considered the corresponding cross section can be schematically written as a convolution,

$$E\,\frac{d^3\sigma^{NS}}{d^3p} \equiv \sigma^{NS} = a(\mu)\{\text{BORN} + a(\mu)[b\,\ln(\mu/\Lambda_{\overline{MS}}) - 2P_{qq}\,\ln(Q^2/\hat{s})\otimes]$$
$$\times\ \text{BORN} + a(\mu)\,K_{q\bar{q}}(\hat{s}/\Lambda^2_{\overline{MS}})\ \otimes\}q_v(Q^2)\otimes q_v(Q^2)\ . \tag{48}$$

Definitions given already in the first lecture are used here. $a(\mu) = a_{\overline{MS}}(\mu)$ to second order, equ. (12), which allows to express $\ln(\mu/\Lambda_{\overline{MS}})$ in terms of $a(\mu)$. $\sigma^{NS}(a(\mu),Q^2)$ is therefore not a polynomial in $a(\mu)$. Concerning the $a(\mu)$ dependence σ^{NS} has the same structure as $R(Q)$ discussed in the first lecture starting with equ. (8). BORN stands for the contribution from $q\bar{q} \to \gamma G$. The function $K_{q\bar{q}}$ does not depend on μ and Q and it is a RS invariant quantity [18], calculated from the diagrams for $q\bar{q} \to \gamma GG$ and the virtual ones for $q\bar{q} \to \gamma G$. The valence quark distributions $q_v(Q^2)$ are to be understood as physical quantities measured by $F_2^{ep}(x,Q^2)$

according to equ.(19). Because of the explicit μ and Q dependent terms in equ.(48) σ^{NS} becomes more stable than the LO result under changes of μ and Q (compare Figs. 23 and 24). Self-consistency of perturbation theory actually requires strong cancellations, such that σ^{NS} of equ.(48) obeys

$$\sigma^{NS}(a(\mu),Q^2) \to \sigma^{NS}(a(\mu'),Q'^2) = \sigma^{NS}(a(\mu),Q^2) + O(a^3) \ , \qquad (49)$$

under the changes $\mu \to \mu'$, $Q \to Q'$. This is easy to verify using equs.(12) and (20):

$$a(\mu) \to \ a(\mu') = a(\mu) + ba^2(\mu) \ln \frac{\mu}{\mu'} \ +O(a^3), \ \Lambda \text{ fixed},$$

$$q_v(Q^2) \to q_v(Q'^2) = q_v(Q^2) + a(\mu) \ln \frac{Q'^2}{Q^2} P_{qq} \otimes q_v(Q^2) + O(a^2) \ . \qquad (50)$$

Since the physical σ^{NS} does not depend on the unphysical variables μ and Q in contrast to its perturbative approximation, equ.(48), an optimized prediction may be obtained demanding the criterion of minimal sensitivity proposed by P.M. Stevenson [18], namely

$$\mu \frac{\partial\sigma^{NS}}{\partial\mu} = 0 \qquad \text{and } Q^2 \frac{\partial\sigma^{NS}}{\partial Q^2} = 0 \qquad (51)$$

at fixed kinematics and $\Lambda_{\overline{MS}}$. Although there is no proof that this principle resolves correctly the ambiguities, it incorporates fully the requirements of renormalization-group invariance [18]. The solution of equ.(51) selects the optimal values for μ and Q at which σ^{NS}_{opt} (equ.(48)) is then evaluated. Because of the complicated analytic structure of σ^{NS} the solution has to be found numerically.

Fig.26 shows the results of such a numerical analysis [75] for the process $(p\bar{p} - pp) \to \gamma X$ at $\sqrt{s} = 63$ GeV, $y = 0$, and the two different values of the $p_T = 6$ and 10 GeV/c, respectively. The cross section $E\, d^3\sigma/d^3p$ (in units of $10^{-36}(10^{-37})\,cm^2/GeV^2$ at $p_T = 6$ (10) GeV/c) for different values of $a(\mu) = \alpha_s/\pi$ and $c_s \equiv Q^2/(p_T^2(1-x_T))$ is presented as a contour plot in Fig.26. It is remarkable that the conditions equ.(51) are satisfied, since there is a stability (saddle) point at physically acceptable values for the coupling and the scale Q. In particular at the larger value of p_T, $p_T = 10$ GeV/c, $E d^3\sigma/d^3p$ does not show any variation with respect to Q^2, when it is calculated at $a = a_{opt} = 0.075$: $E d^3\sigma/d^3p \simeq 8.6\cdot 10^{-37} cm^2/GeV^2$ for $Q^2 \simeq 1$ to $1000\ GeV^2$, indeed an encouraging feature for the optimization program. This contour plot helps to understand the results shown in Fig.24: obviously there the variation is due to the choice $\mu = p_T$, therefore $a(\mu = p_T) < a_{opt}$, and the corresponding cross section is decreasing with increasing Q (compare with Fig.26). In fact the optimum scale $\mu \simeq c\, p_T$ requires a coefficient c less than one. Furthermore it turns out that the HO corrections become almost negligible at the stability point; the corresponding K-factor (equ.41) is $K_{opt} \simeq 0.95$.

One would like to compare the theoretical results (Fig.26) at the saddle point with corresponding experimental values for $(p\bar{p} - pp) \to \gamma X$ at this energy. Since this combination of processes does not contain the gluon distribution, even $\Lambda_{\overline{MS}}$ could be determined, keeping in mind that $(Ed^3\sigma/d^3p)_{opt}$ changes with $\Lambda_{\overline{MS}}$. However, data for $pp \to \gamma X$ and $\bar{p}p \to \gamma X$ [76] are not at all precise enough for this kind of refined analysis of the difference $(\bar{p}p - pp) \to \gamma X$.

Before finishing the discussion on real photons I shortly mention recent results on inclusive photoproduction of hadrons at large p_T, $\gamma N \to \pi^0 X$, and on deep inelastic Compton scattering, $\gamma N \to \gamma X$. At the level of the contributing subprocesses these reactions are related to the ones discussed above, namely to a + b → γ+ X and a + b → γ+γ + X, respectively. Therefore their importance for testing QCD is evident and a consistent description is required. Figs.27 and 28 show the preliminary data for $\gamma N \to \pi^{o} X$ [77]and the final ones for $\gamma N \to \gamma X$ [78] measured by the NA14 Collaboration. Included is the comparison with the corresponding QCD predictions with HO corrections, which are completely calculated at order α_s^2 for photoproduction [79] and at order α_s for the Compton process [80]. The agreement is remarkable. The data are indeed precise enough to state the significant role of these corrections for obtaining the agreement. This point becomes even clearer when Fig.21 for the Drell-Yan pairs is compared with Fig.28 for the deep inelastic Compton process.

CONCLUSION

Up to momentum scales Q ∿ 100 GeV/c, presently accessible, no contradiction to the general statement that QCD is the correct theory of strong interactions is met. Indeed impressive confirmation of the many expectations in different processes is found. In the deep inelastic region a fast moving nucleon behaves as a beam of quarks and gluons: this permits extrapolations to the multi-TeV range [81,82] with the possibility of exploring a new layer of matter [83].

It is also important to note that most of the QCD predictions are still semiquantitative. The progress of improving the predictive quantitative power of QCD is rather slow, mainly due to the fact that except for photons and leptons only hadrons are directly observable but neither quarks nor gluons. In addition the application of QCD perturbation theory, with its preference for almost asymptotic values of Q, is limited to the deep inelastic region, where only very rare events are observed and therefore high precision mesurements are extremely difficult to perform.

ACKNOWLEDGEMENT

I would like to thank the Organizing Committee for arranging this stimulating meeting with its pleasant atmosphere. In particular I am grateful to H. Mitter and W. Plessas for the kind invitation.

FIGURE CAPTIONS

Fig. 1: Inelastic e-α scattering, (a) $Q^2 \simeq 0.08\ GeV^2$ and (b) $Q^2 \simeq 0.1\ GeV^2$. Taken from [2].

Fig. 2: Structure functions measured in lepton-nucleon scattering experiments. Taken from [3].

Fig. 3: Comparison of Q^2 dependence of $F_2(x,Q^2)$ as measured in neutrino and muon scattering. Taken from [4].

Fig. 4: Inclusive π^o cross-sections measured at ISR energies. Taken from [6].

Fig. 5: Effective power n of the scaling behaviour, equ.(4), for π^o production at ISR energies. Taken from [7].

Fig. 6: Scaling behaviour of the single-jet cross section [8].

Fig. 7: The QCD coupling constant as a function of the renormalization mass $\mu(=Q)$ taken from G. Altarelli et al. [81].

Fig. 8: The CDHS xF_3 and F_2 data compared to QCD fits with $\Lambda_{\overline{MS}} = 124$ MeV (dashed curve) and $\Lambda_{\overline{MS}} = 404$ MeV (solid curve), from [24].

Fig. 9: R as a function of x from neutrino scattering [25]. Taken from [4].

Fig. 10: Soft (solid curve) and hard (dashed curve) gluon distribution [26].

Fig. 11: Lego plot of a two-jet event from UA2 [29].

Fig. 12: Phase-space density of large p_T final states averaged over a large number of events at ISR energies. Taken from [6].

Fig. 13: Feynman diagrams in lowest order QCD for two-two scattering of quarks and gluons. Taken from [40].

Fig. 14: Angular distribution of the two-jet cross section, equ.(23). The solid curve is the QCD prediction normalized to the data from UA2 [36].

Fig. 15: The effective structure function F(x), equ.(29), derived from two-jet events at $Sp\bar{p}S$ and computed (dashed curves) from deep inelastic neutrino scattering. Taken from [36].

Fig. 16: Single jet yields at ISR and $Sp\bar{p}S$ compared with QCD predictions in LO from [38].

Fig. 17: Single-jet transverse-momentum distribution of the subprocesses shown in Fig.13 calculated with different sets of structure functions and choices of Λ. Taken from [40].

Fig. 18: The p_{out}-distribution from UA1 compared to a three-jet QCD calculation. Taken from [47].

Fig. 19: Comparison of inclusive π^o production measured at $\sqrt{s}$ = 63 GeV and $\sqrt{s}$ = 540 GeV with corresponding QCD predictions in LO. Taken from [50].

Fig. 20: Comparison of inclusive π^o and charged pion production at $\sqrt{s}$ = 44.8 GeV as function of transverse momentum. The data are from [52].

Fig. 21: Data on production of continuum lepton pairs in pp collisions at two different energies for masses above and below the Υ resonances compared with QCD predictions from [59]. $K \equiv 1$ does not include the HO contributions.

Fig. 22: Inclusive production of prompt photons and π^o's measured at $\sqrt{s}$ = 63 GeV as a function of p_T [10].

Fig. 23: Inclusive production of prompt photons, $pp \to \gamma X$, compared with QCD calculations in LO for two different scales in the struc.funct.[26], $Q^2 = p_T^2/2$ (solid curve) and $Q^2 = 4p_T^2$ (dashed curve). The data are from [10,70].

Fig. 24: Same as Fig.23 except for the QCD calculations, which include all contributions at order α_s^2[71].

Fig. 25: Cross section for $\pi^- p \to \gamma\gamma X$, E_{LAB} = 300 GeV, p_T = 2 GeV/c, y = 0 as a function of z for different bin widths, $\Delta z = 0.2$ and $\Delta z = 0.1$, respectively, as calculated from QCD beyond leading order [73]. The dashed curve is the contribution of $O(\alpha_s)$ in the bin with z = 1.

Fig. 26: Contour plot of $Ed^3\sigma/d^3p$ in units of 10^{-36} (10^{-37}) cm^2/GeV^2 at p_T = 6 (10) GeV/c for prompt photons $(p\bar{p} - pp) \to \gamma X$ at $\sqrt{s}$ = 63 GeV, y = 0, based on QCD calculations [75] at second order with $\Lambda_{\overline{MS}}$ = 200 MeV.

Fig. 27: Inclusive photoproduction of π^o. Data [77] are compared with QCD predictions including HO corrections [79] (full curve).

Fig. 28: Deep inelastic Compton scattering. Data [77,78] are compared with QCD predictions including HO corrections [80](full curve).

REFERENCES

A complete list of references is not attempted. I apologize for omissions. References to recent text books and review papers are included, in which further references may be found.

1. J. Strauss, Results from experiments at the CERN anti-proton-proton collider, contribution to these proceedings.
2. F.E. Close, An Introduction to Quarks and Partons,

Academic, N.Y. 1979.

3. W.B. Atwood, in Lectures on Lepton-Nucleon Scattering and Quantum Chromodynamics, Birkhäuser, 1982.
4. J. Steinberger, preprint CERN-EP/84-153 (1984); CDHS Collab., H. Abramowicz et al., Z. Phys. C12 (1982) 289 and C17 (1983) 283.
5. R.P. Feynman, Photon-Hadron Interactions, Benjamin, 1972.
6. P. Darriulat, preprint CERN-EP/84-152 (1984); and P. Darriulat, Ann. Rev. Nucl. Part. Sci. 30 (1980) 159.
7. R. Stroynowski, in Lectures on Lepton-Nucleon Scattering and Quantum Chromodynamics, Birkhäuser, 1982.
8. E609 Collab., L.R. Cormell et al., Phys. Lett. 150B (1985) 322.
9. T. Ferbel and W.R. Molzon, Rev. Mod. Phys. 56 (1984) 181.
10. C.W. Fabjan, preprint CERN-EP/83-174 (1983).
11. C. Quigg, Gauge Theories of the Strong, Weak and Electromagnetic Interactions, Benjamin, 1983.
12. H.D. Politzer, Phys. Rep. 14C (1974) 129.
13. P. Ramond, Field Theory: A Modern Primer, Benjamin, 1981; C. Itzykson and J.B. Zuber, Quantum Field Theory, McGraw-Hill, 1980.
14. K.G. Wilson, in Recent Developments in Gauge Theories, ed. by G. 't Hooft et al., Plenum, 1980, p.363.
15. G. 't Hooft, Nucl. Phys. B61 (1973) 455; W.A. Bardeen, A.J. Buras, D.W. Duke and T. Muta, Phys. Rev. D18 (1978) 3998.
16. M. Dine and J. Sapirstein, Phys. Rev. Lett. 43 (1979) 668.
17. W. Celmaster and R.J. Gonsalves, Phys. Rev. D20 (1979) 1420.
18. P.M. Stevenson, Phys. Rev. D23 (1981) 2916, and Nucl. Phys. B203 (1982) 472.
19. G. Altarelli, Phys. Rep. 81 (1982) 1; A.J. Buras, Rev. Mod. Phys. 52 (1980) 199; A. Peterman, Phys. Rep. 53C (1979) 157.
20. G. Altarelli, R.K. Ellis and G. Martinelli, Nucl. Phys. B157 (1979) 461.
21. W.J. Marciano, Phys. Rev. D12 (1975) 3861.

22. G. Altarelli and G. Parisi, Nucl. Phys. B126 (1977) 298.
23. F. Dydak, Proc. Int. Symp. on Lepton and Photon Interactions at High Energies, 1983, p.634.
24. A. Devoto, D.W. Duke, J.F. Owens and R.G. Roberts, Phys. Rev. D27 (1983) 508.
25. CDHS Collab., H. Abramowicz et al., Phys. Lett. 107B (1981) 141.
26. D.W. Duke and J.F. Owens, Phys. Rev. D30 (1984) 49.
27. EMC Collab., J.J. Aubert et al., preprint CERN-EP/85-34 (1985).
28. J.D. Bjorken, Phys. Rev. D8 (1973) 4098.
29. UA2 Collab., M. Banner et al., Phys. Lett. 118B (1982) 203; and P. Bagnaia et al., Z. Phys. C20 (1983) 117; UA1 Collab., G. Arnison et al., Phys. Lett. 123B (1983) 115.
30. M. Jacob, in Proc. XXII Int. Conf. High Energy Physics, Leipzig, 1984, Vol.II, p.150; and preprint CERN-TH 4058/84 (1984).
31. L. Di Lella, preprint CERN-EP/85-13 (1985); and Lectures at the JINR-CERN School of Physics, CERN-EP/84-44 (1984).
32. F. Halzen and F. Herzog, in Proc. XV Symp. on Multiparticle Dynamics, Lund, 1984; preprint MAD/PH/195 (1984).
33. R. Møller, in Proc. XV Symp. on Multiparticle Dynamics, Lund, 1984; and preprint NBI-HE 84/13 (1984).
34. M. Greco, preprint LNF-84/61 (P) (1984).
35. B. Combridge, J. Kripfganz and J. Ranft, Phys. Lett. 70B (1977) 234; R. Cutler and D. Sivers, Phys. Rev. D17 (1977) 196.
36. UA1 Collab., G. Arnison et al., Phys. Lett. 136B (1984). 294; UA2 Collab., P. Bagnaia et al., Phys. Lett. 144B (1984) 283.
37. CCOR Collab., A.L.S. Angelis et al., Nucl. Phys. B209 (1982) 284.
38. M. Jacob, in SLAC Summer Institute (1983) and preprint CERN-TH 3693 (1983).
39. R. Horgan and M. Jacob, in Proc. of the CERN School,

Malente 1980, CERN-81-04 (1981); and Nucl. Phys. B179 (1981) 441.
40. B. Humpert, preprint CERN-TH 3934 (1984).
41. for a review, S.L. Wu, preprint DESY 84-028 (1984).
42. UA1 Collab., G. Arnison et al., Phys. Lett. 132B (1983) 132.
43. AFS Collab., T. Åkesson et al., Z. Phys. C25 (1984) 13.
44. G. Sterman and S. Weinberg, Phys. Rev. Lett. 39 (1977) 1436.
45. M.B. Einhorn and B.G. Weeks, Nucl. Phys. B146 (1978) 445.
46. F.A. Berends, R. Kleiss, P. de Causmaecker, R. Gastmans and T.T. Wu, Phys. Lett. 103B (1981) 124.
47. UA1 Collab., G. Arnison et al., preprint CERN-EP-118 (1983).
48. Z. Kunszt and E. Pietarinen, Nucl. Phys. B164 (1980) 45, and preprint CERN-TH 3584 (1983).
49. R.D. Field, in Quantum Chromodynamics,AIP Conf. Proc. No.55 (1978), p.97.
50. UA2 Collab., M. Banner et al., preprint DPhPE 84-07 (1984).
51. R. Baier, J. Engels and B. Petersson, Z. Phys. C2 (1979) 265.
52. F.W. Büsser et al., Nucl. Phys. B106 (1976) 1; C. Kourkoumelis et al., Z. Phys. C5 (1980) 95.
53. S.D. Drell and T.M. Yan, Phys. Rev. Lett. 25 (1970) 316.
54. F. Halzen, in Proc. XXI Int. Conf. High Energy Physics, Paris, 1982, p.C3-381 (1982); E. Reya, Phys. Rep. 69 (1981) 195.
55. R. Baier and R. Rückl, Z. Phys. C19 (1983) 251.
56. H.D. Politzer, Nucl. Phys. B129 (1977) 301; R.K. Ellis, H. Georgi, M. Machacek, H.D. Politzer and G.C. Ross, Nucl. Phys. B152 (1979) 285; D. Amati, R. Petronzio, G. Veneziano, Nucl. Phys. B146 (1978) 29.
57. W. Celmaster and D. Sivers, Ann of Phys. 143 (1982) 1; H.D. Politzer, Nucl. Phys. B194 (1982) 493.
58. L. Beaulieu and C. Kounnas, Nucl. Phys. B141 (1978) 423;

J. Kodaira and T. Uematsu, Nucl. Phys. B141 (1978) 497.

59. G. Altarelli, R.K. Ellis and G. Martinelli, preprint CERN-TH 4049/84 (1984).
60. C. Quigg, Rev. Mod. Phys. 49 (1977) 297.
61. G. Altarelli, R.K. Ellis, M. Greco and G. Martinelli, Nucl. Phys. B246 (1984) 12.
62. UA1 Collab., C. Rubbia, in Proc. of the 4th p$\bar{p}$ Workshop, Bern, 1984.
63. UA2 Collab., J. Schacher, in Proc. of the 4th p$\bar{p}$ Workshop, Bern, 1984.
64. UA1 Collab., G. Arnison et al., Phys. Lett. 122B (1983) 103, Phys. Lett. 126B (1983) 298, Phys. Lett. 129B (1983) 273.
65. UA2 Collab., M. Banner et al., Phys. Lett. 122B (1983) 476; P. Bagnaia et al., Phys. Lett. 129B (1983) 130.
66. S.M. Berman, J.D. Bjorken and J.B. Kogut, Phys. Rev. D4 (1971) 3388.
67. C. Kourkoumelis et al., Z. Phys. C16 (1982) 101.
68. P. Aurenche and J. Lindfors, Nucl. Phys. B168 (1980) 296.
69. E.L. Berger, E. Braaten and R.D. Field, Nucl. Phys. B239 (1984) 52.
70. G. Anassontzis et al., Z. Phys. C13 (1982) 277; CCOR Collab., A.L.S. Angelis et al., Phys. Lett. 98B (1981) 115.
71. P. Aurenche, R. Baier, A. Douiri, M. Fontannaz and D. Schiff, Phys. Lett. 140B (1984) 87.
72. NA24 Collab., W. Heck, talk given at the EP Seminar at CERN, Nov. 1984.
73. P. Aurenche, R. Baier, A. Douiri, M. Fontannaz and D. Schiff, preprint Orsay, to appear.
74. Yu. L. Dokshitzer, D.I. Dyakonov and S.I. Troyan, Phys. Rep. 58 (1980) 269; A. Bassetto, M. Ciafaloni and G. Marchesini, Phys. Rep. 100 (1983) 201.
75. P. Aurenche, R. Baier, M. Fontannaz and D. Schiff, in preparation.

76. AFS Collab., H. Breuker, talk given at the Annecy Workshop, Dec. 1984.
77. G. Wormser, Thèse de Doctorat d'Etat, Orsay (1984).
78. NA14 Collab., P. Astbury et al., preprint CERN/EP 85-07 (1985).
79. P. Aurenche, R. Baier, A. Douiri,M. Fontannaz and D. Schiff, Phys. Lett. 135B (1984) 164.
80. D.W. Duke and J.F. Owens, Phys. Rev. D26 (1982) 1600; Erratum: Phys. Rev. D28 (1983) 1227; P. Aurenche, R. Baier, A. Douiri,M. Fontannaz and D. Schiff, Z. Phys. C24 (1984) 309.
81. A. Ali et al., G. Altarelli et al., in Proc. of the ECFA-CERN Workshop, Lausanne and Geneva, 1984, ECFA 84/85 CERN 84-10 (1984).
82. E. Eichten, I. Hinchliffe, K. Lane and C. Quigg, Rev. Mod. Phys. 56 (1984) 579.
83. W. Buchmüller, contribution to these proceedings.

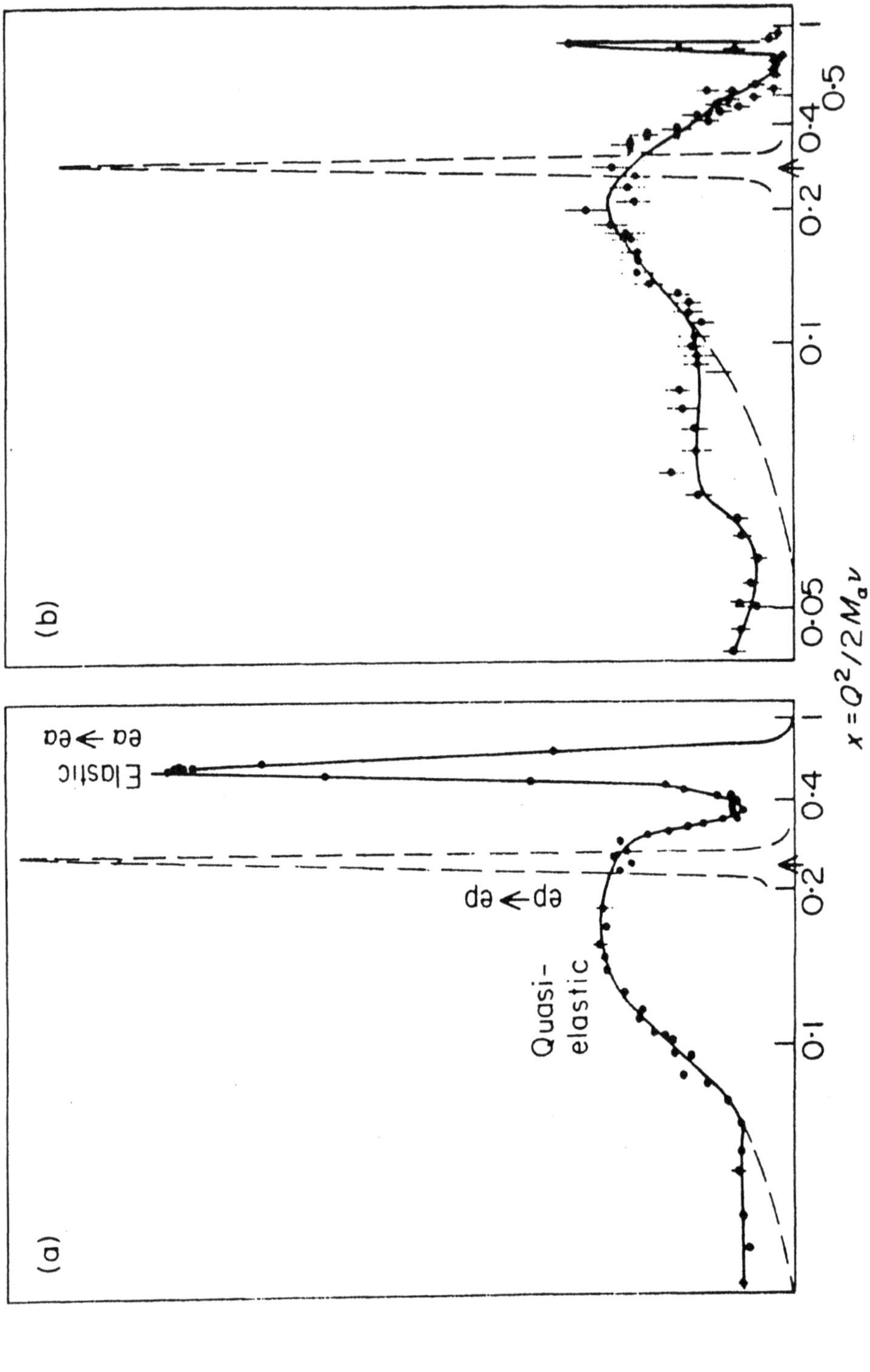

Fig. 1

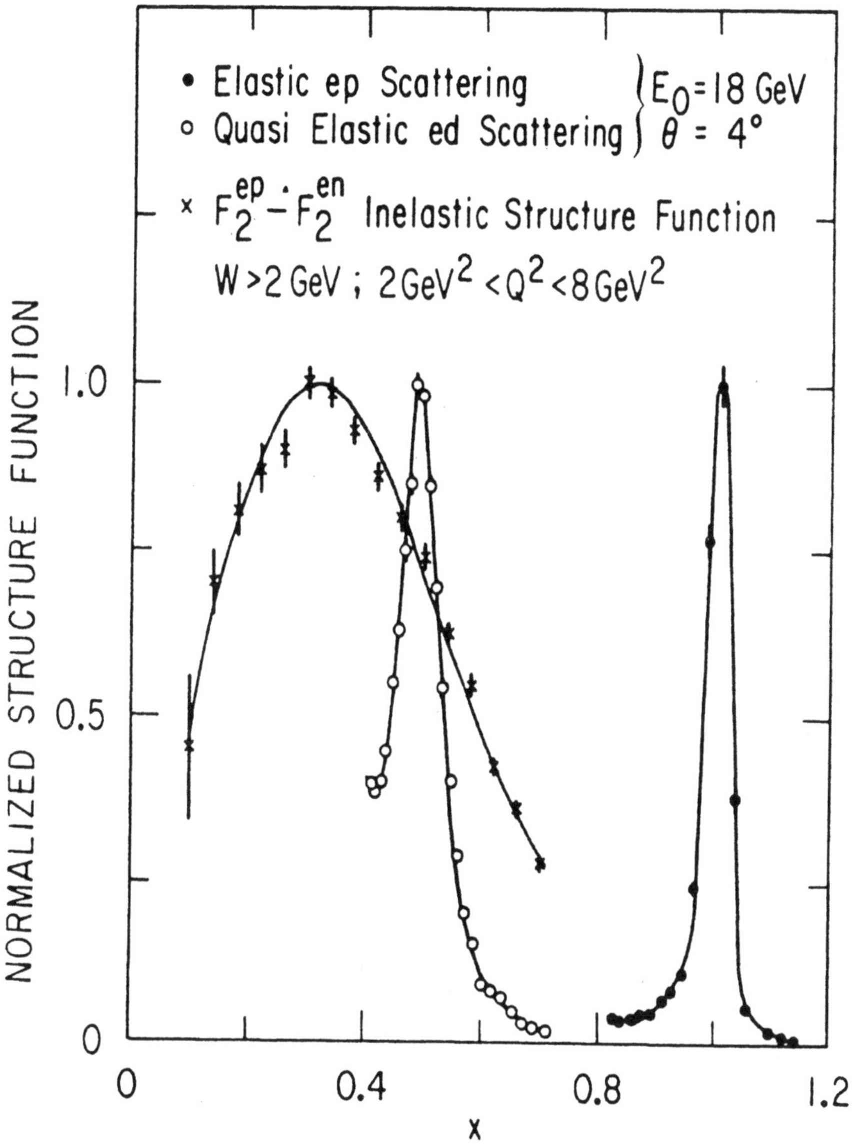

• Elastic ep Scattering
○ Quasi Elastic ed Scattering
E_0 = 18 GeV
θ = 4°
× $F_2^{ep} - F_2^{en}$ Inelastic Structure Function
W > 2 GeV ; 2 GeV^2 < Q^2 < 8 GeV^2
NORMALIZED STRUCTURE FUNCTION
1.0
0.5
0
0
0.4
0.8
1.2
x

Fig. 2

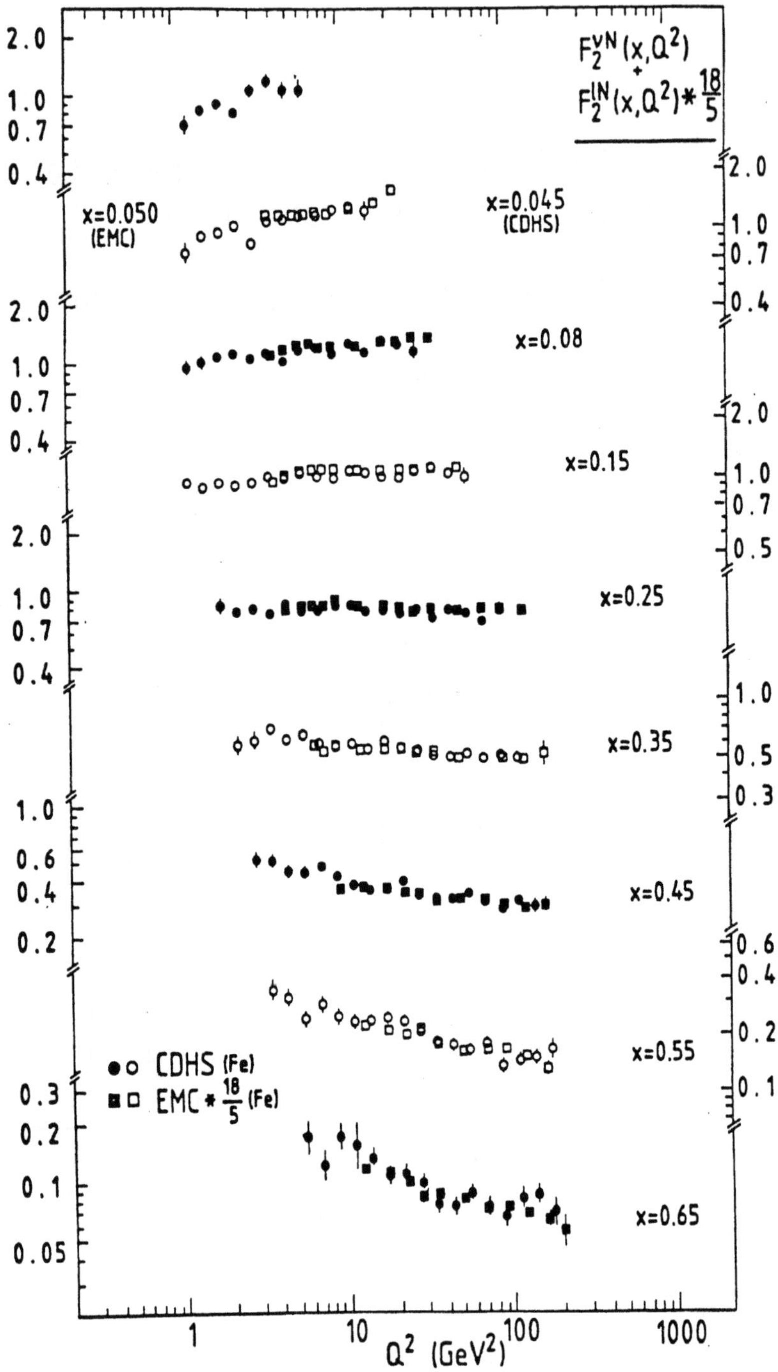

$F_2^{\nu N}(x,Q^2)$
$F_2^{lN}(x,Q^2)*\frac{18}{5}$
x=0.050 (EMC)
x=0.045 (CDHS)
x=0.08
x=0.15
x=0.25
x=0.35
x=0.45
x=0.55
x=0.65
CDHS (Fe)
EMC * 18/5 (Fe)
Q^2 (GeV^2)

Fig. 3

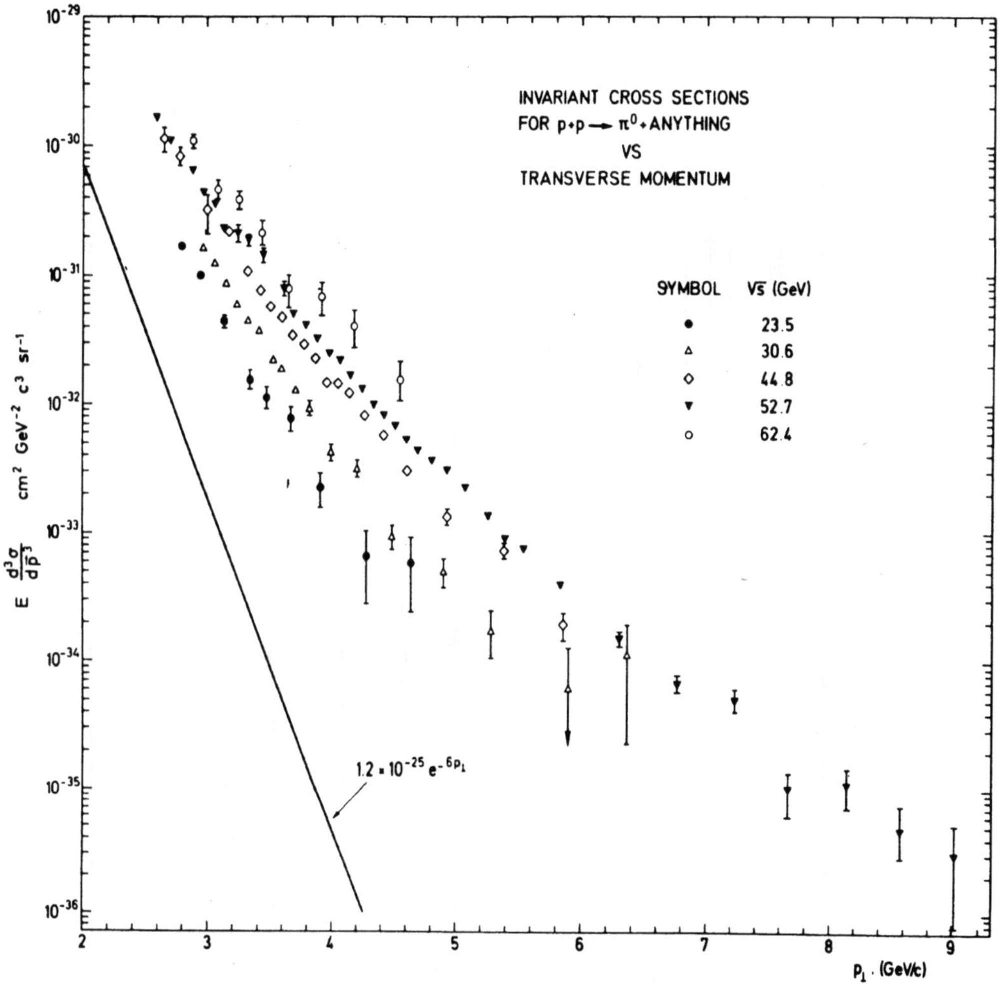

INVARIANT CROSS SECTIONS
FOR p+p → π⁰+ANYTHING
VS
TRANSVERSE MOMENTUM
SYMBOL √s (GeV)
23.5
30.6
44.8
52.7
62.4
$1.2 \times 10^{-25} e^{-6p_\perp}$
$E \frac{d^3\sigma}{dp^3}$ cm² GeV⁻² c³ sr⁻¹
$p_\perp$ (GeV/c)

Fig. 4

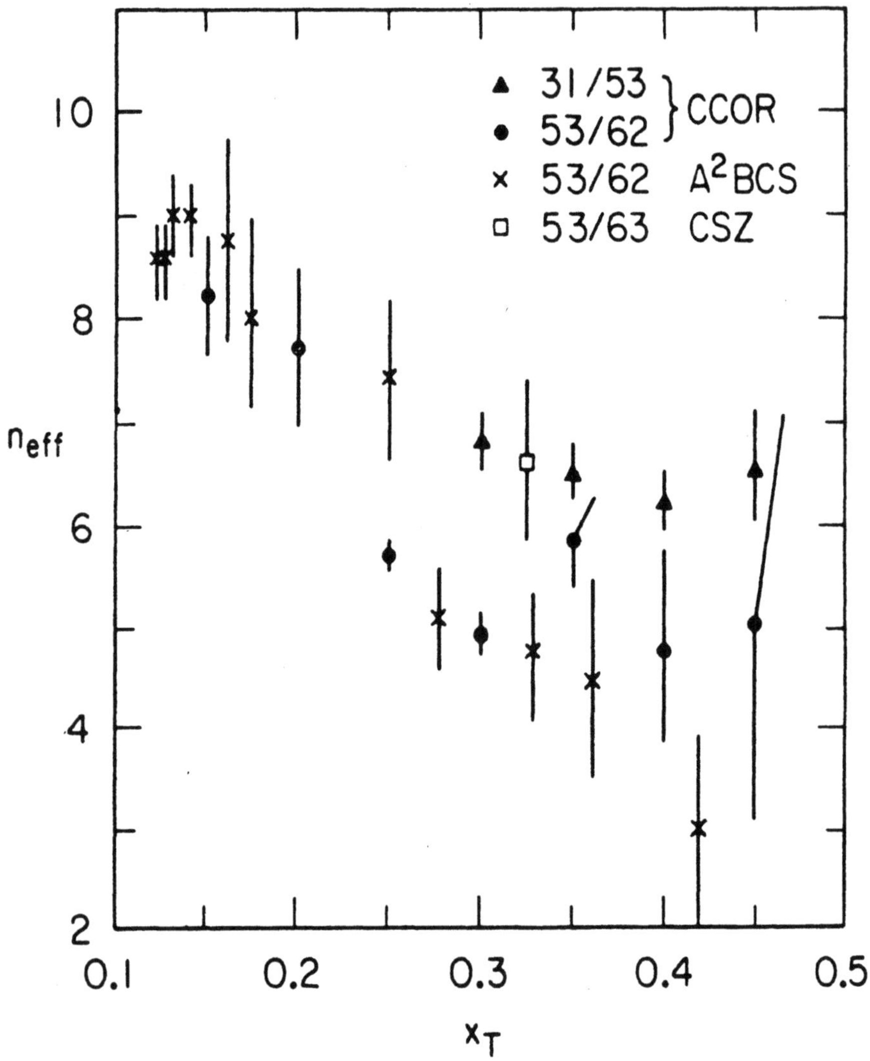
31/53
53/62
CCOR
53/62 A²BCS
53/63 CSZ
10
8
6
4
2
n_eff
0.1
0.2
0.3
0.4
0.5
x_T

Fig. 5

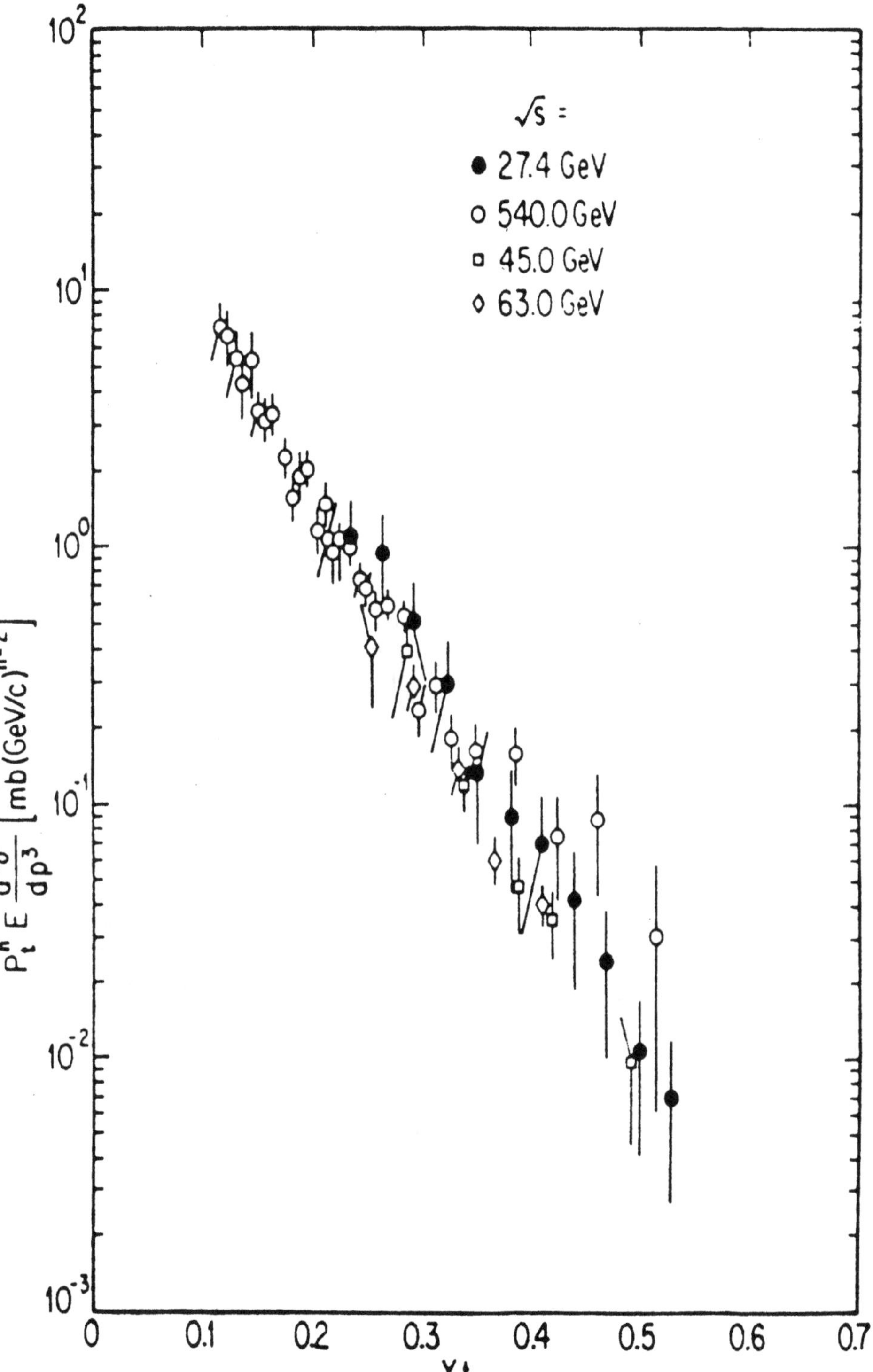

$\sqrt{s}$ =
● 27.4 GeV
○ 540.0 GeV
□ 45.0 GeV
◇ 63.0 GeV
$P_t^n E \frac{d^3\sigma}{dp^3}$ [mb(GeV/c)$^{n-2}$]
10^2
10^1
10^0
10^{-1}
10^{-2}
10^{-3}
0
0.1
0.2
0.3
0.4
0.5
0.6
0.7
Xt

Fig. 6

Fig. 7

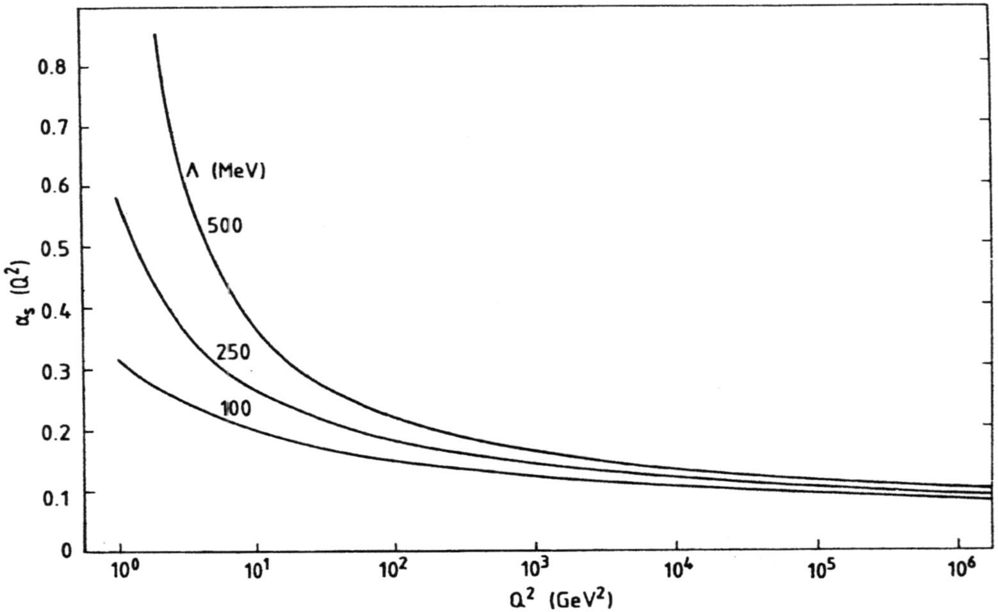

The running coupling constant of QCD in two-loop precision for various values of the scale Λ: $\alpha_s(Q^2) = \alpha_0(Q^2)[1 - b'\alpha_0(Q^2) \ln\ln Q^2/\Lambda^2]$ with $\alpha_0(Q^2) = 1/b \ln Q^2/\Lambda^2$; b and b′ depend on the number n_f of excited flavours. The charm, bottom, and top thresholds ($m_t \sim 40$ GeV) are approximately taken into account by defining Λ at $2m_b \leq Q \leq 2m_t$, where we set $\alpha_s(Q^2) \equiv \alpha_s^{(5)}(Q^2)$ [$\alpha^{(5)}$ meaning that b and b′ are evaluated with $n_f = 5$] and continuing to $2m_c \leq Q \leq 2m_b$ by setting $\alpha_s(Q^2) = [1/\alpha_s^{(5)}(4m_b^2) + 1/\alpha_s^{(4)}(Q^2) - 1/\alpha_s^{(4)}(4m_b^2)]^{-1}$, and so forth, so that the continuity of α_s and the correct large Q^2 behaviour is ensured at each step.

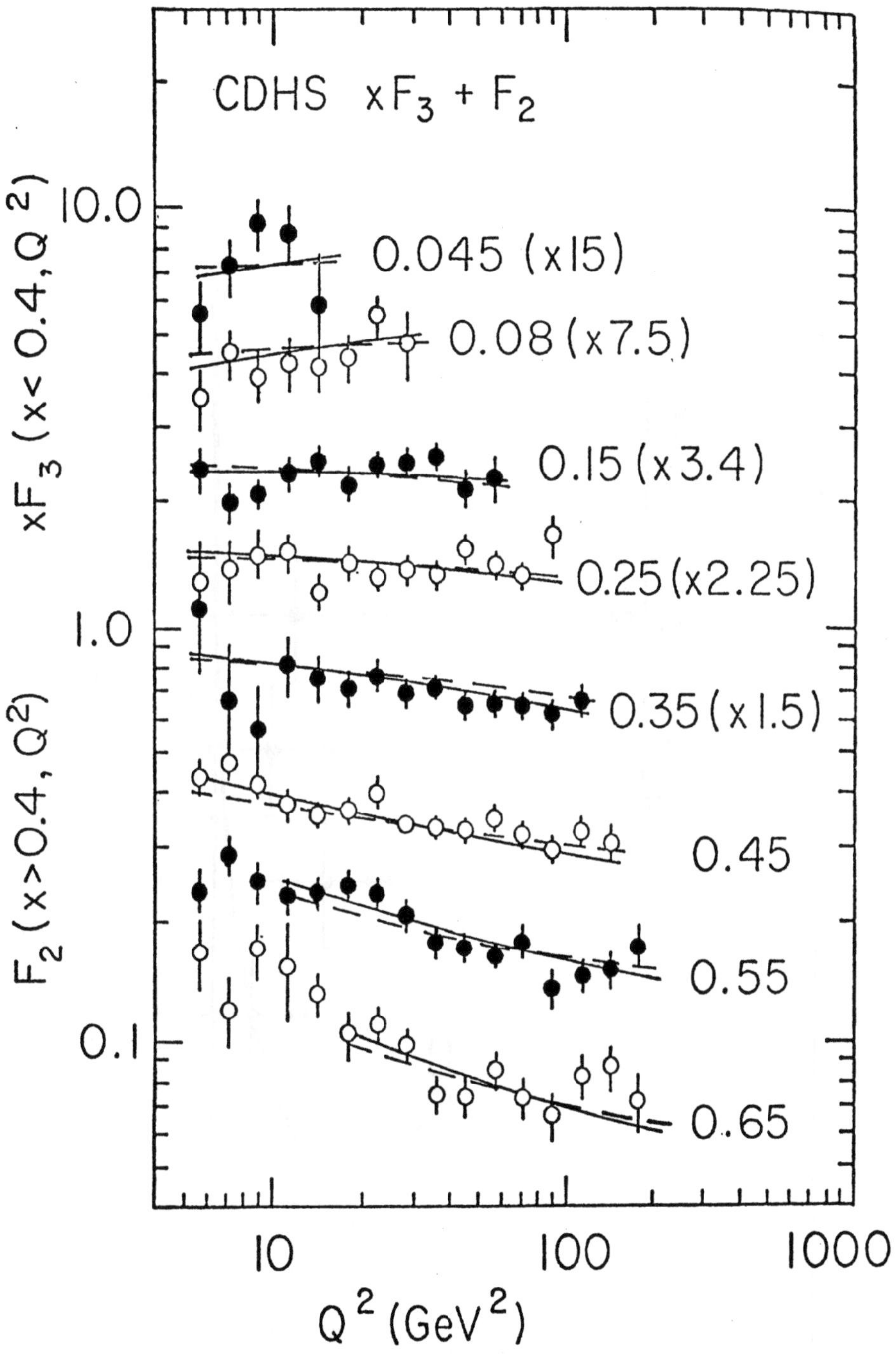

CDHS $xF_3 + F_2$
$xF_3(x<0.4,Q^2)$
$F_2(x>0.4,Q^2)$
10.0
1.0
0.1
0.045 (x15)
0.08 (x7.5)
0.15 (x3.4)
0.25 (x2.25)
0.35 (x1.5)
0.45
0.55
0.65
10
100
1000
Q^2 (GeV^2)

Fig. 8

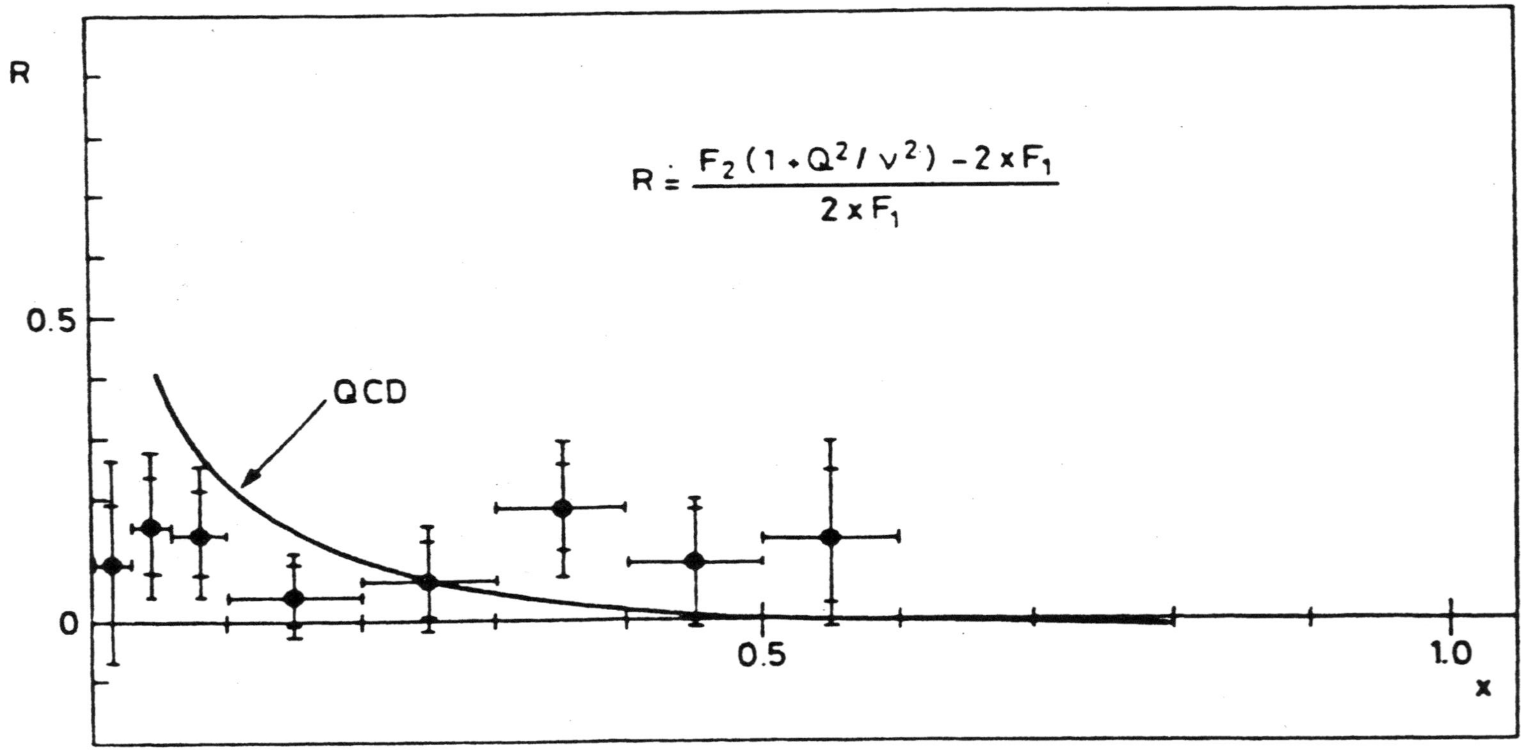

Fig. 9

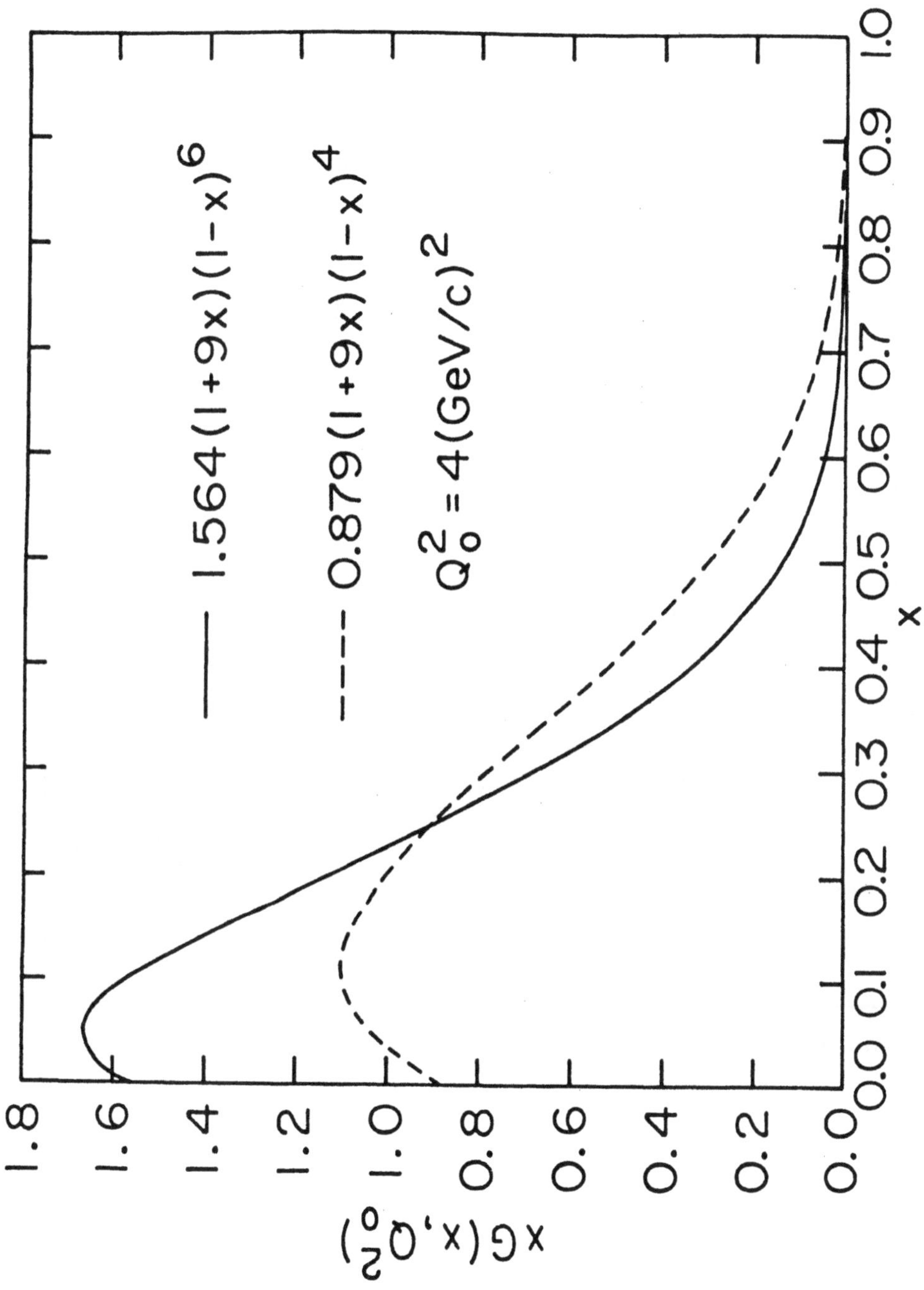
1.564(1+9x)(1-x)^6
0.879(1+9x)(1-x)^4
Q_o^2 = 4(GeV/c)^2
xG(x,Q_o^2)
x
0.0
0.1
0.2
0.3
0.4
0.5
0.6
0.7
0.8
0.9
1.0
0.0
0.2
0.4
0.6
0.8
1.0
1.2
1.4
1.6
1.8

Fig. 10

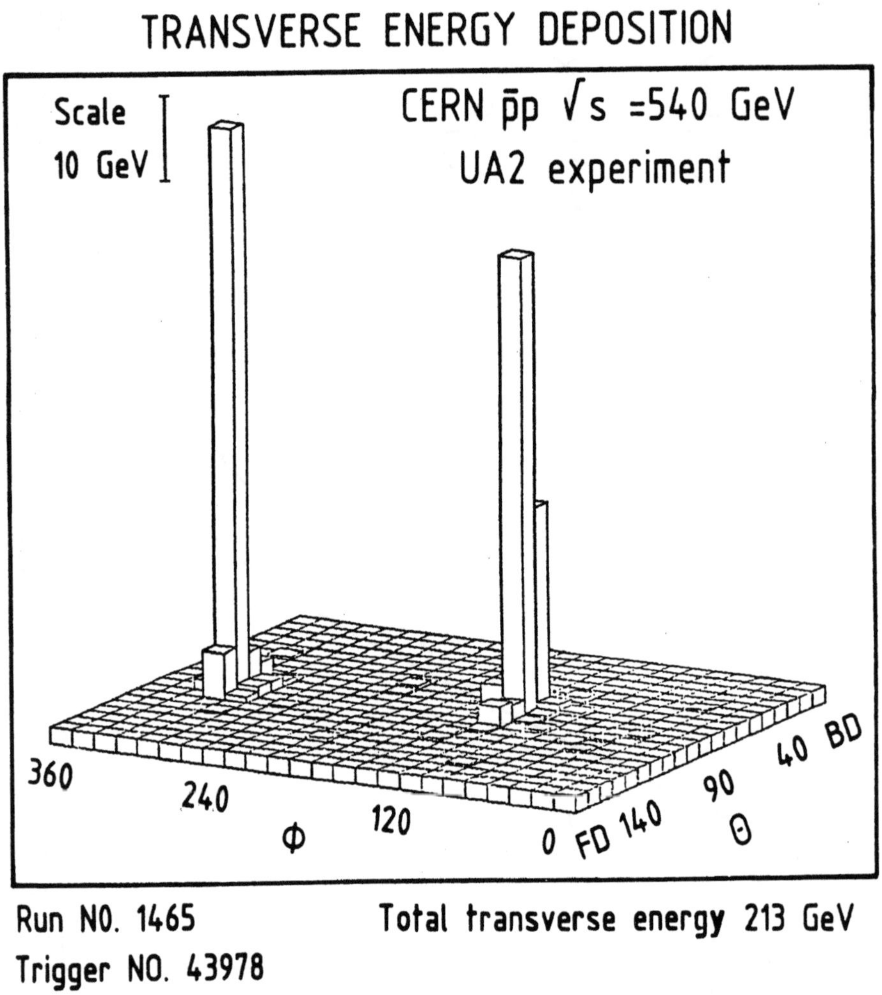
TRANSVERSE ENERGY DEPOSITION
Scale
10 GeV
CERN p̄p √s =540 GeV
UA2 experiment
360
240
Φ
120
0
FD 140
90
40 BD
Θ
Run NO. 1465
Trigger NO. 43978
Total transverse energy 213 GeV

Fig. 11

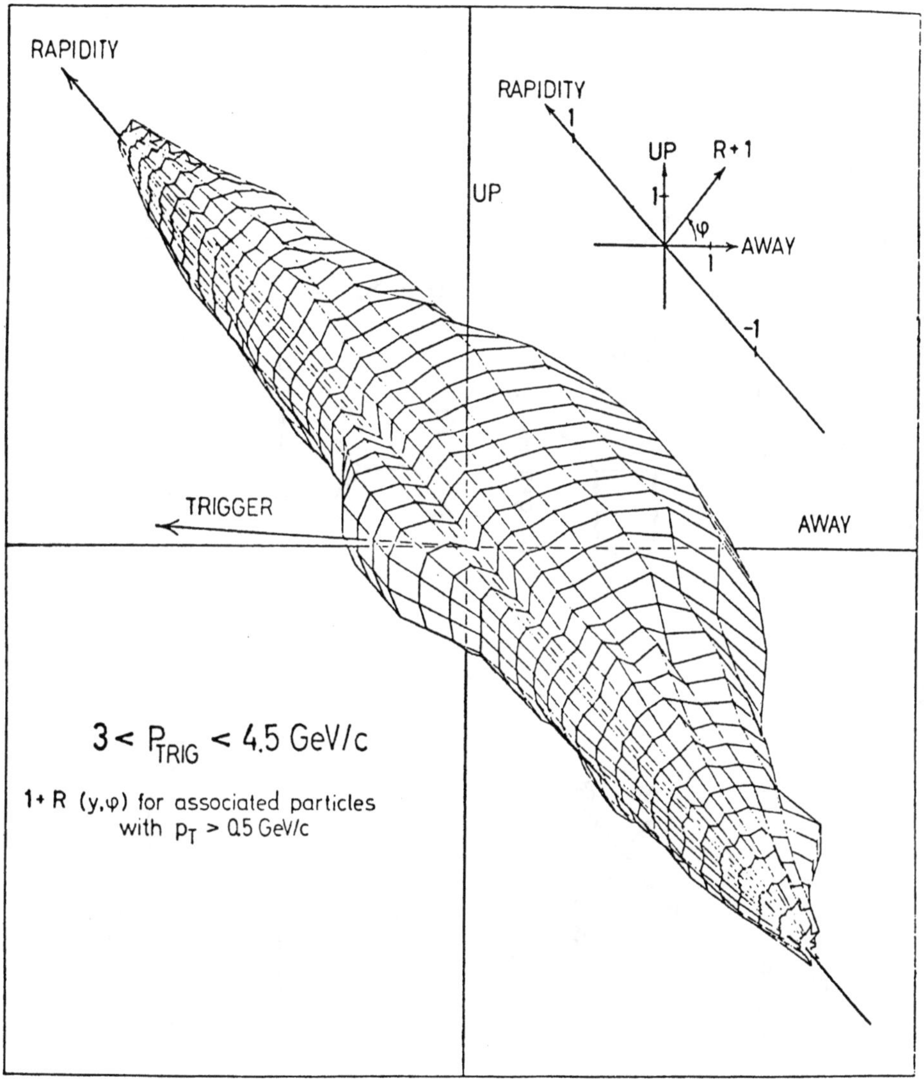
RAPIDITY
RAPIDITY
1
UP
R+1
1
φ
AWAY
1
-1
UP
TRIGGER
AWAY
3< P_TRIG < 4.5 GeV/c
1+ R (y,φ) for associated particles
with p_T > 0.5 GeV/c

Fig. 12

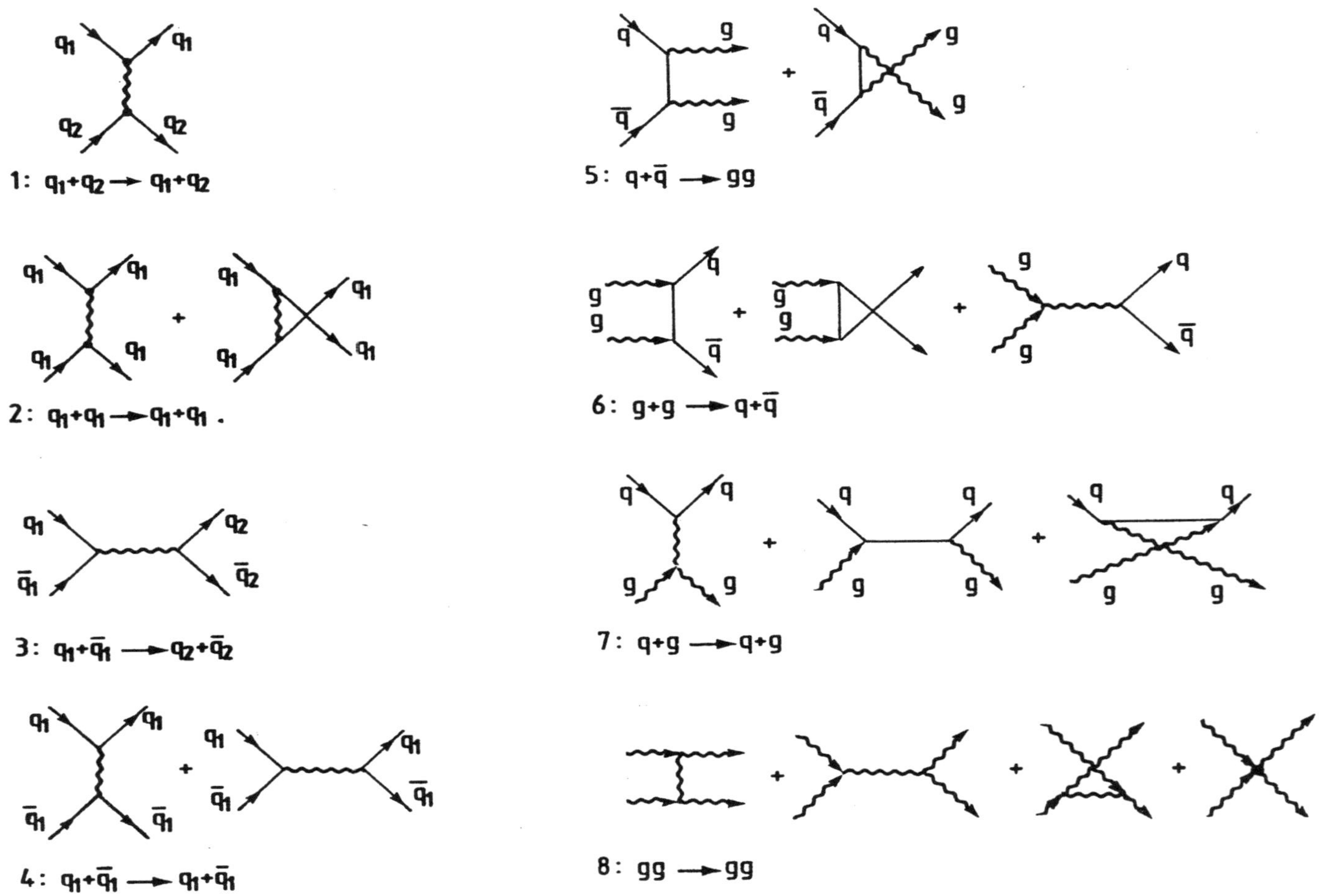
1: q1+q2 → q1+q2
2: q1+q1 → q1+q1 .
3: q1+q̄1 → q2+q̄2
4: q1+q̄1 → q1+q̄1
5: q+q̄ → gg
6: g+g → q+q̄
7: q+g → q+g
8: gg → gg

Fig. 13

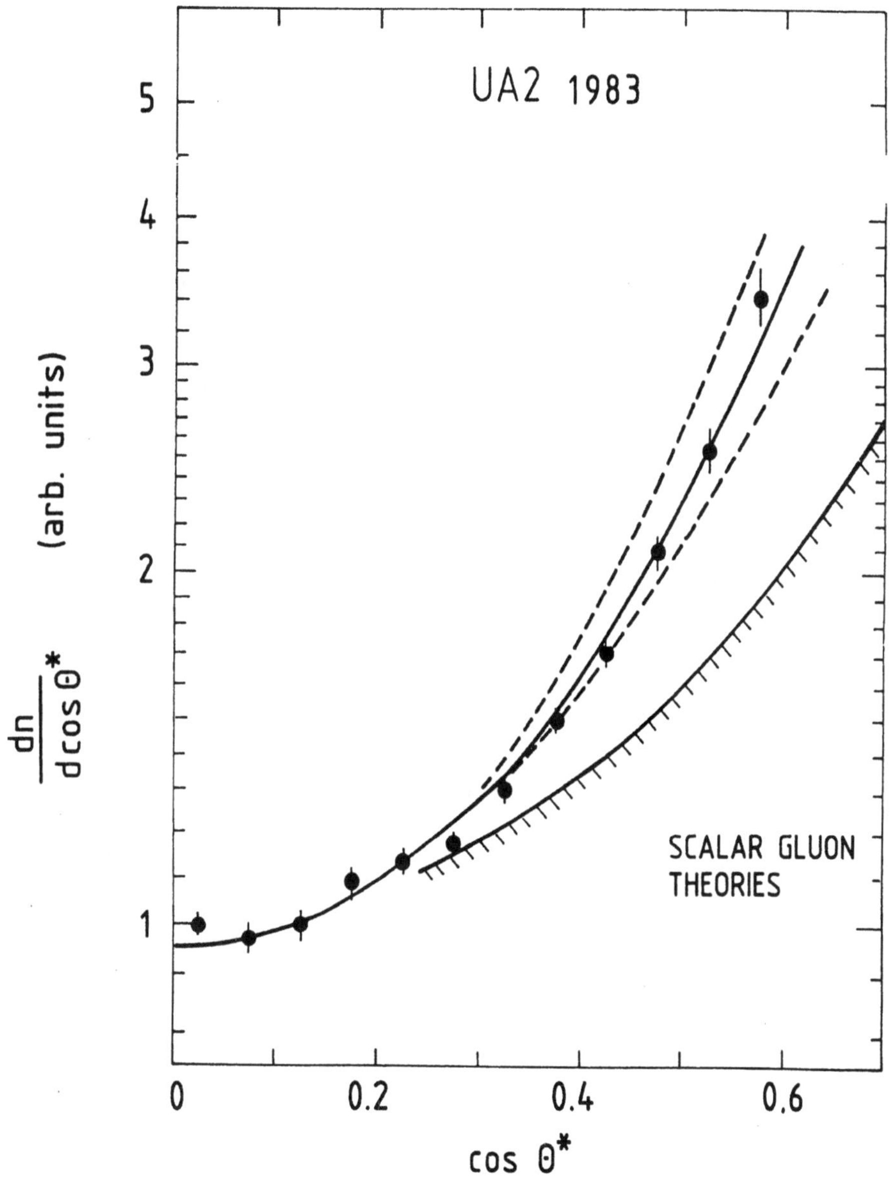
UA2 1983
5
4
3
2
1
(arb. units)
dn/dcos Θ*
SCALAR GLUON
THEORIES
0
0.2
0.4
0.6
cos Θ*

Fig. 14

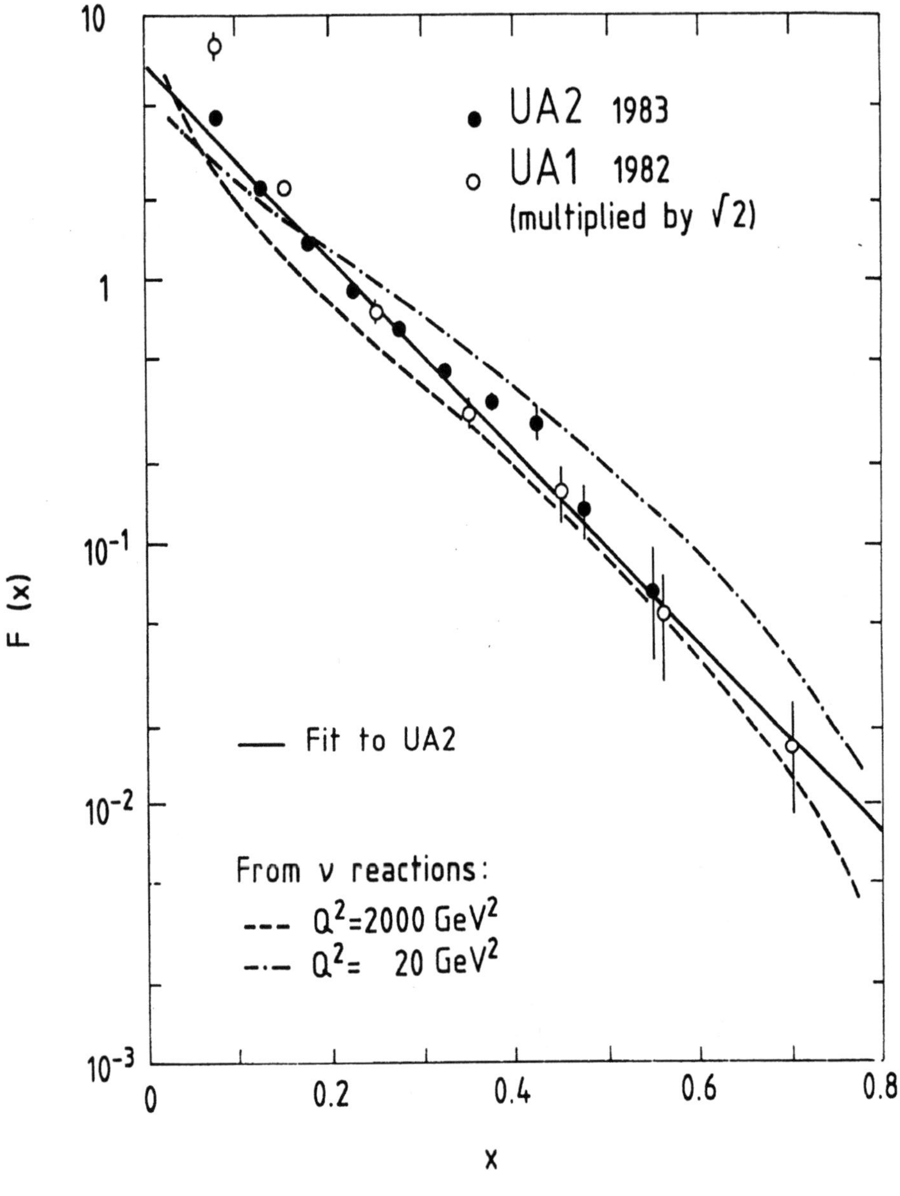

10
1
10^{-1}
10^{-2}
10^{-3}
F (x)
0
0.2
0.4
0.6
0.8
x
UA2 1983
UA1 1982
(multiplied by √2)
Fit to UA2
From ν reactions:
Q^2=2000 GeV^2
Q^2= 20 GeV^2

Fig. 15

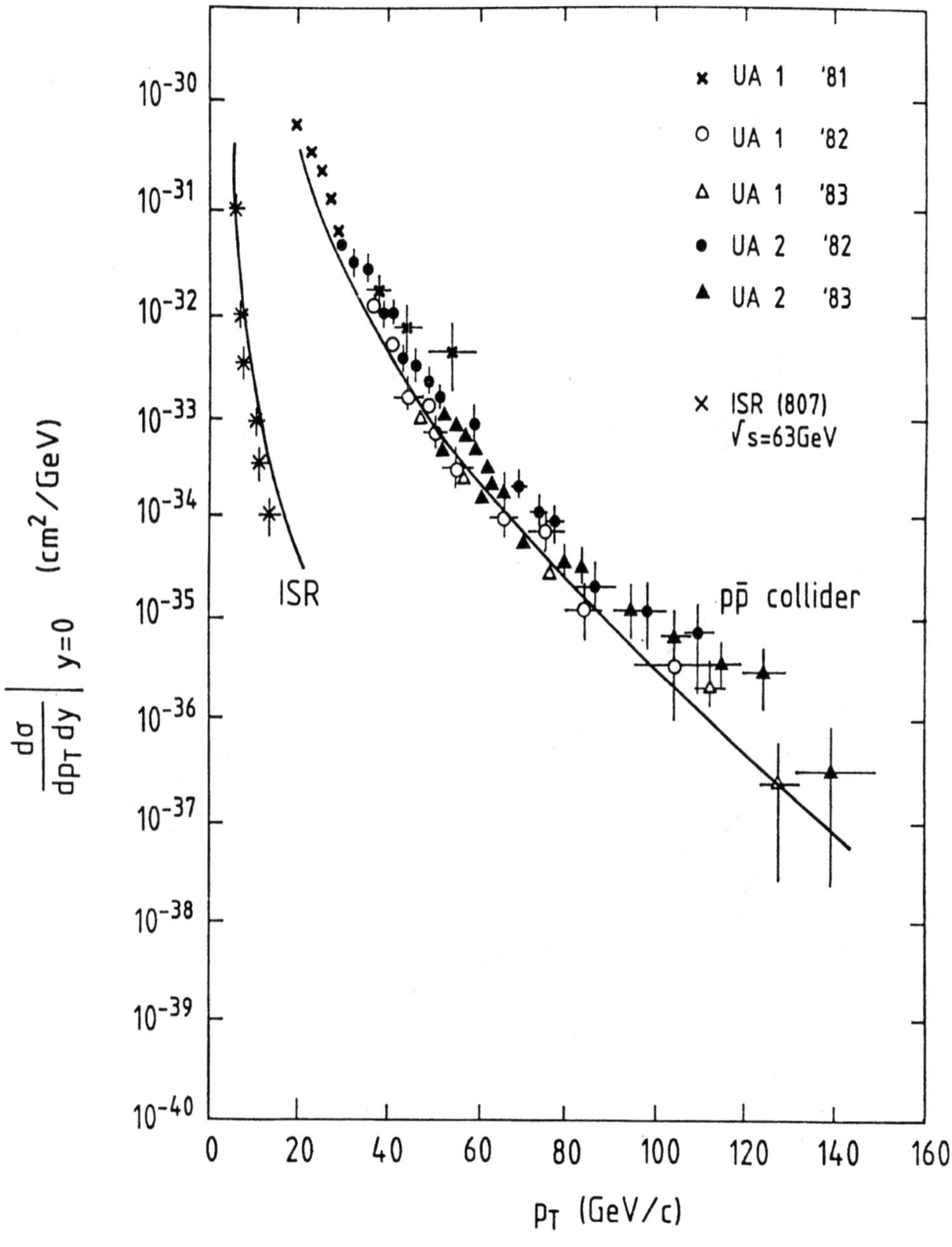
UA 1 '81
UA 1 '82
UA 1 '83
UA 2 '82
UA 2 '83
ISR (807)
√s=63GeV
ISR
pp̄ collider
10^{-30}
10^{-31}
10^{-32}
10^{-33}
10^{-34}
10^{-35}
10^{-36}
10^{-37}
10^{-38}
10^{-39}
10^{-40}
$\frac{d\sigma}{dp_T dy}\Big|_{y=0}$ (cm^2/GeV)
0
20
40
60
80
100
120
140
160
p_T (GeV/c)

Fig. 16

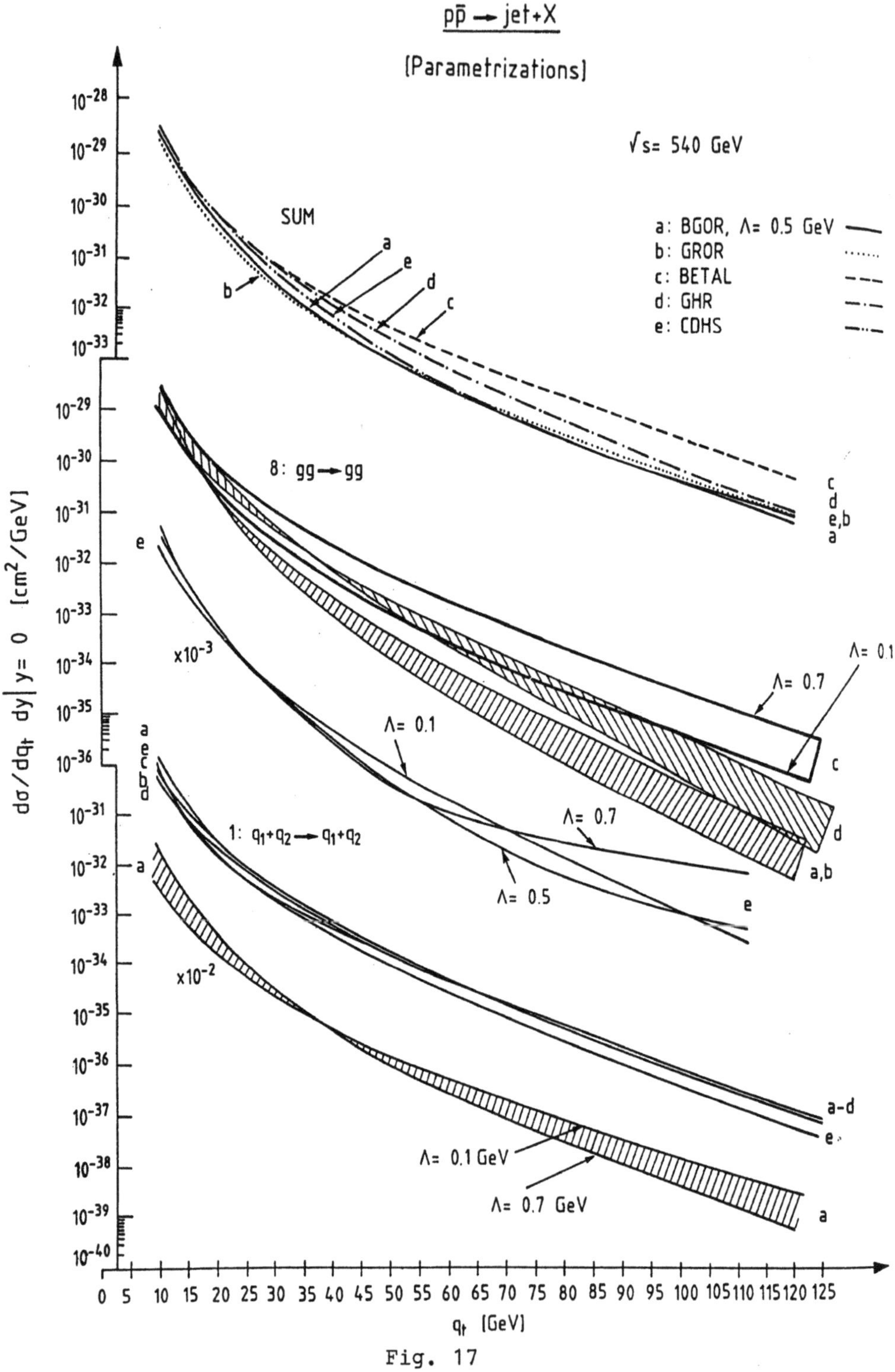
pp̄ → jet+X
(Parametrizations)
√s= 540 GeV
a: BGOR, Λ= 0.5 GeV
b: GROR
c: BETAL
d: GHR
e: CDHS
SUM
8: gg→gg
1: q1+q2→q1+q2
x10^-3
x10^-2
Λ= 0.1
Λ= 0.7
Λ= 0.5
Λ= 0.1 GeV
Λ= 0.7 GeV
dσ/dqt dy| y= 0 [cm²/GeV]
qt [GeV]

Fig. 17

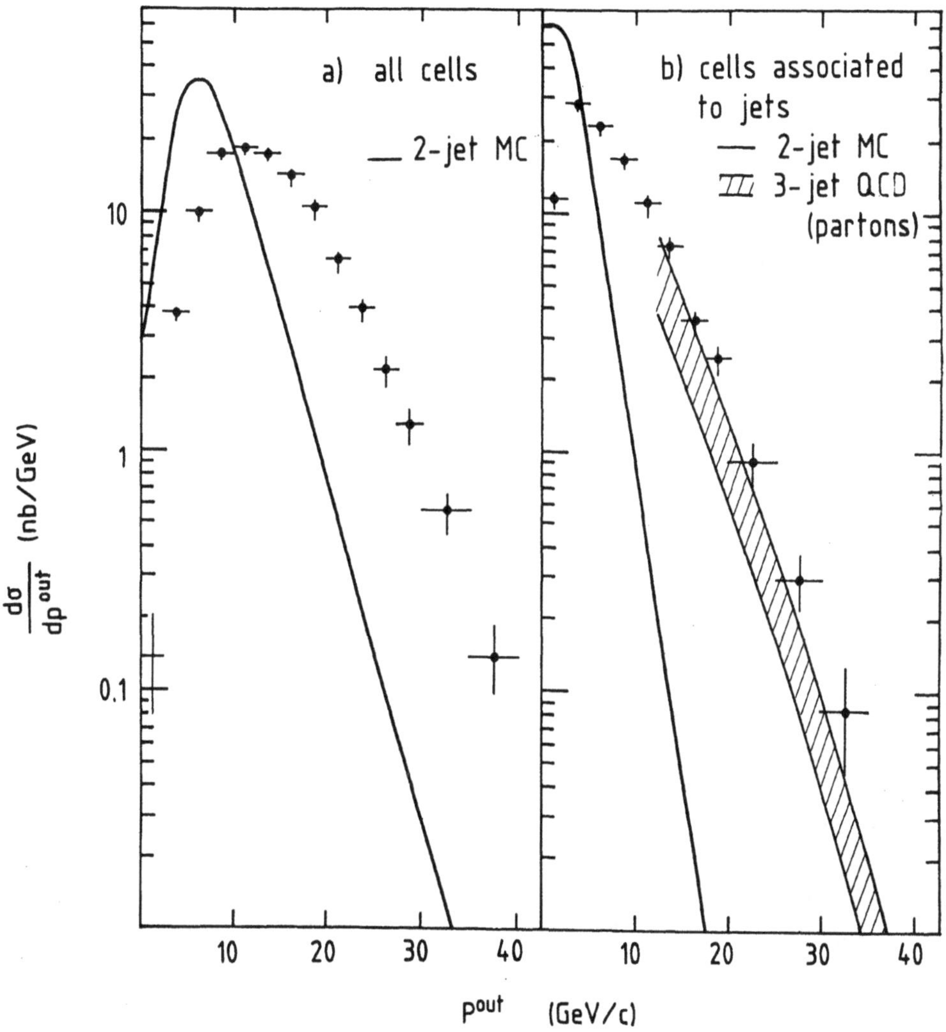
a) all cells
2-jet MC
b) cells associated to jets
2-jet MC
3-jet QCD (partons)
10
1
0,1
dσ/dp^out (nb/GeV)
10
20
30
40
10
20
30
40
p^out (GeV/c)

Fig. 18

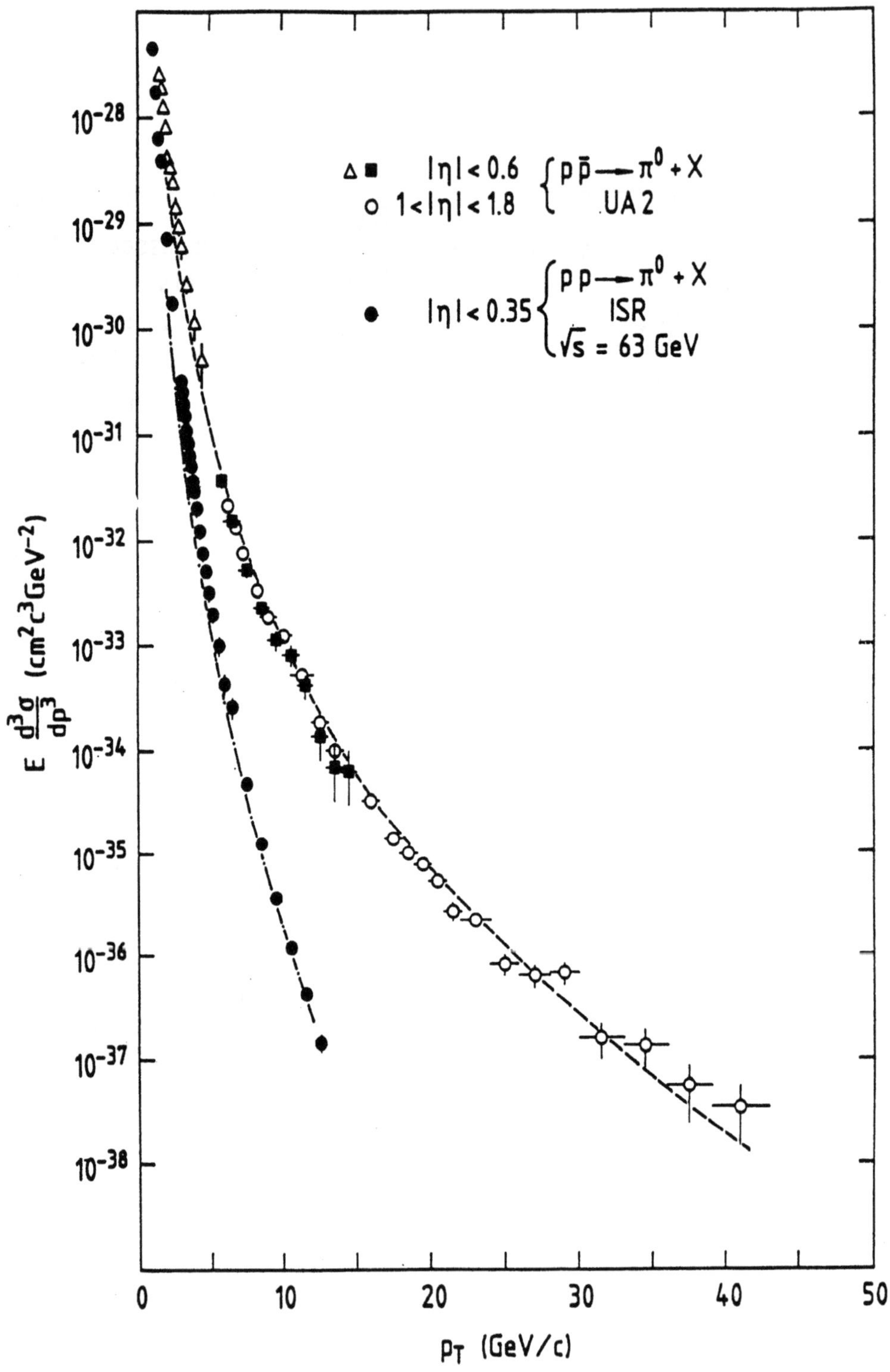

10^-28
10^-29
10^-30
10^-31
10^-32
10^-33
10^-34
10^-35
10^-36
10^-37
10^-38
E d³σ/dp³ (cm²c³GeV⁻²)
△ ■ |η| < 0.6
○ 1<|η|< 1.8
p p̄ → π⁰ + X
UA 2
● |η| < 0.35
p p → π⁰ + X
ISR
√s = 63 GeV
0
10
20
30
40
50
p_T (GeV/c)

Fig. 19

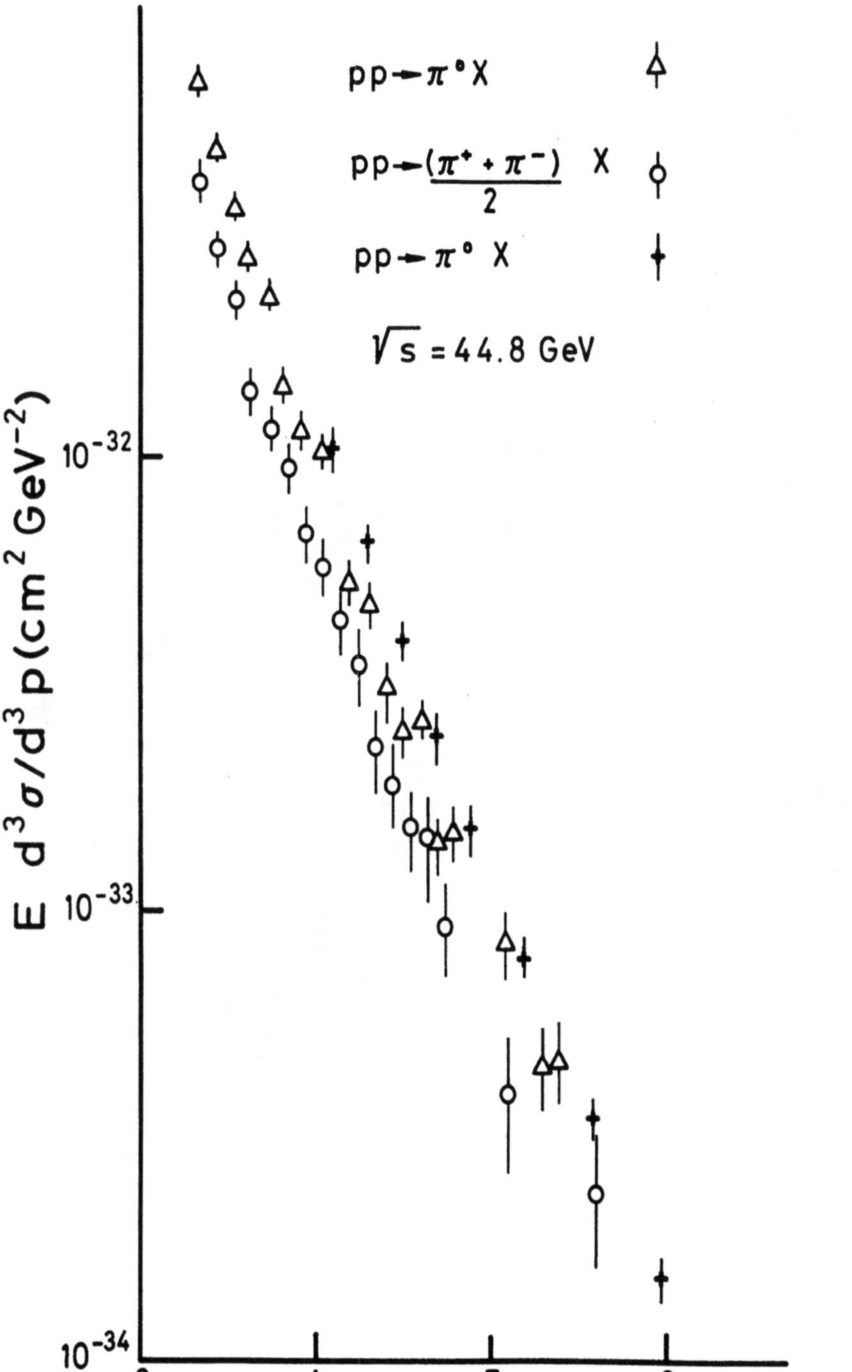
pp → π° X
pp → (π⁺ + π⁻)/2 X
pp → π° X
√s = 44.8 GeV
E d³σ/d³p (cm² GeV⁻²)
10⁻³²
10⁻³³
10⁻³⁴
3
4
5
6
p_T (GeV/c

Fig. 20

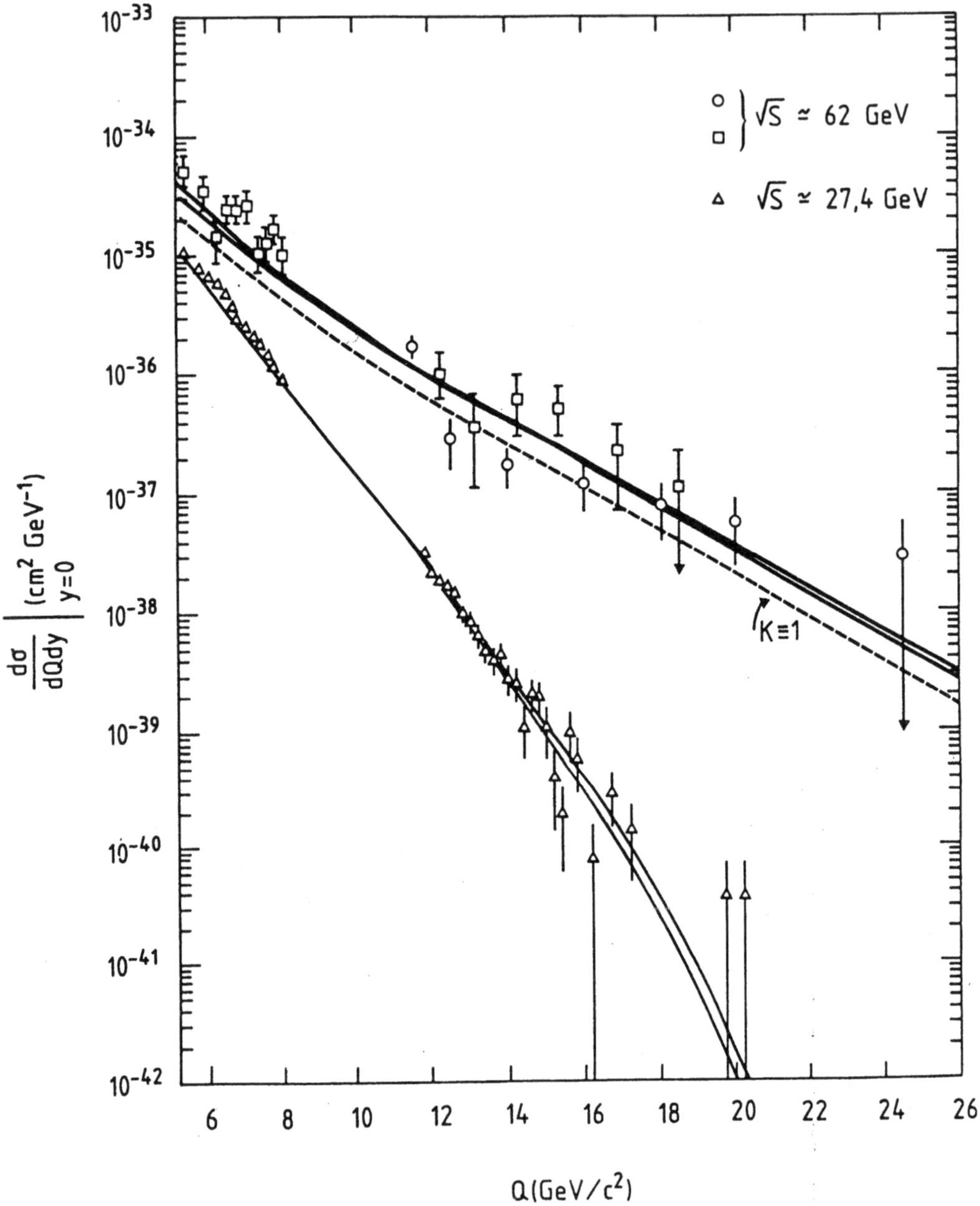

10^{-33}
10^{-34}
10^{-35}
10^{-36}
10^{-37}
10^{-38}
10^{-39}
10^{-40}
10^{-41}
10^{-42}
$\frac{d\sigma}{dQdy}\Big|_{y=0}$ (cm^2 GeV^{-1})
○ □ } $\sqrt{S} \simeq 62$ GeV
△ $\sqrt{S} \simeq 27{,}4$ GeV
K≡1
6 8 10 12 14 16 18 20 22 24 26
Q(GeV/c^2)

Fig. 21

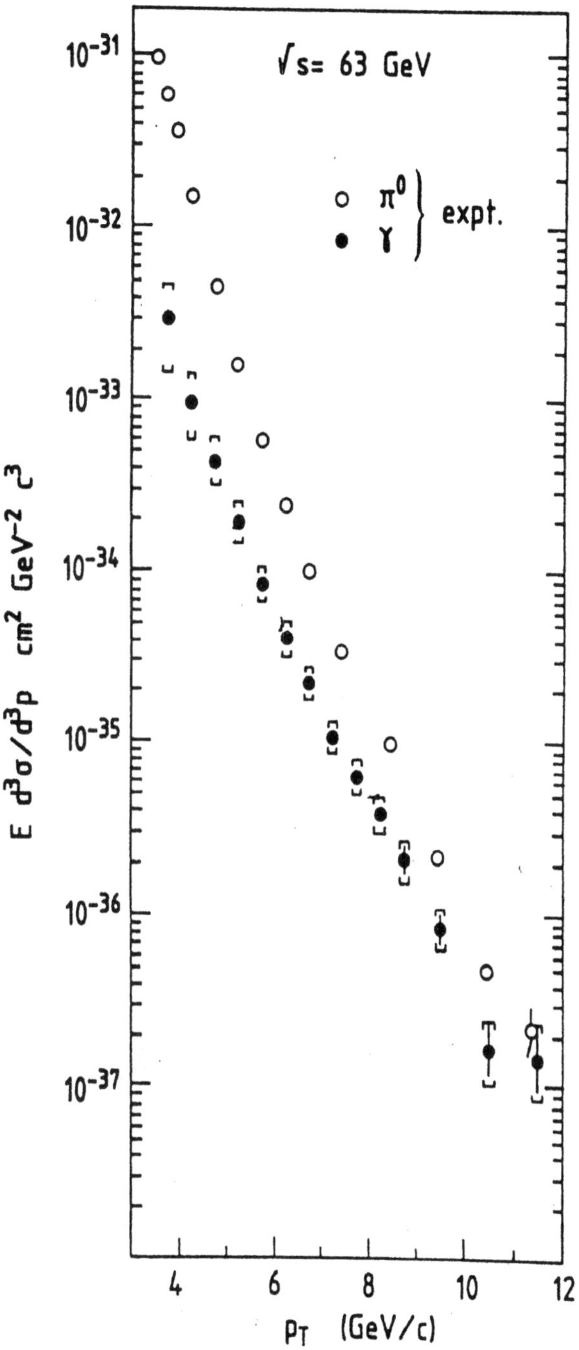

√s= 63 GeV
π⁰
γ
expt.
10^{-31}
10^{-32}
10^{-33}
10^{-34}
10^{-35}
10^{-36}
10^{-37}
E d³σ/d³p cm² GeV⁻² c³
4
6
8
10
12
pT (GeV/c)

Fig. 22

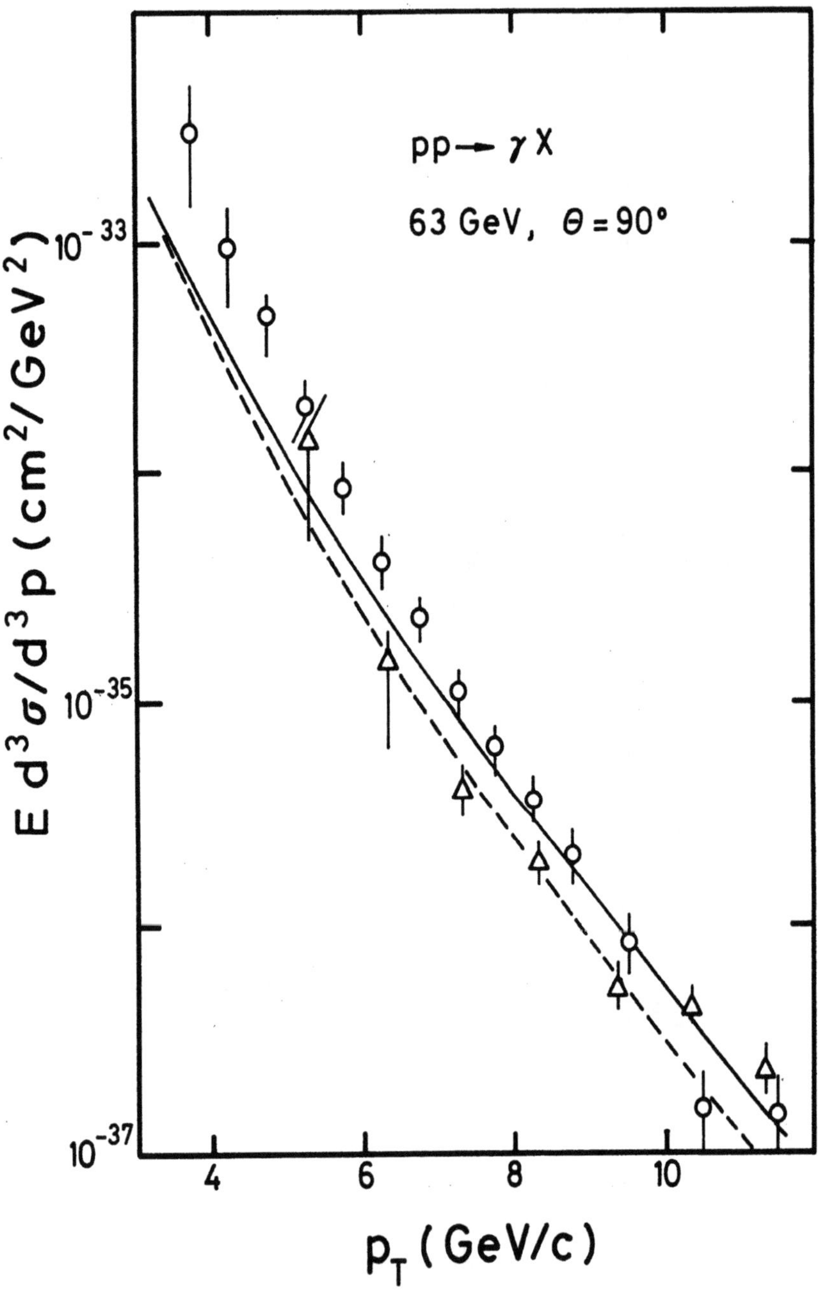

Fig. 23

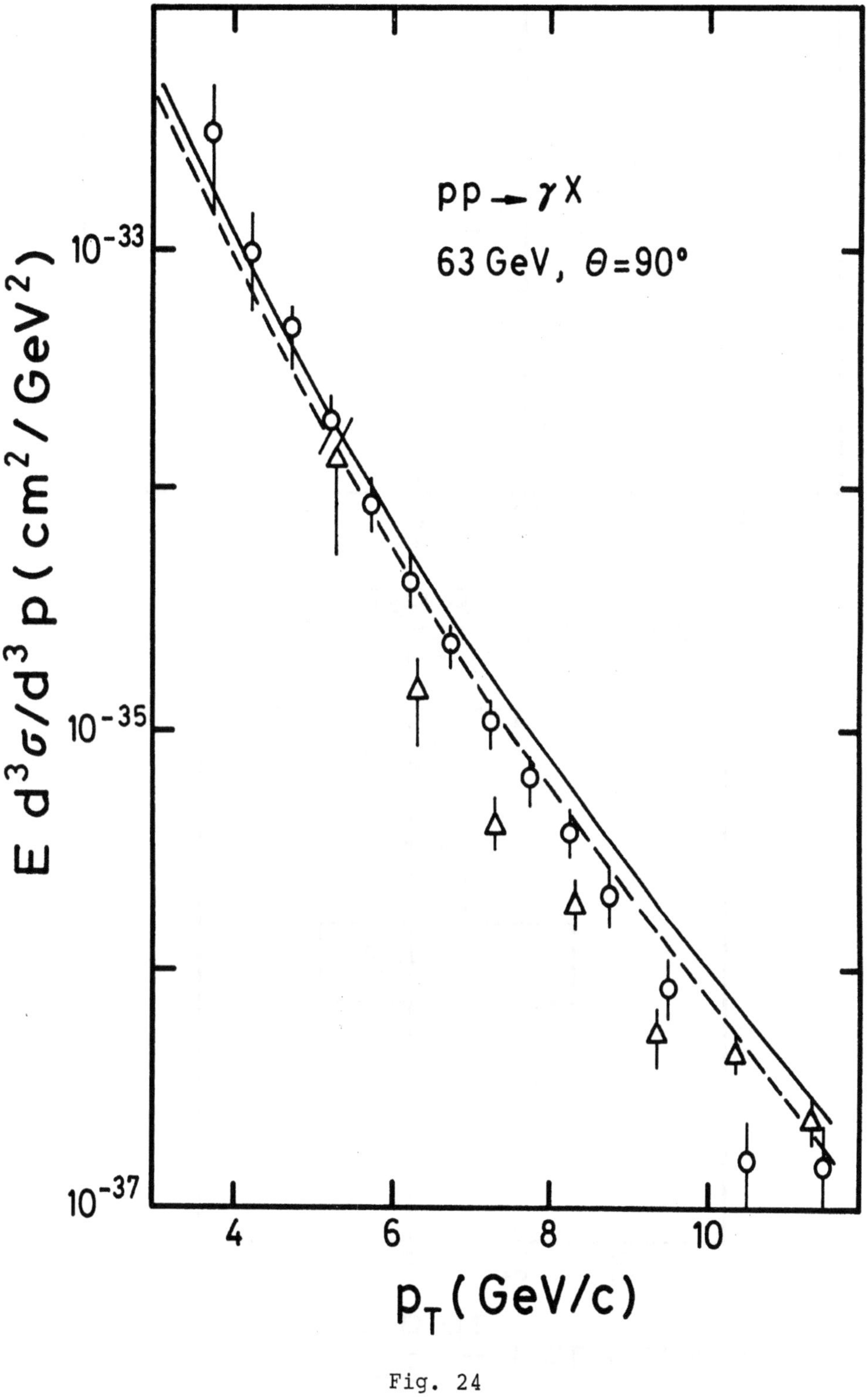

Fig. 24

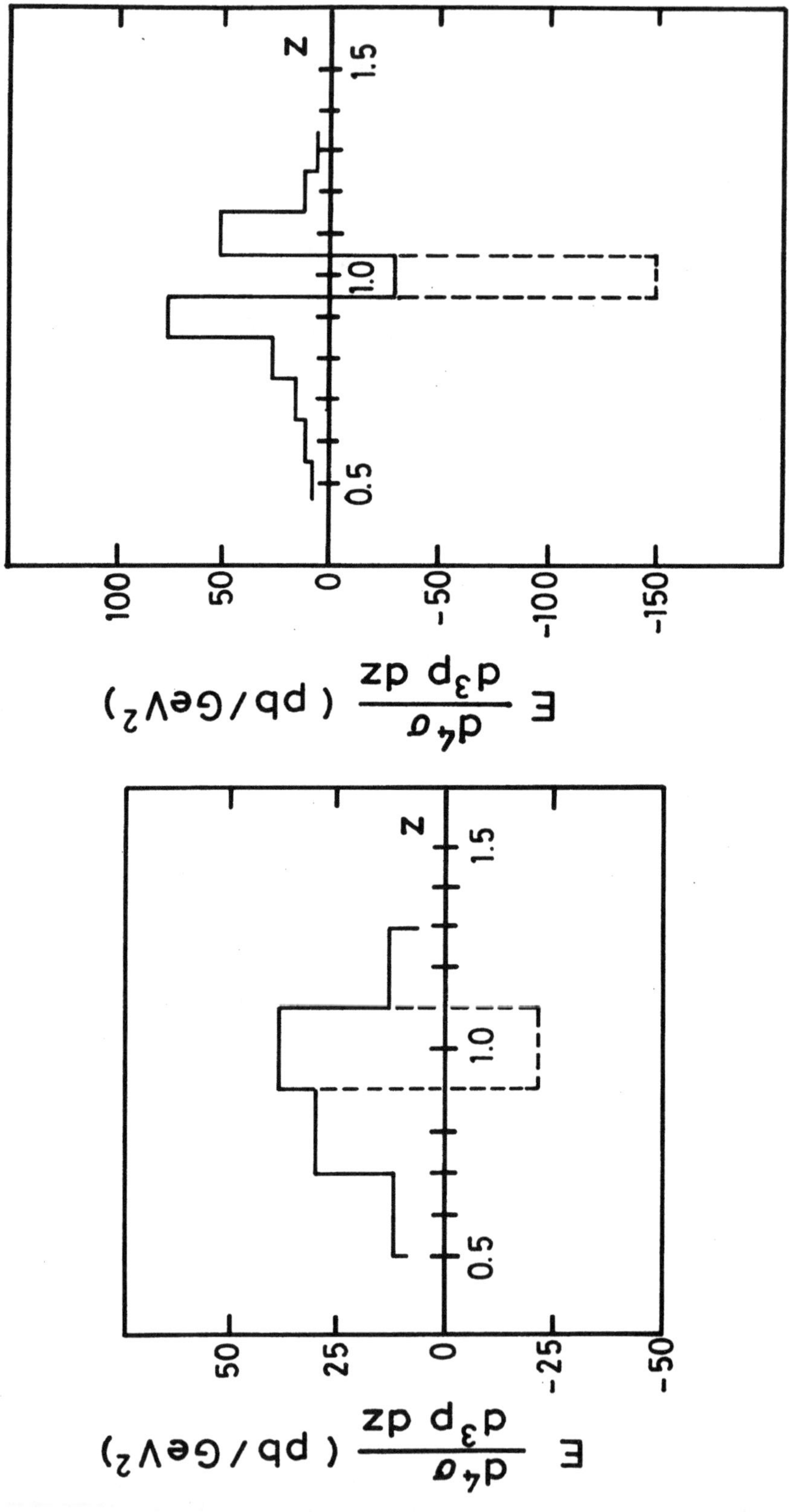

z
0.5
1.0
1.5
100
50
0
-50
-100
-150
$E \frac{d^4\sigma}{d^3p\,dz}$ (pb / GeV^2)
z
0.5
1.0
1.5
50
25
0
-25
-50
$E \frac{d^4\sigma}{d^3p\,dz}$ (pb / GeV^2)

Fig. 25

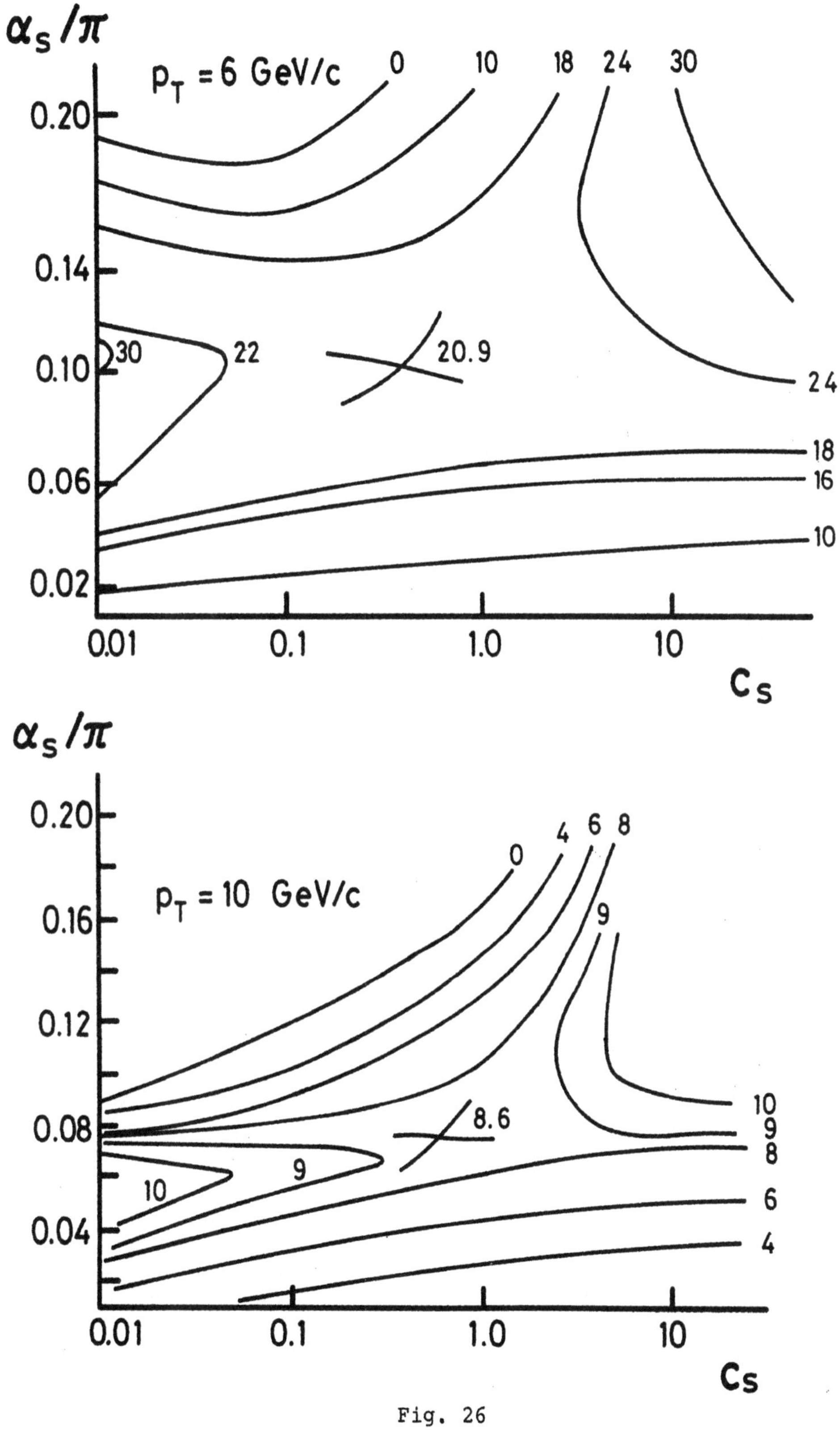

α_s/π
$p_T = 6$ GeV/c
0
10
18
24
30
0.20
0.14
0.10
0.06
0.02
30
22
20.9
24
18
16
10
0.01
0.1
1.0
10
c_s
α_s/π
$p_T = 10$ GeV/c
0.20
0.16
0.12
0.08
0.04
0
4
6
8
9
8.6
10
9
10
9
8
6
4
0.01
0.1
1.0
10
c_s

Fig. 26

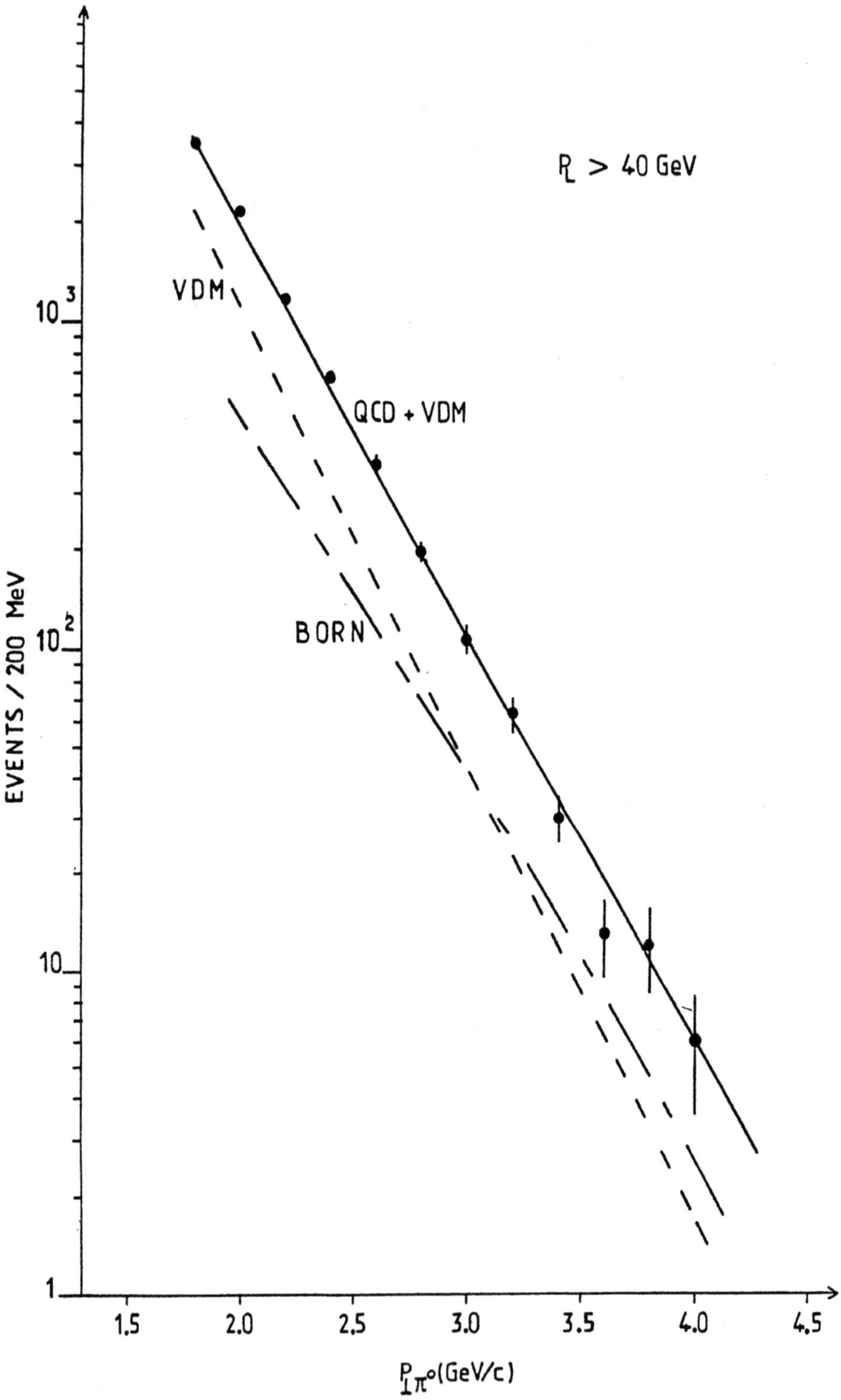

P_L > 40 GeV
VDM
QCD + VDM
BORN
EVENTS / 200 MeV
1
10
10^2
10^3
1.5
2.0
2.5
3.0
3.5
4.0
4.5
$P_{\perp \pi^0}$(GeV/c)

Fig. 27

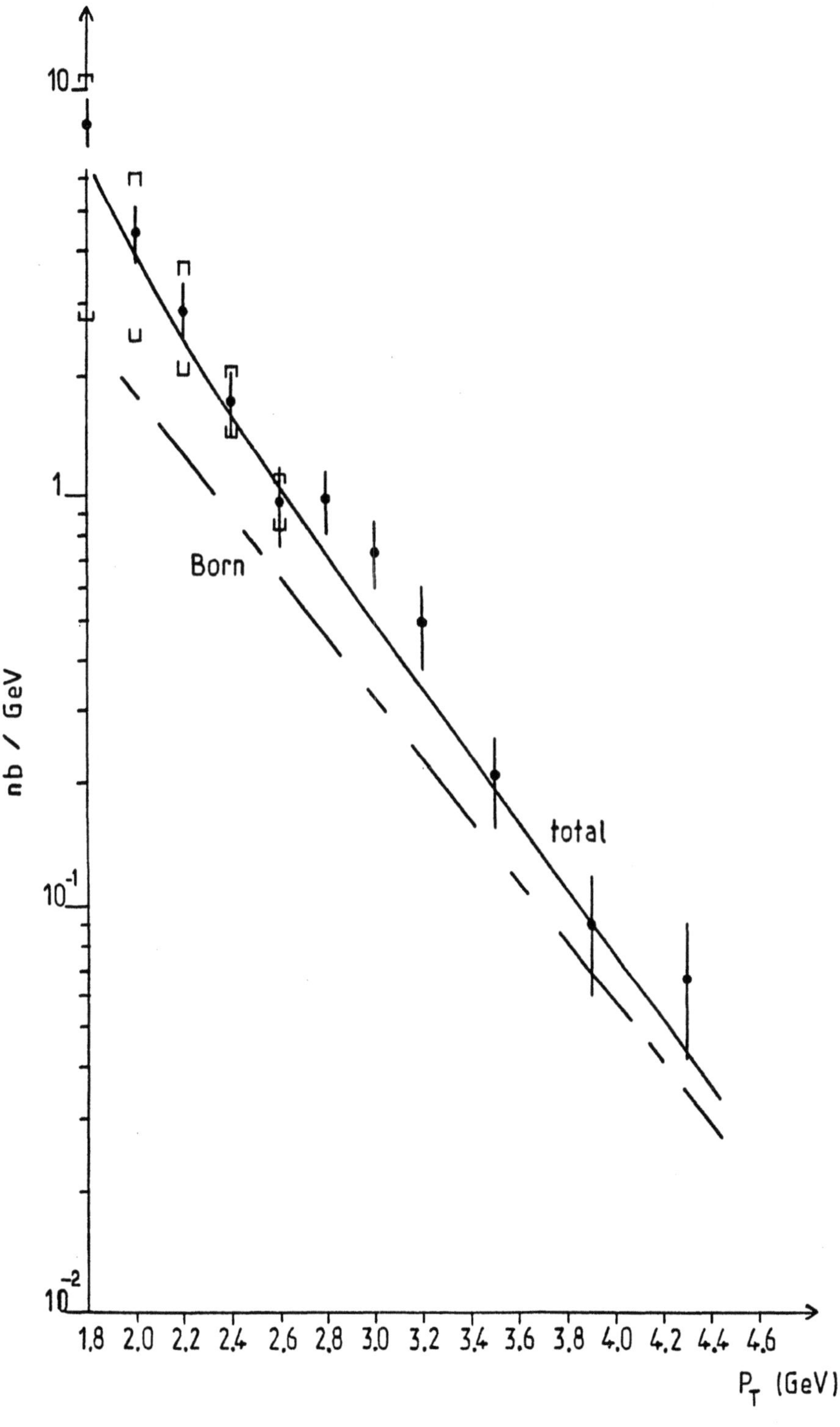
10
1
10^-1
10^-2
nb / GeV
Born
total
1,8 2,0 2,2 2,4 2,6 2,8 3,0 3,2 3,4 3,6 3,8 4,0 4,2 4,4 4,6
P_T (GeV)

Fig. 28

Acta Physica Austriaca, Suppl. XXVII, 517–595 (1985)

COMPOSITE QUARKS AND LEPTONS+

by

W. BUCHMÜLLER
CERN
Geneva

CONTENTS

+) Lectures given at the XXIV. Internationale Universitätswochen für Kernphysik, Schladming, Austria, February 20-March 1, 1985.

1. INTRODUCTION

The present theory of particle physics is the standard model of strong and electroweak interactions [1]. It is a renormalizable quantum field theory which describes sucessfully all our experimental information about the structure of matter down to distances of 10^{-16} cm. Despite its extraordinary success, however, there appears to be a general consensus that the standard model can only be an effective "low energy" Lagrangian of a more fundamental theory. Indeed, the gauge structure of strong, electromagnetic and weak interactions suggests a unification of all forces [2,3], the replication of families and the hierachical pattern of fermion masses point towards a common underlying structure of quarks and leptons, and the "naturalness problem" [4-6] of fundamental scalar fields seems to require a modification of the Higgs sector.

In these lectures we will discuss the current status of attempts to improve some of the unsatisfactory features of the standard model by making quarks, leptons, Higgs bosons and possibly also the vector bosons of the weak interactions composite. Despite considerable efforts [7] in recent years there exists at present no satisfactory "standard composite model" which would provide a convincing explanation of the family replication and the quark-lepton mass sepctrum and which could be distinguished from the standard model through definite, experimentally testable predictions. We will therefore not discuss any specific model in detail but rather concentrate on the general features which characterize the different classes of composite models.

In sec.2 we illustrate the role of global and local symmetries by means of a preon model of Pati-Salam type. Sec.3 is devoted to the bounds which high energy scattering experiments and rare processes impose on the substructure scale Λ. In sec.4 we discuss composite W-bosons, which occur in preon models where the substructure scale is of the order of the Fermi scale, $\Lambda \sim G_F^{-1/2}$, and which represent an alternative to the standard gauge theory of weak interactions. As an example of the rich phenomenology of such models with "nearby compositeness" we describe the signatures of light pseudoscalar leptoquarks. Sec.5 deals with theoretical constraints on the dynamics of (almost) massless composite particles: the effective Lagrangian for Goldstone bosons, the 't Hooft consistency conditions for chiral fermions and the the Case-Gasiorowicz-Weinberg-Witten theorem for massless vector bosons. In sec.6 we discuss the special features of supersymmetric (SUSY) preon models, in particular the natural appearance of massless composite fermions through the Goldstone mechanism. An example of the resulting SUSY σ-models and the particularly interesting coset spaces of exceptional groups are described in sec.7, and finally we present some conclusions in sec.8.

Before we describe attempts to improve unsatisfactory features of the standard model, however, let us recall the basic ingredients of the Glashow-Weinberg-Salam theory [8]. The strong and electroweak interactions are mediated through the exchange of gauge bosons G_μ^A (A=1,..,8), W_μ^I(I=1,..,3) and B_μ which are contained in the gauge group

$$G_{SM} = SU(3)_C \times SU(2)_W \times U(1)_Y \quad . \tag{1}$$

There are three families of quarks and leptons

$$q_L^{\alpha i} = \begin{pmatrix} u_L^{\alpha i} \\ d_L^{\alpha i} \end{pmatrix}, \qquad u_R^{\alpha i} \ , \ d_R^{\alpha i} \ ,$$

$$l_L^{i} = \begin{pmatrix} \nu_L^{i} \\ e_L^{-i} \end{pmatrix}, \qquad (\nu_R^{i}), \ e_R^{-i} \ , \tag{2}$$

with colour, weak isospin and hypercharge quantum numbers

$$q_L \sim (3,2)_{\frac{1}{6}}\,, \qquad u_R \sim (3,1)_{\frac{2}{3}}\,, \qquad d_R \sim (3,1)_{-\frac{1}{3}}\,,$$

$$l_L \sim (1,2)_{-\frac{1}{2}}\,, \qquad (\nu_R \sim (1,1)_{0})\,, \qquad e_R \sim (1,1)_{-1} \tag{3}$$

where $\alpha = 1,\dots,3$ is a colour index and $i = 1,\dots,3$ a generation index for the e-, μ- and τ-families. Note that left-handed fermions are $SU(2)_W$ doublets whereas right-handed fermions are $SU(2)_W$ singlets. So far there is no experimental evidence for righthanded neutrinos ν_R [9] which play a special role because they do not couple to gauge bosons. The spontaneous breaking of $SU(2)_W \times U(1)_Y$ requires scalar fields, such as

$$\phi_1 = \begin{pmatrix} \phi_1^0 \\ \phi_1^- \end{pmatrix} \sim (1,2)_{-\frac{1}{2}}$$

and

$$\phi_2 = \begin{pmatrix} \phi_2^+ \\ \phi_2^0 \end{pmatrix} \sim (1,2)_{\frac{1}{2}}\,, \tag{4}$$

where the superscripts denote the electric charges which are the sum of hypercharge and weak isospin, i.e., $Q = Y + T_3$. In the minimal model the second Higgs doublet is not an independent field but related to ϕ_1 through $\phi_2{}^p = \varepsilon^{pq}\,\phi^*_{1q}$.

Given the $SU(3)_C \times SU(2)_W \times U(1)_Y$ quantum numbers of quark, lepton and Higgs fields their couplings to the gauge bosons G_μ^A, W_μ^I and B_μ are entirely determined up to three universal gauge coupling constants g_3, g_2 and g_1. On the contrary the couplings between fermions and scalars are largely arbitrary. The most general renormalizable, SU(3) x SU(2) x U(1) invariant Lagrangian is given by

$$L_Y = g^{(u)i}_{\ \ j}\ \bar{q}_{Li}\phi_1 u^j_R + g^{(d)i}_{\ \ j}\ \bar{q}_{Li}\phi_2 d^j_R$$

$$(+g^{(\nu)i}_{\ \ j}\bar{\ell}_{Li}\phi_1\nu^j_R) + g^{(e)i}_{\ \ j}\bar{\ell}_{Li}\phi_2 e^j_R \ + \text{c.c.}, \qquad (5)$$

and involves 3 unconstrained complex 3x3 matrices of Yukawa couplings.

In the standard model the effective potential of the Higgs fields ϕ_1 and ϕ_2 has a minimum which braks $SU(2)_W$ x x $U(1)_Y$ to $U(1)_{EM}$:

$$\langle\phi_1\rangle_o = \begin{pmatrix} v_1 \\ 0 \end{pmatrix}, \qquad \langle\phi_2\rangle_o = \begin{pmatrix} 0 \\ v_2 \end{pmatrix}. \qquad (6)$$

This spontaneous symmetry breaking leads to masses for the $W^\pm$ and Z vector bosons:

$$M_W^2 = \frac{1}{2}\, g_2^2 (v_1^2+v_2^2), \qquad M_Z^2 = \frac{1}{2}\, (g_1^2+g_2^2)(v_1^2+v_2^2) \ ,$$

$$v_1^2+v_2^2 = (2\sqrt{2}\ G_F)^{-1} = (174\ \text{GeV})^2\ . \qquad (7)$$

Furthermore one obtaines from equs. (5) and (6) mass matrices for u- and d-type quarks and for the charged leptons.

$$M^{(u)i}_{\ \ \ j} = v_1\ g^{(u)i}_{\ \ \ j}, \quad M^{(d)i}_{\ \ \ j} = v_2\ g^{(d)i}_{\ \ \ j}, \quad M^{(e)i}_{\ \ \ j} = v_2\ g^{(e)i}_{\ \ \ j}, \qquad (8)$$

which yield the mass eigenvalues $m_e, \ldots, m_t$ as well as the parameters $\Theta_1, \Theta_2, \Theta_3$ and δ of the Kobayashi-Maskawa matrix.

The structure of the matter sector of the standard model, as described above, provokes the following questions:

(i) Why is there a threefold replication of quark-lepton generations?

(ii) What is the dynamical origin of symmetry breaking and fermion masses?

It is indeed difficult to believe that the appearence of three quark-lepton families with identical $SU(3)_C \times SU(2)_W \times$ $\times\, U(1)_Y$ quantum numbers is a mere accident and that Yukawa couplings which, from the electron to the b-quark, vary over four orders of magnitude should be considered as fundamental parameters like the gauge couplings. Furthermore it is obvious that an understanding of the Yukawa couplings is impossible within the standard model and that some new structure is required. In analogy to hadron physics, where the pion-nucleon coupling constant can be computed (at least in principle) within QCD, it is natural to investigate the possibility that also quarks, leptons and Higgs bosons are composite. The Lagrangian (5) would then be valid only at momentum transfers small compared to the substructure scale Λ and at distances $r < \frac{1}{\Lambda}$ the full structure of the three-point functions would become visible (cf.Fig.1). It is conceivable that such a preon theory

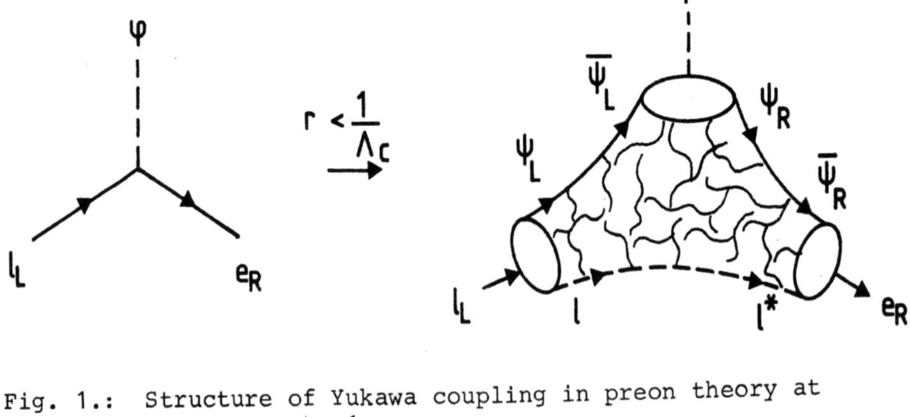

Fig. 1.: Structure of Yukawa coupling in preon theory at distances $r < \frac{1}{\Lambda_C}$.

could explain generations as some sort of excitations and that the fermion mass matrices would be calculable (at least in principle) in terms of a few parameters.

As the desire to understand the fermion mass spectrum is one of the main motivations for quark-lepton substructure, let us briefly recall some basic facts of the phenomenology of quark mass matrices [10]. The mass matrices $M^{(u)}$ and $M^{(d)}$ can be diagonalized by bi-unitary transformations [8]

$$U_L^+ M^{(u)} U_R = M_D^{(u)} = \begin{pmatrix} m_u & & \\ & m_c & \\ & & m_t \end{pmatrix}, \quad U_{L(R)}^+ U_{L(R)} = 1 ,$$

$$V_L^+ M^{(d)} V_R = M_D^{(d)} = \begin{pmatrix} m_d & & \\ & m_s & \\ & & m_b \end{pmatrix}, \quad V_{L(R)}^+ V_{L(R)} = 1, \tag{9}$$

which yield the Kobayashi-Maskawa matrix as

$$K = V_L^+ U_L . \tag{10}$$

Wolfenstein has suggested an interesting parametrization of K [11] which is consistent with all experimental information:

$$K = \begin{array}{c} \\ u \\ c \\ t \end{array} \begin{array}{c} \begin{array}{ccc} d & s & b \end{array} \\ \begin{pmatrix} 1-\frac{\lambda^2}{2} & \lambda & \lambda^3(\rho-i\eta) \\ -\lambda & 1-\frac{\lambda^2}{2} & \lambda^2 A \\ \lambda^2 A(1-\rho-i\eta) & -\lambda^2 A & 1 \end{pmatrix} \end{array} + O(\lambda^4). \tag{11}$$

Here the cosine of the Cabibbo angle $\lambda=\cos\Theta_c \approx 0.22$ plays the role of an expansion parameter, and A,ρ, and η are real numbers of order one.

Recently Frampton and Jarlskog have shown that in a basis where the mass matrices are hermitean they satisfy the relation [12]

$$\frac{1}{m_t} M^{(u)} = \frac{1}{m_b} M^{(d)} + O(\lambda^2) , \tag{12}$$

a regularity pointing towards a common origin of up- and down- quark masses. It is even possible to find a basis where one has

$$\frac{1}{m_t} M^{(u)} = \begin{pmatrix} 0 & & \\ & 0 & \\ & & 1 \end{pmatrix} + O(\lambda^2) = \frac{1}{m_b} M^{(d)} . \tag{13}$$

This is achieved by choosing the fields q_L^i, u_R^i and d_R^i such that $V_L = V_R = U_R = 1$, which implies

$$K = U_L . \tag{14}$$

In this basis $M^{(d)}$ is diagonal and one has $(\lambda^2 \gtrsim \frac{1}{20})$

$$\frac{1}{m_b} M^{(d)} = \begin{pmatrix} \frac{m_d}{m_b} & & \\ & \frac{m_s}{m_b} & \\ & & 1 \end{pmatrix} = \begin{pmatrix} 0 & & \\ & 0 & \\ & & 1 \end{pmatrix} + O(\lambda^2) , \tag{15}$$

where we have used $m_d/m_b \gtrsim 1/600$, $m_s/m_b \gtrsim 1/30$ [13]. For the up-quark mass matrix one obtains from equs. (9)-(11) and with $m_u/m_t < 1/4000$, $m_c/m_t < 1/18$ (cf. [13], $m_t > 20$ GeV):

$$\frac{1}{m_t} M^{(u)} = K \begin{pmatrix} \frac{m_u}{m_t} & & \\ & \frac{m_c}{m_t} & \\ & & 1 \end{pmatrix} = \begin{pmatrix} 0 & & \\ & 0 & \\ & & 1 \end{pmatrix} + O(\lambda^2) . \tag{16}$$

Equs.(12) and (13) suggest that, as many authors have speculated, the quark mass matrices may be calculable in some kind of perturbation theory, starting from

$$\frac{1}{m_b} M^{(d)}_{(0)} = \frac{1}{m_t} M^{(u)}_{(0)} = \begin{pmatrix} 0 & & \\ & 0 & \\ & & 1 \end{pmatrix} . \qquad (17)$$

The different mass scales for u- and d-type quark masses could be due to different vacuum expectation values of the two Higgs doublets H_1 and H_2:

$$\frac{m_b}{m_t} \approx \frac{v_2}{v_1} \approx \frac{1}{10} . \qquad (18)$$

The goal of composite models is to understand the structure of the quark and lepton mass matrices and to compute eventually the parameters $v_1, v_2, \lambda, \ldots$ from a more fundamental preon theory.

2. A PREON MODEL OF PATI-SALAM TYPE

Pati and Salam were the first to speculate that quarks and leptons may have a further substructure [14]. They were led to this conjecture by the symmetries of the quark-lepton spectrum and in this section we will follow their approach. Early preon models can be found in [15]-[17] and a more complete list of references in the reviews [18]-[20]. There is an interesting class of composite models with only fermionic preons, such as the models of Harari [21], Shupe [22], Harari and Seiberg [23], Bars [24] and Schrempp and Schrempp [25], as well as the recent work on "composite Higgs models" [26], which we will not discuss. We will focus on fermion-boson models which have a simple quantum number structure. One may object that fundamental scalar fields are "unnatural" [4-6] but, as we will see in sections 6 and 7, this problem does not exist in supersymmetric models which we consider to be the most promising candidates for a theory of quark-lepton substructure. The content of this section is closely related to the model of Barbieri, Masiero and Mahapatra [27], similar models can be found in [28]-[31].

It is very instructive to study the dependence of the global symmetries of the standard model on the Higgs part of the Lagrangian (cf.(5)). For arbitrary Yukawa couplings $g^{(u,d,\nu,e)i}{}_j \neq 0$ one has

$$G = SU(3)_C \times SU(2)_W \times U(1)^3 ,$$

with

$$U(1)^3 = U(1)_q \times U(1)_\ell \times U(1)_\phi \quad . \tag{19}$$

The U(1) charges of quarks, leptons and Higgs fields are given in table I. One linear combination, the hypercharge Y, is gauged together with

Table I.: U(1) charges of quarks, leptons and Higgs scalars.

	q_L	ℓ_L	ϕ_1	u_ℓ	d_R	ν_R	e_R	ϕ_2
Q_q	1	0	0	1	1	0	0	0
Q_ℓ	0	1	0	0	0	0	1	0
Q_ϕ	0	0	1	-1	1	-1	1	-1

colour and weak isospin. The two remaining U(1) factors correspond to baryon and lepton number:

$$Y = \frac{1}{6} Q_q - \frac{1}{2} Q_\ell - \frac{1}{2} Q_\phi \; , \tag{20a}$$

$$B = \frac{1}{3} Q_q \; , \tag{20b}$$

$$L = Q_\ell \; . \tag{20c}$$

It is a veryremarkable feature of the standard model Lagrangian that it contains no global symmetries beyond baryon and lepton number. This changes drastically in the extreme case of vanishing Yukawa couplings, $g^{(u,d,\nu,e)i}{}_j=0$. In this case there is no difference between the three families and one gains a global U(3) invariance for each irreducible $SU(3)_C \times SU(2)_W \times U(1)_Y$ representation (cf.(3)), i.e.,

$$G' = SU(3)_C \times SU(2)_W \times U(1)_Y \times U(3)^6 \; . \tag{21}$$

Note that in the absence of Yukawa couplings the right handed neutrinos are free particles. The huge symmetry G' has to be broken to G in order to generate masses for quarks and leptons. To understand how this happens is one of the big problems of composite models where one wants to calculate fermion masses starting, in first approximation, from three massless families of quarks and leptons.

Many preon models are based on the global symmetry which one obtains for equal Yukawa couplings, $g^{(u)} = g^{(d)} = g^{(\nu)} = g^{(e)} \equiv g$, in the case of one family. Then the Lagrangian (5) becomes

$$L_H = g^{(u)} \bar{q}_{Lp} \phi_1^p u_R + g^{(d)} \bar{q}_{Lp} \phi_2^p d_R \qquad +$$

$$+ g^{(\nu)}\bar{\ell}_{Lp}\phi_1{}^p\nu_R + g^{(e)}\bar{q}_{Lp}\phi_2^p e_R \quad + \text{c.c.}$$

$$= g\ \bar{f}_{Lp}\phi_s{}^p f^s{}_R + \text{c.c.}\ , \qquad (22)$$

with

$$f_L^p = (q_L^{\alpha p},\ \ell_L^p)\ ,\quad f_R^s = \begin{pmatrix} u_R^\alpha & \nu_R \\ d_R^\alpha, & e_R^- \end{pmatrix}\ ,$$

$$\phi_S^p = (\phi_1{}^p,\ \phi_2{}^p)\ .$$

Obviously the Lagrangian (22) has the symmetry

$$G'' = SU(2)_L \times SU(2)_R \times SU(4)_{PS} \times U(1)^2\ , \qquad (23)$$

under which fermions and scalars transform as

$$f_L \sim (2,1,4),\quad f_R \sim (1,2,4),\quad \phi \sim (2,2,1)\ . \qquad (24)$$

The Pati-Salam symmetry $SU(4)_{PS}$ contains lepton number as "fourth colour" [2]. It appears here as a global symmetry which, in the limit of vanishing gauge couplings g_1 and g_3, is an exact invariance of the whole Lagrangian. An attractive feature of the $SU(4)_{PS}$ symmetry is that it contains B-L. The three diagonal SU(4) λ-matrices are λ_3 and λ_8, which are contained in the $SU(3)_C$ subgroup, and

$$\lambda_{15} = \frac{1}{\sqrt{6}} \begin{pmatrix} 1 & & & \\ & 1 & & \\ & & 1 & \\ & & & -3 \end{pmatrix} = \sqrt{\frac{3}{2}}\ (B-L)\ . \qquad (25)$$

The electric charge takes the symmetric form

$$Q = \frac{1}{2}(B-L) + T_{3L} + T_{3R} \quad , \tag{26}$$

where $T_{3L(R)}$ are the diagonal generators of $SU(2)_{L(R)}$.

The fermion representation (24) is chiral (i.e., different for left and right handed fermions) with respect to $SU(2)_L \times SU(2)_R$, and vectorlike (i.e., identical for left and right handed fermions) with respect to $SU(4)_{PS}$. Therefore it is natural, to associate in a fermion-boson composite model the $SU(2)_L \times SU(2)_R$ subgroup with fermionic preons and the SU(4) subgroup with scalar preons [14]. We thus introduce as preons the fermions

$$\psi_L^P \sim (2,1,1) \quad , \qquad \psi_R^S \sim (1,2,1) \tag{27a}$$

and the scalars

$$S = (h^\alpha, \ell) \sim (1,1,4) \tag{27b}$$

with $SU(3)_C \times SU(2)_W \times U(1)_Y$ gauge quantum numbers

$$\psi_L \sim (1,2)_0, \qquad \psi_R^1 \sim (1,1)_{\frac{1}{2}}, \qquad \psi_R^2 \sim (1,1)_{-\frac{1}{2}} \quad ,$$

$$h \sim (3,1)_{\frac{1}{6}}, \qquad \ell \sim (1,1)_{-\frac{1}{2}} \quad . \tag{28}$$

The most common choice for the binding force is a confirning "hypercolour" gauge interaction with gauge group $SU(n)_{HC}$ under which the preons transform like

$$\psi_{L,R} \sim \underline{n} \quad , \qquad s \sim \underline{n}^* \quad . \tag{29}$$

Equs.(27) - (29) define a preon theory whose bound state spectrum is in principle calculable. Unfortunately the dynamics of confining gauge theories is very complicated and currently no rigorous results are known for theories with fermions and scalars in the fundamental representation of the gauge group. In analogy with QCD one expects that the simplest hypercolour singlet composite operators with different "flavour" quantum numbers create different physical states from the vacuum. This yields quarks and leptons as fermion-boson bound states,

$$q_L = (\psi_L h) \;, \qquad u_R = (\psi_R^1 h) \;, \qquad d_R = (\psi_R^2 h) \;,$$

$$\ell_L = (\psi_L \ell) \;, \qquad \nu_R = (\psi_R^1 \ell) \;, \qquad e_R^- = (\psi_R^2 \ell) \;, \tag{30}$$

and the Higgs scalars as fermion-antifermion bound states,

$$\phi_S^P = (\bar{\psi}_{RS} \psi_L^P) \; . \tag{31}$$

Vacuum expectation values of the composite scalars ϕ_S^P,

$$\langle \phi_S^P \rangle = \langle \bar{\psi}_{RS} \psi_L^P \rangle = v \; \delta_S^P \; , \tag{32}$$

can lead to the technicolour-type [5] symmetry breaking $SU(2)_L \times SU(2)_R \rightarrow SU(2)_D$ and thereby give masses to the W-bosons.

In addition to the wanted bound states of equs.(30) and (31) one expects composite vector and scalar bosons corresponding to the following operators ($SU(2)_L \times SU(2)_R \times$ $\times\ SU(4)_{PS}$ quantum numbers):

$$W^I_{L\mu} = (\bar{\psi}_L \gamma_\mu \frac{1}{2}\tau^I \psi_L) \sim (3,1,1) \quad , \tag{33a}$$

$$W'_{L\mu} = (\bar{\psi}_L \gamma_\mu \psi_L) \quad \sim \quad (1,1,1) \; , \tag{33b}$$

$$W^I_{R\mu} = (\bar{\psi}_R \gamma_\mu \frac{1}{2}\tau^I \psi_R) \sim \quad (1,3,1) \quad , \tag{33c}$$

$$W'_{R\mu} = (\bar{\psi}_R \gamma_\mu \psi_R) \quad \sim \quad (1,1,1) \, , \tag{33d}$$

$$V^A_\mu = (s^\dagger \overleftrightarrow{D}_\mu \frac{1}{2}\lambda^A s) \sim \quad (1,1,15) \; , \tag{33e}$$

$$V'_\mu = (s^\dagger D_\mu s) \quad \sim \quad (1,1,1) \; , \tag{33f}$$

$$\chi^A = (s^\dagger \frac{1}{2}\lambda^A s) \quad \sim \quad (1,1,15) \; , \tag{33g}$$

$$\chi^I = (s^\dagger s) \quad \sim \quad (1,1,1) \; . \tag{33h}$$

Furthermore there will be bound states of higher spin and towers of excitations for all of the states of equs.(30) - (33). The appearance of bound states in addition to quarks, leptons and Higgs bosons is an unavoidable feature of preon models. Some of these new staes have an interesting phenomenology. The 15-plet χ^A, for instance, contains scalars with the following colours and electric charges:

$$\{\chi^A\} = \{\underline{8}_0, \; \underline{3}_{\frac{2}{3}} \; , \; \underline{3}^*_{-\frac{2}{3}}, \; \underline{1}_0\} \; . \tag{34}$$

The colour octet couples to quarks, the colour triplet to a quark-lepton pair (cf. Fig.2). It is the scalar leptoquark [2] which, as we will see in the following section, could cause rather spectacular effects.

In the preon model described above the strong and electroweak interactions are fundamental gauge interactions. There is an interesting alternative version [27] in which not the full $SU(2)_L \times U(1)_Y$ symmetry group is gauged but only the U(1) subgroup generated by the electric charge. The weak interactions are mediated by the composite vector bosons W_L and W_R (cf. (33a), (33c)) and parity is assumed to be spontaneously broken, yielding $M_{W_R} >> M_{W_L}$. The correct neutral current structure can be obtained through a dynamical $W_L^3 - W_R^3 - \gamma$ mixing. In a more radical approach even gluons and photon are considered as composite fields [29,32].

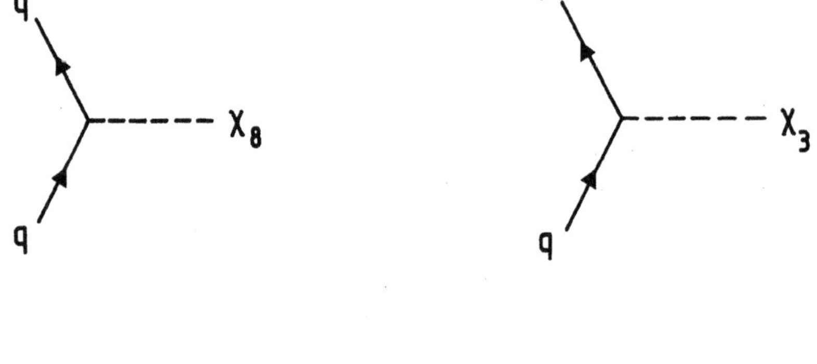

Fig. 2.: Couplings of colour octet and triplet (leptoquark) scalars to quarks and leptons.

One of the main motivations for composite models is the replication of quark-lepton families. The bound states (30),however, represent only one family. Where do the others come from? Are they "radial excitations" or "pair excitations"? Should one introduce an additional preon carrying "generation number" or enlarge the global symmetry, for instance from SU(4) to $SU(4n_f)$ for n_f generations? Is the family replication connected with the center Z_n of the hypercolour group $SU(n)_{HC}$ or enforced by anomaly matching? A number of interesting possibilities has been discussed in more detail by Harari [33], but a fully satisfactory solution of the generation puzzle still remains to be found.

What is the mass scale Λ of the hypercolour interaction and the corresponding size $r \sim \frac{1}{\Lambda}$ of the composite quarks and leptons? As the preon theory has to explain the mass of the weak vector bosons (independent of whether they are elementary or composite) Λ ,being the only mass scale of the theory, cannot be much larger than the Fermi scale $G_F^{-1/2} \sim 300$ GeV. As we will see in the following section such a "nearby compositeness" is barely compatible with experimental bounds on Λ.

So far we have only considered an example of the "quantum numerology" of preon models. The main dynamical problem of quark-lepton substructure, the appearance of almost massless composite fermions, will be discussed in sec.5.

3. BOUNDS ON THE SUBSTRUCTURE SCALE Λ

The substructure scale of the preon model considered in the previous section was the Fermi scale, i.e., $\Lambda \sim G_F^{-1/2} \sim 300$ GeV. There exists, however, at present no "standard composite model" and hence also no theoretically preferred value of Λ. Indeed, preon models have been constructed with values of Λ covering the whole range from $G_F^{-1/2}$ up to the Planck mass $M_{PL} \sim 10^{19}$ GeV [18]. In this section we will discuss the experimental bounds on Λ which have been obtained from scattering experiments and rare processes.

(i) Scattering experiments

The classical process which probes the elementarity of the electron is Bhabba scattering (cf.Fig.3). A finite size of the electron would lead to an electromagnetic formfactor and a deviation from the pointlike coupling to the photon [34]:

$$1 \to F(q^2) = 1 + \frac{q^2}{\Lambda_F^2} + O(\frac{1}{\Lambda_F^4}), \quad |q^2| << \Lambda_F^2 . \tag{35}$$

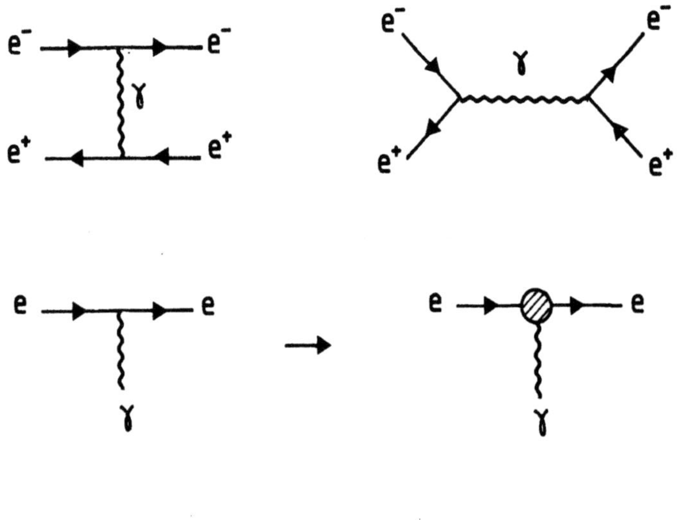

Fig. 3.: Bhabba scattering with formfactor modification.

The current lower bound on the formfactor scale Λ_F reads [34]

$$\Lambda_F > 200 \text{ GeV} , \tag{36a}$$

which corresponds to the upper bound for the electron size

$$r_e \stackrel{\sim}{\sim} \frac{1}{\Lambda_F} < 10^{-16} \text{ cm} . \tag{36b}$$

If electrons are composite objects they will have contact interactions in addition to the standard model interactions. These contact interactions are caused by the interchange of constutuents or, equivalently, the exchange of heavy reaonances of various spins (cf. Fig.4). They are described by an effective Lagrangian of the form ($|q^2| << M_V^2, M_S^2$)

$$L_{eff} = - \frac{g_V^2}{2M_V^2} \bar{e}\gamma^\mu e \, \bar{e}\gamma_\mu e - \frac{g_S^2}{2M_S^2} \bar{e}e \, \bar{e}e + \ldots , \tag{37}$$

where g_V, g_S, M_V and M_S are the coupling constants and masses of a heavy vector V and a heavy scalar S. For a strong binding force at the preon level one may expect, in analogy to QCD, also large values for the coupling constants g_V and g_S, i.e., $g_{V,S}^2/4\pi = O(1)$. It is therefore natural to define a substructure scale Λ_c [35] as

$$\frac{g^2}{M^2} = \frac{4\pi}{\Lambda_c^2} \quad . \tag{38}$$

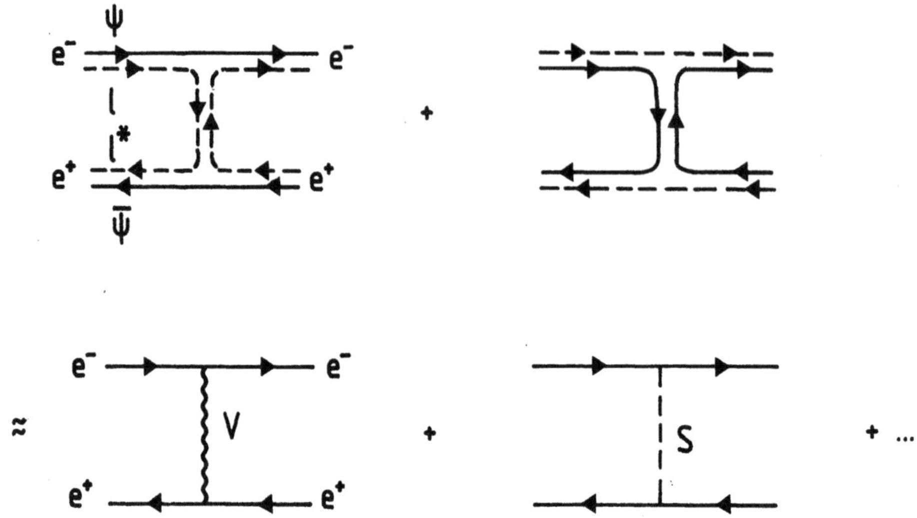

Fig. 4.: Dynamical origin of contact interactions: constituent exchange or, equivalently, exchange of heavy resonances.

The so defined scale Λ_c corresponds physically to the mass of a strongly interacting resonance. Obviously, the form-factor correction (35) also leads to a contact term. From

$$2\,\frac{e^2}{q^2}\,\frac{q^2}{\Lambda_F^2} = \frac{4\pi}{\Lambda_c^2} \tag{39}$$

one obtains for the relation between the two scales Λ_F and Λ_c:

$$\Lambda_c = \frac{1}{\sqrt{2\alpha}}\,\Lambda_F \approx 8\Lambda_F\ . \tag{40}$$

Contact interactions may have various Lorentz structures. Of particular interest are the chirality conserning left-left, right-right, vector-vector and axial vector-axial vector interactions (LL-, RR-,VV- and AA-"models"). Eichten, Lane and Peskin have studied the effect of the contact interactions on the angular distribution in Bhabba scattering (cf. Fig.5). Their analysis of PETRA data at $\sqrt{s}$ = 35 GeV led to the bounds [35]

$$\Lambda_c > \begin{cases} 0.75 \text{ TeV}, & \text{for LL-, RR-models} \\ 1.5 \text{ TeV}, & \text{for VV-, AA-models} \end{cases} \quad . \tag{41}$$

The analysis of Bhabba scattering at $\sqrt{s}$ = 100 GeV at SLC and LEP (cf. Fig.6) is expected to yield the bounds [35]

$$\Lambda > \begin{cases} 2 \text{ TeV}, & \text{for LL-, RR-models} \\ 5 \text{ TeV}, & \text{for VV-, AA-models} \end{cases} \quad . \tag{42}$$

Rückl has studied the effect of contact interactions in ep-scattering [37,38]. For the LL-model

$$L_{eff} = \frac{4\pi}{\Lambda_c} \bar{e}_L\gamma^\mu e_L\bar{q}_L\gamma_\mu q_L \tag{43}$$

he finds that the structure function F_2 at x = 0.5 and $Q^2 = (100 \text{ GeV})^2$ charges by 25 % (60 %) for Λ_c = 2 TeV (1.5 TeV) (cf. Fig.7), which illustrates the sensitivity of HERA experiment to substructure effects.

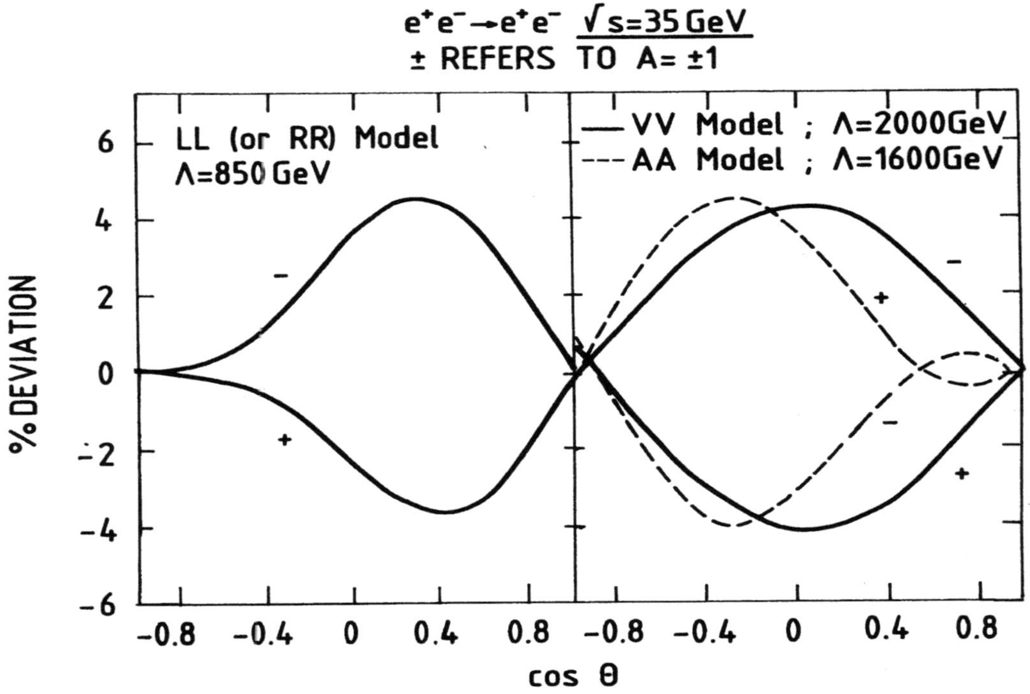

Fig. 5.: Deviation of the cross section for $e^+e^- \to e^+e^-$ from the standard model prediction due to contact interactions at $\sqrt{s}$ = 35 GeV (see text). The +(-) sign refers to a contact term with opposite (equal) sign compared to the weak interaction current Lagrangian (from ref. 35).

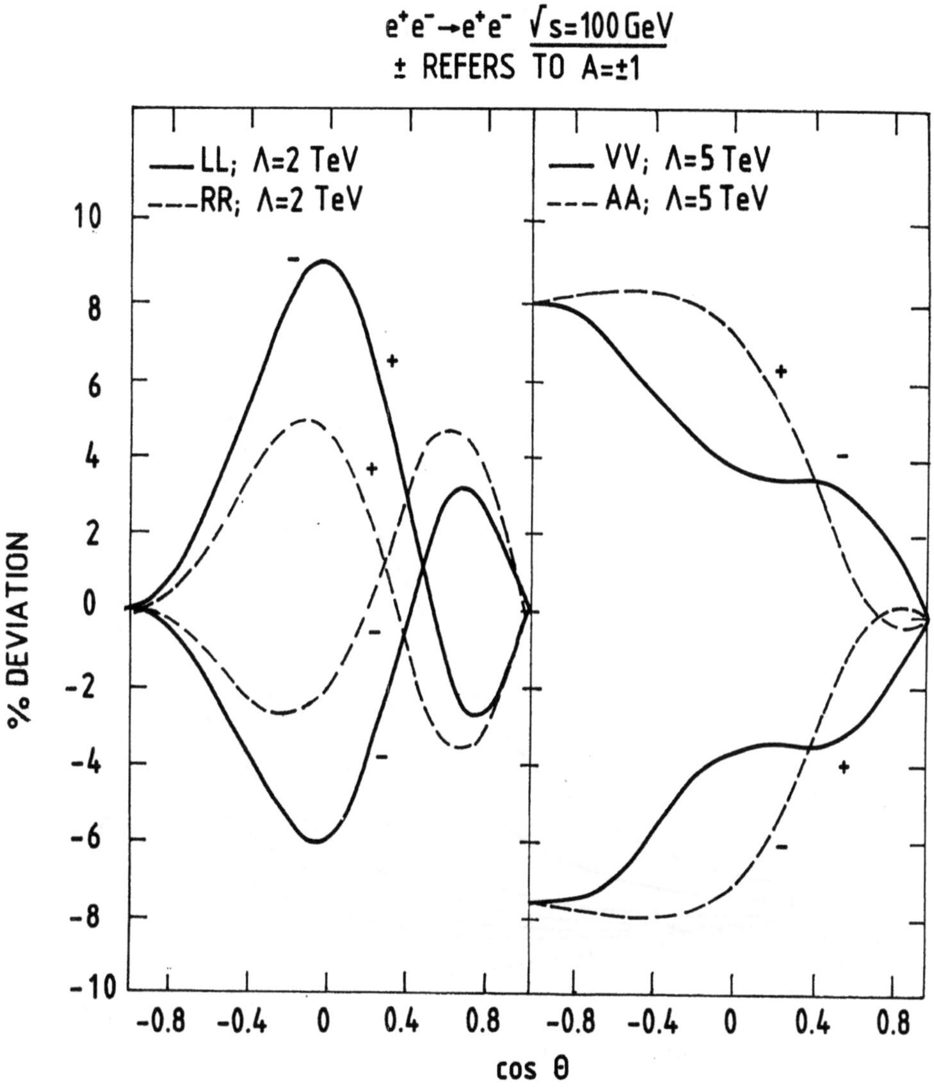

Fig. 6.: Deviation of the cross section for $e^+e^- \to e^+e^-$ from the standard model prediction due to contact interactions at $\sqrt{s}$ = 100 GeV. (cf. Fig.5, from ref.35).

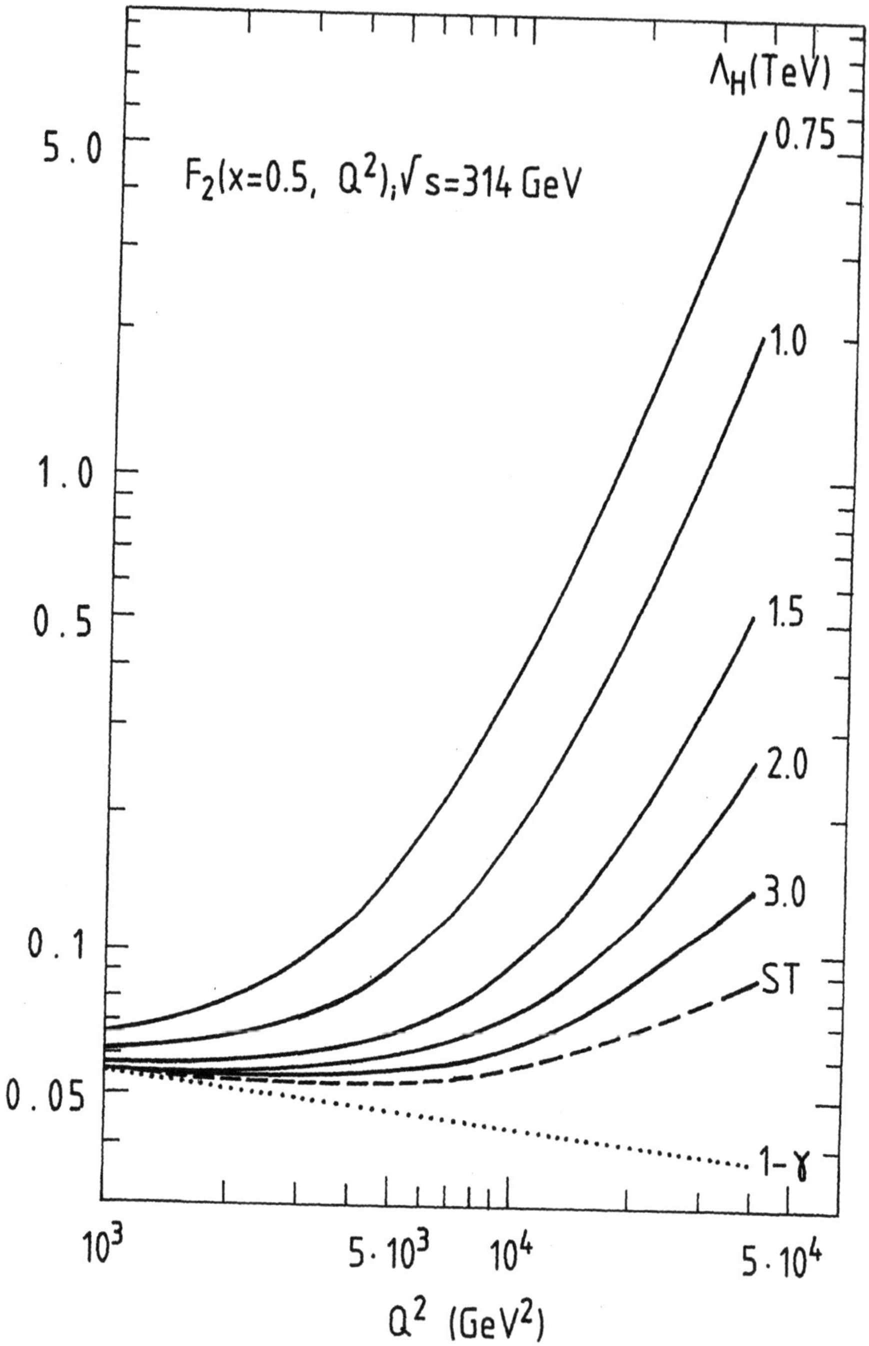

Fig. 7.: Deviation of the proton structure function F_2 at $x = 0.5$, $\sqrt{s} = 314$ GeV, from the standard model prediction due to the contact interaction (43) (from ref.37).

Following Abolins et al. [36] we have listed in table II the bounds on the substructure scale Λ_c and the (electromagnetic or colour) formfactor scales Λ_F for various processes which can be obtained at the next generation of accelerators. Despite the considerable increase in energy the bounds will be improved by at most a factor of three compared to the current bound (41).

Table II: Expected bounds from future colliders on the mass scales Λ_c and Λ_F for various processes. In the second column the assumed integrated luminosity per year is listed (from ref.36).The values for Λ_F and estimated according to equ.(40) with α_{EM} =0.008 and α_s = 0.1.

process $\sqrt{s}$ [GeV]	$\int L dt/yr$	Λ_c [TeV]	Λ_F [GeV]	subprocess tested
ep, 10 x 1000	10^{39}	4	500	$eq \to eq$
e^+e^-, $\leq$ 100	10^{38}	2	250	$e^+e^- \to e^+e^-$
e^+e^-, $\geq$ 200	10^{38}	3	380	
pp, 2000	10^{40}	2	890	$qq \to qq$
$p\bar{p}$, 2000	10^{37}	1.5	700	
$p\bar{p}$, 2000	10^{37}	4	500	$q\bar{q} \to \mu^+\mu^-$

(ii) Rare processes

Rare processes provide stringent constraints on possible extensions of the standard model. They have been studied in connection with horizontal interactions [39,40], extended technicolour theories [41,42] as well as composite models [43-46]. A systematic study of possible manifestations of physics beyond the standard model can be carried out by examining all higher dimensional operators [44] which can be constructed from the fields of the standard model. In preon models such operators, which contain the contact interactions, will be generated through the exchange of heavy resonances. There exist more than 100 operators of dimension 6 [44,46]. Some interesting examples are:

$$O_1 = \frac{1}{2} g_2 \Gamma^{i}_{1j} \; \bar{\ell}_{Li} \sigma^{\mu\nu} \tau^I \phi e^j_R W^I_{\mu\nu} + \text{c.c.} ,$$

$$W^I_{\mu\nu} = \partial_\mu W^I_\nu - \partial_\nu W^I_\mu + g_2 \varepsilon_{IJK} W^I_\mu \, W^K_\nu , \tag{44a}$$

$$O_2 = \frac{1}{2} g_1 \Gamma^{i}_{2j} \; \bar{\ell}_{Li} \sigma^{\mu\nu} \phi e^j_R B_{\mu\nu} + \text{c.c.} ,$$

$$B_{\mu\nu} = \partial_\mu B_\nu - \partial_\nu B_\mu , \tag{44b}$$

$$O_3 = \Gamma^{i}_{3j} \phi^\dagger \phi \bar{\ell}_{Li} \phi e^j_R + \text{c.c.} , \tag{44c}$$

$$O_4 = \Gamma^{ij}_{4kl} \, \bar{q}_{Li} \gamma^\mu q^k_L \bar{\ell}_{Lj} \gamma_\mu \ell^\ell_L , \tag{44d}$$

$$O_5 = \Gamma^{ij}_{5kl} \bar{q}_{Li} \gamma^\mu q^k_L \bar{q}_{Lj} \gamma_\mu q^\ell_L ; \tag{44e}$$

the matrices $\Gamma^{i}_{1j}, \ldots, \Gamma^{ij}_{5kl}$ contain a priori arbitrary flavour mixing.

The flavour diagonal part of the operators O_1 and O_2 contributes, after spontaneous symmetry breaking, to the anomalous magnetic moments of the leptons. Following the convention of the contact interactions in defining a substructure scale, one obtains (cf. Fig.8)

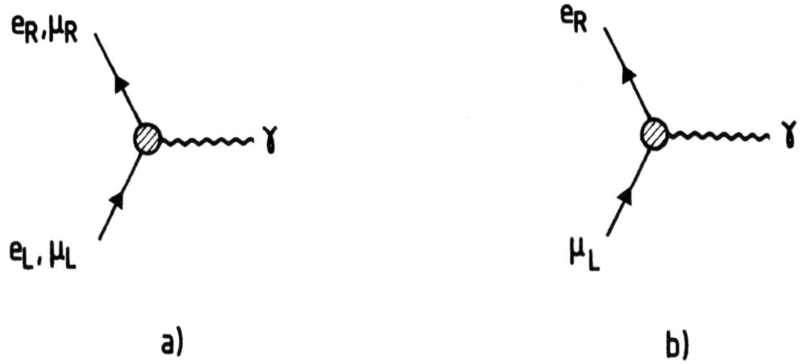

Fig. 8.: Magnetic dipole couplings of leptons causing anomalous magnetic moments (a) and the $\Delta G = 1$ transition $\mu \to e\gamma$ (b).

$$O_a = e \frac{4\pi}{\Lambda^2} v \, \bar{\ell}_{Li} \bar{\sigma}^{\mu\nu} e^i_R F_{\mu\nu} + \text{c.c.} , \tag{45}$$

where $F_{\mu\nu}$ is the electromagnetic field strength tensor and v the vacuum expectation value of the Higgs field. Recalling the Gordon decomposition

$$\bar{\psi}\gamma_\mu\psi \, A^\mu = \frac{1}{2m} \bar{\psi} \overset{\leftrightarrow}{iD}_\mu \psi A^\mu + \frac{1}{4m} \bar{\psi}\bar{\sigma}^{\mu\nu}\psi F_{\mu\nu} + \partial_\mu \, (\ldots) , \tag{46}$$

$$D_\mu = \partial_\mu + ie \, A_\mu ,$$

one immediately obtains for the anomalous magnetic moment corresponding to (45):

$$\delta a_\ell = 16 \, \frac{m_\ell v}{\Lambda^2} . \tag{47}$$

Usually anomalous magnetic moments due to compositeness [47,48] are estimated as

$$\delta a'_\ell = \left(\frac{m_\ell}{\Lambda}\right)^2 , \tag{48}$$

which leads to weaker bounds on Λ than (47). In order to obtain (48) from the operators O_1 and O_2 one has to assume that the matrices $\Gamma_{1,2}$ are proportional to the Yukawa coupling matrix $g^{(e)}$. For arbitrary matrices Γ a stringent bound follows already from O_3 which, after spontaneous symmetry breaking, leads to the mass term

$$O_m = \frac{4\pi}{\Lambda^2} v^3 (\bar{\ell}_{Li} e_R{}^i + \bar{e}_{Ri} \ell_L{}^i) . \tag{49}$$

Requiring the corresponding lepton mass to be smaller than the electron mass one finds

$$\Lambda > 360 \text{ TeV} . \tag{50}$$

Agreement between QED calculations and experiment allows an anomalous magnetic moment $|\delta a| < 2 \times 10^{-10}$ (3×10^{-8}) [49] for the electron (muon). Using equ.(47) ((48)) this implies the bounds on Λ

$$\Lambda > \begin{cases} 150 \text{ TeV} \quad (40 \text{ GeV}) , \text{ electron} \\ 170 \text{ TeV} \quad (580 \text{ GeV}) , \text{ muon} . \end{cases} \tag{51}$$

Bounds much stronger than (50) or (51) can be obtained from flavour-changing processes. It is convenient to classify them by means of a generation number [40]. Transitions between the $e(\mu)$- and $\mu(\tau)$-families are $\Delta G = 1$ processes, transitions between the e- and τ-families $\Delta G=2$ processes.

The flavour off-diagonal part of the operators O_1 and O_2 contains a term causing the $\Delta G=1$ process $\mu\to e\gamma$ (cf.Fig.8):

$$O_{\mu e\gamma} = e\,\frac{4\pi}{\Lambda^2}\, v\, \bar{\mu}_L \bar{\sigma}^{\mu\nu} e_R F_{\mu\nu} + \text{c.c.} \quad . \tag{52}$$

From the experimental upper bound for the branching ratio,

$BR(\mu\to e\gamma) < 1.7 \times 10^{-10}$ [50], one obtains

$$\Lambda > 3 \times 10^4 \text{ TeV } (200 \text{ TeV}) \; , \tag{53}$$

where the number in bracket has been obtained by replacing in (52) $4\pi v$ by m_μ.

A paticularly dangerous process for preon models of Pati-Salam type is the decay $\bar{K}^0 \to \mu^- e^+$ (cf. Fig.9).

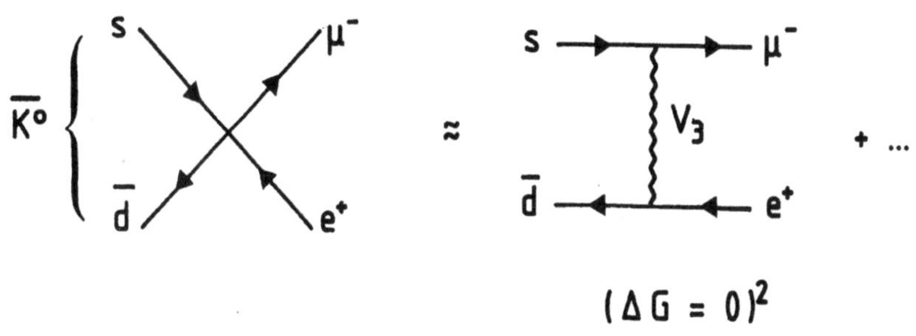

Fig. 9.: The $\Delta G=0$ contact interaction for $\bar{K}^0\to\mu^- e^+$ induced by the exchange of a leptoquark vector boson.

The corresponding operator O_4 can be generated through the exchange of a leptoquark V_3 with two $\Delta G=0$ vertices. From

$$O_{K^0\mu e} = \frac{4\pi}{\Lambda^2} \bar{d}_L \gamma^\mu s_L \bar{\mu}_L \gamma_\mu e_L + \text{c.c.} \quad (54)$$

and the bound on the ratios of partial widths (cf.[40],[50])

$$\frac{\Gamma(K_L^0 \to \mu^\pm e^\mp)}{\Gamma(K^+ \to \mu^+ \nu_\mu)} < 2.3 \times 10^{-6} \quad (55)$$

one obtains

$$\Lambda > 110 \text{ TeV} . \quad (56)$$

The strongest bound follows from the $\Delta G = 2$ process of K^0-$\bar{K}^0$ mixing. Comparing the contribution of the operator O_5 (cf. Fig. 10),

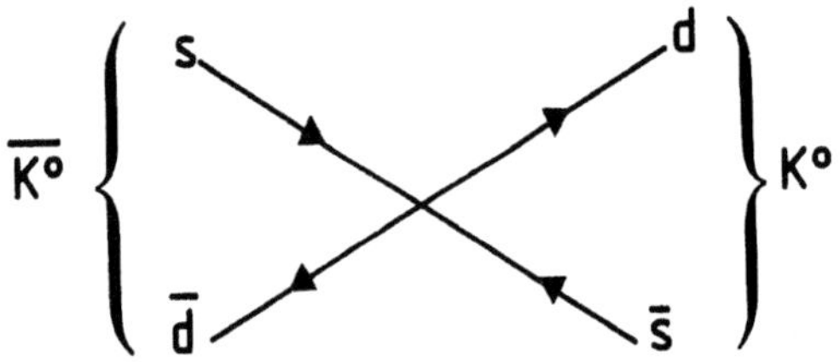

Fig. 10.: The contact interaction for K^0-$\bar{K}^0$ mixing.

$$O_{K^0\bar{K}^0} = \frac{2\pi}{\Lambda^2} \bar{s}_L \gamma^\mu d_L \bar{s}_L \gamma_\mu d_L + \text{c.c.} , \quad (57)$$

with the standard model box graph contribution of Gaillard and Lee one finds (cf.[40])

$$\Lambda > 4300 \text{ TeV} . \quad (58)$$

Clearly the bounds (53), (56) and (58) are incompatible with "nearby compositeness", i.e., $\Lambda < 1$ TeV. It is conceivable however, that the flavour changing processes are forbidden by appropriate symmetries. Electron or muon number conservation, for instance, which are symmetries of the standard model without right-handed neutrinos, would forbid the decay $K_L^o \to \mu^\pm e^\mp$. An alternative dynamical mechanism to prevent this decay has been suggested by Greenberg, Mohapatra and Nussinov [51]. As can be seen from Fig.11, the process $\bar{K}^o \to \mu^- e^+$ is not allowed, if a generation quantum is associated with the scalar preons, i.e., instead of 4 preons $\chi = (h^\alpha, \ell)$ one introduces $4n_f$

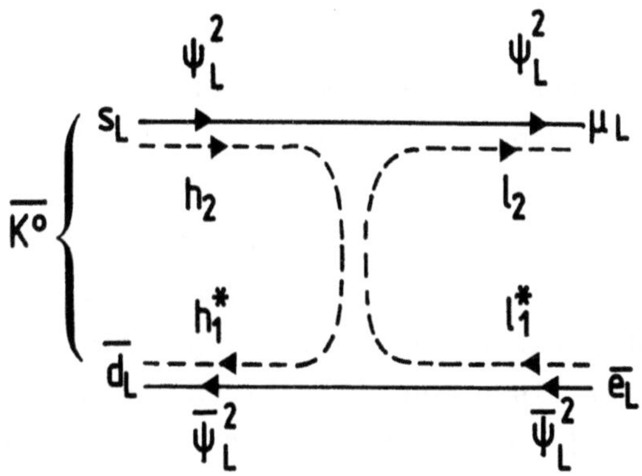

Fig. 11.: Preon diagram for the decay $\bar{K}^o \to \mu^- e^+$. For unbroken generation symmetry the scalar lines cannot be connected and the process is forbidden (cf. ref.51).

preons $\chi_i = (\chi_i^\alpha, \ell_i)$, $i = 1 \ldots n_f$. This, however, shifts the problem of family replication to the preon level which is contrary to the motivation for quark-lepton substructure.

A more detailed discussion of dynamical mechanisms

which can guarantee "natural flavour conservation" will be given in ref.46. The only unavoidable low-energy effects of compositeness appear to be anomalous magnetic moments of quarks and leptons, which, according to (51), are marginally consistent with a substructure scale below 1 TeV.

4. COMPOSITE W-BOSONS

In preon models with "nearby compositeness", i.e., $\Lambda \sim G_F^{-1/2} \sim 300$ GeV, the weak intermediate vectors bosons appear generally as bound states and not as fundamental gauge bosons. The idea of composite W-bosons is compatible with all present experimental results and represents a consistent alternative to the gauge theory of weak interactions. The main ingredients in models with composite W-bosons are the dynamical mixing of the photon with the neutral W-boson, first considered by Bjorken [52] and Hung and Sakurai [53] in connection with the neutral current structure of weak interactions, and the principle of vector meson dominance, as discussed by Fritzsch, Kögerler and Schildknecht [54,55]. This picture of composite W-bosons appears to be dynamically realized in theories with scalar condensates [56-59]. The effective Lagrangian of composite W-bosons is that of a massive Yang-Mills theory [60].

The effective Lagrangian of a preon theory with composite W-boson contains kinetic terms for the SU(2) triplet of W-bosons and the photon, the mass term for the W-bosons and the couplings of the vector bosons to the conserved SU(2) and electromagnetic currents:

$$L = -\frac{1}{4} W^I_{\mu\nu} W^{I\mu\nu} - \frac{1}{4} F_{\mu\nu} F^{\mu\nu} + \frac{1}{2} M_W^2 W^I_\mu W^{I\mu} -$$

$$- gW^{I\mu} j^{I}_{\mu} - e\, A^{\mu} j^{EM}_{\mu} \,, \qquad (59)$$

with

$$W^{I}_{\mu\nu} = \partial_{\mu} W^{I}_{\nu} - \partial_{\nu} W^{I}_{\mu} + g\, \varepsilon_{IJK} W^{J}_{\mu} W^{K}_{\nu} \,,$$

$$F_{\mu\nu} = \partial_{\mu} A_{\nu} - \partial_{\nu} A_{\mu} \,, \qquad \partial^{\mu} j^{I}_{\mu} = \partial^{\mu} j^{EM}_{\mu} = 0 \,.$$

The dynamical mixing of the elementary photon with the composite neutral W^3-boson is described by the $U(1)_{EM}$ gauge invariant "current mixing" [61]

$$L_{\gamma W} = - \frac{1}{2} \lambda\, F_{\mu\nu} W^{3\mu\nu} = \lambda A_{\nu} \Box\, W^{3\nu} + \partial_{\mu}\, (\ldots) \,. \qquad (61)$$

From equs. (59) and (60) one reads off the propagators and vertices shown in Fig.12.

Using the Feynman rules of Fig.12 one can easily construct the effective current - current Lagrangian as a power series in λ:

$$i\, L_{eff} = \frac{1}{2}\, j^{EM} \overset{\gamma}{\times\!\sim\!\sim\!\sim\!\times} j^{EM} + \frac{1}{2}\, j^{EM} \underset{(-\lambda q^2)^2}{\overset{\gamma \quad W^3 \quad \gamma}{\times\!\sim\!\sim\!\times\!\sim\!\sim\!\times\!\sim\!\sim\!\times}} j^{EM} + \ldots$$

$$+\, j^{EM} \underset{-\lambda q^2}{\overset{\gamma \quad W}{\times\!\sim\!\sim\!\times\!\sim\!\sim\!\times}} j^{3} + j^{EM} \underset{(-\lambda q^2)^3}{\overset{\gamma \quad W^3 \quad \gamma \quad W^3}{\times\!\sim\!\sim\!\times\!\sim\!\sim\!\times\!\sim\!\sim\!\times\!\sim\!\sim\!\times}} j^{3} + \ldots$$

$$+ \frac{1}{2}\, j^{I} \overset{W^I}{\times\!\sim\!\sim\!\sim\!\times} j^{I} + \frac{1}{2}\, j^{3} \underset{(-\lambda q^2)^2}{\overset{W^3 \quad \gamma \quad W^3}{\times\!\sim\!\sim\!\times\!\sim\!\sim\!\times\!\sim\!\sim\!\times}} j^{3} + \ldots . \qquad (61)$$

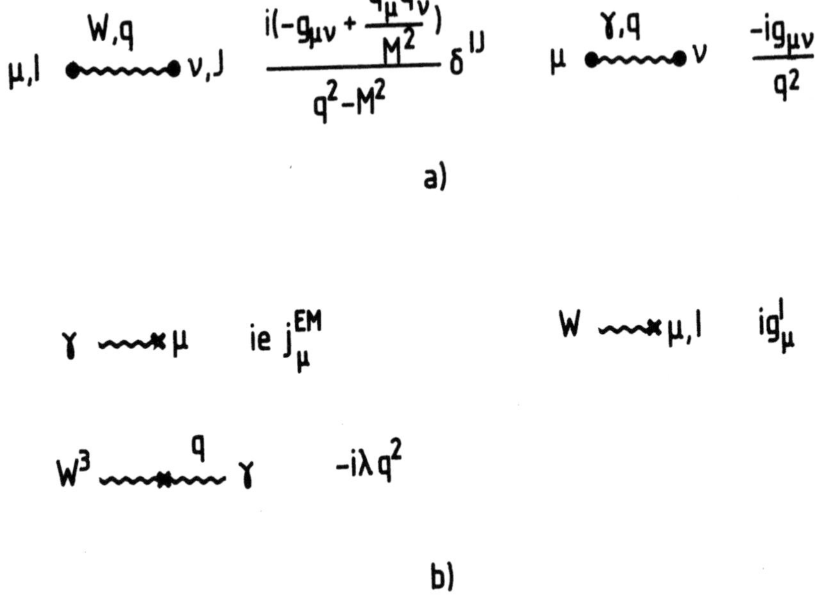

Fig. 12.: Propagators (a) and vertices (b) from the Lagrangian (59).

For $|q^2| \ll M_W^2$ only four terms remain:

$$L_{eff} \underset{|q^2| \ll M_W^2}{\approx} \frac{1}{2}\left(\frac{e^2}{q^2} - \frac{e^2\lambda^2}{M_W^2}\right) j_\mu^{EM} j^{EM\mu} - \frac{1}{2}\frac{g^2}{M_W^2} j_\mu^I j^{I\mu}$$

$$+ \frac{e\lambda g}{M_W^2} j_\mu^{EM} j^{3\mu}$$

$$= \frac{1}{2}\frac{e^2}{q^2} j_\mu^{EM} j^{EM\mu} - \frac{1}{2}\frac{g^2}{M_W^2} j_\mu^+ j^{-\mu}$$

$$- \frac{1}{2}\frac{g^2}{M_W^2}\left(j_\mu^{EM} - \frac{e\lambda}{g} j_\mu^3\right)^2 \quad , \tag{62}$$

where

$$j^{\pm}_{\mu} = \sum_i \bar{f}_{Li} \gamma_{\mu} \tau^{\pm} f_{Li} \ .$$

The result (62) is precisely the standard model weak interaction Lagrangian which, as we have seen, follows simply from global SU(2) invariance and current mixing. The Weinberg angle is a function of e,g and λ:

$$\sin^2\theta_W = \frac{e\lambda}{g} \ . \tag{63}$$

The γ-W^3 mixing also causes a mass shift of the neutral W-boson. The kinetic terms of (59) and (60) correspond to the propagator matrix

$$\Delta^{-1}(q^2) = \begin{pmatrix} q^2 & \lambda q^2 \\ \lambda q^2 & q^2 - M_W^2 \end{pmatrix} , \tag{64}$$

whose determinant,

$$\det \Delta^{-1}(q^2) = q^2[q^2(1-\lambda^2) - M_W^2] \ , \tag{65}$$

has the eigenvalues $q^2 = 0$ and $q^2 = M_W^2/(1-\lambda^2)$ which correspond to the massless photon and the massive Z-boson with

$$M_Z = \frac{M_W}{\sqrt{1-\lambda^2}} \ . \tag{66}$$

From equ.(63), (66) and $\frac{g^2}{M_W^2} = 4\sqrt{2}\, G_F$ M_Z can be determined as function of M_W (cf. Fig.13). Weinberg's mass prediction is obtained, if one supplements (63) by the so called

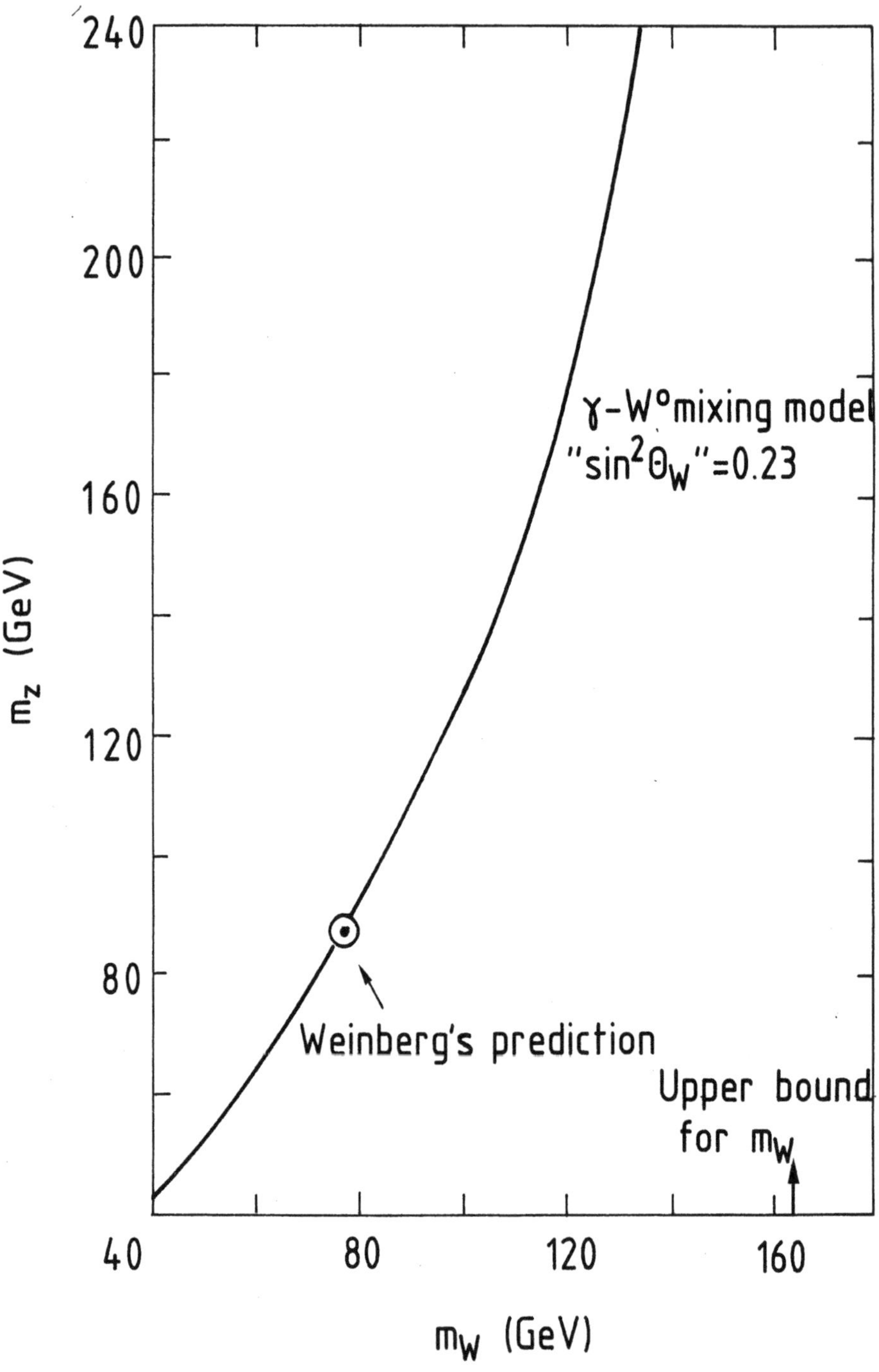

Fig. 13.: M_Z as function of M_W in the W^3-γ mixing scheme (from ref.62).

"unification condition" [53]

$$e = g\lambda \ , \tag{67}$$

which implies

$$\lambda = \sin\theta_W , \qquad M_Z = \frac{M_W}{\cos\theta_W} \quad . \tag{68}$$

One easily verifies that, if the "unification condition" holds, the effective current-current Lagrangian becomes SU(2) invariant at large momentum transfers. Contrary to (61) one has to sum up a geometric series in λ for $q^2 >> M_W^2$, yielding the result ($e = g\lambda$)

$$L_{eff} \underset{|q^2|>>M_W^2}{\approx} \frac{1}{1-\lambda^2} \frac{1}{q^2}\{\frac{1}{2} e^2 j_\mu^{EM} j^{EM\mu} - \lambda eg \, j_\mu^{EM} j^{3\mu}$$

$$+ \frac{1}{2} g^2 (1-\lambda^2) j_\mu^I j^{I\mu} + \frac{1}{2} g^2 \lambda^2 j_\mu^3 j^{3\mu}\}$$

$$= \frac{1}{2q^2} \{g'^2 j_\mu^Y j^{Y\mu} + g^2 j_\mu^I j^{I\mu}\} \ , \tag{69}$$

where j_μ^Y and g' are the standard model hypercharge current and coupling constant,

$$j_\mu^Y = j_\mu^{EM} - j_\mu^3 \ , \qquad g' = \frac{e}{\sqrt{1-\lambda^2}} = \frac{e}{\cos\theta} \quad . \tag{70}$$

The Lagrangian (69) corresponds to the exchange of massless B- and W-bosons as in the standard model for large-momentum transfers.

The "unification condition" (67) can also be understood as a consequence of vector meson dominance [54,55]. The experimental value of $\sin^2\theta_W \approx 0.23$ requires a large γ-W^3 mixing parameter λ, whereas in analogy to the ρ-γ mixing in hadron physics one would expect $\lambda = O(\alpha_{EM})$. The required large dynamical γ-W^3 mixing can be understood, if the excitation energy of the composite W-bosons is of the order of 1 TeV, i.e., much larger than the mass of the lowest vector boson [63,64] (cf. Fig.14). In this case, however,

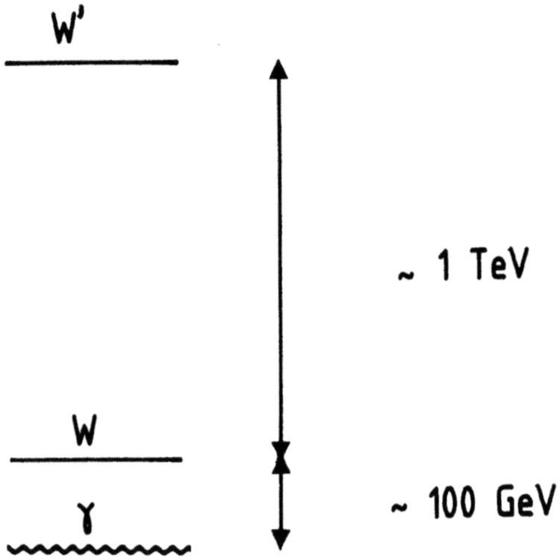

Fig. 14.: Pattern of vector boson masses required for a large dynamical W^3-γ mixing.

one expects W-dominance [54,55], i.e., the formfactor of the isovector part of the electromagnetic current should be dominated by the isovector boson of lowest mass (cf. Fig.15). This implies precisely the "unification condition" (67).

The extension of vector meson dominance beyond the isovector part of the electromagnetic current requires at least one additional SU(2) singlet vector boson V, which corresponds to the B-boson of the standard model.

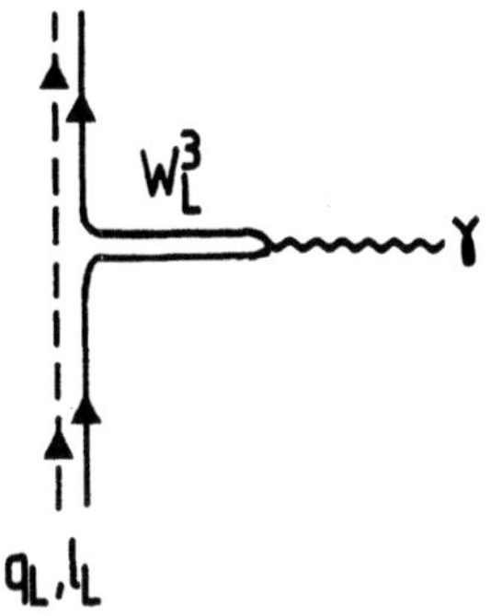

Fig. 15.: W-dominance of the isovector part of the electromagnetic form factor.

In Pati-Salam type preon models the required isosinglet V is part of a $SU(4)_{PS}$ 15-plet which contains in addition leptoquarks V_3 and V_3* and a colour octet vector boson V_8 which should dominate the colour formfactor of quarks [60,65] (cf. Fig. 16). A phenomenological analysis of W^3-γ-V mixing [66] yields for the isosinglet mass the lower bound

$$M_V > 250 \text{ GeV} . \qquad (71)$$

If one requires also V_8-dominance of the quark colour form factor one obtains [65]

$$M_V > 600 \text{ GeV}, \qquad M_{V_8} > 1 \text{ TeV} . \qquad (72)$$

There exists a vast literature on the phenomenology of "nearby compositeness", which has partly been reviewed in [67] and[68]. An interesting example are light pseudoscalar leptoquarks, which have been discussed by Schrempp and Schrempp [69] as a possible explanation of three

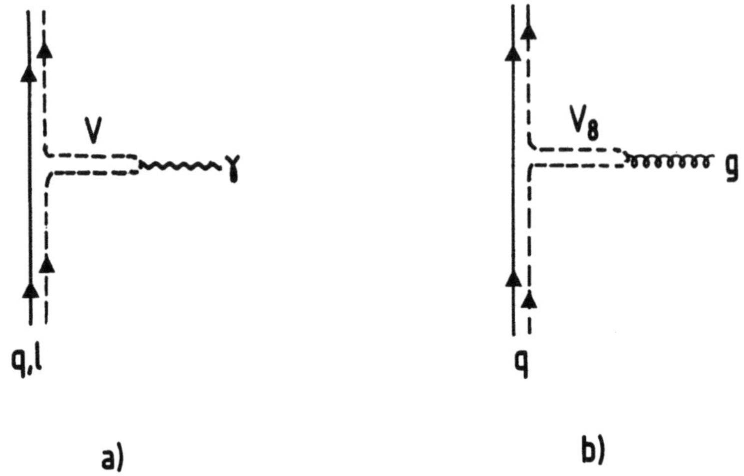

Fig. 16.: V-dominance of the isoscalar part of the electromagnetic form factor (a), and V_8-dominance of the colour form factor (b).

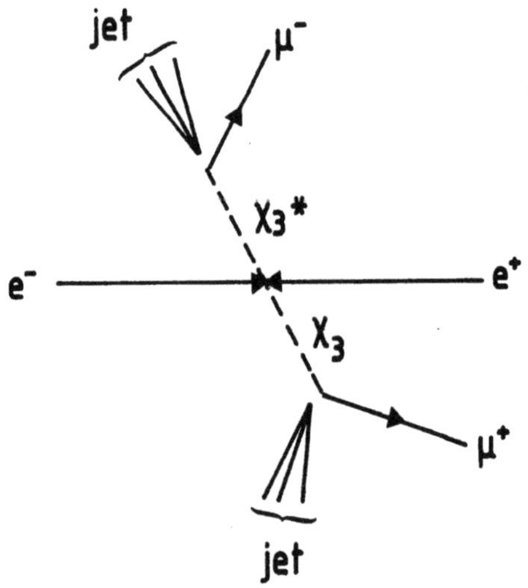

Fig. 17.: Leptoquark production and decay in e^+e^- collision.

striking events, containing two muons of opposite charge and two jets, which were reported by the CELLO collaboration [70] and the UA1 collaboration [71].

Light pseudoscalar leptoquarks (cf.(34))

$$\chi_3 = (h\ell^*) \sim \underline{3}\ \tfrac{2}{3}$$

appear naturally in preon models of Pati-Salam type as a consequence of the spontaneous symmetry breaking

$$SU(4)_{PS} \to SU(3)_C \times U(1)_{B-L} \quad . \tag{73}$$

They are pseudo Goldstone bosons with a mass [69]

$$m_{\chi_3} = O(\sqrt{\frac{\alpha_s}{\pi}}\Lambda) = O(40 \text{ GeV}) \quad . \tag{74}$$

If there are separate leptoquarks for each generation, the leptoquark of the μ-family will have the decay modes

$$\chi_3 \to s\mu^+ \quad , \tag{75a}$$

$$\chi_3 \to c\nu_\mu \quad . \tag{75b}$$

The first decay mode could explain the CELLO event (cf.Fig.17) which would imply $m_{\chi_3} \gtrsim 20$ GeV. As a consequence of this explanation one expects $\mu^+\mu^-$ jet events in $p\bar{p}$ collisions with an invariant mass of two μ-jet combinations of 20 GeV (cf. Fig.18). This signature as well as the expected production rates are indeed compatible [69] with two events reported by the UA1 collaboration [71] (cf. Fig. 19). Obviously, no firm conclusions can be drawn from three events, but their qualitative features are very intriguing.

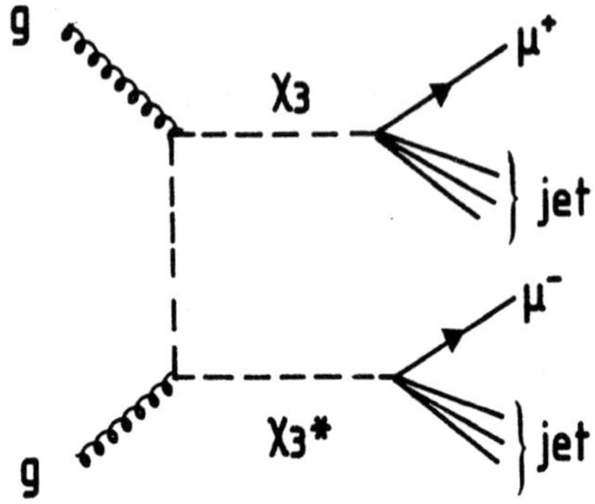

Fig. 18.: Leptoquark production and decay in $p\bar{p}$ collisions.

5. MASSLESS COMPOSITE SCALARS, FERMIONS AND VECTOR BOSONS

The main problem of quark-lepton substructure is the dynamics of bound states whose size is very small compared to their Compton wave length, i.e.,

$$\frac{1}{m_f} \frac{1}{\langle r_f \rangle} >> 1 . \tag{76}$$

Bound states familiar from atomic and nuclear physics, such as the deuteron or positronium, are of opposite type. They are non relativistic, and from the uncertainty relation

$$\frac{1}{\langle r \rangle} \sim \langle p \rangle = m \langle \frac{v}{c} \rangle \tag{77}$$

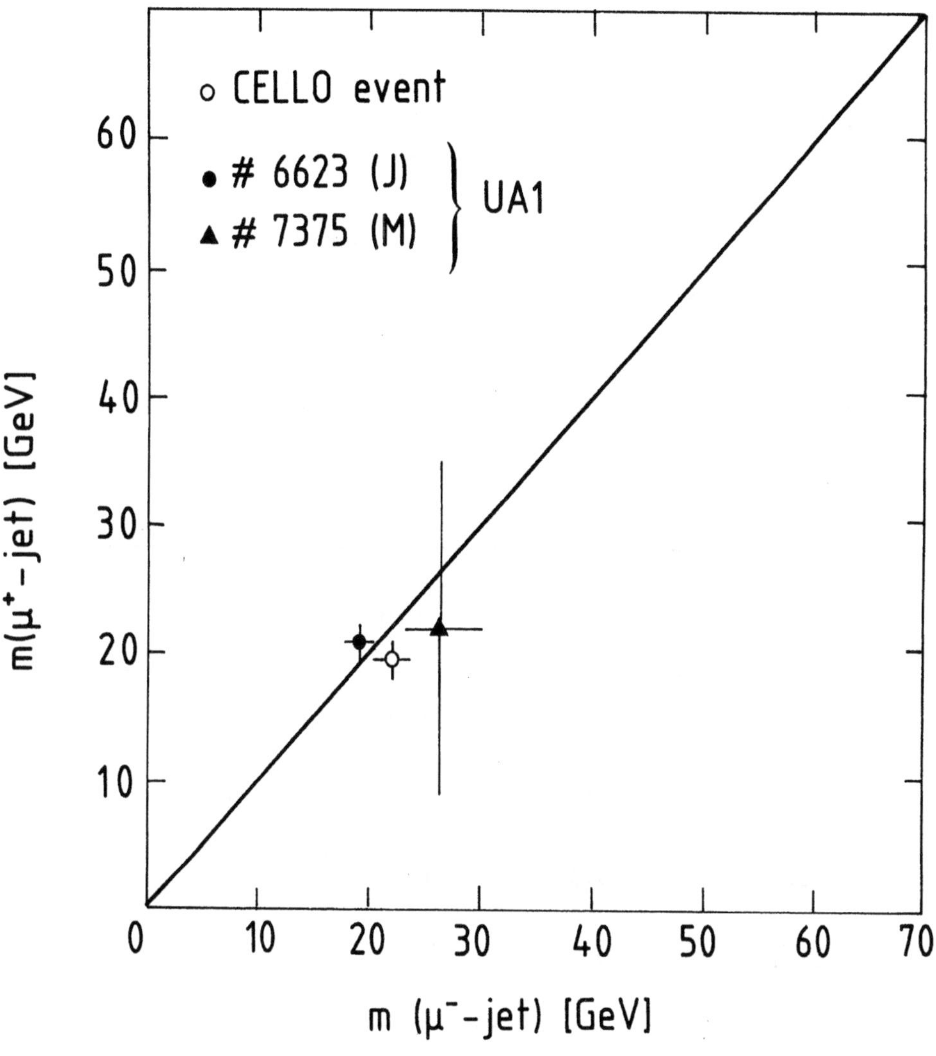

Fig. 19.: m(μ^+-jet) versus m(μ^--jet) for the CELLO [70] and UA1[71] events (from ref.69).

one obtains

$$\frac{1}{m}\frac{1}{<r>} = <\frac{v}{c}> < 1 \quad . \tag{78}$$

Even the relativistic QCD bound states satisfy (78) (cf. table III), except for the pion which, as a pseudo-Goldstone boson, plays a special role. It is clear from table III that, if quarks and leptons are composite, their structure is likely to be very different from the bound states we know.

Table III: Masses and sizes of some "elementary" particles (cf. [50], [72])

particle	mass m [MeV]	size <r> [fm]	$\frac{1}{m}\frac{1}{<r>}$
proton	938	0.86	0.25
pion	135	0.75	2.0
electron	0.5	$< 4 \times 10^{-4}$	$> 10^{6}$
neutrino	0	>0	∞

A crucial ingredient in the description of (almost) massless fermions is chiral symmetry [73]. Let us consider the Lagrangian of a set of Dirac fermions $\psi^{i} = (\psi^{\alpha i})$, $i = 1 \ldots N$, which interact with massless nonabelian vector bosons V^{A}_{μ} :

$$L = i\bar{\psi}_i \not{D}\psi^i - m\bar{\psi}_i\psi^i ,$$

$$\not{D}^\alpha_\beta = \gamma^\mu(\delta^\alpha_\beta\partial_\mu + ig\,\frac{1}{2}(\lambda^A)^\alpha_\beta V^A_\mu) . \tag{79}$$

In the case of the gauge group SU(3) the Lagrangian (79) corresponds to QCD with N flavours. In terms of the chiral projections

$$\psi_{\substack{L\\R}} = \frac{1}{2}(1\mp\gamma_5)\psi , \qquad \gamma_5\psi_{\substack{L\\R}} = \mp\psi_{\substack{L\\R}} , \tag{80}$$

the Lagrangian (79) reads

$$L = i\bar{\psi}_{Li}\not{D}\psi^i_L + i\bar{\psi}_{Ri}\not{D}\psi^i_R - m(\bar{\psi}_{Li}\psi^i_R + \bar{\psi}_{Ri}\psi^i_L) . \tag{81}$$

In the chiral limit $m \to o$ the classical theory has the conserved currents

$$J^A_{L(R)\mu} = \bar{\psi}_{L(R)i}\gamma_\mu(T^A)^i_j\psi^j_{L(R)} , \qquad \partial^\mu J^A_{L(R)\mu} = 0 , \tag{82}$$

where T^A are the generators of U(N); the corresponding charges

$$Q^A_{L(R)} = \int d^3x\; J^A_{L(R)o}(x) \tag{83}$$

generate the chiral symmetry $U(N)_L \times U(N)_R$. In the quantum theory only the currents of the $SU(N)_L \times SU(N)_R \times U(1)_{L+R}$ subgroup are conserved, whereas the $U(1)_{L-R}$ symmetry is broken due to instanton effects.

Massless composite particles are very special objects and, as we will discuss in the following, some of their properties can be determined from general theoretical principles without solving the complete bound state problem.

(i) Massless composite scalars

The only known composite particles, whose size is small compared to their Compton wave length, are the pions. They are very light because they are pseudo-Goldstone bosons associated with the spontaneous breaking of chiral symmetry in QCD. It is an important consequence of their Goldstone nature that their self interactions at distances

$$r > \frac{1}{f_\pi} \sim \frac{1}{\Lambda_{QCD}}$$

are entirely determined by the symmetry breaking pattern.

The QCD Lagrangian with two flavours $u_{L,R}$ and $d_{L,R}$ has the global symmetry

$$G = SU(2)_L \times SU(2)_R \times U(1)_V , \tag{84}$$

which is generated by the charges $Q^A_{L,R}$ $(A = 1,\ldots,3)$ and Q_V. One easily verifies that the vacuum expectation values

$$\langle 0|\bar{u}u|0\rangle = \langle 0|\bar{d}d|0\rangle \sim \Lambda^3_{QCD} \tag{85}$$

break G to the subgroup

$$H = SU(2)_V \times U(1)_V , \tag{86}$$

generated by the charges $T^A = Q^A_L + Q^A_R$ and Q_V. The vacuum is no longer invariant under the axial $SU(2)_A$ transformations generated by $X^A = Q^A_R - Q^A_L$. The corresponding Goldstone bosons are the $SU(2)_V$ triplet of pions which are created from the vacuum by the axial current:

$$\langle 0|\bar{q}(0)\gamma_\mu\gamma_5 \tfrac{1}{2}\tau^A q(0)|\pi^B(p)\rangle = ip_\mu \delta^{AB} f_\pi \ . \tag{87}$$

An effective Lagrangian for the pion self interactions at momentum transfers smaller than f_π can be obtained by constructing first a nonlinear realization of the full chiral group G on the fields $\pi^A(x)$ which is linear for the subgroup H [74]

$$\frac{1}{i}[T^A,\pi^B] = \varepsilon_{ABC}\ \pi^C \ ,$$

$$\frac{1}{i}[X^A,\pi^B] = -f_\pi[\tfrac{1}{2}(1-\frac{1}{f_\pi^2}\vec{\pi}^2)\delta^{AB} + \frac{1}{f_\pi^2}\pi^A\pi^B] \ . \tag{88}$$

The requirement of invariance under (88) leads to a non-polynomial Lagrangian which is unique up to redefinition of the pion field :

$$L_{eff} = \frac{1}{2(1+\frac{1}{f_\pi^2}\vec{\pi}^2)}\ \partial^\mu\pi^A\partial_\mu\pi^A \ . \tag{89}$$

An expansion in powers of $1/f_\pi^2$ yields interaction terms, where always two pions couple with derivatives, which is a consequence of the low-energy theorems for Goldstone particles.

(ii) Massless composite fermions

Spontaneously broken chiral symmetry leads to massless composite scalar particles as Goldstone bosons. Unbroken chiral symmetry requires massless composite fermions whose spectrum is strongly restricted through 't Hooft's consistency conditions [6].

Let us consider the vector and axial vector currents for a gauge theory of the type defined in (79),

$$J^A_\mu(x) = \bar{\psi}_i(x)\gamma_\mu (T^A)^i_j\psi^j(x) \ ,$$

$$J^A_{5\mu}(x) = \bar{\psi}_i(x)\gamma_\mu\gamma_5(T^A)^i_j\psi^j(x) \ , \tag{90}$$

and compute the 3-point functions of one axial and two vector currents [75]

$$\Gamma^{ABC}_{\mu\nu\lambda}(k_1,k_2) = \int d^4x_1 d^4x_2 e^{i(k_1x_1+k_2x_2)} \langle 0| T(J^A_{5\mu}(0)J^B_\nu(x_1)J^C_\lambda(x_2))|0\rangle$$

$$= d^{ABC}(R)\Gamma_{\mu\nu\lambda}(k_1,k_2) \ ,$$

$$d^{ABC}(R) = \mathrm{tr}[T^A\{T^B,T^C\}] \ . \tag{91}$$

The tensor $d^{ABC}(R)$ depends on the representation R of the fermions ψ^i and is related to the corresponding tensor of the fundamental representation R_o through the anomaly coefficient:

$$d^{ABC}(R) = A(R)\ d^{ABC}(R_o) \ . \tag{92}$$

Note that the tensor d^{ABC} changes its sign, if the representation R is changed to its complex conjugate R^*:

$$d^{ABC}(R^*) = (-)^3\mathrm{tr}[T^{A*}\{T^{B*},T^{C*}\}]$$

$$= -\ \mathrm{tr}[T^{AT}\{T^{BT},T^{CT}\}]$$

$$= -\mathrm{tr}[\{T^C,T^B\}T^A] = -d^{ABC}(R) \ . \tag{93}$$

The 3-point function $\Gamma^{ABC}_{\mu\nu\lambda}$ can be computed perturbatively by evaluating the graphs shown in Fig.20.

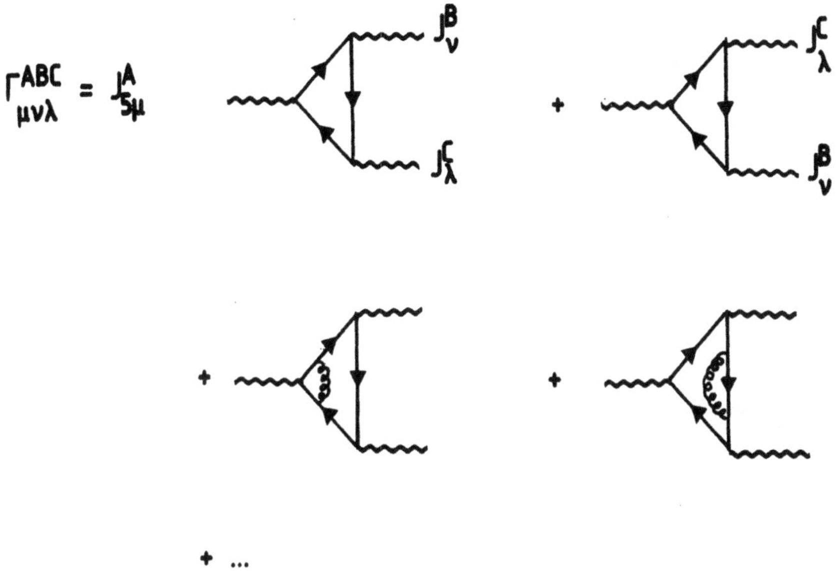

Fig. 20.: Contributions to the three-point function of one axial and two vector currents in perturbation theory (cf.(91)).

Classically the vector and axial vector currents (90) are conserved in the chiral limit,

$$\partial^{\mu} J^{A}_{\mu} = 0 \ , \qquad \partial^{\mu} J^{A}_{5\mu} = 0 \ , \tag{94}$$

and thus one expects naively ($q = k_1 + k_2$):

$$k_1^{\nu}\ \Gamma^{ABC}_{\mu\nu\lambda} = k_2^{\lambda}\ \Gamma^{ABC}_{\mu\nu\lambda} = 0 \ , \tag{95a}$$

$$q^{\mu}\ \Gamma^{ABC}_{\mu\nu\lambda} = 0 \ . \tag{95b}$$

It turns out, however, that there exists no regularization scheme, which is needed to give a meaning to the divergent quantity $\Gamma^{ABC}_{\mu\nu\lambda}$, which preserves (95a) and (95b). Choosing (95a) to be valid, i.e., the conservation of the vector current, the divergence of the axial current does not vanish but is given by the Adler-Bell-Jackiw (ABJ) anomaly:

$$q^{\mu}\,\Gamma^{ABC}_{\mu\nu\lambda}(k_1,k_2) \underset{m\to 0}{\to} \frac{1}{2\pi^2}\, d^{ABC}(R)\,\varepsilon_{\nu\lambda\rho\sigma}\,k_1^{\ \rho} k_2^{\sigma} \quad . \tag{96}$$

This result, which is obtained from a 1-loop calculation, is not modified through higher orders in perturbation theory, as Adler and Bardeen have shown. It is therefore natural to assume that (96) is an exact property of the theory.

Let us consider now a preon theory with some exact chiral flavour symmetry G and let $A(R_p)$ be an anomaly coefficient calculated from the fermionic preons. The composite fermions fall into two representations of G,

$$\text{a)}\ R_h\ , \qquad R_h = R_h^{*}\ ,$$

$$\text{b)}\ R_\ell\ , \qquad R_\ell \neq R_\ell^{*}\ . \tag{97}$$

Fermions in the <u>real</u> representation R_h are expected to have masses of the order of the hypercolour scale Λ, for fermions in the <u>chiral</u> representation R mass terms are forbidden.

If the ABJ anomaly is an exact property of the theory, it must be possible to calculate the anomaly coefficient in terms of the fundamental as well as the composite fields. If G is spontaneously broken the preon anomaly can be reproduced at the bound state level through the contribution of Fig.21, which involves the direct coupling of the Goldstone bosons to the broken currents. If G is unbroken, the preon anomaly must be reproduced by the composite fermions.

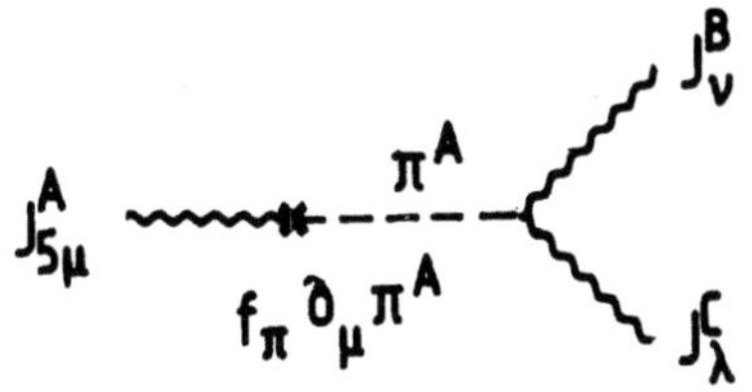

Fig. 21.: Direct coupling of Goldstone boson to current of spontaneously broken symmetry.

As the heavy fermions contained in R_h do not contribute (cf. (92), (93)) we obtain 't Hooft's consistency condition [6]

$$A(R_p) = A(R_\ell) \ , \tag{98}$$

which represents a very stringent constraint on preon theories.

It is very instructive to study the implications of (98) for the preon model of Pati-Salam type described in sec.2. Associated with the fermionic preons $\psi^p = \psi^p_L + \psi^p_R$ (i = 1,2) there are the following classically conserved currents (α=1,...,n for $SU(n)_{HC}$) :

$$J^{(F)}_\mu(x) = \bar{\psi}_\alpha(x)\gamma_\mu\psi^\alpha(x) \ , \tag{99a}$$

$$J^A_\mu(x) = \bar{\psi}_\alpha(x)\gamma_\mu\frac{1}{2}\tau^A\psi^\alpha(x) \ , \tag{99b}$$

$$J^A_{5\mu}(x) = \bar{\psi}_\alpha(x)\gamma_\mu\gamma_5 \ \frac{1}{2}\tau^A\psi^\alpha(x) \ ; \tag{99c}$$

the charge F = 3B+L corresponds to fermion number. In terms of composite quarks and leptons the currents (99) read

$(i = 1,\ldots,n_F)$:

$$J_\mu^{(F)}(x) = \bar{q}_i(x)\gamma_\mu q^i(x) + \bar{\ell}_i(x)\gamma_\mu \ell^i(x) \ , \tag{100a}$$

$$J_\mu^A(x) = \bar{q}_i(x)\gamma_\mu \frac{1}{2}\tau^A q^i(x) + \bar{\ell}_i(x)\gamma_\mu \frac{1}{2}\tau^A \ell^i(x) \ , \tag{100b}$$

$$J_{5\mu}^A(x) = \bar{q}_i(x)\gamma_\mu \gamma_5 \frac{1}{2}\tau^A q^i(x) + \bar{\ell}_i(x)\gamma_\mu \gamma_5 \frac{1}{2}\tau^A \ell^i(x) \ , \tag{100c}$$

with

$$q^i = q_L^i + q_R^i \ , \qquad \ell^i = \ell_L^i + \ell_R^i \ ,$$

$$q_R^i = \begin{pmatrix} u_R^i \\ d_R^i \end{pmatrix} \ , \qquad \ell_R^i = \begin{pmatrix} \nu_R^i \\ e_R^i \end{pmatrix} \ .$$

The anomaly coefficient for the 3-point function of $J_\mu^{(F)}$, J_ν^A and $J_{5\lambda}^A$ is easily obtained. As the preons ψ^p are in the fundamental representation of SU(2) with F-charge 1, $A(R_p)$ is simply the multiplicity of SU(2) doublets, i.e.,

$$A(R_p) = n \tag{101}$$

for the hypercolour group $SU(n)_{HC}$. In the same way one obtains for the composite fermions

$$A(R_\ell) = 4n_F \ , \tag{102}$$

because each generation contains 4 SU(2) doublets. From 't Hooft's consisting condition (98) we thus obtain

$$n = 4n_f \ , \tag{103}$$

i.e., the matching of anomalies requires the hypercolour group $SU(12)_{HC}$ [27], if the bound state spectrum shall contain three generations of (almost) massless quarks and leptons! This result is not particularly encouraging. It means that the number of preons in the model of sec.2 has to be larger than the number of quarks and leptons whose spectrum we want to explain. As a consequence of 't Hooft's consistency conditions it appears indeed unavoidable that a theory, whose bound-state spectrum contains massless fermions, is much more complicated than this sector of massless bound states. Thus we cannot hope for a simple preon theory. This, however, may not be too surprising since the theory for pions at distances larger than $1/f_\pi$ is also much simpler than QCD.

(iii) Massless composite vector bosons

One could expect that a preon theory with massless bound states of spin 0 or $\frac{1}{2}$ may also lead to massless composite vector bosons. It has been shown, however, by Case and Gasiorowicz [76], and Weinberg and Witten [77] that massless vector bosons with global quantum numbers cannot exist. The theorem [77] reads as follows: "A theory that allows the construction of a Lorentz-covariant conserved four-vector current J_μ cannot contain massless particles of spin > $\frac{1}{2}$ with nonvanishing values of the conserved charge $\int J_0 d^3x$."

The proof is based on the comparison of two matrix elements of the current J_μ, which correspond to "forward" and "backward scattering", in the limit of zero momentum. Due to Lorentz invariance one has

$$\langle p'|J_\mu|p\rangle \underset{p'\to p}{\to} g\,\frac{p_\mu}{|\vec{p}|}\,, \tag{104}$$

with $g \neq 0$ because the charge of the massless particle is assumed to be nonvanishing. On the other hand, helicity conservation implies

$$\langle p',\lambda|J_\mu|p,\lambda\rangle = 0 \tag{105}$$

for $p = (|\vec{p}|,\vec{p})$, $p' = (|\vec{p}|,-\vec{p})$ and $2\lambda>1$. Considering the limit $|\vec{p}|\to 0$ leads to a contradiction between (104) and (105) for $\lambda > \frac{1}{2}$ which proves the proposition.

Similarly one can prove that particles with spin $j > 1$ cannot exist in theories with a Lorentz-covariant energy-momentum tensor [77]. These theorems do not exclude Yang-Mills gauge theories or supergravity theories because these theories do not have Lorentz-covariant conserved currents or energy-momentum tensors, respectively.

We conclude this section by emphasizing again that some of the properties of composite massless particles can be determined from general theoretical principles without solving the difficult problem of relativistic bound states. Massless particles with nonvanishing quantum numbers with respect to global symmetries have to have spin 0 or $\frac{1}{2}$. Massless composite scalars can be obtained as Goldstone bosons; their self interactions at distances larger than the inverse symmetry breaking scale are completely determined by the pattern of symmetry breaking. Massless composite fermions have to satisfy 't Hooft's consisting conditions, which appear to lead unavoidably to rather complicated preon theories.

6. QUASI-GOLDSTONE FERMIONS

Interesting candidates for a theory of quark-lepton substructure are supersymmetric (SUSY) confining gauge theories. Supersymmetric preon models [78-82] have been suggested in order to avoid the "naturalness problem" [4-6] for scalar preons whose presence appears to be necessary in phenomenologically acceptable models. Furthermore supersymmetric theories provide new mechanisms to generate massless composite fermions: supersymmetry may protect chiral symmetry from being dynamically broken [83-85], a spontaneous breakdown of extended supersymmetry can lead to Goldstone fermions [86], or quasi-Goldstone fermions [87] can appear as SUSY partners of Goldstone bosons due to the spontaneous breaking of global symmetries in supersymmetric theories.

Let us consider, for example, a SUSY confining gauge theory, such as SQCD, with some global chiral symmetry G. The bound state spectrum of such a theory will unavoidably contain massless composite fermions because either

(i) G is unbroken, which implies massless "baryons" (cf. sec.5(ii)), or

(ii) G is spontaneously broken to H with unbroken SUSY, which yields quasi-Goldstone fermions as SUSY partners of "pions", or

(iii) SUSY is spontaneously broken with G broken or unbroken, which leads at least to one massless fermion, the Goldstino.

The building blocks of supersymmetric gauge theories are chiral superfields $\chi_a = (\tilde{\chi}_a, \eta_{La})$, which contain complex scalars and Weyl fermions as physical degrees of freedom, and vector superfields $V^A = (\lambda_L^A, V_\mu^A)$ containing Weyl fermions and massless vector fields [88]. In strongly interacting confining theories it is likely that a global symmetry G of the Lagrangian is spontaneously broken to a subgroup H through vacuum expectation values of the lowest components of composite chiral and vector superfields,

$$<0|\chi_a \chi_b|0> \sim \Lambda_{HC}^2 ,$$

$$<0|\bar{\chi}_a e^{2gV} \chi_b|0> \sim \Lambda_{HC}^2 , \qquad (106)$$

where Λ_{HC} is the hypercolour scale. Due to supersymmetry the massless composite Goldstone bosons will be part of chiral superfields $\phi_i = (\tilde{\phi}_i, \psi_{Li})$.

A natural question to ask is whether the Goldstone mechanism in a supersymmetric theory could be the dynamical origin of composite quarks and leptons which are (almost) massless with respect to their substructure scale Λ. In such a picture masses of scalar quarks and leptons much smaller than Λ could arise from the breaking of supersymmetry and some explicit breaking of the original global symmetry G, for instance through gauge interactions, which would turn the massless Goldstone bosons into massive pseudo-Goldstone bosons (cf. Fig.22). The masslessness of the quasi-Goldstone fermions could be guaranteed by some chiral symmetry contained in the unbroken subgroup $H \subset G$ [89]. In order to realize this idea, one has to solve a group theoretical problem: the unbroken subgroup H has to be identified with a physical group which contains at least the group of unbroken gauge interactions,

$$H \supset SU(3)_C \times U(1)_{EM} \times \ldots . ; \qquad (107)$$

furthermore, H must be embedded into a larger group G such that the broken generators in G/H have the quantum numbers of quarks and leptons with respect to H.

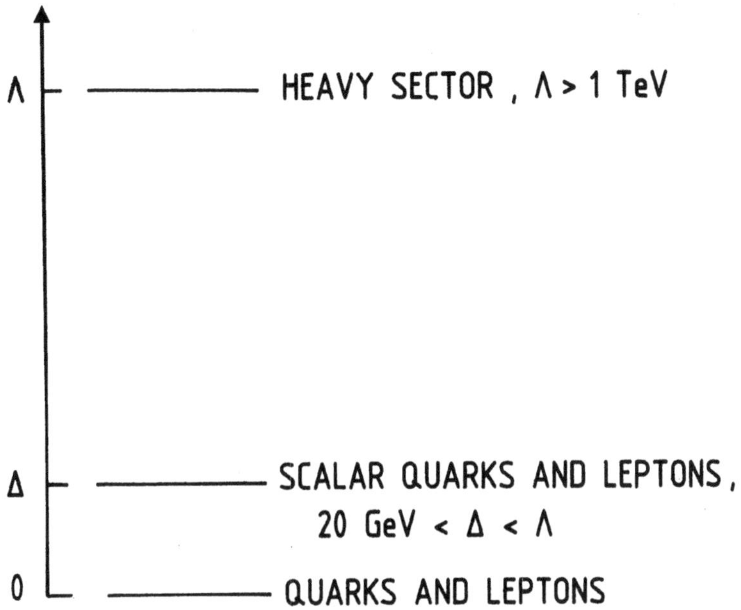

Fig. 22.: Qualitative picture of mass spectrum in supersymmetric preon models.

A simple example is the coset space U(6)/SU(2) x U(4) [82,89-91], where the SU(2) and SU(4) subgroups are identified as weak isospin and Pati-Salam symmetry (cf. sec.2). From the adjoint representation of U(6),

$$U(6) : \quad \left(\begin{array}{c|c} U(4) & (4,2^*) \\ \hline (4^*,2) & U(2) \end{array}\right) , \tag{108}$$

one reads off that the Goldstone bosons (GB's) corresponding to U(6)/U(2)xU(4) transform as

$$\text{GB's} \sim (4,2^*) + (4^*,2) , \tag{109}$$

i.e., they form a real representation of U(2) x U(4). Their

SUSY partners, however, the quasi-Goldstone fermions (QGF's), may be chosen to transform as a chiral representation of $U(2) \times U(4)$,

$$QGF's \sim (4,2^*) \ . \qquad (110)$$

The Goldstone bosons and the quasi-Goldstone fermions form together a chiral multiplet $\phi_p^a \sim (4,2^*)$ $(a = 1,\ldots,4;$ $p = 1,2)$ which corresponds to the left-handed sector of one supersymmetric quark-lepton family. The possibility to obtain chiral fermions [89,92] as SUSY partners of real Goldstone bosons is of crucial importance in model building [93-96]. The coset space $U(6)/U(2) \times U(4)$ can also be associated with the real representation $(4,2^*)+(4^*,2)$ of QGF's [90]. It is a dynamical question which case is realized.

The coset space necessary to accommodate one quartet of left-handed quarks and leptons is $U(6)/U(2) \times U(4)$. The study of explicit models as well as general arguments [97-99] show, however, that the unbroken subgroup of $U(6)$ is smaller than $U(2) \times U(4)$. In particular the breaking of $U(2)$ to $SU(2)$ gives rise to a Goldstone chiral multiplet, the "novino" [93] , which contains only one Goldstone boson and one additional scalar (quasi-Goldstone boson). The presence of their $SU(2) \times SU(4)$ neutral chiral multiplet ϕ has important consequences for the self-interactions of all Goldstone multiplets ϕ_p^a and ϕ [59,93,100].

The interactions of Goldstone superfields at distances $r < \frac{1}{f}$, where f is the mass scale of symmetry breaking, are described by supersymmetric σ-models [102]. They have the special property that the complex scalar fields can be interpreted as coordinates of a Kähler manifold. This means that the metric, which determines the σ-model Lagrangian, can locally be expressed as the second derivative of a potential:

$$L_s = \partial^\mu \tilde{\phi}_i^* g_{ij}(\tilde{\phi}^*,\tilde{\phi}) \partial_\mu \tilde{\phi}_j \quad ,$$

$$g_{ij}(\tilde{\phi}^*,\tilde{\phi}) = \frac{\partial^2}{\partial \tilde{\phi}_i^* \partial \tilde{\phi}_j} K(\tilde{\phi}^*,\tilde{\phi}) \ . \tag{111}$$

The supersymmetric extension of the Lagrangian for the scalar fields takes a simple form in terms of the Kähler potential K [103]:

$$L = \int d^4\theta K(\bar{\phi},\phi)$$

$$= L_{kin} - \frac{1}{f^2} \tilde{\phi}_i^* \tilde{\phi}_i \partial^\mu \tilde{\phi}_j^* \partial_\mu \tilde{\phi}_j - \frac{1}{2f^2} \bar{\psi}_{Li}\gamma^\mu \psi_{Li} \bar{\psi}_{Lj}\gamma_\mu \psi_{Lj} + \cdot\cdot \tag{112}$$

The residual interactions of the quasi-Goldstone fermions are of current-current type which is of interest in models with composite W-bosons [90,91]. This is in contrast to Goldstone fermions where the residual interactions involve derivatives [86]:

$$L = L_{kin} - \frac{1}{2f^4} \bar{\psi}_i \gamma^\mu \partial^\nu \psi_j \bar{\psi}_i \gamma_\nu \partial_\mu \psi_j + \cdots\ . \tag{113}$$

The Kähler potentials for the coset spaces based on U(6), which we discussed above, are exactly known. For U(6)/U(2) x U(4) one has [103]

$$K = f^2 \ln(\det(\delta_q^p + \frac{1}{f^2} \bar{\phi}_a^p \phi_q^a)) \ . \tag{114}$$

In the case of U(6)/SU(2) x U(4) the Kähler potential is given by [100,101]

$$K = F[e^{i/f_1(\phi-\bar{\phi})} \det(\delta_q^p + \frac{1}{f_2^2}\bar{\phi}_a^p \phi_q^a)] \ , \tag{115}$$

where F is an arbitrary function up to the constraints

$$\frac{\partial^2 F}{\partial\bar{\phi}\partial\phi} = 1 \ , \quad \frac{\partial^2 F}{\partial\bar{\phi}^p_a \partial\phi^b_q} = \delta^a_b \delta^q_p \ . \tag{116}$$

The arbitrariness of the Kähler potential (115) compared to (114) is due to the presence of one quasi-Goldstone boson in the case of U(6)/SU(2) x U(4) which implies that the metric of the Kähler manifold is not determined entirely by the U(6) symmetry.

The coset space U(6)/SU(2) x U(4), which accommodates one quartet of left-handed quarks and leptons, is the main building block of more realistic models which are based on U(6) x U(6)'. In such models the W-bosons may be composite [91,95] or fundamental [89,95,96]. In the latter case the local $SU(2)_W \times U(1)_Y$ symmetry is broken through the technicolour type breaking [104]of SU(2) x SU(2)' to the diagonal SU(2) subgroup. The class of U(6) x U(6)' models has interesting theoretical and phenomenological features, but so far no completely satisfactory preon model has been found. With respect to the question of family replication it is an interesting possibility [91] that the number of quark-lepton generations is a consequence of the dimension of the hypercolour group and 't Hooft's anomaly matching conditions.

7. COMPOSITE HIGGS FIELDS AND EXCEPTIONAL COSET SPACES

As we saw in the previous section the effective Lagrangian for Goldstone superfields is a supersymmetric σ-model, i.e., a pure D-term. The absence of a superpotential raises the question how fermion masses can be generated in such theories. In this section we will discuss a nonperturbative mechanism for fermion mass generation [105], which is a consequence of supersymmetry breaking. The questions of supersymmetry breaking and its implications have been discussed in a similar context by Masiero and Veneziano [106] and Achiman, Aoyama and van Holten [107].

Let us consider a simple SUSY σ-model for "leptons" based on the coset space $U(3)xU(2)/U(2)xU(1)^2$. The unbroken subgroup $U(2) \times U(1)^2$ contains the electroweak gauge group $SU(2)_W \times U(1)_Y$ and the global $U(1)_\ell$ of lepton number. The Goldstone superfields with their $SU(2)_W \times U(1)_Y \times U(1)_\ell$ quantum numbers are [105]

$$L^i \sim (2, -\frac{1}{2}, 1), \; E_c \sim (1,1,-1) \; , \; S \sim (1,0,0) \; ; \qquad (117)$$

they correspond to a left-handed doublet of leptons, a left-handed positron and an additional singlet containing one quasi-Goldstone boson.

An exact Kähler potential for the coset space $U(3) \times U(2)/$ $/U(2) \times U(1)^2$, which can be obtained by standard methods, reads

$$K_o = f^2 + f(S+\bar{S}) + \bar{L}_i L^i + \bar{E}_c E_c + \bar{S}S$$

$$+ \frac{1}{f^2} \bar{L}_i L^i \bar{E}_c E_c [(1+ \frac{1}{f}S)(1 + \frac{1}{f}\bar{S})]^{-1} \; . \qquad (118)$$

Adding a scalar mass $\Delta \ll f$ for the scalars $\tilde{L}^i$ and $\tilde{E}_c$, and gauging $SU(2)_W \times U(1)_Y$ [104,108] one obtains (the anomalies of the gauged currents are irrelevant in the following discussion and will be absent in a realistic theory including quarks):

$$L_\sigma = \int d^4\theta[\bar{L}_i(e^{2gW-g'B})^i_j L^j + \bar{E}_c\, e^{2g'B} E_c](1-\delta) + \bar{S}S$$

$$+ \frac{1}{f^2}\bar{L}_i\bar{E}_c(e^{2gW+g'B})^i_j L^j E_c[(1+\frac{1}{f}S)(1+\frac{1}{f}\bar{S}]^{-1} ,$$

$$\delta = \Delta^2\theta\theta\bar{\theta}\bar{\theta}, \qquad \Delta << f . \tag{119}$$

The Lagrangian (119) is similar to the supersymmetric Nambu-Jona Lasinio (SNJL) model [84]. Therefore one may expect that the soft supersymmetry breaking in (119) induces a further symmetry breaking through vacuum expectation values of composite fields describing bound states of L^i and E_c [109]. In analogy to the SNJL model such composite fields can be found by rewriting (119) with the aid of auxiliary superfields H_1 and H_2:

$$L_t = \int d^4\theta\{[\bar{L}_i(e^{2gW-g'B})^i_j L^j + \bar{E}_c e^{2g'B} E_c](1-\delta) + \bar{S}S$$

$$+ \bar{H}_{1i}(e^{2gW+g'B})^i_j H^j_i \}$$

$$+\int d^2\theta[fH^i_1 H_{2i} + H^i_1 H_{2i} S + L^i H_{2i} E_c] + \text{c.c.} \tag{120}$$

In (120) H_2 appears without kinetic term, i.e., it acts as a Lagrange multiplier field. Inserting the corresponding constraint for H_1 in (120) yields the old Lagrangian (119). Therefore, although H_1 does appear with kinetic term in (120), neither H_1 nor H_2 correspond to independent physical degrees of freedom. From the equations of motion one obtains:

$$H^i_1 = - \frac{1}{f}(1 + \frac{1}{f}S)^{-1} L^i E_c ,$$

$$H_{2i} = \frac{1}{f^2}(1+\frac{1}{f}S)^{-1}(-\frac{1}{4})\bar{D}^2[(1+\frac{1}{f}\bar{S})^{-1}\bar{L}_j\bar{E}_c(e^{2gW+g'B})^j_i] . \tag{121}$$

It is remarkable that H_1 and H_2 have the quantum numbers of the Higgs fields in the supersymmetric standard model [110], and that they interact with the other chiral superfields through a superpotential. It is indeed a non-trivial consequence of the Kähler potential (118) that the auxiliary fields are chiral and not vector superfields!

As in the SNJL model one finds that one-loop radiative corrections generate a kinetic term for H_2 (cf. Fig.23):

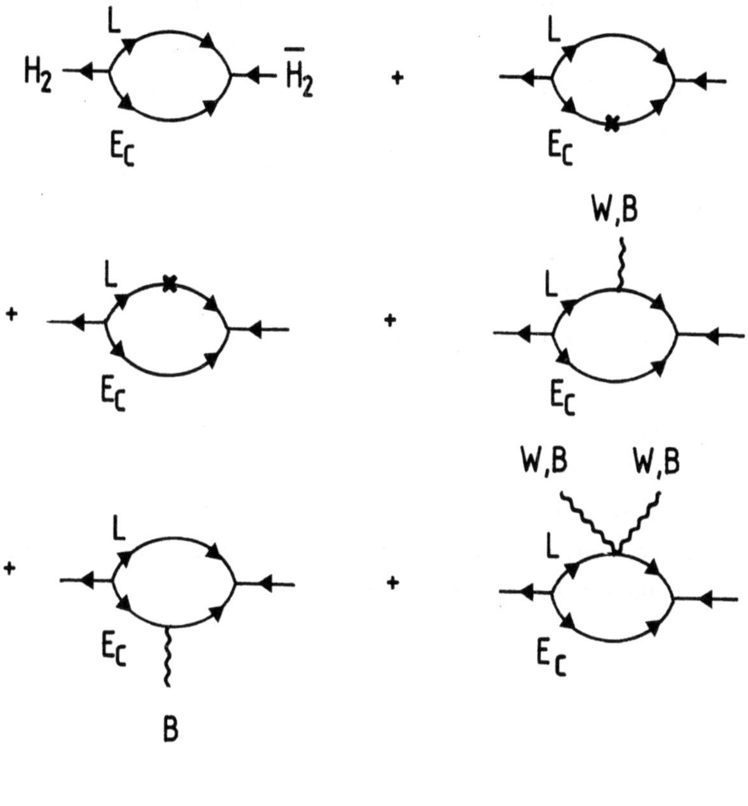

Fig. 23.: Supergraphs contributing to the kinetic term of H_2.

$$L_{I\ell} = Z\int d^4\theta \bar{H}_2^i (e^{-2gW-g'B})_i^{\ j} H_{2j}(1+2\delta) + O\left(\frac{1}{\ln\frac{f^2}{\Delta^2}}\right) , \qquad (122a)$$

$$\text{with } Z = \frac{1}{16\pi^2}\left(\ln\frac{f^2}{\Delta^2} + O(1)\right) . \qquad (122b)$$

Therefore in the one-loop approximation, which may be justifiable by means of a $\frac{1}{N}$ expansion in an extended version of the model, we obtain the Higgs fields H_1 and H_2 as bound states of the fields L^i, E_c and S. The supersymmetry breaking term in (122) corresponds to a negative mass squared for $\tilde{H}_2$. This leads to a vacuum expectation value of H_2 and thereby via the superpotential in (120) to fermion masses whose existence has already been anticipated as an infrared cutoff in some of the one-loop integrals of Fig.23. Given the composite Higgs fields, the described mechanism for $SU(2)_W \times U(1)_Y$ breaking is similar to the radiative symmetry breaking proposed by Ibáñez and Ross [111].

From the Lagrangian (120), (122) the entire mass spectrum of the model can be calculated in the standard manner [110] as function of the σ-model scale f and the SUSY breaking scale Δ. Minimization of the scalar potential yields the vacuum expectation values

$$<\tilde{H}_1> = \begin{pmatrix} 0 \\ V_1 \end{pmatrix} , \qquad \sqrt{Z}<\tilde{H}_2> = \begin{pmatrix} 0 \\ V_2 \end{pmatrix} , \qquad (123a)$$

with

$$V_1^2 = \frac{2(\frac{1}{4}(g^2+g'^2)Z-1)Z\Delta^2}{\frac{1}{2}(g^2+g'^2)Z-1}$$

$$v_2^2 = \frac{(g^2+g'^2)Z^2\Delta^2}{2(\frac{1}{2}(g^2+g'^2)Z-1)} \quad . \tag{123b}$$

The physical states are the vector bosons $W^{\pm}$, Z and γ, the neutrino ν and the Dirac fermions e, h^+, h^-, h and h' (which are combinations of Weyl fermions from H_1, H_2, L, E_c, S and gauginos), the complex sclars $\tilde{N}, \tilde{E}, \tilde{E}_c$, $\tilde{S}$ and $\tilde{H}^{\mp}$, and the real scalars $\tilde{H}_o$, $\tilde{H}_1^o$ and $\tilde{H}_2^o$. The corresponding masses are given by

$$(\cos\theta_W = \frac{g}{\sqrt{g^2+g'^2}}, \ \cos\phi = \frac{v_2}{\sqrt{v_1^2+v_2^2}}) :$$

$$M_W^2 = \frac{1}{2}g^2(v_1^2+v_2^2), \qquad M_Z^2 = \frac{1}{2}(g^2+g'^2)(v_1^2+v_2^2), \qquad M_\gamma^2 = 0 , \tag{124a}$$

$$m_{h^+} = \sqrt{2}\ \sin\phi\ M_W, \qquad m_{h^-} = \sqrt{2}\ \cos\phi\ M_W , \tag{124b}$$

$$m_h = \sqrt{2}\ \Delta, \qquad m_{h'} = M_Z , \tag{124c}$$

$$m_e = \sqrt{2}\ \Delta\ \cos\phi , \qquad m_\nu = 0 , \tag{124d}$$

$$M^2_{\tilde{N}} = \Delta^2(1+2\cos^2\phi) , \qquad M^2_{\tilde{E}} = \Delta^2(1+4\cos^2\phi\sin^2\theta_W) , \tag{124e}$$

$$M^2_{\tilde{E}_c} = \Delta^2(1+2\cos^2\phi(1-2\sin^2\theta_W)) , \qquad M^2_{\tilde{S}} = 2\Delta^2\cos^2\phi, \tag{124f}$$

$$M^2_{\tilde{H}^+} = M^2_W - 2\Delta^2, \qquad M^2_{\tilde{H}_o} = 0 , \tag{124g}$$

$$M^2_{\tilde{H}^o_{1,2}} = \Delta^2[x \pm (\frac{x(x^2-5x+8)}{x-1})^{1/2}], \qquad x = \frac{1}{2}(g^2+g'^2)Z = \frac{M_Z^2}{2\Delta^2} \gtrsim 2. \tag{124h}$$

So far we have only discussed a model for "leptons". A more realistic theory including quarks and more than one generation will have a much more complicated set of composite Higgs fields. In such a theory the one-loop graphs of the type shown in Fig.23 will be dominated by the t-quark, i.e., the multiplet which couples most strongly to the composite Higgs fields. Identifying m_e with the t-quark mass yields indeed a phenomenologically consistent mass spectrum. Choosing for instance as input $M_W = 81$ GeV, $M_Z = 92$ GeV and $m_e = 38$ GeV, in order to fix the parameters Δ, f and $\sin^2\theta_W$, one obtains for the remaining masses (in GeV) from equs.(124):

$$m_{h^+} = 66, \qquad m_{h^-} = 94, \qquad m_h = 46\ ,$$

$$m_{h'} = 92, \qquad m_\nu = 0, \qquad M_{\tilde{N}} = 50,$$

$$M_{\tilde{E}} = 41, \qquad M_{\tilde{E}_c} = 43, \qquad M_{\tilde{S}} = 38,$$

$$M_{\tilde{H}^+} = 67, \qquad M_{\tilde{H}^o} = 0, \qquad M_{\tilde{H}_1} = 82,$$

$$M_{\tilde{H}_2^o} = 42\ . \tag{125}$$

The σ-model scale f can be expressed in terms of the Fermi constant G_F and the supersymmetry breaking scale Δ (cf. (122), (123)):

$$\ln\frac{f^2}{\Delta^2} = 8\pi^2\,\frac{v_1^2+v_2^2}{\varepsilon^2} = \frac{2\sqrt{2}\pi^2}{\Delta^2 G_F}\ . \tag{126}$$

From (126) it is clear that Δ has to be of the order of the Fermi scale $G_F^{-1/2}$, and that f is naturally much larger than $G_F^{-1/2}$, i.e., the substructure scale has to be identified with the unification mass M_{GUT} or the Planck mass M_{PL} rather than the Fermi scale (in fact for the mass spectrum (125) one obtains $f > M_{PL}$, but equ.(126) is likely to be modified in a more realistic theory).

The possibility that Higgs fields may be described as bound states of quark and leptons has been considered before by Harari and Seiberg [112]. More recently Napoly has studied dynamical symmetry breaking through fermion condensates in a grand-unified SO(10) model [113]. If quarks and leptons are themselves composite, vacuum expectation values of composite Higgs fields correspond to higher dimensional preon condensates which are currently also being studied by Mizrachi and Peccei in connection with fermion mass generation [114]. Of course, Higgs fields should be considered as bound states of quarks and leptons only in an effective sense, i.e., in the same way as nucleons [115] and ρ-mesons [116] can be obtained from a σ-model of pions.

What is the correct σ-model for (at least) three generations of quarks and leptons? The most interesting coset spaces are clearly those involving exceptional groups [87,90,92,117-121], and σ-models based on E_7 [118], E_6 [119] and E_8 [121,122] have already been constructed. From the SU(5) decomposition of the adjoint representations [123] it is abvious that a spontaneous symmetry breaking of E_6, E_7 or E_8 will yield Goldstone multiplets with the quantum numbers of quarks and leptons:

$$E_6 : \underline{78} = \underline{24} + \underline{5} + \underline{\bar{5}} + 2\mathrm{x}(\underline{10} + \underline{\overline{10}}) + 4 \times \underline{1} ,$$

$$E_7 : \underline{133} = \underline{78} + 3\mathrm{x}(\underline{5} + \underline{\bar{5}}) + \underline{10} + \underline{\overline{10}} + 5 \times \underline{1} ,$$

$$E_8 : \underline{248} = \underline{133} + 6\mathrm{x}(\underline{5} + \underline{\bar{5}}) + 2\mathrm{x}(\underline{10} + \underline{\overline{10}}) + 15 \times \underline{1} , \qquad (127)$$

where $\underline{24}$, $\underline{78}$, $\underline{133}$ and $\underline{248}$ denote the adjoint representation of SU(5), E_6, E_7 and E_8 respectively.

In order to accomodate three quark-lepton families in a coset-space G/H the dimension of G has to be sufficiently large. As each generation contains 15 Weyl fermions one has

$$n_G \geq n_H + 2 \times 15 \times 3 = n_H + 90 \,. \tag{128}$$

Thus for $H \supset SU(5)$ E_7 and E_8 are possible choices for G whereas for $H \supset SO(10)$ or E_6 only E_8 is large enough. Appropriate Kähler manifolds can be found by means of a recipe recently given by Bordemann, Forger and Römer [124]. It states that G/H is a Kähler manifold, if H takes the form $H = H' \times U(1)^p$ where H' is a semi-simple subgroup of G which one obtains from the Dynkin diagram of G by crossing out p roots; $U(1)^p$ (p = rank (G) -rank (H')) are the U(1) factors of G which are not contained in H'. Some examples for $G = E_6, E_7, E_8$ are shown in Fig.24.

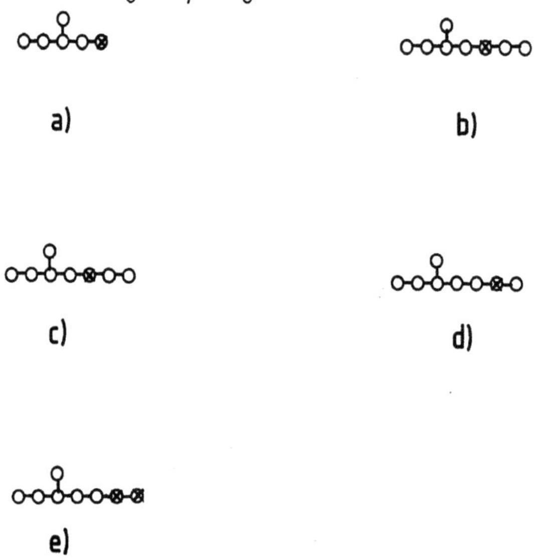

Fig. 24.: Dynkin diagrams for the Kähler manifolds $E_6/$ $/SO(10) \times U(1)$ (a), $E_7/SU(5) \times SU(3) \times U(1)$ (b), $E_8/SO(10) \times SU(3) \times U(1)$ (c), $E_8/E_6 \times SU(2) \times$ $\times U(1)$ (d), $E_8/E_6 \times U(1)^2$ (e).

Given a Kähler manifold G/H one has to find the G-invariant complex structures Ω which determine the H-quantum numbers of the chiral superfields of the supersymmetric σ-models. The complex structures can be easily found by decomposing the adjoint representation of G with respect to H and selecting those roots of G/H whose charges with respect to a U(1) factor contained in $U(1)^p$ have the same sign.

For the five cases of Fig.24 one finds

a) $E_6/SO(10) \times U(1)$ [119]:

$$\Omega = \underline{16}^{1} . \tag{129}$$

b) $E_7/SU(5) \times SU(3) \times U(1)$ [118]:

$$\Omega = (5,1)^{6} + (\bar{5},3)^{4} + (10,3)^{2} . \tag{130}$$

c) $E_8/SO(10) \times SU(3) \times U(1)$ [121, 122]:

$$\Omega = (16,3)^{1} + (\overline{16},1)^{3} + (1,3)^{4} + (10,\bar{3})^{2} . \tag{131}$$

d) $E_8/E_6 \times SU(2) \times U(1)$ [124]:

$$\Omega = (27,2)^{1} + (\overline{27},1)^{2} + (1,2)^{3} . \tag{132}$$

e) $E_8/E_6 \times U(1)^2$ [124]:

$$\Omega = \underline{27}^{1,1} + \underline{27}^{1,-1} + \underline{\overline{27}}^{2,0} + \underline{1}^{3,1} + \underline{1}^{3,-1} + \underline{1}^{0,2} . \tag{133}$$

In the last case there also exists a unique complex structure, although H contains two U(1) factors.

Only the coset spaces b) and c) contain three families. As in case b) SU(5) cannot be gauged because of anomalies, the coset space c) appears to be the most promising candidate for a description of quarks and leptons. Its characteristic feature is the prediction of a fourth mirror family.

It is conceivable that the investigation of supersymmetric σ-models for quarks and leptons will lead to an understanding of their quantum numbers (and the prediction of some specific new particles), and that the nonrenormalizable interactions of the Goldstone multiplets, together with gauge interactions and supersymmetry breaking, can eventually explain the quark-lepton mass spectrum. It is certainly a very remarkable phenomenon, that the coset spaces of successive exceptional groups (cf. [126]), which include the interesting gauge groups, contain precisely the quantum numbers of quarks and leptons. We conclude this section by listing this "exceptional sequence":

		$E_{n+1}/E_n \times U(1)$:
$E_3 = SU(3) \times SU(2)$	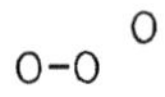	
		$(3,2)$ + c.c.
$E_4 = SU(5)$	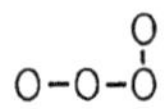	
		10 + c.c.
$E_5 = SO(10)$	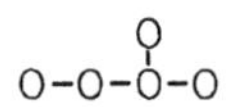	
		$16(=\bar{5}+10+1)$ + c.c.
E_6	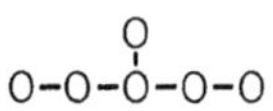	
		$27(=16+10+1)$ + c.c.
E_7	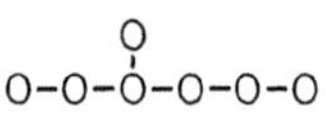	
		$56(=27+\overline{27}+1+1)+1$ + c.c.
E_8	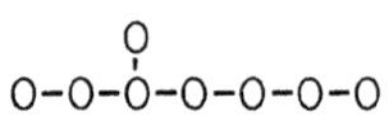	

8. CONCLUSIONS

In these lectures we have discussed various aspects of quark-lepton substructure. Starting from the standard gauge theory of strong and electroweak interactions we have argued that the replication of quark-lepton generations and the hierachical pattern of their masses suggest a further substructure of matter. We have studied a representative preon model of Pati-Salam type and examined experimental bounds on the compositeness scale Λ. It is phenomenologically possible, although unlikely, that Λ is as small as the Fermi scale $G_F^{-1/2} \sim 300$ GeV. In this case the W-vector bosons of the weak interactions would be composite and their effective Lagrangian would be a massive Yang-Mills theory.

A central question of quark-lepton substructure is the dynamics of massless bound states. The only known composite particles, whose mass is small compared to their inverse size, are the pions which are the pseudo-Goldstone bosons of the spontaneously broken axial isospin symmetry of QCD. Massless composite fermions which originate from an asymptotically free preon theory have to satisfy 't Hooft's anomaly matching conditions. An apparently unavoidable consequence of these consistency conditions is that preon theories have to be very complicated and that the number of preons is generally larger than the number of massless composite fermions. A further important constraint is that massless particles with global quantum numbers have to have spin 0 or spin $\frac{1}{2}$.

The most interesting candidates for a theory of quark-lepton substructure are confining supersymmetric gauge theories which contain naturally massless composite fermions. They can appear because of unbroken chiral symmetries or as a consequence of symmetry breaking: Goldstone fermions arise from the spontaneous breaking of supersymmetry, and quasi-Goldstone fermions from the spontaneous breaking of global symmetries in the case of unbroken supersymmetry. The group structure of supersymmetric preon models is similar to non-supersymmetric models, the amount of complexity is even larger in the supersymmetric case.

The main ingredients of preon models have been borrowed from QCD: the concepts of a confining, asymptotically free hypercolour force and of chiral symmetry. It is disappointing that so far attempts based on these ideas have not led to a "standard composite model" and that in particular no convincing explanation has been found for the main puzzle of the standard model, the replication of quark-lepton families. Therefore we conclude that the QCD-motivated approach described in most of these lectures is presumably inappropriate and that novel ideas are needed.

In connection with supersymmetric preon models supersymmetric σ-models play an important role. It is conceivable that these σ-models are more interesting than the renormalizable theories which motivate them. The existence of Kähler manifolds whose linear symmetries are the relevant gauge groups and whose coordinates carry quark-lepton quantum numbers is a remarkable phenomenon. It is an intriguing possibility that the role of supersymmetric σ-models for the dynamics of quarks and leptons could be analogous to the role of ordinary σ-models for the dynamics of hadrons before the discovery of QCD.

ACKNOWLEDGEMENTS

These lectures are an extended version of lectures given at he Herbstschule für Hochenergiephysik, Maria-Laach, September 1984. I thank Professors J.K. Bienlein, H. Mitter and W. Plessas for the invitations to Maria-Laach and Schladming. I am grateful to U. Ellwanger, O. Napoly, M.G. Schmidt, D. Wyler, T. Yanagida and, especially, R.D. Peccei for a stimulating collaboration on some aspects of quark-lepton substructure.

REFERENCES

1. S.L. Glashow, Nucl. Phys. 22 (1961) 579.
   S. Weinberg, Phys. Rev. Lett. 19 (1967) 1264.
   A. Salam, in Elementary Particle Theory, ed. N. Svartholm (Almqvist and Wiksell, Stockholm, 1968) 367.
2. J.C. Pati and A. Salam, Phys. Rev. D10 (1974) 275.
3. H. Georgi and S.L. Glashow, Phys. Rev. Lett. 32 (1974) 438.
4. K.G. Wilson, Phys. Rev. D3 (1971) 1818.
5. L. Susskind, Phys, Rev. D20 (1979) 2619.
6. G. 't Hooft, in Recent Developments in Gauge Theories, Cargèse 1979, eds. G. 't Hooft et al. (Plenum, New York and London).
7. For a review and an extensive list of references, see:
   L. Lyons, Progress in Particle and Nuclear Physics 10 (1983) 227; Oxford preprint 2/84 (1984).
8. For an introduction, see, for instance: G.G. Ross, Grand unified theories (The Benjamin/Cummings Publishing Company, Inc., 1985).
9. For a recent review, see:
   D. Wyler, Lectures given at the Escuela Mexicana de Particulas y Campos, Oaxtebec, Mexico, December 1984, ETH preprint (1985).
10. There exists a vast literature on this subject. For recent reviews, see:
   G. Ecker, Acta Phys. Pol. B15 (1984) 179; B. Stech, in Proceedings of the Europhysics Topical Conference

on Flavor Mixing in Weak Interactions, Erice, 1984, ed. by L.L. Chau (Plenum Press, New York, 1984).

11. L. Wolfenstein, Phys. Rev. Lett. 51 (1983) 1945.
12. P.H. Frampton and C. Jarlskog, Phys. Lett. 154B (1985) 421.
13. J. Gasserand H. Leutwyler, Phys. Rep. 87C (1982) 77.
14. J.C. Pati and A. Salam, ref.2, footnote 7.
15. O.W. Greenberg and C.A. Nelson, Phys. Rev. D10 (1974) 2567.
16. K. Matumoto, Prog. Theor. Phys. 52 (1974) 1973.
17. J.C. Pati, A. Salam and J. Strathdee, Phys. Lett. 59B (1975) 265.
18. M.E. Peskin, in Proc. of the Lepton Photon Symposium, Bonn University, ed. W. Pfeil (1981), p.880.
19. R.D. Peccei, in Proc. of the Fourth Topical Workshop on Proton-Antiproton Collider Physics, Bern, eds. H. Hänni and J. Schacher (1984), p.483.
20. H. Terazawa, in Proc. of the XXII International Conference on High Energy Physics, Leipzig, eds. A. Meyer and E. Wieczorek (1984), p.63.
21. H. Harari, Phys. Lett. 86B (1979) 83.
22. M.A. Shupe, Phys. Lett. 86B (1979) 87.
23. H. Harari and N. Seiberg, Phys. Lett. 98B (1981) 269; Nucl. Phys. B204 (1982) 141.
24. I. Bars, Nucl. Phys. B208 (1982) 77.
25. B. Schrempp and F. Schrempp, Nucl. Phys. B231 (1984) 109; ibid. B242 (1984) 203.
26. H. Georgi and D.B. Kaplan, Phys. Lett 145B (1984) 216; H. Georgi, Phys. Lett. 151B (1985) 57.
27. R. Barbieri, A. Masiero and R.N. Mohapatra, Phys. Lett. 105B(1981) 369.
28. L. Abbott and E. Farhi, Phys. Lett. 101B (1981) 69.
29. O.W. Greenberg and J. Sucher, Phys. Lett. 99B (1981) 339.
30. H. Fritzsch and G. Mandelbaum, Phys. Lett. 102B (1981) 369.

31. R. Casalbuoni and R. Gatto, Phys. Lett. 103B (1981) 113.
32. H. Terazawa and K. Akama, Phys. Lett. 96B (1980) 276.
33. H. Harari, Phys. Rep. 104 (1984) 159.
34. For a more complete discussion, see:
S. Yamada, in Proc. of the XXII International Conference on High Energy Physics, Leipzig, eds. A. Meyer and E. Wieczorek (1984) p.72.
35. E.J. Eichten, K.D. Lane and M.E. Peskin, Phys. Rev. Lett. 50 (1983) 811.
36. M. Abolins et al., in Proc. of the 1982 Summer Study on Elementary Particles and Fields, Snowmass, eds. R. Donaldson et al. (Amer. Inst. Phys. N.Y., 1983) p.274.
37. R. Rückl, Phys. Lett. 129B (1983) 363.
38. R. Rückl, Nucl. Phys. B234 (1984) 91.
39. T. Maehara and T. Yanagida, Prog. Theor. Phys. 61 (1979) 1434.
40. R.N. Cahn and H. Harari, Nucl. Phys. B176 (1980) 135.
41. G.L. Kane and R. Thun, Phys. Lett. 94B (1980) 513.
42. S. Dimopoulos and J. Ellis, Nucl. Phys. B182 (1981) 505.
43. I. Bars, in Proc. Moriond Workshop on Quarks, Leptons and Supersymmetry, ed. J. Tran Thanh Van (1982) p.541.
44. C.N. Leung, S.T. Love and S. Rao, preprint Fermilab-Pub-84/74-T (1984).
45. Ch. Kopper, Phys. Lett. 155B (1985) 409.
46. W. Buchmüller and D. Wyler, in preparation.
47. R. Barbieri, L. Maiani and R. Petronzio, Phys. Lett. 96B (1980) 63.
48. S. Brodsky and S.D. Drell, Phys. Rev. D22 (1980) 2236.
49. T. Kinoshita and W.B. Lindquist, Phys. Rev. Lett. 47 (1981) 1573; T. Kinoshita, B. Nižić, and Y. Okamoto, Phys. Rev. Lett. 52 (1984) 717.
50. Review of Particle Properties, Rev. Mod. Phys. 56 (1984) No.2, Part II.

51. O.W. Greenberg, R.N. Mohapatra and S. Nussinov, Phys. Lett. 148B (1984) 465.
52. J.D. Bjorken, Phys. Rev. D19 (1979) 335.
53. P.Q. Hung and J.J. Sakurai, Nucl. Phys. B143 (1978) 81.
54. R. Kögerler and D. Schildknecht, preprint CERN-TH 3231 (1982).
55. H. Fritzsch, R. Kögerler and D. Schildknecht, Phys. Lett. 114B (1982) 157.
56. H.G. Dosch, M. Kremer and M.G. Schmidt, Phys. Lett. 137B (1984) 88; Z. Phys. C26 (1985) 569.
57. V.A. Matveev, M.E. Shaposhnikov and A.N. Tavkhelidze, Moscow preprint P-0325 (1983).
58. S. Narison, Montpellier preprint PM 84/23 (1984).
59. W. Buchmüller and M.G. Schmidt, Nucl. Phys. B285 (1985) 230.
60. B. Schrempp and F. Schrempp, preprint DESY 84-055 (1984).
61. N.M. Kroll, T.D. Lee and B. Zumino, Phys. Rev. 157 (1967) 1376.
62. J.J. Sakurai, Acta Phys. Aust. Suppl. XXIV (1982) 275.
63. P. Chen and J.J. Sakurai, Phys. Lett. 110B (1982) 481.
64. H. Fritzsch and G. Mandelbaum, Phys. Lett. 109B (1982) 224.
65. W. Buchmüller, Phys. Lett. 145B (1984) 151.
66. M. Kuroda, D. Schildknecht and K.-H. Schwarzer, Bielefeld preprint BI-TP 84/22 (1984).
67. R.D. Peccei, in Proc. of the ECFA-CERN Workshop, Lausanne, ed. M. Jacob, (1984) p.329.
68. F.M. Renard, Montpellier preprint PM/84/4 (1984).
69. B. Schrempp and F. Schrempp, Phys. Lett. 153B (1985) 101.
70. CELLO collaboration, H.J. Behrend et al., Phys. Lett. 141B (1984) 145.
71. UA1 collaboration, G. Arnison et al., Phys. Lett. 155B (1985) 442.
72. O. Dumbrajs et al., Nucl. Phys. B216 (1983) 304.
73. For a thorough review of "chiral symmetry and chiral

symmetry breaking", see: M.E. Peskin, Lectures given at the 1982 Los Houches Summer School of Theoretical Physics, preprint SLAC-PUB-3021 (1982).

74. S. Weinberg, Phys. Rev. 166 (1968) 1568.
75. See, for instance, C. Itzykson and J.-B. Zuber, Quantum Field Theory (McGraw-Hill Inc., 1980).
76. K.M. Case and S. Gasiorowicz, Phys. Rev. 125 (1962) 1055.
77. S. Weinberg and E. Witten, Phys. Lett. 96B (1980) 59.
78. J.-M. Gérard, J. Govaerts, Y. Meurice and J. Weyers, Phys. Lett. 116B (1982) 29; Nucl. Phys. B234 (1984) 138.
79. R. Barbieri, Phys. Lett. 121B (1983) 43.
80. J.C. Pati and A. Salam, Nucl. Phys. B214 (1983) 109.
81. Y.J. Ng and B.A. Ovru, Phys. Lett. 125B (1983) 147; Phys. Rev. D29 (1984) 138.
82. F. Bordi, R. Casalbuoni, D. Dominici and R. Gatto, Phys. Lett. 127B (1983) 419.
83. H.P. Nilles, Phys. Lett. 112B (1982) 455.
84. W. Buchmüller and S.T. Love, Nucl. Phys. B204 (1982) 213.
85. G. Veneziano, Phys. Lett. 124B (1983) 357.
86. W.A. Bardeen and V. Višnjić, Nucl. Phys. B194 (1982) 422.
87. W. Buchmüller, S.T. Love, R.D. Peccei and T. Yanagida, Phys. Lett. 115B (1982) 233; W. Buchmüller, R.D. Peccei and T. Yanagida, Phys. Lett. 124B (1983) 67.
88. We use the conventions of J. Wess and J. Bagger, Supersymmetry and Supergravity (Princeton University Press, Princeton, 1983).
89. R. Barbieri, A. Masiero and G. Veneziano, Phys. Lett. 128B (1983) 493.
90. W. Buchmüller, R.D. Peccei and T. Yanagida, Nucl. Phys. B227 (1983) 503.
91. O.W. Greenberg, R.N. Mohapatra and M. Yasue, Phys. Lett. 127B (1983) 419; Phys. Rev. Lett. 51 (1983) 1737.
92. C.L. Ong, Phys. Rev. D27 (1983) 911.
93. W. Buchmüller, R.D. Peccei and T. Yanagida, Nucl. Phys. B231 (1984) 53.

94. W. Lerche and D. Lüst, Nucl. Phys. B244 (1984) 157.
95. W. Buchmüller, R.D. Peccei and T. Yanagida, Nucl. Phys. B244 (1984) 186.
96. A. Masiero, R. Pettorino, M. Roncadelli and G. Veneziano, preprint CERN-TH. 4166 (1985)
97. W. Lerche, Nucl. Phys. B238 (1984) 582.
98. G. Shore, Nucl. Phys. B248 (1984) 123.
99. W. Lerche, thesis, Munich 1985.
100. T. Kugo, S. Uehara and T. Yanagida, Phys. Lett. 147B (1984) 321.
101. M. Bando, T. Kuramoto, T. Maskawa and S. Uehara, Phys. Lett. 138B (1984) 94.
102. For a review,see: J. Bagger, Lectures given at the NATO Advanced Study Institute on Supersymmetry, Bonn 1984, preprint SLAC-PUB-3461 (1984).
103. B. Zumino, Phys. Lett. 87B (1979) 203.
104. A. Buras and W. Słominski, Nucl. Phys. B223 (1983) 157.
105. W. Buchmüller and U. Ellwanger, preprint CERN-TH 4113 (1985).
106. A. Masiero and G. Veneziano, preprint CERN-TH 3950 (1984).
107. Y. Achiman, S. Aoyama and J.W. van Holten, Phys. Lett. 150B (1985) 153.
108. J. Bagger and E. Witten, Phys. Lett. 118B (1982) 103.
109. W. Buchmüller and U. Ellwanger, Nucl. Phys. B245 (1984) 237.
110. For a review, see: H.P. Nilles, Phys. Rep. 110 (1984).
111. L.E. Ibáñez and G.G. Ross, Phys. Lett. 110B (1982) 215.
112. H. Harari and N. Seiberg, Phys. Lett. 100B (1981) 41.
113. O. Napoly, Saclay preprint SPhT-84-51 (1984).
114. R.D. Peccei, Lectures given at CERN, February 1985.
115. T.H.R. Skyrme, Proc. Roy. Soc. A260 (1961) 127; E. Witten, Nucl. Phys. B223 (1983) 433.
116. M. Bando, T. Kugo, S. Uehara, K. Yamawaki and T. Yanagida, Phys. Rev. Lett. 54 (1985) 1215.

117. C.L. Ong, Phys. Rev. D27 (1983) 3044.

118. T. Kugo and T. Yanagida 134B (1984) 313.

119. Y. Achiman, S. Aoyama and J.W. van Holten, Phys. Lett. 141B (1984) 64; Wuppertal preprint WU B 85-2 (1985).

120. L.E. Ibáñez, Phys. Lett. 150B (1985) 127.

121. C.L. Ong, preprint SLAC-PUB-3459 (1984).

122. S. Irié and Y. Yasui, Tohoku preprint TU/85/278 (1985).

123. R. Slansky, Phys. Rep. 79 (1981) 1.

124. M. Bordemann, M. Forger and H. Römer, preprint CERN-TH 4111 (1985).

125. W. Buchmüller and O. Napoly, preprint CERN-TH 4197 (1985)

126. D.I. Olive, in Unification of the Fundamental Particle Interactions II, Erice 1981, edited by J. Ellis and S. Ferrara (Plenum Press, New York, 1983).

Acta Physica Austriaca, Suppl. XXVII, 597–604 (1985)

A CONSISTENT MESON-EXCHANGE MODEL FOR THE NN INTERACTION+

by

Ch. ELSTER, K. HOLINDE*), R.MACHLEIDT++)
Institut für Theoretische Kernphysik d. Universität Bonn
Nussallee 14-16, D-5300 Bonn

*) Institut f. Kernphysik der KFA-Jülich, D-5170 Jülich

++) Theory group, TRIUMF
4004 Wesbrook Mall, Vancouver, B.C. Canada V6T2A3

Quantum chromodynamics (QCD) is the fundamental theory of strong interactions. Therefore, the nucleon-nucleon (NN) interaction is completely determined by the underlying quark-gluon dynamics. However, partly due to our present lack of knowledge about the details of the confinement mechanism, we are far away from a quantitative understanding of the NN force over the whole range in this way. Moreover, there is a good chance that conventional hadrons like nucleons, mesons and Δ-isobars, while being manifestations of the quark structure of hadrons, remain the relevant degrees of freedom for a wide range of nuclear physics phenomena. In that case, the overwhelming part of the nuclear force can be constructed in terms of meson-nucleon and meson-isobar vertices, which represent a natural and effective

+Seminar given at the XXIV. Internationale Universitätswochen für Kernphysik,Schladming,Austria, February 20-March 1, 1985

description of a complicated n-quark reaction.

Genuinely new quark-gluon exchange processes, not being taken into account by meson exchange, occur only for overlapping hadrons. Therefore, their role depends critically on the hadron size, which has not yet been determined unambiguously from QCD. By studying in detail how far meson exchange works and where it breaks down, one might get some idea about this quantity. In any case, effective meson-nucleon vertices should contain form factors, which suppress meson-exchange for small distances. In fact, historically, such form factors were introduced in a purely ad hoc way, supplying a sufficient fall off for high momenta, necessary to get a solution of the scattering equation. Now, due to the extended quark structure of hadrons, the form factor has become an important physical concept, being directly related to the hadron size. We mention in this connection attempts to understand the EMC-effect by increasing the nucleon size in the medium. If this picture is confirmed, the form factor would be <u>the</u> essential degree of freedom when going from the free NN interaction to that in a nuclear environment. We have constructed a meson-exchange model of the NN interaction which is solely based on suitable meson-nucleon and meson-nucleon-Δ-isobar vertices W_α (note that Δ-Δ transitions are not included). Antinucleons are left out from the beginning since it is known that the contribution of negative-energy states is quite small provided pseudovector coupling is used for the NNπ vertex, which is suggested as an effective coupling (Weinberg Lagrangian) starting from chiral invariance. We treat the Hamiltonian $H = H_o + W$ in time-ordered (non-covariant) perturbation theory. This treatment corresponds to standard many-body theory and thus leads to a well-defined transition from the two- to the many-body problem. Since one of our main goals is to provide a meson-exchange NN interaction suitable for application in nuclear structure, this procedure should therefore be adequate for that purpose.

A suitably defined transition matrix $\langle\alpha'\beta'|T|\alpha\beta\rangle$ can be represented by a series expansion defined by all diagrams containing two ingoing (α,β) and two outgoing (α',β') nucleon lines. This series can be partially summed by solving a (three-dimensional) integral equation

$$T(z) = V(z) + V(z)\,\frac{1}{z-h_o^{(N)}}\,T(z)$$

where $h_o^{(N)}$ is the free Hamiltonian involving the (relativistic) kinetic energies of the nucleons and z is their starting energy. In principle, the quasipotential V(z) contains the (infinite) sum of all diagrams involving at least one meson or one Δ-isobar in each intermediate state. (Note that each time-ordering has to be taken into account separately). In practice, however, one should make a choice according to the following guidelines: First, since the total meson mass exchanged between two nucleons determines the range of the NN force (higher mass implying shorter range), one should restrict oneself to meson exchanges with a mass below the cutoff masses Λ_α, which parametrize the formfactor and are directly related to the bag size. It turns out that, in order to obtain a quantitative fit to the NN data, Λ_α must be in the range of 1.2 - 1.6 GeV, i.e. a hadron extension of the order of 0.5 fm has to be implemented. Consequently, it is reasonable and consistent to include all relevant exchange up to a mass of about 1 GeV.

Furthermore, one should take advantage of the antagonist structure of π and ρ; namely, the corresponding tensor forces have opposite sign. Thus, there is a strong cancellation between explicit 2π-exchange diagrams and those involving $\pi\rho$-exchange. This leads us to the set of diagrams

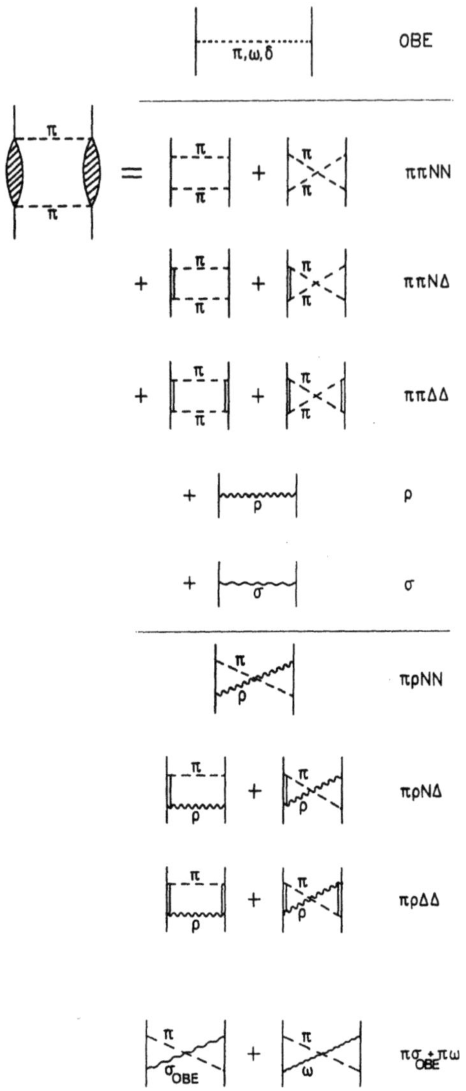

Fig.1: Set of irreducible diagrams building up V(z). Each box resp. crossed-box diagram contains the various time orderings of the noncovariant perturbation theory apart from the nucleon box, where the iterative pieces have to be left out as they are generated by the scattering equation. A full line represents a nucleon and a double line a Δ(1232)-isobar.

in Fig.1 building up V(z). The evaluation of the crossed-box diagrams is much more complicated than for the box diagrams and for that reason, they have sometimes been neglected. However, they are non-negligible and help to provide a scalar-isoscalar character of the 2π-exchange contribution (at least in higher partial waves), which is suggested by results from dispersion theory.
We view the Δ-isobar not as the empirical P_{33}-resonance, built up by Chew-Low-type diagrams in πN scattering, but as a genuine particle in the quark model sense, mainly because we want (and have) to include corresponding NΔπρ(and ΔΔπρ) contributions. Since we stay below pion-production threshold, part of the the dressing of the Δ can be conveniently taken into account by using the empirical Δ-mass. Renormalization of the πNΔ-vertex has been left out, i.e. we take the quark-model value for the bare πNΔ-coupling. Since we leave out corresponding terms at the NΔρ vertex, the combined effect of these vertex renormalizations is probably small, due to the strong cancellation between ππ and πρ contributions.
The correlated 2π-exchange in the P-wave state gives rise to the well known ρ-meson. On the other hand, the ππ-S-wave interaction does not lead to a resonance. However, Durso et al. [1]have shown that the diagrams of Fig.2 may be well approximated by the exchange of a scalar-isoscalar boson with a definite strength and mass distribution, which we will denote by σ. It turns out that this correlated 2π-exchange contribution provides about 2/3 of the total 2π-exchange. Note that in ref. 1 the ππ-S-wave interaction entering into the diagrams of Fig.2 has been determined from the empirical ππ-S-wave phase shifts. σ is thus introduced in a well defined way in contrast to the σ_{OBE} used in former one-boson-exchange models.
Additional (correlated and uncorrelated) 2π-exchange processes involving higher-mass πN resonances have been left out. As shown in ref. 1 these contributions are probably small, especially since we stay below the pion-production threshold

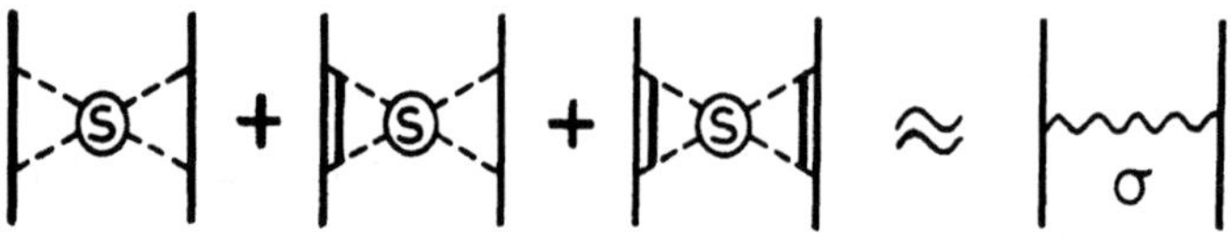

Fig.2: Correlated 2π-exchange contributions considered in the work of Durso et al.[1]. The circled S symbolizes the ππ-S-wave interaction adjusted to empirical ππ-S-wave scattering.

in our model.

In order to compare our 2π-exchange contribution with the corresponding result obtained by dispersion theory [2], in which empirical information of πN- and ππ-scattering is used, we consider sufficiently high partial-wave phase shifts of NN scattering, such that there is no cutoff dependence, either due to form factors in our model or cutting off dispersion in ref. 2 . It turns out that there is in fact good agreement between both models.

Concerning the 3π-exchange contribution, it is necessary to include the (repulsive)πρ terms in order to obtain a quantitative fit to the low partial waves, which are sensitive to the short-range part of the interaction. We stress, however, that we still have to use a rather large ω-coupling ($g_\omega^2 \sim 10$), comparable to OBE-models, since the πρ contribution is completely needed to cut down the overattraction of the $\pi\pi$ contribution.

There are further 3π-exchange contributions of comparable ranges, which are however to a large extent cancelled by corresponding 4π-exchange terms. For example, noniterative $\pi\sigma$-exchange is almost completely cancelled by $\pi\omega$-exchange. Thus, perhaps unexpectedly, there seems to be a strong convergence in the expansion of mesonic diagrams, provided they are grouped together in a suitable way.

A characteristic feature of our model is the small contribution of the intermediate-range tensor force, characterized by a low deuteron D-state probability of 4.3 %. This is because we keep the recoil effects in the meson propagator $[(z-E'-E-\omega_\pi)\omega_\pi]^{-1}$ of the pion exchange. Neglecting these retardation effects, i.e. putting $z = E' + E$, we obtain the conventional static OPE-Propagator, i.e. $[(\vec{q}\,'-\vec{q})^2+m^2]^{-1}$. Note that the recoil effects tend to suppress the iterations of $V_\pi(z)$, since large intermediate momenta are involved. For example, the second iteration is reduced almost by a factor of 2. Consequently, in our model, the effect of the OPE tensor force, which mainly builds up the higher iterations, is relatively small compared to models using the static propagator, e.g. those based on the Blanckenbecler-Sugar (BBS) reduction of the Bethe-Salpeter (BS) equation. (Models avoiding the reduction by using the full Bethe-Salpeter equation [3] keep these retardation effects and thus have a comparable small tensor force contribution as in our model.) Since the amount of tensor force has an enormous impact on results in nuclear physics, it is of utmost importance to keep these recoil terms. Whether this is done in a covariant way like in the BS-equation or noncovariantly as in our model is of negligible practical relevance. In fact, from this point of view (i.e. getting the correct amount of tensor force), the BBS reduction (if not corrected by higher order contributions in the interaction kernel) is to be rejected, inspite of being covariant.

Our model provides a quantitative description of the NN data. For details we refer the reader to ref.[4].

REFERENCES

1. J.W. Durso, A.D. Jackson, B.J. Verwest, Nucl. Phys. A345 (1980) 471.
2. R. Vinh Mau, The Paris Nucleon-Nucleon Potential, in Mesons in Nuclei, Vol.I (North Holland, A'dam 1979) p.151.
3. M.J. Zuilhof, J.A.Tjon, Phys. Rev. C24 (1981) 736.
4. R. Machleidt, Lecture Notes in Physics 197 (1984), p. 352, Ch. Elster, K. Holinde, R. Machleidt, in preparation.

Acta Physica Austriaca, Suppl. XXVII, 605–613 (1985)

ANTINUCLEON-NUCLEON INTERACTION IN A COUPLED-CHANNEL APPROACH WITH SEPARABLE POTENTIALS+

by

W. SCHWEIGER, W. PLESSAS, and J. HAIDENBAUER
Institut für Theoretische Physik
Karl-Franzens-Universität Graz
Universitätsplatz 5, A-8010 Graz
Austria

The antinucleon-nucleon ($\bar{N}$-N) interaction is in general approached via two conceptually different techniques: the optical-potential method and the coupled-channel formalism. Each one has its own-mostly technical-advantages; furthermore there is a unique transition from the coupled-channel to the optical-potential framework.

The coupled-channel formalism is most suitable especially for the use of separable interactions. Since in this case the underlying dynamical equations (integral equations of Lippmann-Schwinger type, say) are solvable by algebraic means, one can include a large number of channels explicitly without complicating the problem too much. Of course, this would be very appropriate for the $\bar{N}$-N system with its many decay channels.

+) Seminar given at the XXIV. Internationale Universitätswochen für Kernphysik, Schladming, Austria, February 20-March 1, 1985

There is a long tradition of separable interactions above all in the N-N system [1,2]. By now the development has matured such that in practice any two-body interaction can be represented in an on-shell and off-shell equivalent way through a sufficiently accurate separable approximation [3]; as an example we mention the separable representation of the Paris N-N potential [4]. Evidently the most important usefulness of separable interactions turns out in few-body applications, where they presently already allow to incorparate two-body subsystem properties in a rather refined and complete way [5].

With respect to the $\bar{N}$-N system there is in addition the evidence from microscopic annihilation models [6] that a significant part of the interaction is nonlocal and in particular separable [7,8]. Similarly the elastic interaction can be expected to be nonlocal at shorter distances, where heavy(multi)-meson exchanges or quark degrees-of-freedom play a role.

Due to these considerations we developed our separable coupled-channel model for the $\bar{N}$-N interaction [9]. We regard the long-range elastic interaction to be well-established, since it can be derived from the meson-exchange part of the N-N interaction. In fact, we applied a G-parity transformation to the Paris potential [10] and provided for a smooth cut-off at $r \lesssim 0.8$ fm. The elastic $\bar{N}$-N interaction obtained in this manner was then cast into a separable form by means of the Ernst-Shakin-Thaler (EST) method [11], which guarantees for a reliable on-shell and off-shell equivalence of the resulting separable potential [3,4,9]. The parametrization of the elastic interaction fixed by this way can be found in Ref.9.

For practical purposes but also in resemblance to microscopic theories [6 - 8] the annihilation interaction was also taken of separable form.In our first attempt and, of course, since the $\bar{N}$-N annihilation is far from being completely understood, we provided only one effective 2-particle decay channel per isospin state. In addition we assumed spinless, non-interacting decay products of equal masses. For a definite isospin channel the strong interaction thus takes the form

$$\mathcal{V}_s = \begin{bmatrix} V_s & V_a^+ \\ V_a & 0 \end{bmatrix}, \qquad (1)$$

where all operators are separable. V_s represents the elastic interaction, while V_a causes the coupling to the annihilation channels.

Our model was designed for all partial waves up to L = 2, where angular-momentum coupling was included up to J = 3. The only open parameters, which are contained in the annihilation part V_a, namely, the coupling strengths and ranges were fitted to experimental data (integrated elastic, charge-exchange, and total cross sections at E_{Lab} = 50 and 100 MeV). The necessary calculations were performed in the framework of the non-relativistic Lippmann-Schwinger equation with rigorous inclusion of Coulomb effects [12] and also taking into account the n-p mass differences [9].

Here we demonstrate the results of our model for elastic differential cross sections below $E_{Lab} \gtrsim 100$ MeV (the quality of the fit to integrated cross section data can be estimated from Fig.3 of Ref.9). The lowest energy so far accessible by experiments is E_{Lab} = 20 MeV. In Fig.1 we show the differential cross section after subtraction of the pure Coulomb amplitude (solid curve). The agreement with the experimental data is remarkably good (note that this a prediction of our model and not a fit). In Fig.1 also the result after subtraction of Coulomb-nuclear interference is shown (dashed-dotted curve) together with the purely nuclear result (dotted curve). From comparing the latter curves it is seen that the Coulomb-distortion effect on the strong interaction is by no means negligible, it amounts to about 10 %. Figs. 2-4 reveal the situation at the energies E_{Lab} = 30,50,and 70 MeV, respectively. The reproduction of the data by our model is still satisfactory everywhere. The influence of the Coulomb-distortion effect is seen to be diminishing with increasing energy.

The separable interaction model as developped so far is not adequate for laboratory kinetic energies beyond $E_{Lab} \gtrsim 150$ MeV. In order to be valid up to higher energies the annihilation has to be modified so as to account for the energy dependence induced by the opening of additional decay channels. Corresponding work paying regard also to fundamental dynamical concepts for the annihilation interaction is presently under way.

This work was supported by Fonds zur Förderung der Wissenschaftlichen Forschung in Österreich, project 5234.

REFERENCES

1. W. Plessas, L. Mathelitsch, F. Pauss, and H.F.K. Zingl, Nonlocal Separable Interactions in the Two-Nucleon System (Univ. of Graz, Graz, 1975).
2. W. Plessas, Acta Phys. Austriaca 54 (1982) 305.
3. J. Haidenbauer and W. Plessas, Phys. Rev. C27 (1983) 63 and references therein.
4. J. Haidenbauer and W. Plessas, Phys. Rev. C30 (1984) 1822.
5. Y. Koike, W. Plessas, and H. Zankel, Preprint UNIGRAZ-UTP 02/85, to appear in Phys. Rev. C.
6. A.M. Green, this Proceedings.
7. M. Maruyama and T. Ueda, Nucl. Phys. A364 (1981) 297;

Phys. Lett. 124B (1983) 121; ibid. 149B (1984) 436; M. Maruyama, Prog. Theor. Phys. 69 (1983) 937.

8. A.M. Green, J.A. Niskanen, and J.M. Richard, Phys. Lett. 121B (1983) 101; A.M. Green and J.A. Niskanen, Nucl. Phys. A412 (1984) 448; ibid., A430 (1984) 605.
9. W. Schweiger, J. Haidenbauer, and W. Plessas, Preprint UNIGRAZ-UTP 01/85, to appear in Phys. Rev. C.
10. W.N. Cottingham, M. Lacombe, B. Loiseau, J.M. Richard, and R. Vinh Mau, Phys. Rev. D8 (1973) 800.
11. D.J. Ernst, C.M. Shakin, and R.M. Thaler, Phys. Rev. C8 (1973) 507.
12. W. Schweiger, W. Plessas, L.P. Kok, and H. van Haeringen, Phys. Rev. C27 (1983) 515; ibid. 28 (1983) 1414.
13. D. Spencer and D.N. Edwards, Nucl. Phys. B19 (1970) 501.

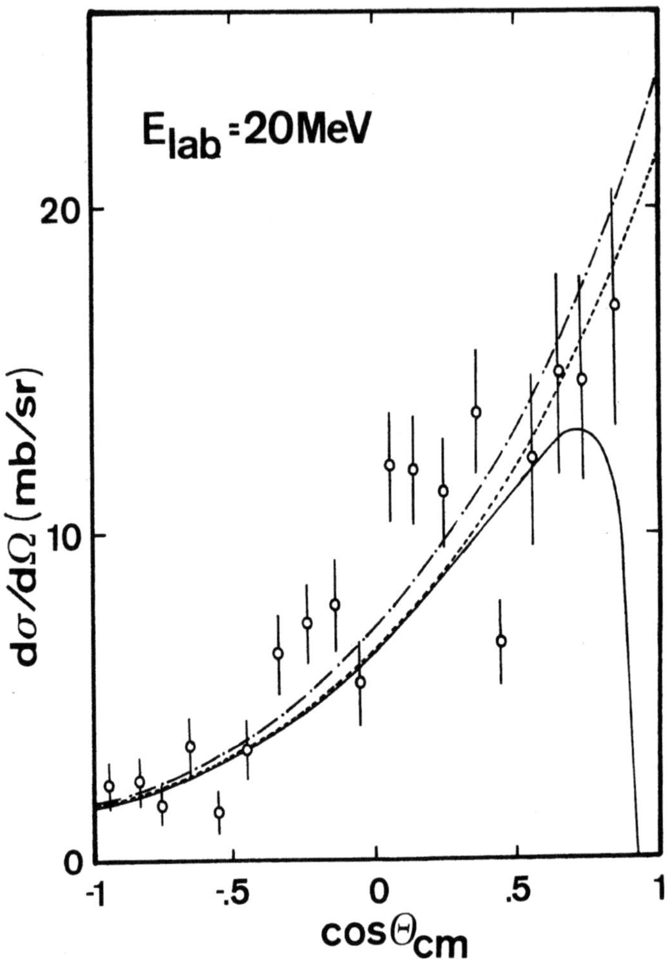

Fig. 1.: Predictions of our separable $\bar{N}$-N interaction model (Ref.9) for elastic $\bar{p}$-p differential cross section at E_{Lab} = 20 MeV. The solid line represents the result after subtraction of the pure Coulomb amplitude, i.e. with Coulomb-nuclear interference and Coulomb distortion. In the dashed-dotted curve only Coulomb-distortion effects are contained, while the dotted line is the purely nuclear result.. Experimental data are from Ref.13.

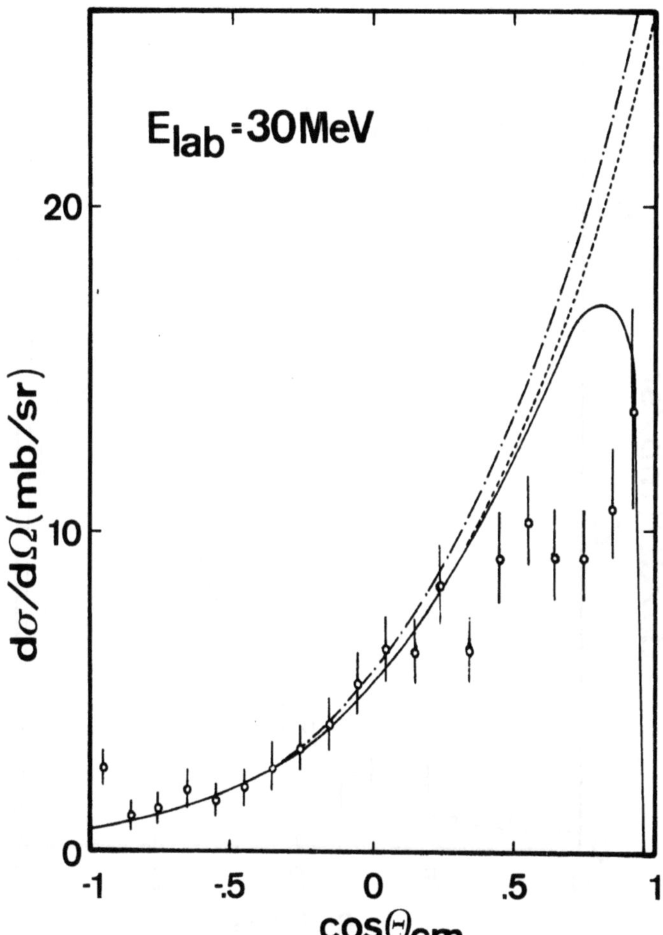

Fig. 2.: Same as Fig.1 for E_{Lab} = 30 MeV.

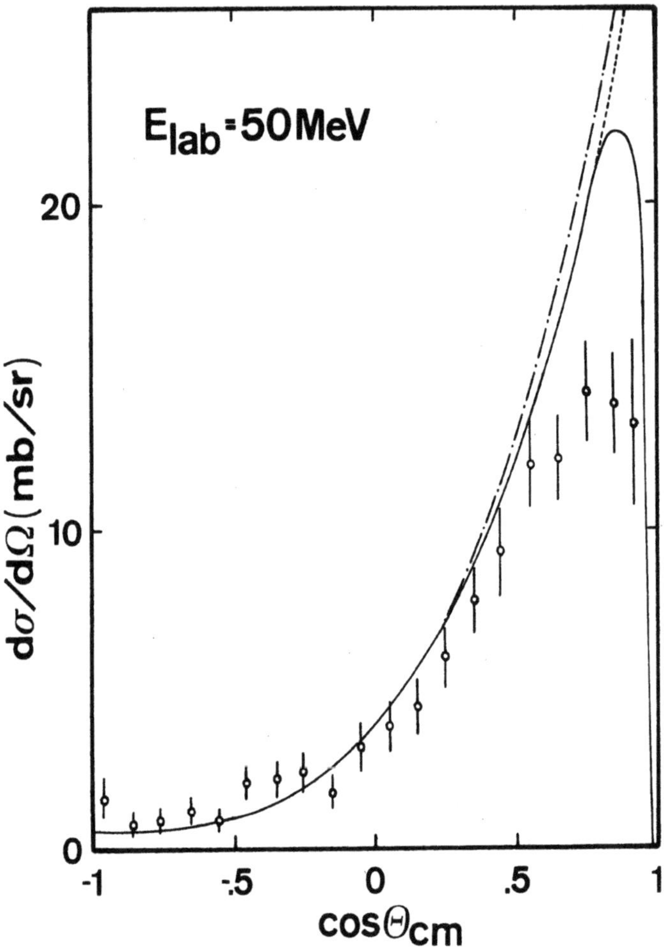

Fig. 3.: Same as Fig.1 for E_{Lab} = 50 MeV.

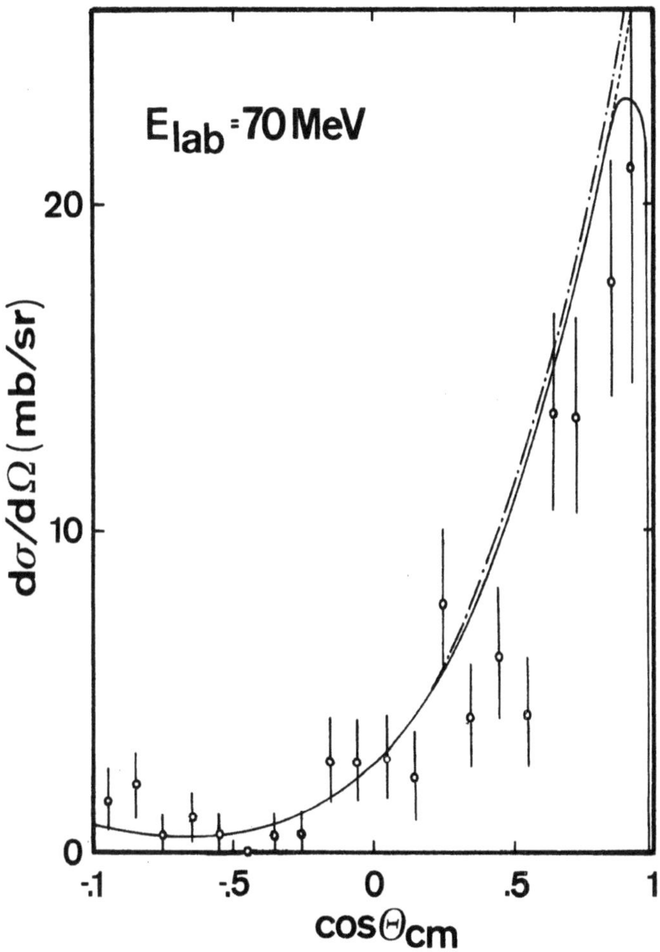

Fig. 4.: Same as Fig.1 for E_{Lab} = 70 MeV.

Acta Physica Austriaca, Suppl. XXVII, 615–625 (1985)

SU(3) ANALYSIS OF $p\bar{p}$ DECAYS INTO TWO MESONS⁺

by

U. HARTMANN
Institut für Physik
Johannes Gutenberg-Universität
D-6500 Mainz, West Germany

ABSTRACT

The process $p\bar{p} \to MM'$ is analyzed in terms of 8 SU(3) invariant amplitudes with definite charge conjugation properties. The amplitudes are constructed such that they either correspond to quark annihilation or rearrangement diagrams. For the C'-odd case the process $p\bar{p} \to PS+PS$ is examined, whereas for the C'-even case the decay $p\bar{p} \to PS+V$ is analyzed. It will be shown that for C'-odd both annihilation and rearrangement contribute, whereas rearrangement is dominant in the C'-even case.

⁺Seminar given at the XXIV. Internationale Universitätswochen für Kernphysik, Schladming, Austria, February 20-March 1, 1985.

1. INTRODUCTION

With LEAR being in operation since 1982 we can expect - and already got some - new data on $p\bar{p}$ decay at rest. This prospect renewed the theoretical interest in the field of $N\bar{N}$ - and especially $p\bar{p}$ - interactions and already led to a variety of papers which can be split into two categories. The authors of the first approach [1-5] derive various annihilation potentials in order to explain the $p\bar{p}$ data, whereas the authors of ref. 6-9 use group theoretical methods.

Furthermore, the authors of ref. 1-6 rely on specific models for the $q\bar{q}$ annihilation vertex whereas the results quoted in ref. 9 are independent of certain models such as $^3P_0$, using as minimal ingredients just SU(3), C-invariance and - at one point - the quark line rule[+].

2. CONSTRUCTION OF SU(3) INVARIANTS

From the point of SU(3) we are dealing with the problem of $8 \otimes 8$ ($p\bar{p}$) $\rightarrow 8 \otimes 8$ (MM'). In general we have, because of $8 \otimes 8 = 27 \oplus 10 \oplus \overline{10} + 8 \oplus 8 \oplus 1$, 8 invariant amplitudes (or reduced matrix elements).

In order to include pseudoscalar (PS), vector (V) and tensor (T) mesons in one framework a new quantum number C' is defined via:

$$C' = C(p\bar{p})C(M)C(M') \qquad (1)$$

where $C(p\bar{p})$ is the charge conjugation eigenvalue of the protonium atom and

[+]For a review of the various approaches see ref. 10.

$$C(M) = \begin{cases} +1 & \text{if M PS- or T-meson} \\ -1 & \text{if M V-meson.} \end{cases} \qquad (2)$$

With this new quantum number we have five C'-even amplitudes, namely 27, $10 - \overline{10}$, 8_{ss}, 8_{sa}, 1 and three C'-odd ones: $10 + \overline{10}$, 8_{as} and 8_{aa}. The indices a and s denote the symmetricity or antisymmetricity of the amplitudes with respect to the exchange of the two mesons (first index) and the two baryons (second index).

Knowing this one can switch to a more convenient set of SU(3) invariants which are easily related to quark diagrams. The mesons are denoted by M_i^j, i.e. their usual octet representation and the baryons are also represented by their quark content, i.e.

$$B_{i[jk]} = \varepsilon_{mjk}\, B_i^m \qquad (3)$$

where B_i^m stands for the usual baryon-octet representation. The new set of SU(3) invariants is now defined by:

$$\langle MM'|T|p\bar{p}\rangle_{C'=\pm} = \sum_i q_i^{\pm}\, Q_i^{\pm} . \qquad (4)$$

I. C' = +

i) annihilation type (see Fig. 1a)

$$q_{2s}^{+} = \bar{B}^{a[bc]} \bar{B}_{a[bd]} (\bar{M}_c^{\ e} \bar{M}_e'^{\ d} + \bar{M}_c'^{\ e} \bar{M}_e^{d})$$

$$q_{3s}^{+} = \bar{B}^{c[ab]} B_{d[ab]} (\bar{M}_c^{\ e} \bar{M}_e'^{\ d} + \bar{M}_c'^{\ e} \bar{M}_e^{d}) .$$

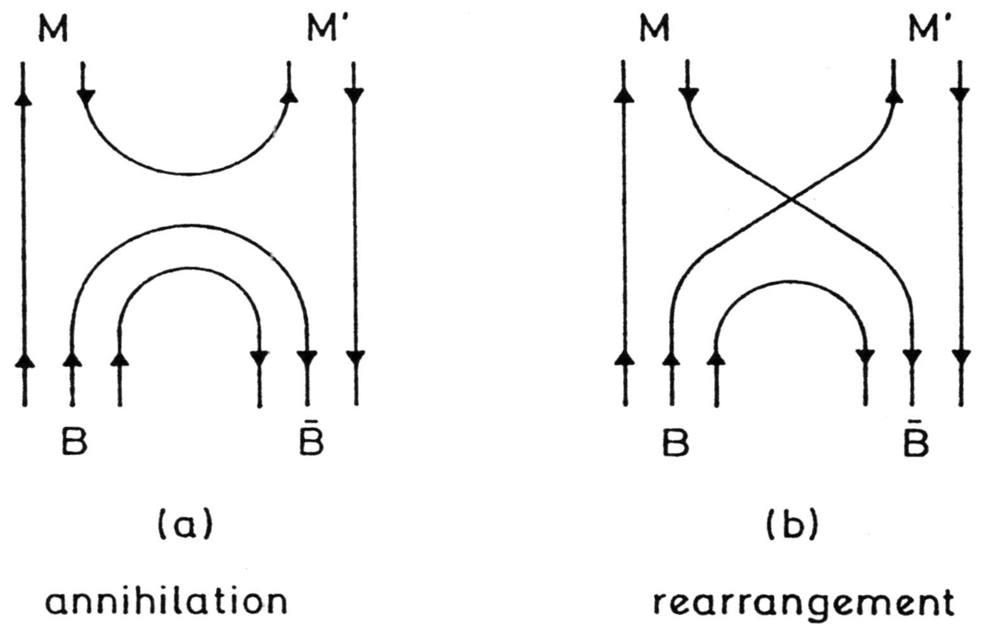

<u>Fig. 1.:</u> Diagrams representing the two types of connected SU(3) tensor contractions for $\bar{B}B \rightarrow MM'$, a) annihilation type,
b) rearrangement type

ii) rearrangement type (see Fig. 1b)

$$q_{4s}^{+} = \bar{B}^{a[bc]} B_{a[de]} \bar{M}_b{}^d \bar{M}'_c{}^e$$

$$q_{5s}^{+} = \bar{B}^{b[ac]} B_{d[ae]} (\bar{M}_b{}^d \bar{M}'_c{}^e \pm \bar{M}'_b{}^d \bar{M}_c{}^e) \qquad (5)$$

II. C' = -

i) annihilation type (see Fig. 1a)

$$q_{2a}^{-} = \bar{B}^{a[bc]} B_{a[bd]} (\bar{M}_c{}^e \bar{M}'_e{}^d - \bar{M}'_c{}^e \bar{M}_e{}^d)$$

$$q_{3a}^{-} = \bar{B}^{c[ab]} B_{d[ab]} (\bar{M}_c{}^e \bar{M}'_e{}^d - \bar{M}'_c{}^e \bar{M}_e{}^d)$$

ii) rearrangement type (see Fig. 1b)

$$q_{6a}^{-} = \bar{B}^{a[bc]} B_{d[ae]} \bar{M}_b{}^d \bar{M}'_c{}^e - \bar{B}^{d[ae]} B_{a[bc]} \bar{M}_d{}^b \bar{M}'_e{}^c$$

The Q's refer to the invariant amplitudes and have to be deduced from experiment (or calculated by using a certain model) whereas the q's represent the various projections.

The inclusion of singlet mesons gives in general five more C-even amplitudes which however correspond to disconnected diagrams and should therefore, according to the quark line rule, be discarded. Thus, the use of the above (5+3) invariant amplitudes automatically incorporates the predictions of the quark line rule.

In Table 1 the values of the quark model couplings for $p\bar{p}$ into various two-body meson final states are listed.

Table 1: Values of quark model couplings for $\bar{p}p$ into various two-body meson states. Particle symbols are used as generic SU(3) symbols.

	C'-even					C'-odd		
	ann. type		rearr.type			ann.type		rearr.type
	q^+_{2s}	q^+_{3s}	q^+_{4s}	q^+_{5s}	q^+_{5a}	q^-_{2a}	q^-_{3a}	q^-_{6a}
$\bar{p}p \rightarrow$ K^+K^-	1	2	0	0	0	-1	-2	0
$K^0\overline{K^0}$	1	0	0	0	0	-1	0	0
$\pi^+\pi^-$	2	2	-1	0	0	0	-2	-1
$\pi^0\pi^0$	2	2	-1	0	0	0	0	0
$3\eta_8\eta_8$	2	2	1	2	0	0	0	0
$\sqrt{3}\pi^0\eta_8$	0	2	0	1	1	0	0	0
$3\sqrt{2}\eta_1\eta_8$	4	4	2	4	0	0	0	0
$3\eta_1\eta_1$	4	4	2	4	0	0	0	0
$\sqrt{6}\pi^0\eta_1$	0	4	0	2	2	0	0	0

3. COMPARISON WITH EXPERIMENT AND CONCLUSIONS

In order to compare the results presented in Table 1 with experiment, i.e. the experimental observed branching ratio (B.R.), phase space and statistical factors have to be taken into account. Denoting with S the spin of the $p\bar{p}$-atom, with p and ℓ the c.m. momentum and lowest angular momentum of the two mesons we define a - what we call - reduced branching ratio C_{eff} via

$$B.R. = C_{eff}\ (25 + 1)\ \left(\frac{p}{M_p}\right)^{2\ +1\ell} \qquad M_p\text{: proton mass} \tag{6}$$

The square root of C_{eff} is then analyzed in terms of the SU(3) quark model invariant amplitudes , i.e.

$$|C_{eff}|^{1/2}_{C'=\pm} = |\sum_i q_i^{\pm}\ Q_i^{\pm}|\ . \tag{7}$$

Sometimes (e.g.,$p\bar{p} \rightarrow \omega f$) C_{eff} is a strongly varying function of the actual meson mass; in these cases C_{eff} is calculated using Monte-Carlo integration.

Table 2 shows B.R. taken from bubble chamber experiments [11] and the calculated C_{eff} for various $p\bar{p}$ decay modes. For the C'-even case we can now easily deduce the following relations for the decay $p\bar{p}(^3S_1)\rightarrow V + PS$:

$$\begin{aligned}
&|Q^+_{2s}(V,PS)| \leq 0.01 \\
&|Q^+_{3s}(V,PS)| \leq 0.01 \\
&|Q^+_{4s}(V,PS)| = 0.11(3) \\
&|Q^+_{5s}(V,PS) + Q^+_{5a}(V,PS)| = 0.10(3)\ .
\end{aligned} \tag{8}$$

A comparison of equations (5) with figure 1 shows that Q^+_{2s} and Q^+_{3s} correspond to annihilation diagrams whereas Q^+_{4s}, Q^+_{5s}, Q^+_{5a} correspond to rearrangement diagrams. The con-

Table 2: Branching ratios for $\bar{p}p \to M + M'$

$\bar{p}p$	C'	M	M'	B.R. ($\times 10^3$)	ℓ	$C_{eff}^{1/2}$	Ref.
$^3S_1$	-	PS	PS				
		K^+	$\overline{K^-}$	0.96(8)	1	0.023(1)	11a
		K^0	K^0	0.80(5)	1	0.021(1)	11a
		π^+	π^-	3.7(3)	1	0.036(2)	11a
$^3S_1$	+	PS	V				
		K^+	$\overline{K^{*-}}$	<0.07	1	$\leq$0.009	11a
		K^0	K^{*0}	<0.09	1	$\leq$0.011	11a
		π^+	ρ^-	19.3(1.0)	1	0.011(1)	11a
		η	ρ^0	5.2(1.4)	1	0.07(1)	11b
		η'	ρ^0	1.4(7)	1	0.08(2)	11c
$^1S_0$	+	V	V				
		K^{0*}	$\overline{K^{0*}}$	<2.9	1	$\leq$0.3	11d
		ρ^0	ρ^0	1.2(1.2)	1	0.08(8)	11a
		ρ^0	ω	23.3(2.3)	1	0.35(2)	11a

Table 2: continued

$\bar{p}p$	C'	M	M'	B.R. ($\times 10^3$)	ℓ	$C_{eff}^{1/2}$	Ref.
$^1S_0$	+	PS	T				
		π^+	A_2^-	14(2)	2	0.62(5)	11e
		π^0	f	4.3(1.5)	2	0.27(6)	11a
		η	f	0.13(12)	2	0.12(11)	11b
$^3S_1$	-	PS	T				
		π^+	A_2^-	6(1)	2	0.24(2)	11e
$^3S_1$	+	V	T				
		ρ^0	f	16(4)	1	0.65(9)	11a
		ω	f	34(4)	1	1.6(1)	11a

clusion to be drawn is therefore, that in the above decay mode rearrangement is dominating annihilation by a factor of ten in amplitudes.

In the C'-odd case the decay $p\bar{p}(^3S_1) \rightarrow PS + PS$ is examined and leads to the equations:

$$|Q^-_{2a}(PS,PS)| = 0.021(1)$$
$$|Q^-_{2a}(PS,PS) + 2\, Q^-_{3a}(PS,PS)| = 0.023(3)$$
$$|2\, Q^-_{3a}(PS,PS) + Q^-_{6a}(PS,PS)| = 0.036(2)\ . \qquad (9)$$

Unfortunately,the data are now not sufficient to determine the amplitudes, nevertheless these relations show that both annihilation and rearrangement contribute equally in this C'-odd channel.

With the prospect of new data [12] , especially those of $p\bar{p}$ annihilations from $^1P_1$ and $^3P_{0,1,2}$ states, one will be able to make further checks on the rearrangement dominance in the C'-even channel.

ACKNOWLEDGEMENT

I would like to thank E. Klempt and J.G. Körner for their participation in this project.

REFERENCES

1. H. Maruyama and T. Ueda, Nucl. Phys. A364 (1981) 297
2. A.M. Green, J.A. Niskanen and J.M. Richard, Phys. Lett. 121B (1983) 101.
3. A.M. Green and J.A. Niskanen, Nucl. Phys. A412 (1984) 448.
4. S. Furui and A. Faessler, Nucl. Phys. A424 (1984) 525.
5. S. Furui, Orsay preprint (1984).
6. C. Dover and P. Fishbane, Nucl. Phys. B244 (1984) 349.
7. H. Genz, Phys. Rev. D28 (1983) 1094.
8. H. Genz, "Comment Concerning Annihilation Diagrams in $\bar{p}p$ Annihilation into two Pseudoscalar Mesons", Karlsruhe preprint (1984).
9. U. Hartmann, E. Klempt, J. Körner, "Proton-Antiproton into two Mesons: Charge Conjugation, Quark Rule and SU(3)", (in preparation)
10. A.M. Green, "Mechanisms for $N\bar{N}$ Annihilation", Lectures given at the "XXIV. Internationale Universitätswochen für Kernphysik", Schladming (1985), within this proceeding.
11. a) R. Armenteros and B. French, "$N\bar{N}$ Interactions", in High Energy Physics, Vol.4, ed. E.H.S. Burhop, Academic Press Inc., New York (1969) and references therein.
    b) P. Espigat et al., Nucl. Phys. B36 (1968) 174.
    c) M. Foster et al. Nucl. Phys. B8 (1972) 93.
    d) S. Baltay et al., Phys. Rev. Lett. 15 (1965) 532.
    e) J. Diaz et al., Nucl. Phys. B16 (1970) 239.
12. A. Ahmad et al., "$\bar{p}p$ Annihilation at Rest From Atomic P States", presented by C. Amsler at the 7th European Symposium on Antiproton Interactions, Durham, 1984.

Acta Physica Austriaca, Suppl. XXVII, 627–644 (1985)

ANTIPROTON-NUCLEUS SCATTERING AND ITS RELATION TO ANTINUCLEON-NUCLEON POTENTIAL+

by

S. ADACHI
and
H.V. VON GERAMB
Theoretische Kernphysik
Universität Hamburg
Luruper Chaussee 149, 2000 Hamburg 50, W.-Germany

ABSTRACT

Elastic and inelastic antiproton-nucleus scattering is analysed with the microscopic optical model potential and the distorted wave g-matrix approximation based on the effective interaction obtained from the Paris antinucleon-nucleon potential. The experimental data from LEAR are well reproduced on the expense of an uncertainty in the final imaginary central potential.

+Seminar given at the XXIV. Internationale Universitätswochen für Kernphysik, Schladming, Austria, February 20-March 1, 1985.

1. INTRODUCTION

Since 1983, a fair amount of data has been accumulated for antiproton-nucleus scattering with the antiproton beam from the LEAR facility at CERN [1][2]. They are summarized in Table 1. The data were preferentially taken from ^{12}C where elastic and inelastic data exist in the momentum transfer region up to 3 fm^{-1}. Now is the time to perform systematic analyses with a microscopic approach successfully used in the nucleon-nucleus scattering. In the analyses we refer to the total accumulation of data.

Table 1: Experimental data base for antiproton-nucleus elastic and inelastic scattering.

nucleus	p(MeV/c)	kind of data
^{12}C	300,600	elastic scattering
^{12}C	550	elastic polarization
^{12}C	300	inelastic scattering 0^+ to 2^+ (4.44 MeV)
^{12}C	600	inelastic scattering 0^+ to 2^+ (4.44 MeV) 0^+ to 0^+ (7.65 MeV) 0^+ to 3^- (9.64 MeV)
^{16}O	600	elastic scattering
^{18}O	600	elastic scattering
^{40}Ca	300,600	elastic scattering
^{208}Ph	300,600	elastic scattering

2. THE EFFECTIVE INTERACTION AND THE MICROSCOPIC OPTICAL MODEL POTENTIAL

Elastic scattering of nucleons from nuclei have been analysed systematically with the optical model potential (OMP). Recently the microscopic OMP based on the Brückner reaction matrix convoluted with the target density appears to be a successful approach [3][4][5]. In this approach the OMP for a nucleon/antinucleon with energy E and momentum k is given by the lowest order term of the Bethe-Brückner expansion of the mass operator

$$U(k,E;k_F) = \sum_{|\vec{p}|\leq k_F} \langle \phi_{kp} | g(\omega) | \phi_{kp} \rangle \tag{1}$$

where the matrix element is antisymmetrised in the case of nucleons but not for antinucleons. The reaction matrix $g(\omega)$ is a solution of the integral equation

$$g(\omega) = V + VG_Q^{(+)}(\omega)g(\omega) \tag{2}$$

with the Green function

$$G_Q^{(+)}(\omega) = \sum_{\vec{p}',\vec{k}'} \frac{Q(k',p';k_F)}{\omega-e(k')-e(p')+\varepsilon}, \tag{3}$$

where ω denotes the starting energy and e's are single particle energies. We take the on-shell value for the parameter ω

$$\omega = e(k) + e(p), \tag{4}$$

where e(k) and e(p) are single-particle energies of the projectile and a hit particle in the target respectively. The single particle energy is given by a sum of a kinetic energy and a potential energy. The Pauli operator Q assures an intermediate excitation of a nucleon above the Fermi level. The nuclear medium corrections appear in the Pauli blocking effect by the Pauli operator Q and the binding energy effect by the potential part in the single particle energies. Defining the correlated wave function ψ by the relation

$$g(\omega)\,|\phi_{kp}\rangle = V|\psi_{kp}\rangle \ , \tag{5}$$

where ϕ is an uncorrelated wave function, the equivalent equation to equ.(2) is

$$\psi_{kp} = \phi_{kp} + G_Q^{(+)}(\omega)V\psi_{kp} \ . \tag{6}$$

The equation (2) or (6) is solved in nuclear matter to obtain the reaction matrix. The standard technique uses a spherically averaged Pauli operator. There are some alternatives for the choice of the potential energy. We solved equ.(6) in coordinate space with partial wave decomposition neglecting the potential energy part of e.

To use the reaction matrix in finite nuclei, we have to construct an effective interaction. This effective interaction is based on the assumption that there exists a local potential which reproduces in an average way the reaction matrix

$$\langle\phi|g(\omega)|\phi\rangle \approx \langle\phi|\ V_{eff}(r;E,k_F)\,|\phi\rangle \ . \tag{7}$$

It consists of central, spin-orbit and tensor parts

$$V_{eff}(r) = \sum_{ST} V_c^{ST}P^SP^T + \sum_T V_{LS}^T P^T(\vec{L}.\vec{S}) + \sum_T V_{tensor}^T P^T S_{12} \ , \tag{8}$$

where S and T show spin and isospin channels respectively. To remove the dependence of the effective interaction on the momentum of the nucleon in the target, we perform an average over the momentum $\vec{p}$ in the reaction matrix

$$(LSJT|g(\vec{k})|L'SJT) = \frac{\sum_{L''} \int_{|\vec{p}|\leq k_F} d\vec{p}\, j_L(qr) V^{J,ST}_{LL''}(r)\, u^{J,ST}_{L'L''}(r;k_F,\vec{k},\vec{p})}{\int_{|\vec{p}|\leq k_F} d\vec{p}\, j_L(qr)\, j_{L'}(qr)} \tag{9}$$

In addition to this average the angular momentum average is performed to eliminate the angular momentum dependence of the effective interaction. The details are given in the previous paper [4]. The obtained effective interaction $V_{eff}(r;E,k_F)$ is local, complex, energy-denpendent and density-dependent. The density-dependence comes from the Pauli operator and the momentum averaging procedure and is included in the form of the local density approximation

$$k_F(r) = [\frac{3\pi^2}{2}\rho(r)]^{1/3} .$$

An alternative representation to equ.(8) of the effective interaction is its classification with respect to the operator nature

$$\begin{aligned} V_{eff} &= V_c^{o} + V_c^{\sigma}(\vec{\sigma}.\vec{\sigma}) + V_c^{\tau}(\vec{\tau}.\vec{\tau}) + V_c^{\sigma\tau}(\vec{\sigma}.\vec{\sigma})(\vec{\tau}.\vec{\tau}) \\ &+ (V_{LS}^{o} + V_{LS}^{\tau}(\vec{\tau}.\vec{\tau}))(\vec{L}.\vec{S}) \\ &+ (V_{tensor}^{o} + V_{tensor}^{\tau}(\vec{\tau}.\vec{\tau}))S_{12} . \end{aligned} \tag{10}$$

To get some insight into this interaction, we show in Fig.1

the absolute magnitude squared of the Fourier transform of the four components of the antinucleon-nucleon central interaction for three different energies. These effective interactions are obtained from the Paris antinucleon-nucleon potential [6] in the limit of $k_F \to 0$. From this plot, one can deduce some magnitude effect: Transition without spin or/and isospin flip dominate. Charge-exchange and spin-flip transitions are an order of magnitude smaller. The Gamow-Teller transitions appear somewhat suppressed.

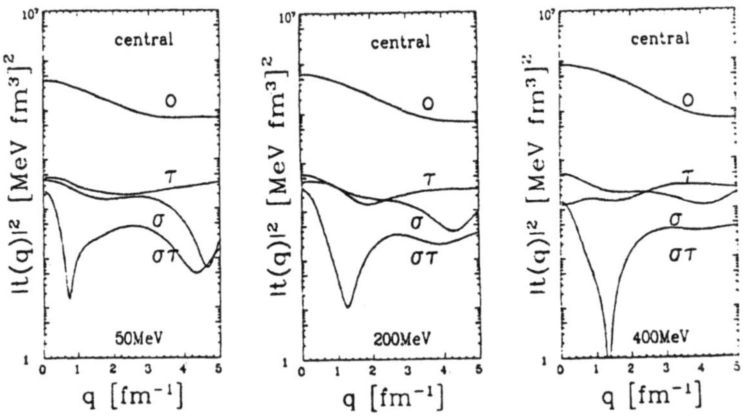

Fig.1.: The absolute magnitude squared of the Fourier transform of the central effective interaction for 50, 200 and 400 MeV.

The folding expression of the OMP is given for the central and spin-orbit interactions

$$V_c(r)+iW_c(r) = \frac{2}{\pi}\int k^2 dk j_o(kr)[(\tilde{\rho}_n(k)+\tilde{\rho}_p(k))\int s^2 ds j_o(ks) V_o^c(s) + (\tilde{\rho}_n(k)-\tilde{\rho}_p(k))\int s^2 ds j_o(ks) V_\tau^c(s)], \quad (11)$$

$$V_{LS}(r)+iW_{LS}(r) = \frac{1}{2\pi}\,\frac{1}{r}\int k^2 dk j_1(kr)[(\tilde{\rho}_n(k)+\tilde{\rho}_p(k))\int s^3 ds j_1(ks) V_o^{LS}(s) + (\tilde{\rho}_n(k)-\tilde{\rho}_p(k))\int s^3 ds j_1(ks) V_\tau^{LS}(s)], \quad (12)$$

where

$$\tilde{\rho}_\alpha(k) = 4\pi\int x^2 dx j_o(kx)\,\rho_\alpha(x) \qquad (\alpha = p \text{ or } n) . \quad (13)$$

In case of antinucleon scattering, the antisymmetrization is suppressed and only the direct term remains. In Fig.2 the OMP obtained for antiproton scattering from ^{40}Ca is shown for various energies. Throughout the folding procedure for the antiproton OMP we use the effective interaction obtained in the limit of $k_F \to 0$ with the energy parameter E corresponding to the projectile at $r \to \infty$. In the final calculation, we need the imaginary part of the central OMP increased by 20 $\sim$ 50 % to obtain a perfect fit. This reflects some inappropriateness of the prescription to generate the effective interaction and shows the necessity for the proper treatment in solving the Brückner-Bethe-Goldstone equation.

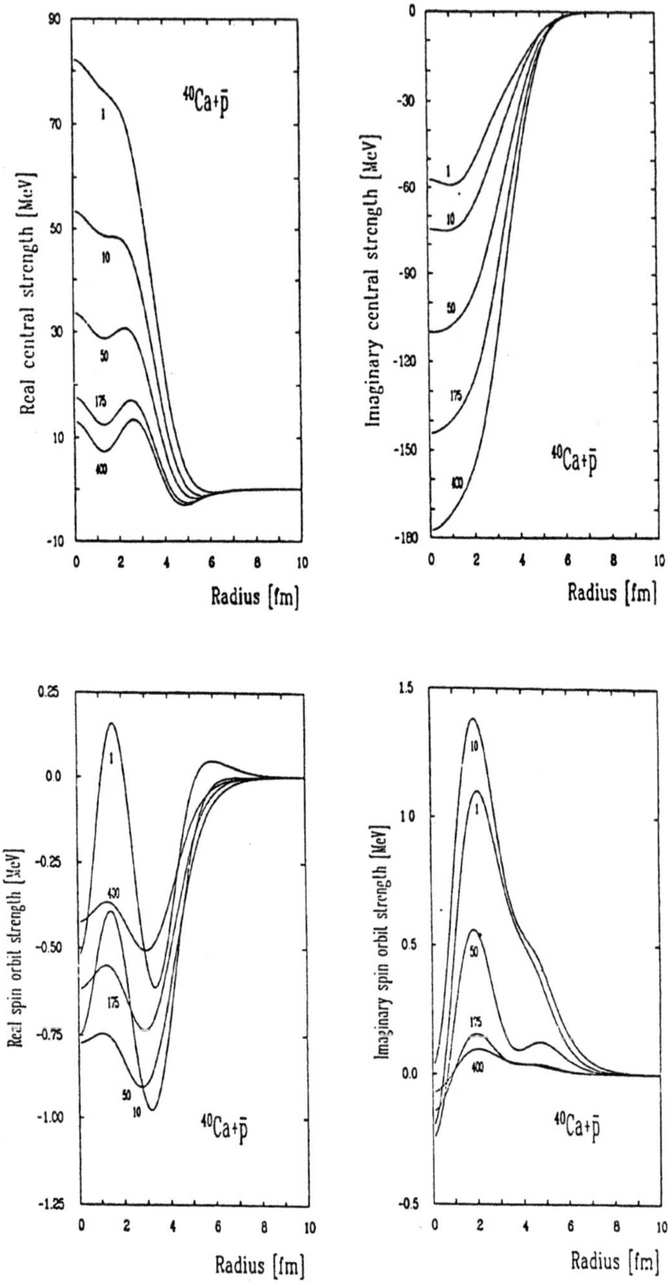

Fig.2.: The microscopic OMP for antiproton scattering from ^{40}Ca for various energies.

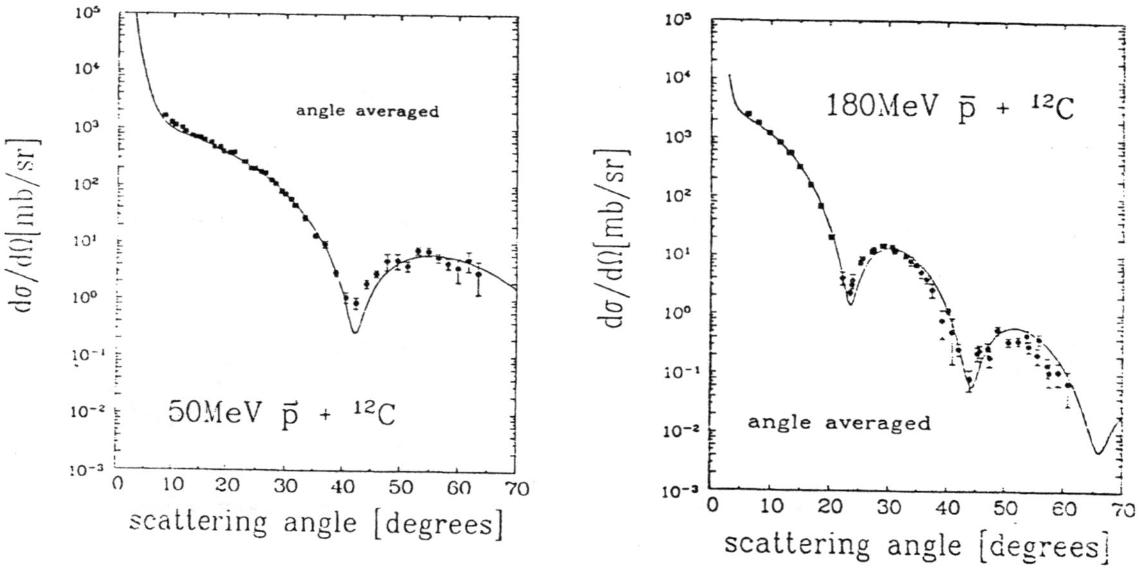

Fig. 3.: The differential cross sections for elastic scattering. The data were taken with 300 and 600 MeV/c antiprotons.

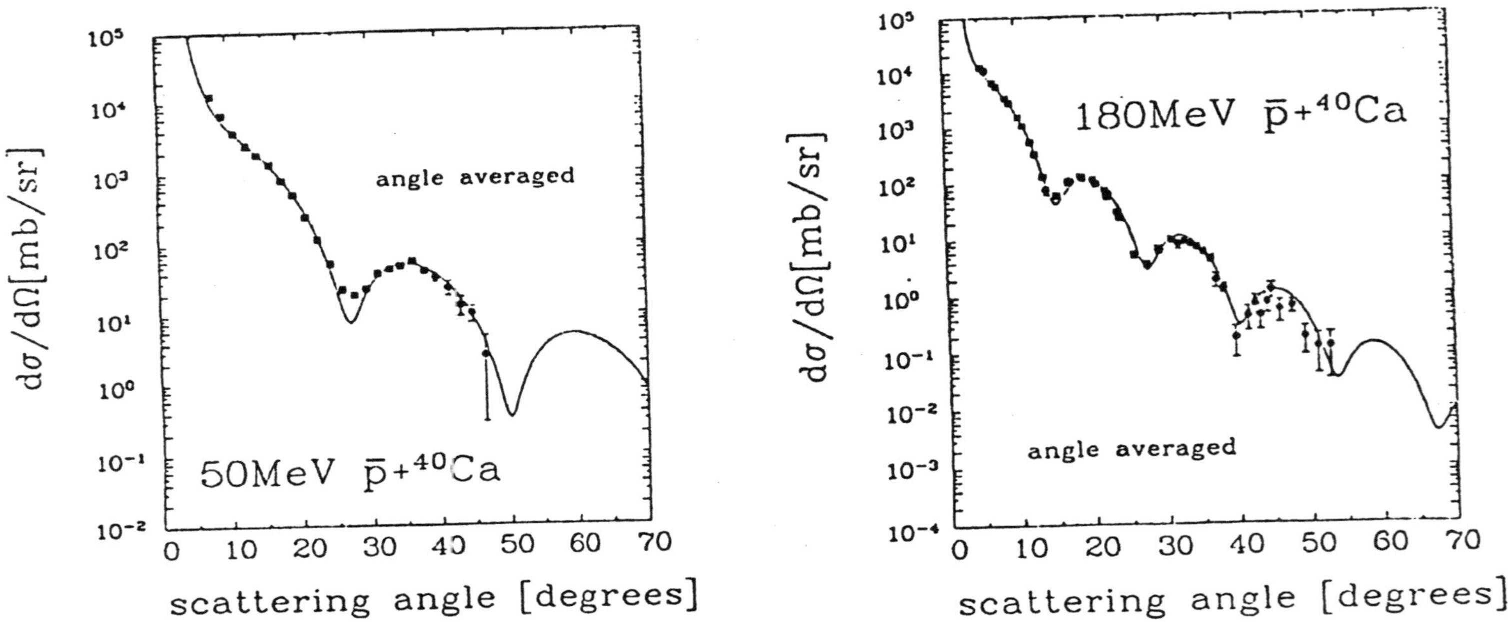

Fig. 3.: continued

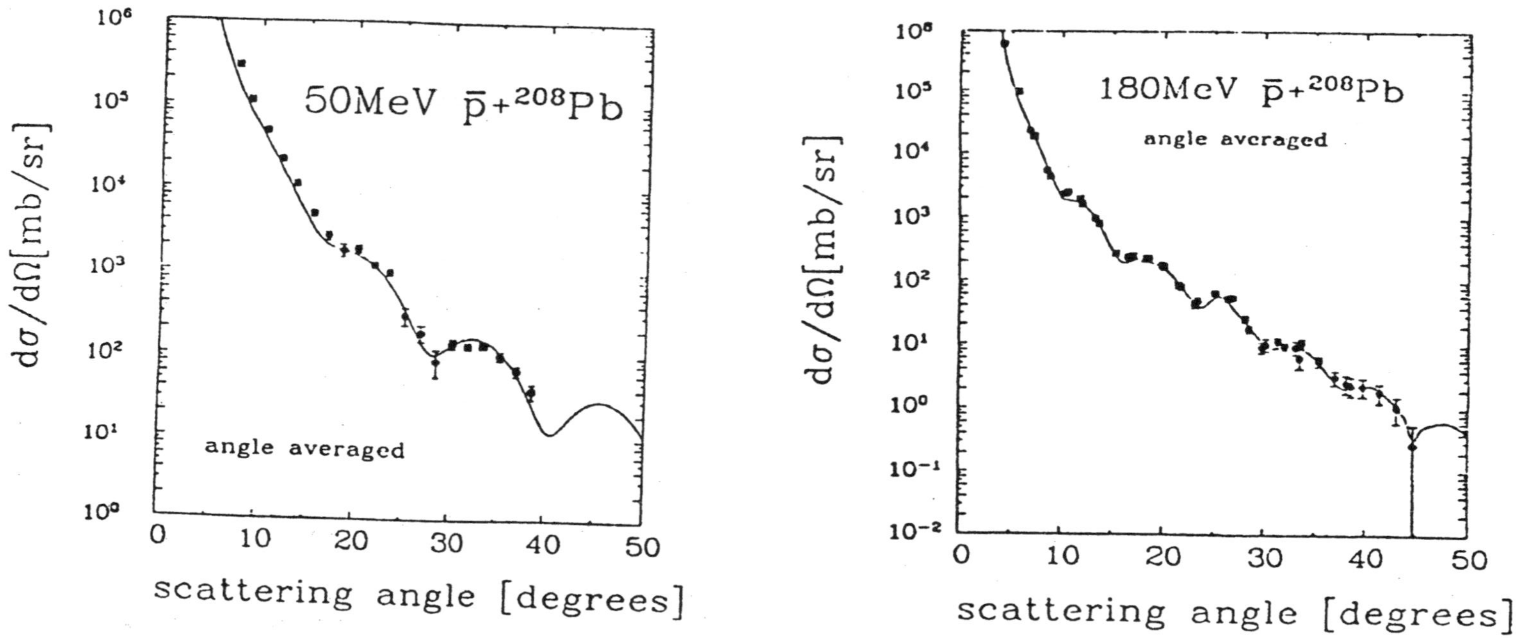

Fig. 3.: continued

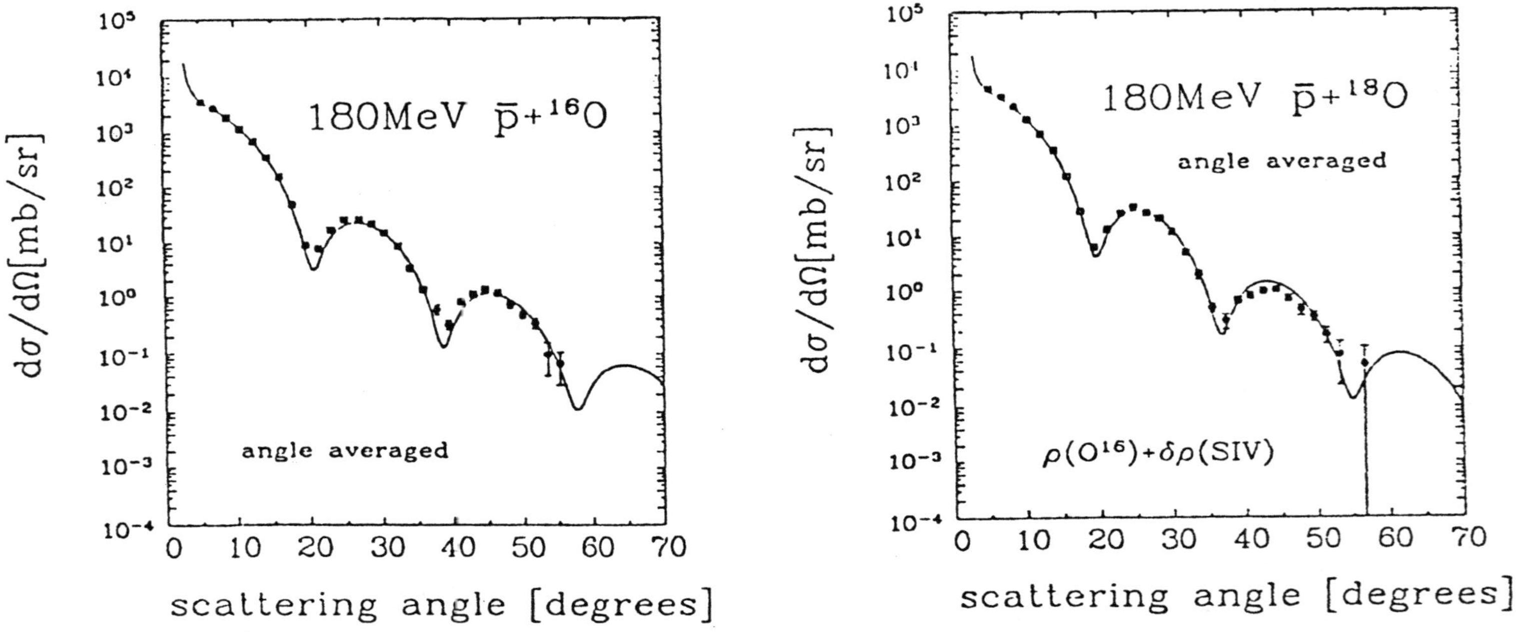

Fig. 3.: continued

3. ELASTIC SCATTERING

The cross sections calculated with the microscopic OMP by the computer code ECIS [8] are shown in Fig.3. In the calculation the angle-average was performed with a typical experimental angular width from 1 to 2 degrees. The agreement with the experimental data [1] is very good. In the folding procedure of the OMP, we use the proton and neutron densities supplied by L. Ray [9] except for ^{18}O. The densities of ^{18}O are obtained by adding the density difference between ^{16}O and ^{18}O from Hartree-Fock calculation with the Skyrme IV interaction [10]to Ray's proton and neutron density of ^{16}O.

A small polarization value is obtained in all cases. In Fig.4 the polarization of 550 MeV/c antiproton scattering from ^{12}C is shown. The agreement with the experimental data [2] is excellent.

4. INELASTIC SCATTERING

The inelastic transition amplitude is given by the distorted wave g-matrix approximation

$$T_{fi} = \sum_{j_1 j_2} S(j_1 j_2 J_i J_{fj} I) \cdot \langle \chi^{(-)}(0)\phi_{j_2}(1) | g(|\vec{r}_1 - \vec{r}_0| ; k_F(r_1)) | \chi^{(+)}(0)\phi_{j_1}(1) \rangle \tag{14}$$

where S denotes spectroscopic amplitudes from RPA or shell model calculations [11], ϕ shell model wave functions and χ distorted waves computed with the microscopic OMP. We therefore have consistent calculations for inelastic as well as elastic scattering. In Fig.5 the cross sections are shown for the transitions to 2^+(4.44 MeV) and 3^-(9.64 MeV) states in ^{12}C. The cross section of the transition to the second

0^+(7.65 MeV) state in ^{12}C is shown in Fig.6. This calculation is based on the Cranell transition density deduced from electron scattering [12]. This transition density has a zero at 2.6 fm and this point corresponds to 25 % target saturation density. The cross section from the calculation including the outer part or the inner part of the transition density is shown by a dashed line and a dash-dotted line respectively in Fig.6. The smallness of the inner part contribution, which is smaller than the outer part contribution by two orders of magnitude, gives a measure from the region within 2.6 fm. Since the annihilation dominates in the surface as expected, the potential feature in the inner region is difficult to verify.

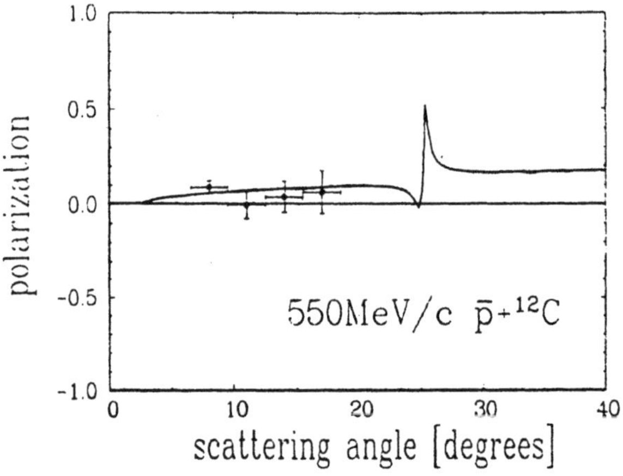

Fig.4.: The polarization for elastic scattering of 550 MeV/c antiprotons from ^{12}C.

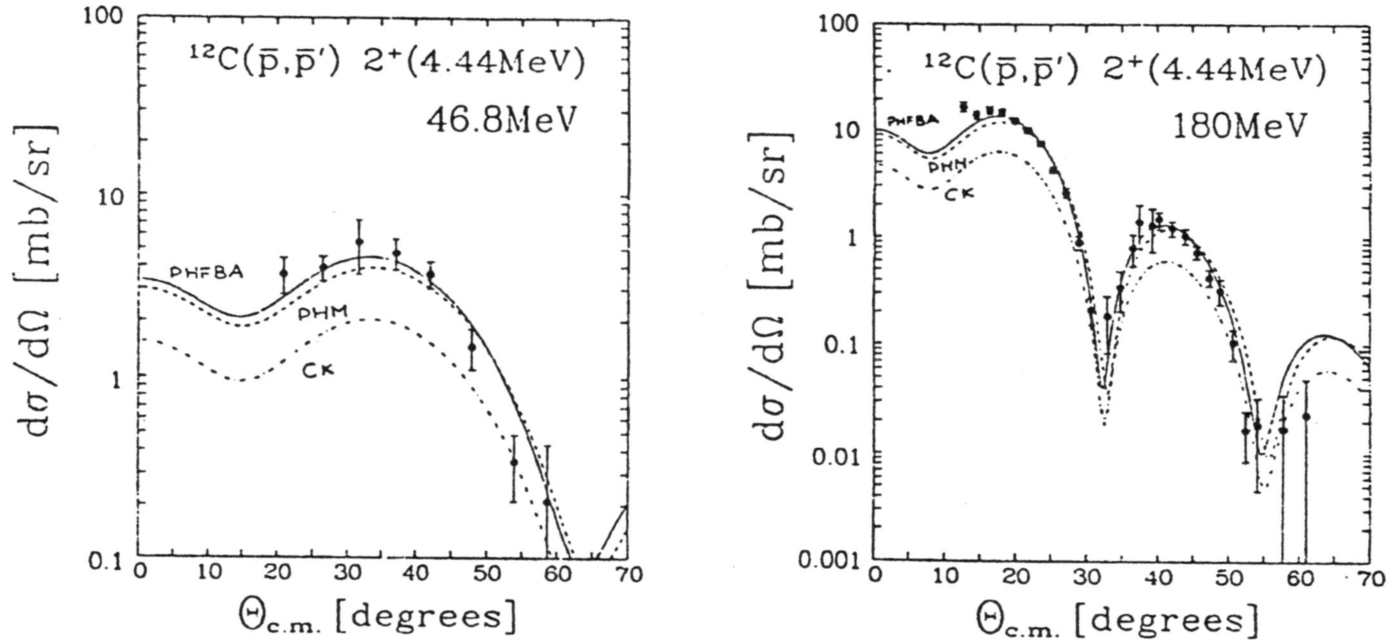

Fig. 5.: The differential cross sections for inelastic scattering of 300 and 600 MeV/c antiprotons exciting the 2^+ and 3^- states in ^{12}C. The nuclear structure models used in DWGA are refered to by CK, MK, PHFBA and PHM [11].

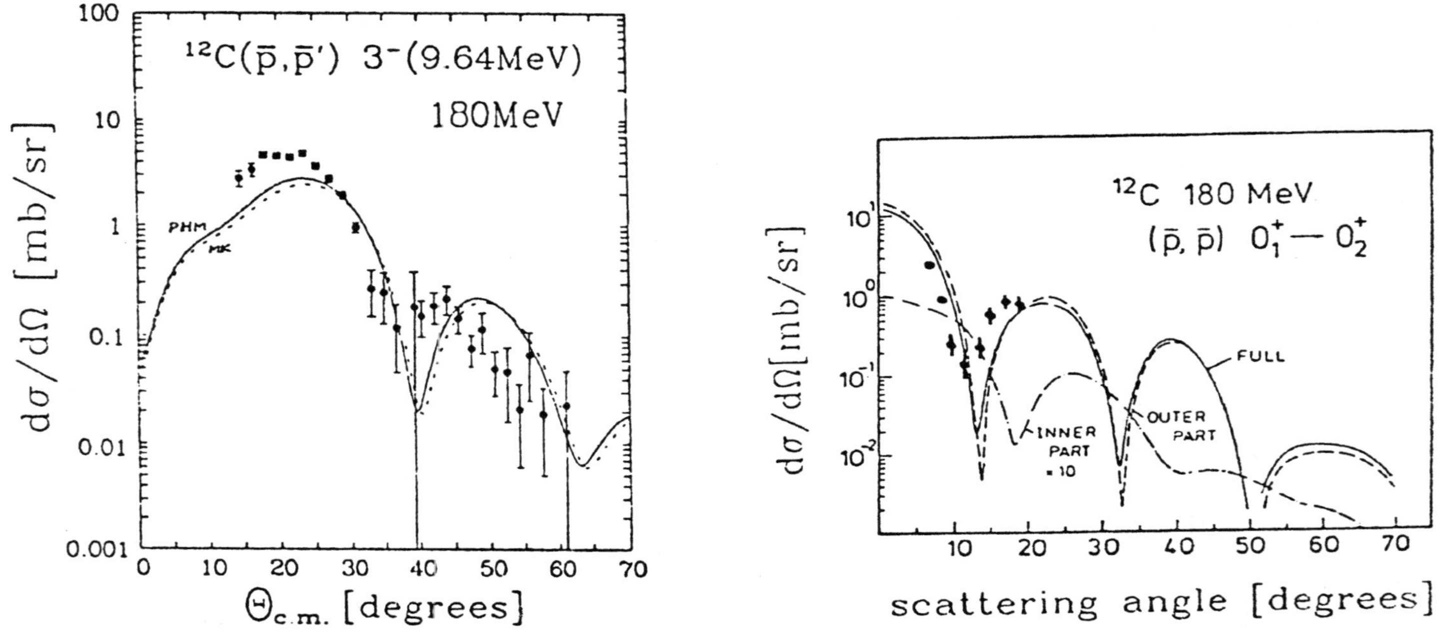

Fig. 6.: The differential cross sections for inelastic scattering of 600 MeV/c anti-protons exciting the second 0^+ state in ^{12}C.

5. SUMMARY

(1) All experimental data for antiproton-nucleus scattering from LEAR (PS184) are analysed with a microscopic approach connecting the nucleon-antinucleon Paris potential with the OMP and the DWGA. The essential ingredient is the effective interaction identified with the reaction matrix.

(2) The data for 300 MeV/c and 600 MeV/c are well reproduced on the expense of the 20 ∿ 50 % uncertainty in the final imaginary central OMP which we attribute to the procedure used to generate the effective interaction.

(3) The polarization is small in all cases. Our prediction of typically 8 % polarization for 550 MeV/c antiproton scattering from ^{12}C is in agreement with the experimental data from LEAR (PS172).

(4) The normal parity transitions dominate the spin or/and isospin flip transitions. The advantage of the antiproton-nucleus scattering is that the nucleus acts as an excellent filter for spin or/and isospin flip of the effective interaction weighted with the target density. Exchange does not obscure this statement as in the case of nucleon scattering. Because of the dominance of the annihilation in the surface, it is difficult to verify the features of the OMP or nuclear structures in the inner region. On the other hand, antiprotons are a good tool to verify the density tail.

REFERENCES

1. D. Garreta et al., Phys. Lett. 135B (1984) 266.
2. R. Birsa et al., preprint CERN-EP/85-28.
3. J. Hüfner and C. Mahaux, Ann. Phys. (N.Y.) 73 (1972) 525.
4. J.P. Jeukenne, A. Lejeune and C. Mahaux, Phys. Rep. 25C (1976) 83.

5. H.V. von Geramb, in The Interaction between Medium Energy Nucleons in Nuclei, AIP Conference, Proceedings no.97, ed. H.O. Meyer (AIP,New York, 1983).
6. L. Rikus, K. Nakano and H.V. von Geramb, Nucl. Phys. A414 (1984) 413.
7. J. Coté, M. Lacombe, B. Loiseau, B. Moussallam and R. Vinh Mau, Phys. Rev. Lett. 48 (1982) 1319.
8. J. Raynal, in Computing as a Language of Physics (IAEA, Vienna, 1972).
9. L. Ray, private communication.
10. M. Beiner, H. Flocard, Nguyen Van Giai and P. Quentin, Nucl. Phys. A238 (1975) 29.
11. W. Bauhoff et al., Nucl. Phys. A410 (1983) 180.
12. H. Cranell, private communication.

Acta Physica Austriaca, Suppl. XXVII, 645–654 (1985)

AN OPTICAL MODEL FROM THE MULTI-CHANNEL FINITE-RANK POTENTIAL+

by

L. CANTON and G. PISENT
Dipartimento di Fisica dell'Università di Padova
Istituto Nazionale di Fisica Nucleare,
Sezione di Padova

In references [1], [2], [3] the authors have outlined a detailed analysis of the two-body, multi-channel scattering process by a finite-rank interaction. As is known from the general theory of multi-channel reactions, the general case, when some channels are open and some remain closed, is rather cumbersome.

In this paper an alternative approach to this general treatment is given, by application of the Feshbach operators $\mathcal{P}$ and $\mathcal{Q}$, projecting on the open and closed channel subspaces, respectively.

Thus, we have obtained a definition of the Jost matrix $\underline{f}$ (in the open channel subspace) which confirms in a more fundamental way the result obtained through a rather complicated matrix algebra, in ref.2. Following Feshbach [4], [5], we introduce the projectors $\mathcal{P}$ and $\mathcal{Q}$, on the subspace of open and closed channels, respectively.

+) Seminar given at the XXIV. Internationale Universitätswochen für Kernphysik, Schladming, Austria, February 20-March 1, 1985.

The effective potential in the open channel subspace assumes the following expression:

$$V^{eff} = PVP + PV\mathcal{Q}g^{(Q)}\mathcal{Q}VP \ , \tag{1}$$

where V is the two-body, multi-channel interaction, and

$$g^{(Q)} = (E-\mathcal{Q}H\mathcal{Q})^{-1} \tag{2a}$$

is the total propagator in the closed channel subspace. We indicate by H the total hamiltonian, by E the total energy. The free propagator in the closed channel subspace is

$$g_o^{(Q)} = (E-\mathcal{Q}Ho\mathcal{Q})^{-1} \ , \tag{2b}$$

where Ho is the unperturbed hamiltonian (H=Ho+V). The two propagators are connected by the following Lippmann-Schwinger equation:

$$g^{(Q)} = g_o^{(Q)} + g_o^{(Q)}\mathcal{Q}V\mathcal{Q}g^{(Q)} \ . \tag{3}$$

We go now to our model, which is based on a channel-dependent, finite-rank interaction, which is written, e.g. in the configuration space r, as follows:

$$V_{cc'}(r,r') = \Sigma_{nn'}\, v_{cn}(r)\lambda_{cn,c'n'}v_{c'n'}(r') \ . \tag{4}$$

Label c=1,2,..$\mathcal{C}$ runs over all (open and closed) channels and n=1,2,...$\mathcal{N}$ is the rank index. Expansion in angular momentum, spin states and target states is assumed as in paper [1] section 2. All channel quantum numbers are included in c, while conserved quantum numbers are omitted. For the properties of strength λ and form factor v, and for all other details, see references [1] and [2].

We stress the fact that, when $r \to \infty$, only the assumption $v(r) \to \exp(-\beta r)$ is needed to guarantee the existence of all the integrals given in the following.

Now it is convenient to distinguish between open channels (labelled throughout by index $p=1,2,\ldots P$) and closed channels ($q=1,2,\ldots Q$). Obviously $P+Q=C$. We will indicate with C (or P,Q) the dimension of the space and the space itself.

By application of equ. (1) to the interaction (4), we find the effective interaction in the open channel subspace, namely:

$$V^{eff}_{pp'}(r,r') = \Sigma_{nn'} v_{pn}(r) \Lambda_{pn,p'n'} v_{p'n'}(r') \ , \qquad (5)$$

where Λ is an effective strength in the space $P \otimes N$, namely:

$$\Lambda_{pn,p'n'} = \lambda_{pn,p'n'} + \Sigma_{q_1n_1q_2n_2} \lambda_{pn,q_1n_1} G^{(Q)}_{q_1n_1q_2n_2} \lambda_{q_2n_2p'n'} \ . \qquad (6)$$

The matrix $G^{(Q)}$ is defined as follows:

$$G^{(Q)}_{qn,q'n'} = \int_0^\infty v_{qn}(r) r dr g^{(Q)}_{qq'}(r,r') r' dr' v_{q'n'}(r') \ , \qquad (7)$$

$g^{(Q)}_{qq'}(r,r')$ being the coordinate representation of the total propagator (2a).

From these equations it is seen that the spaces of the problem are the space $C \otimes N$, and its subspaces $P \otimes N$ and $Q \otimes N$. Consequently it is convenient to redefine the projectors as follows:

$$\mathcal{P}_{cn,c'n'} = \delta_{cc'}\delta_{nn'}\delta_{c\epsilon\mathcal{P}} , \tag{8a}$$

$$\mathcal{Q}_{cn,c'n'} = \delta_{cc'}\delta_{nn'}\delta_{c\epsilon\mathcal{Q}} . \tag{8b}$$

We introduce now, besides equ.(7), the following matrix

$$[G_o^{(Q)}]_{qn,q'n'} = \delta_{qq'}\int_0^\infty v_{qn}(r)rdrg_{oq}^{(Q)}(r,r')r'dr'v_{qn'}(r') , \tag{9}$$

relative to the free propagator in the closed channel subspace. In fact $g_{oq}^{(Q)}(r,r')$ in equ.(9) is the coordinate representation of the propagator (2b), and may be calculated through the following expression:

$$g_{oq}^{(Q)}(r,r') = -\frac{2m}{\hbar^2} F_q(r_<)O_q(r_>) . \tag{10}$$

For further details on this formula, see the equ.(3.7) of reference[2].

It is easily verified that the connection (3) between complete and free propagator leads in our case to the following matrix equation:

$$G^{(Q)} = [(G_o^{(Q)})^{-1} - \mathcal{Q}\underline{\lambda}\,\mathcal{Q}]^{-1} . \tag{11}$$

This equation needs some comments:

- Observe that $G_o^{(Q)} \equiv \mathcal{Q}G_o\mathcal{Q}$; $G^{(Q)} \equiv \mathcal{Q}G\mathcal{Q}$.
- Given any matrix $\underline{A}$ in space $C\otimes N$, $\mathcal{Q}\underline{A}\mathcal{Q}$ is the projection of $\underline{A}$ into $\mathcal{Q}\otimes N$, and is a singular matrix in space $C\otimes N$.
- Here and in the following, the operation $(\mathcal{Q}\underline{A}\mathcal{Q})^{-1}$ is intended to be defined by the following identity:

$$(\mathcal{Q}\underline{A}\mathcal{Q})^{-1}\,(\mathcal{Q}\underline{A}\mathcal{Q}) = \mathcal{Q}. \qquad (12)$$

Summarizing: in a C-channel (only $P<C$ open), N-rank interaction process, the strength $\underline{\lambda}$ is defined in space $C\otimes N$. As seen in equations (5,6,11), the process has been reduced to an effective "optical" interaction in the P open channels, with an effective strength Λ defined in space $P\otimes N$.

We apply now to this effective interaction, the techniques outlined in references 2,3.

The effective Fredholm matrix, relative to the physical solution, is:

$$\underline{M}^{eff} = \underline{\Lambda} - G_o^{(P)(\pm)} , \qquad (13)$$

where the matrices are in space $P\otimes N$, and where:

$$[G_o^{(P)(\pm)}]_{pn,p'n'} = \delta_{pp'} \int_0^\infty v_{pn}(r)rdrg_{op}^{(P)(\pm)}(r,r')r'dr'v_{pn'}(r'), \qquad (14)$$

$$g_{op}^{(P)(\pm)}(r,r') = -\frac{2m}{\hbar^2} F_p(r_<)O_p^{(\pm)}(r_>) . \qquad (15)$$

Note that equations (14,15) are very similar to (9,10), but refer to open (instead of closed) channels. In spite of this formal analogy, the detailed analytical forms relative to real and imaginary momenta are obviously very different (more details are contained in ref.2).

Separation into real and imaginary parts, by means of standard techniques, gives:

$$\underline{M}^{eff(\pm)}_{pn,p'n'} = \underline{P}^{eff}_{pn,p'n'} \pm ic\delta_{pp'}w_{pn}w_{pn'} , \qquad (16)$$

where

$$\underline{P}^{eff}_{pn,p'n} = [\underline{\Lambda}^{-1}]_{pn,p'n'} - \delta_{pp'} PP[G_o^{(P)}]_{pn,pn'} \quad , \tag{17}$$

$$PP[G_o^{(P)}]_{pn,pn'} = -\frac{2m}{\hbar^2} PP\int_o^\infty \frac{w_{pn}(x) w_{pn'}(x) x dx}{x^2 - k_p^2} \quad , \tag{18}$$

$$w_{pn} = \sqrt{2\pi/k_p} \int_o^\infty F_p(r) v_{pn}(r) r dr \; ; \qquad c = \frac{2m}{\hbar^2}\frac{\pi}{2} \; . \tag{19}$$

The symbol PP means Cauchy principal part. Calculate now the image of the effective Fredholm matrix, relative to the physical solution, namely:

$$\underline{N}^{eff(\pm)}_{pp'} = \delta_{pp'} \pm ic \Sigma_{n,n'} w_{pn} [(\underline{P}^{eff})^{-1}]_{pn,p'n'} w_{p'n'} . \tag{20}$$

Then let us write down the effective Fredholm matrix and its image, relative to the regular solution ϕ, namely:

$$\underline{M}^{eff}_{\phi\; pn,p'n'} = \underline{P}^{eff}_{pn,p'n'} - c\delta_{pp'} w_{pn} \hat{w}_{pn'} \tag{21}$$

$$\underline{N}^{eff}_{\phi\; pp'} = \delta_{pp'} - c\Sigma_{n,n'} \hat{w}_{pn} [(\underline{P}^{eff})^{-1}]_{pn,p'n'} w_{p'n'} \; , \tag{22}$$

where

$$\hat{w}_{pn} = \sqrt{2\pi/k_p} \int_o^\infty G_p(r) v_{pn}(r) r dr \; . \tag{23}$$

For further details on equations (19,23), see equations (2.4a) and (3.12b) of ref.2. Note that $[\underline{N}^{eff(+)}]^* = \underline{N}^{eff(-)}$ while $\underline{N}^{eff}$ is real (all the analysis is restricted to the physical momentum space).

Finally we are able to extract the Jost matrix and the scattering matrix in the open channel subspace, namely:

$$\underline{f}^{(\pm)} = \underline{N}^{eff(\pm)} \, [(\underline{N}_{\phi}^{eff})^{-1}] \, , \tag{24}$$

$$\underline{S} = \underline{f}^{(-)} [(\underline{f}^{(+)})^{-1}] = \underline{N}^{eff(-)} \, [(\underline{N}^{eff(+)})^{-1}] \, . \tag{25}$$

We have treated the scattering problem relative to open channels, here from the viewpoint of Feshbach's theory, and in ref.2 through complicated matrix algebras. Now we have to compare equations (20,22) with equations (5.2b, 5.12) of ref.2 , which are written in the following for easy reference:

$$N^{(\pm)}_{pp'} = \delta_{pp'} \pm ic\Sigma_{n,n'} w_{pn} [\underline{P}^{-1}]_{pn,p'n'} w_{p'n'} \, , \tag{26}$$

$$N_{\phi pp'} = \delta_{pp'} - c\Sigma_{n,n'} \hat{w}_{pn} [\underline{P}^{-1}]_{pn,p'n'} w_{p'n} \, . \tag{27}$$

The expression of $\underline{P}$, derived from equ.(4.8b) of paper [2], with reference to the symbols here used, is the following:

$$P_{cn,c'n'} = [\underline{\lambda}^{-1}]_{cn,c'n'} - \delta_{cc'} (\delta_{c\epsilon\mathcal{Q}} [G_o^{(Q)}]_{cn,c'n'} + \\ + \delta_{c\epsilon P} PP [G_o^{(P)}]_{cn,c'n'}) \, . \tag{28}$$

Comparison between equations (17) and (28) clearly shows that

$\underline{P}$ and $\underline{P}^{eff}$ not only are different, but are defined in different spaces. Nevertheless it will be shown in the Appendix that the following identities hold:

$$[(\underline{P}^{eff})^{-1}]_{pn,p'n'} = [\underline{P}^{-1}]_{pn,p'n'}, \tag{29a}$$

$$[(\underline{P}^{eff})^{-1}] = P\,\underline{P}^{-1}P\ . \tag{29b}$$

Equation (29a) is between matrix elements in space $P \otimes N$. Equation (29b) is also between matrices, and it is intended that in the right member we invert $\underline{P}$ in space $C \otimes N$, then project on subspace $P \otimes N$, and finally extract the open channel submatrix.

If identity (29) is assumed, one immediately recognizes that

$$\underline{N}^{eff(\pm)} = \underline{N}^{(\pm)}\ ;\quad \underline{N}_{\phi}^{eff} = \underline{N}_{\phi}\ . \tag{30}$$

It follows that the Jost and scattering matrices, defined here through Feshbach's theory, are the same we obtained in ref.2.

In particular in reference 2 the matrix $\underline{f}$ has been defined in the complete channel space, and then reduced to the open channel subspace, by means of a technique due to Newton [6]. Here we define $\underline{f}$ directly in the open channel subspace through the "effective" interaction, as defined by Feshbach. The final result, that the two definitions coincide, is not at all trivial.

APPENDIX

Here we give the proof of equ.(29). Let $\underline{A}$ be any non-singular matrix and P a projector. Then, calling $Q = \mathbf{1}-P$, the following identity holds:

$$P\underline{A}\, P - P\underline{A}Q(Q\underline{A}Q)^{-1}Q\underline{A}P = (P\underline{A}^{-1}P)^{-1} \ . \tag{A1}$$

We apply this identity to the P-matrix:

$$P\underline{P}^{-1}\, P = (P\underline{P}P - P\underline{P}Q(Q\underline{P}Q)^{-1}\, Q\underline{P}P)^{-1} \ , \tag{A2}$$

and, by means of equ.(28), we obtain the following expression:

$$P\underline{P}^{-1}P = (P\underline{\lambda}^{-1}P - PPP\, G_o^{(P)}P - P\underline{\lambda}^{-1}Q(Q\underline{\lambda}^{-1}Q - G_o^{(Q)})^{-1}Q\underline{\lambda}^{-1}P)^{-1} . \tag{A3}$$

Now, remembering also equ.(17), it can be easily seen that equ.(29) holds if we prove the following identity:

$$\underline{\Lambda}^{-1} = P\underline{\lambda}^{-1}P - P\underline{\lambda}^{-1}Q(Q\underline{\lambda}^{-1}Q - G_o^{(Q)})^{-1}Q\underline{\lambda}^{-1}P \ . \tag{A4}$$

To do this, we apply again identity (A1) to the right term of equ.(A4), and find:

$$P\underline{\lambda}^{-1}P - P\underline{\lambda}^{-1}Q(Q\underline{\lambda}^{-1}Q - G_o^{(Q)})^{-1}Q\underline{\lambda}^{-1}P = (P(\underline{\lambda}^{-1} - G_o^{(Q)})^{-1}P)^{-1} . \tag{A5}$$

Furthermore, it could be demonstrated that:

$$(P(\underline{\lambda}^{-1} - G_o^{(Q)})^{-1}P)^{-1} = P\, (\underline{\lambda} - (G_o^{(Q)})^{-1})^{-1}P \tag{A6}$$

(we omit this lengthy proof for brevity).

Finally, if we apply the identity (A1) to the right term of equ.(A6) (see also equ.(6,11)), we obtain the following equality:

$$P(\underline{\lambda}-(G_o^{(Q)})^{-1})P=\{P\underline{\lambda}P+P\underline{\lambda}Q((G_o^{(Q)})^{-1}-Q\underline{\lambda}Q)^{-1}Q\underline{\lambda}P\}^{-1}=\underline{\Lambda}^{-1} \qquad \text{(A7)}$$

and consequently, the proof of equ.(A4).

REFERENCES

1. G. Pisent, Nuovo Cimento 72A (1982) 303.
2. G. Pisent, L. Canton, DFPD 14/85 and to be submitted to Nuovo Cimento.
3. G. Pisent, L. Canton, DFPD 3/85 and Lettere al Nuovo Cimento, in press.
4. H. Feshbach, Ann. Phys. (NY) 5 (1958) 357.
5. H. Feshbach, Ann. Phys. (NY) 19 (1962) 287.
6. R.G. Newton, "Scattering theory of waves and particles", (Mc Graw-Hill,1966).

Acta Physica Austriaca, Suppl. XXVII, 655–660 (1985)

NUCLEAR SATURATION IN A RELATIVISTIC BRUECKNER-HARTREE-FOCK APPROACH+

by

R. BROCKMANN
Institut für Theoretische Physik
Universität Regensburg
D-8400 Regensburg, W. Germany

R. MACHLEIDT
TRIUMF
Vancouver, B.C., Canada V6T 2A3

Within the framework of a relativistic Brueckner-Hartree-Fock (RBHF) approach the equation of state for nuclear matter has been determined. The relativistic extension of the nonrelativistic Brueckner-Hartree-Fock method effectively leads to more repulsion in nuclear matter. This additional repulsion is necessary to describe quantitatively the empirical saturation point in nuclear matter. Our parameterfree RBHF approach agrees very well with a recent determination of the equation of state at higher densities by Stock et al., too.

+) Seminar given at the XXIV. Internationale Universitätswochen für Kernphysik, Schladming, Austria, February 20-March 1, 1985

In most cases the starting point of a description of nuclear matter is a one-boson-exchange potential (OBEP), which fits phase shifts. Then this OBEP is used to derive a non-relativistic G-matrix. It turns out that all calculations yield values for the binding energy and saturation density which are located on the Coester band. However the Coester band does not pass through the empirically determined saturation point ($E/A = 15.6$ MeV, $K_F = 1.36$ fm^{-1}). In these calculations the binding energy has been worked out only up to second order in the hole-line expansion. Roughly speaking, this means that only terms to second order in the density have been taken into account.

Day [1] could show that the inclusion of higher orders in the hole-line expansion (third order and the most important set of fourth order diagrams) does not lead off the Coester band. In addition he showed that the hole-line expansion converges rapidly. Consequently, one has to conclude that nuclear matter cannot be described in the framework of conventional nonrelativistic nuclear physics.

One way out of this deficiency might be the explicit introduction of the internal degrees of freedom for the nucleon. This is suggested by the fact that the intermediate range of the nucleon-nucleon interaction is due to two-pion exchange including the excitation of a nucleon into a $\Delta(1232)$. Detailed investigations show, however, that the inclusion of isobar degrees of freedom again hardly leads the saturation point of the equation of state off the Coester band [2].

Another aspect of the nuclear matter problem is the inclusion of relativistic effects. The main difference of these relativistic calculations compared to nonrelativistic ones is the explicit treatment of the lower components of the nucleon Dirac wave functions. It turns out that the ratio of the upper to the lower component for a free nucleon is different from that for a nucleon in medium. The lower component of the nucleon spinor in medium is determined by the Dirac Hartree-Fock equations. The change of the small component for the nucleon wave function induced by medium effects leads to changes in the binding energy per nucleon and the density of nuclear matter. These effects do not exist in a nonrelativistic description of nuclear matter.

In ref.3 details are discussed how to set up the relativistic Bethe-Goldstone equation for the relativistic G-matrix, which plays the role of the nucleon-nucleon interaction in the nuclear matter medium. Using this relativistic G-matrix the binding energy per nucleon has been worked out within the Dirac-Hartree-Fock approach. The self-consistency for the nucleon wave function effectively yields more repulsion in nuclear matter. This leads to smaller values for both the density and the magnitude of the binding energy per nucleon. The self-consistency of the nucleon spinors moves the Coester band in the direction towards the empirical saturation point. Therefore it is possible to construct an OBEP (see table 1), which fits the nucleon-nucleon phase shifts on one hand and to obtain in the relativistic Brueckner-Hartree-Fock approach the empirically known saturation point of nuclear matter on the other hand. This was for a long time an outstanding problem of nuclear physics.

In figure 1 the results of the Brueckner-Hartree-Fock calculations with the OBEP of table 1 are summarized. The dashed and the full line display the binding energy per nucleon as a function of Fermi momentum K_F with plane wave spinors and with self-consistently determined nucleon spinor wave functions, respectively. The empirical values at the saturation point are within the rectangle. The dashed curve corresponds roughly to a nonrelativistic Brueckner-Hartree-Fock calculation, while the full line displays the analogous

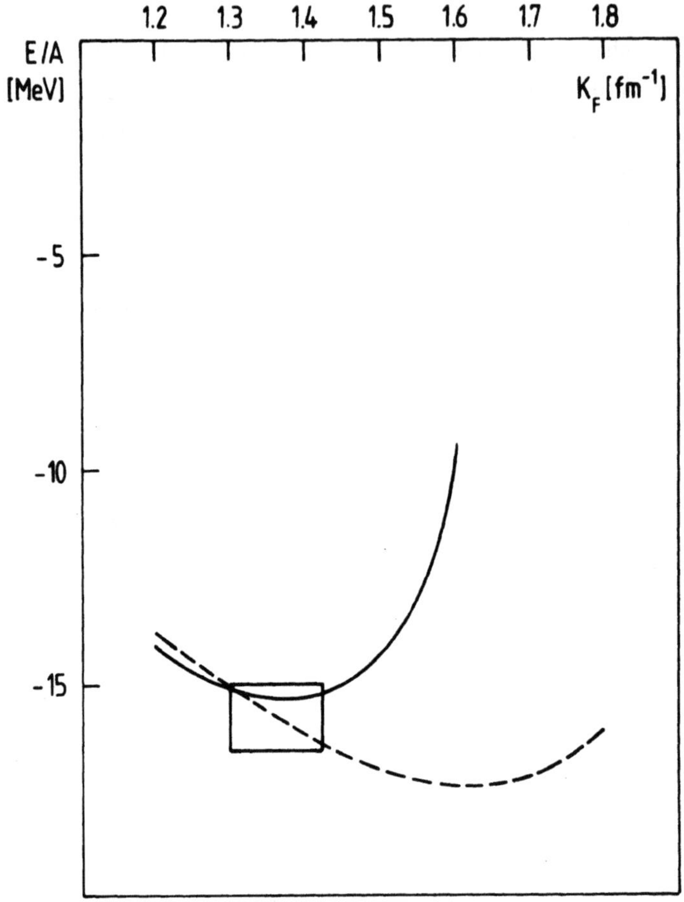

Fig.1.: The dashed and the full line display the binding energy per nucleon as a function of Fermi momentum K_F with plane wave spinors and with self-consistently determined nucleon spinor wave functions, respectively. The empirical values are within the rectangle.

relativistic result. It is important to note that these nuclear matter calculations are parameterfree. Only the free nucleon-nucleon force has been parametrized. No additional parameters arise for the nuclear matter calculation.

Another interesting point is the behaviour of nuclear matter at high densities. At three times nuclear matter density we obtain 65 MeV for the binding energy per nucleon. From a recent relativistic heavy ion experiment Stock et al. [4] concluded that the binding energy per nucleon is 50 ± 10 MeV at three times nuclear matter density. Sano et al. [5] argued that with the addition of missing kinetic energy terms the high density nuclear matter equation of state would be even stiffer ($E/A = 70 \pm 10$ MeV at $\rho = 3\rho_o$, ρ_o is the nuclear matter density). The stiffest equation of state in a nonrelativistic approach delivers only $E/A = 10$ MeV at $\rho = 3\rho_o$.

In conclusion we are able to describe the nuclear matter equation of state up to four times nuclear matter density quantitatively within the relativistic Brueckner-Hartree-Fock approach, while there is no nonrelativistic parameterfree model.

Table 1: Properties of the exchanged mesons for a relativistic OBEP, which fits NN phase shifts and reproduces saturation properties of nuclear matter. Dipole form factors have been used.

Meson	T	J^P	m (MeV)	$g^2/4\pi$	f/g	Λ
π	1	0^-	138	14.6		1200
η	0	0^-	548.8	3.0		1500
σ	0	0^+	570	8.58		2000
δ	1	0^+	983	3.3		1500
ω	0	1^-	782.6	20.0		1500
ρ	1	1^-	769	0.8	6.1	1500

REFERENCES

1. B.D. Day, Phys. Rev. C24 (1981) 1203; Phys. Rev. Lett. 47 (1981) 226.
2. K. Holinde, Phys. Rep. 68 (1981) 121.
3. R. Brockmann and R. Machleidt, Phys. Lett. 149B (1984) 284.
4. R. Stock et al., Phys. Rev. Lett. 49 (1982) 1236.
5. M. Sano et al., Berkeley preprint 1984.

Acta Physica Austriaca, Suppl. XXVII, 661–666 (1985)

INTERACTION AND STRUCTURE OF NUCLEONS IN DEUTERIUM+

by

St. GLAZEK
Institut für Theoretische Physik
Universität Regensburg
W.-Germany
and
Institute of Theor. Physics
Warsaw University
Poland

The following results are sketched in this contribution.

1. The high Q^2 predictions for the elastic electromagnetic deuteron form factors based on the nucleon-nucleon interaction via exchange of mesons $\pi,\sigma,\eta,\rho,\omega,\phi,\delta$ compared with QCD predictions and improved by QCD constraints on the nucleon-meson vertices.

2. The unique nonstandard solution to the problems of the conventional approach to the deuteron smearing corrections in the deep inelastic electron-deuteron scattering, including consideration of dynamical off-shell corrections to the nucleon structure and indicating out that these dynamical corrections are much bigger than the properly calculated Fermi motion effects, for $x \lesssim 0.6$.

+Seminar given at the XXIV. Internationale Universitätswochen für Kernphysik, Schladming, Austria, February 20-March 1, 1985.

1. HIGH Q^2 DEUTERON FORM FACTOR

Using formulae described in ref. 1 we calculate the asymptotic behaviour of the Rosenbluth form factor $A(Q^2)$. The standard one-meson exchanges, with a) nucleon-meson form factors $F = \Lambda^2(\Lambda^2-t)^{-1}$ for scalar pseudoscalar and vector mesons without the Pauli coupling ($\kappa = 0$), and with b) $F = [\Lambda^2(\Lambda^2-t)^{-1}]^{3/2}$ for vector mesons with $\kappa \neq 0$, give the same asymptotic prediction $A \sim (Q^2)^{-10}$ as that obtained from QCD in ref. 2.

The relevance of quark degrees of freedom in the deuteron at experimentally available momentum transfers was inferred from the fact that the reduced form factor picture [2] reproduces well the shape of A down to $Q^2 \sim 1\ GeV^2$, although its normalization is unknown. Therefore, for comparison, we calculated the shape of A in our collinear approximation for five different one meson exchange models [3]. All of them reproduce the shape of A just as well as the reduced form factor formula does. The normalization is given in our case by the magnitude of the two nucleon wave function of the deuteron $u'(0)$. The deuteron wave function is introduced into our collinear model [1] using light front dynamics and it has a unique connection with the non-relativistic wave function thanks to the resemblance between the Weinberg and the Schrödinger equations [4]. To saturate A we need a value of $u'(0)$ almost an order of magnitude larger than that obtained from the reasonable value corresponding to the Reid soft-core potential.

The mesonic theory of the nucleon-nucleon interaction has to take into account that QCD predicts asymptotically $F \sim t^{-3}$ (in contrast to standard F above). Including this additional suppression as suggested in ref. 5 we get a steeper slope of A, in disagreement with experiment and yielding A too small by another order of magnitude for $Q^2 \sim 6\ GeV^2$.

The simple conclusion from collinear approximation

that two nucleons exchanging mesons are not sufficient to explain the data for A is consistent with results of model calculations with full deuteron wave function [6]. Using results of Gross et al. from refs. 3,6 we have performed a test of our collinear model for the deuteron current. In the interesting region of $Q^2 \sim$ few GeV^2 the collinear model does not reproduce the results of impulse approximation with full wave function. This is a caution for all collinear models, particulary QCD collinear projection which effectively describes confining forces. The deuteron is loosely bound but even for $Q^2 \sim$ few GeV^2 it can not be treated as a pair of nucleons each carrying half of the deuteron momentum. This would mean that the motion of nucleons inside the deuteron plays an essential role and provides important effects at relativistic momentum transfers. The fast fall off of the meson-nucleon form factors required by QCD could be a remedy for high momenta, validate a nonrelativistic approach and the collinear approximation. But then to explain A at high Q^2 one has to include new mechanisms.

Our conjecture is that the proper description of the deuteron structure contains nucleons exchanging mesons at large distances and quarks at short distances. Sharp nucleon-meson form factors and quark components at short distances are simulated for observables intensitive to short distance structure by the Schrödinger equation with Yukawa potential and repulsive core. But for high Q^2 observables the repulsive core has to be filled in with quarks and gluons.

2. DEEP INELASTIC DEUTERON STRUCTURE

The current status of theoretical approaches being actually applied to deep inelastic electron deuteron scattering is reviewed in ref. 7. The main question is how to calculate the deuteron structure if one assumes that the deuteron is a proton-neutron bound state. The usually discussed ambiguities are of the order of few percent. However already 3 % ambiguity in the bound nucleon structure gives more than 6 % error in the neutron structure function extracted from deuteron data and almost 20 % ambiguity in the d/u ratio at $x = 0.6$. Note that the EMC effect is of the order of 10 %.

The deep inelastic kinematics singles out z axis in the target rest system, directed along the spatial momentum transfer. We develop the light front dynamics with this z axis defining the light front vector η.

The nucleon is described as a bound state of the active quark a and the passive core c, with the nucleon-quark-core coupling proportional to $\not\eta = \gamma^+$ and $(p^- - c^- - a^-)^{-2}$ which is square of the two-body propagator on the light front motivated by counting rules. p^- is the $p^o - p^3$ component of the momentum of the target, to be continued off shell in the deuteron D to $D^- - n^-$, where n^- is the minus component of the momentum of the spectator nucleon. This simple model gives the following physical properties of the nucleon: spin conservation in hard elastic scattering, asymptotic behaviour of the elastic form factor $\sim(Q^2)^{-2}$, structure function F_2 behaviour $(1-x)^3$ for $x \to 1$ with the active quark carrying spin of the nucleon, satisfaction of charge sum rule and fitting the shape of F_2 for $x > 0.2$.

We imbed this nucleon model into the deuteron. The deuteron wave function is introduced on the light front and we are able to express the deuteron structure by the off-shell continued nucleon structure functions smeared with the probability density ρ. Using the resemblance between the Schrödinger and the Weinberg equations the density ρ is shown to be equal to the Schrödinger $|\psi|^2$ divided by $\sqrt{m^2+\bar{k}^2}$. The same ρ is normalized to unity by the deuteron charge. The fermi motion smearing for $x \lesssim 0.7$ is less than 1 % while the dynamical off-shell correction produces effects of the order of $3 < \bar{k}^2 > m^2$ which is about 5 % in the deuteron. The factor

3 comes from the $(1-x)^3$ behaviour of the nucleon structure function. The dynamical effect in the ratio F_{2N}/F_{2D} is opposite to the observed effect in ratio F_{2O}/F_{2Fe}, what suggests that the EMC effect may be larger than as it is usually interpreted neglecting dynamical corrections.

Our calculation provides also a universal prescription how to continue the on-shell structure functions $W(q^2,\nu)$ off shell in the smearing expression for the bound-state structure functions. One has to treat $W(q^2,\nu)$ as a function of $W(q^2,(p+q)^2)$, preserving proper threshold behaviour in $W^2 = (p+q)^2$. The product pq is independent of the mass shell condition because $q^+ = q\eta = 0$ and the off-shell continuation is done along the vector η, having $\eta^2 = 0$. Gauge invariance is not violated. The longitudiual deuteron cross section vanishes in the limit $Q^2 \to 0$ without ad hoc corrections plaguing conventional approaches [7]. The proton-like sum rules are satisfied but the above mentioned effects complicate the expression for moments of the bound-state structure functions.

The author greately acknowledges invitation to Schaldming and asking for this contribution by Dr. W. Plessas, The author is grateful to Prof. Werner for his valuable comments on the manuscript.

REFERENCES

1. St. Glazek, in "Few Body Problems in Physics" Vol. II - Contributed Papers, B. Zeitnitz (editor) Elsevier Science Publishers B.V., 1984, p.307.
2. S.J. Brodsky, Chueng-Ryong Ji and G.P. Lepáge, Phys. Rev. Lett. 51 (1983) 83.
3. K. Holinde, K. Erkelenz, R. Alzetta, Nucl. Phys. A194 (1972) 161; K. Holinde, Phys. Rep. 68 (1981) 123; W.W. Buck, F. Gross, Phys. Rev. D20 (1979) 2361.

4. J.M. Namyslowski, Progress in Particle and Nuclear Physics, Vol. 14, no.2 (1984) p.1.
5. S.J. Brodsky, SLAC-PUB-3256, November 1983; U. Kaulfuss and M. Gari, Ruhr-Universität Bochum preprint, 1983. The fall off obtained by Kaulfuss and Gari is much flatter than that suggested by Brodsky.
6. St. Glazek, Acta Physica Polonica, B14 (1983) 893; R.G. Arnold, C.E. Carlson, F. Gross, Phys. Rev. C21 (1980) 1426.
7. e.g. D. Kusno, M.J. Moravcsik, Phys. Rev. C27 (1983) 2173.

Acta Physica Austriaca, Suppl. XXVII, 667–676 (1985)

SELECTED TOPICS IN THE EMC EFFECT[+]

by

J. SZWED
Jagellonian University
Kraków

ABSTRACT

The explanations of the EMC ratio of the bound to free nucleon structure functions are reviewed. A general scheme shared by most of the models is presented. The model with nucleons, Δ isobars and pions is described in detail. Other applications of the EMC effect are also given.

I would like to present a short review on the EMC effect. Since the measurement of the ratio of the structure functions

$$R^A(x) = \frac{\frac{1}{A}F_2^A(x)}{\frac{1}{2}F_2^D(x)}$$

[+]Seminar given at the XXIV. Internationale Universitätswochen für Kernphysik, Schladming, Austria, February 20-March 1, 1985.

for iron [1] (Fig. 1) and subsequently for other elements [2] many models attempted to describe the deviation of the ratio $R^A(x)$ from unity. The Fermi motion, important for larger nuclei, causes quite different behaviour (Fig. 1) and plays a role only at Bjorken x close to unity [3]. Consequently one attributes the effect to a significant change of quark distribution inside the nucleon due to nuclear interactions. The surprise was, at least to particle physicists, that these nuclear effects do not vanish at highest momentum transfers Q^2.

Let me start with presenting a framwork shared by most of the models. It expresses the nuclear structure function by a set of structure functions which are better known. This decomposition formula reads

$$F_2^A(x,Q^2) = \sum_a \int_x^A dy\ f^a(y) F_2^a(\frac{x}{y},Q^2) \qquad (1)$$

where $f^a(y)$ is the density of objects "a" inside the nucleus and $F_2^a(z)$ is the structure function of "a". Although the above formula has been never proven it is believed to work at large Q^2 under the condition that the time necessary for the object "a" to be created and interact with the photon is shorter than the average nuclear interaction time which is proportional to $1/m_\pi$[5]. Since the first time depends on the momentum distribution $f^a(z)$, the consistency has to be checked each time the function $f^a(z)$ is constructed. Models of the EMC effect differ among themselves by the choice of the objects "a". Below I review several possibilities.

a = nucleon. As already said this naive expectation is wrong, even after the Fermi motion has been included.

$a = \alpha(3q)+\beta(6q)+ \ldots + \omega(3Aq)$. [6]. This linear combination of multiquark states is usually saturated with the first two terms although a model with 3A quarks [7] or an average

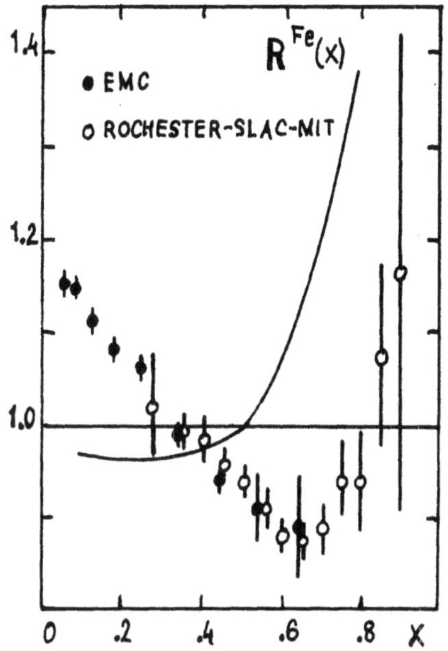

Fig. 1.: The EMC ratio R defined in equ.(1) for iron [1,2]. The curve shows the Fermi motion effect [3].

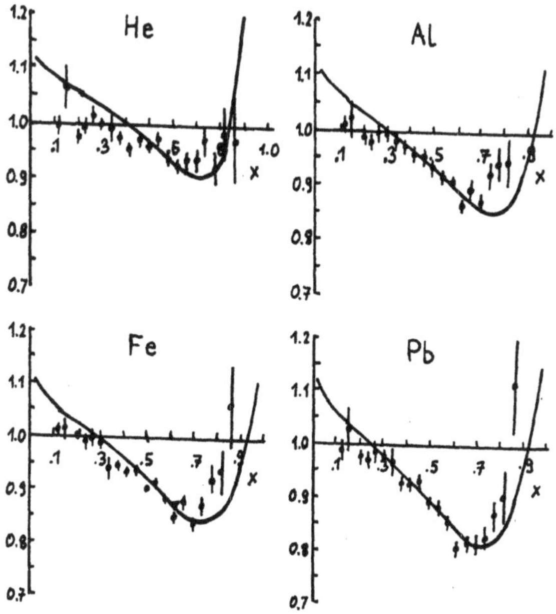

Fig. 2.: The data [1,2] and the N-Δ-π model results [13] on the EMC ratio for different targets.

number of quarks per nucleon [8] are also considered. In each case the distributions $f^a(y)$ of multiquark states and their structure functions have to be constructed. In the latter case the dimensional counting rules or the bag model arguments are usually applied.

a = distorted nucleon. This usually means a nucleon of slightly larger radius or smaller mass [9]. In this class of models a very interesting suggestion connects the EMC effect with the QCD scaling violations [5,10]. Let me describe the last possibility in more detail. The nucleon structure function $F_2^N(x,Q^2)$ depends on the momentum transfer Q^2 through the dimensionless ratio Q^2/μ_N^2 where μ_N sets up the scale. This scale when connected with the confinement radius $\lambda_N \sim 1/\mu_N$ is obviously A-dependent since in the nucleus there are, in addition to the quark modes which propagate across the nucleon, possible modes of the range of the nucleus size λ_A. One can thus write $\lambda_A^2/\lambda_N^2 = \mu_N^2/\mu_A^2$ and construct a model for the change of the confinement radius $\lambda_N \to \lambda_A$. As already argued increasing A means increasing λ_A and can be translated to decreasing μ_A. Since the last scale shows up only in the ratio Q^2/μ_A^2 this effectively means increasing Q^2 when a common scale μ_N is used for all elements. The EMC ratio can be thus written as a ratio of free nucleon structure functions measured at different Q^2

$$R = \frac{\frac{1}{A}F_2^A(x,Q^2/\mu_A^2)}{F_2^N(x,Q^2/\mu_N^2)} = \frac{F_2^N(x,\xi Q^2/\mu_N^2)}{F_2^N(x,Q^2/\mu_N^2)}$$

with ξ depending on λ_A^2/λ_N^2 and Q^2. The model reproduces well the x-dependence of R in the range $0.2 < x < 0.6$ and gives also a correct A-dependence.

a = N,Δ,π,[11-13]. I concentrate on this model not only because I have worked with my collaborators on it [11,13] but also because it seems to be the closest to low energy nuclear physics. It approximates the complicated structure resulting from the nuclear interactions by the presence, in addition to free nucleons, of Δ isobars and π mesons. One writes thus the nuclear structure function as

$$F_2^A(x,Q^2) = \int_x^A dy\ f^N(y) F_2^N(\frac{x}{y},Q^2) + \int_x^A dy\ f^\Delta(y) F_2^\Delta(\frac{x}{y},Q^2)$$

$$+ \int_x^A dy\ f^\pi(y) F_2^\pi(\frac{x}{y},Q^2)\ . \qquad (2)$$

I shall discuss each contribution to the three integrals separately. The free nucleon structure function is measured in deep inelastic scattering off proton and deuteron targets. A parametrisation [13] of these data [14] gives us $F_2^N(x)$ in Equ. (2). The distribution of nucleons inside nucleus A is given in the approximation of the free Fermi gas by [12]

$$f^N(y) = \begin{cases} \frac{3}{4}(\frac{m_N}{k_F})^3 \left[(\frac{k_f}{m_N})^2 - (z-\eta)^2 \right] & \text{for } |z-\eta| < \frac{k_F}{m_N} \\ 0 & \text{for } |z-\eta| \geq \frac{k_F}{m_N} \end{cases}$$

where m_N, k_F and $\eta = \int_0^A dy\ f^N(y)y$ are the nucleon mass, the Fermi momentum and the momentum fraction carried by the nucleons, respectively.

The Δ isobar structure function, which is not measured directly, has been constructed [11] by making use of the spin-isospin structure of the Δ isobars and nucleons. Whereas in the nucleon case the structure function consists of three terms, e.g. for the proton

$$F_2^P(x,Q^2) = \frac{4}{9} A_o(x,Q^2) + \frac{2}{9} A_1(x,Q^2) + S(x,Q^2)$$

with $A_I(x,Q^2)$ being the valence quark contribution with the spectator quarks in spin-isospin I state and $S(x,Q^2)$ denoting the sea quark contribution - in the case of Δ isobars only $A_1(x,Q^2)$ and $S(x,Q^2)$ contribute, allowing us to write e.g. for Δ^+

$$F_2^{\Delta+}(x,Q^2) = \frac{2}{3} A_1(x,Q^2) + \beta S(x,Q^2)$$

where β is fixed by the momentum sum rule. The distribution of Δ isobars $f^{\Delta}(z)$ is taken again to be that of the Fermi gas.

The pion structure function $F_2^{\pi}(x,Q^2)$ can be measured experimentally in the massive (Drell-Yan) lepton pair production, so we use a parametrisation [13] of these data [15]. The pion distribution $f^{\pi}(z)$ is taken from a nuclear matter calculation [16] which uses the many-body Schrödinger equation with the potential being a sum of a phenomenological, short range repulsion term and a standard term resulting from the hamiltonian of n-Δ-π interactions. This last calculation gives in particular the amount of Δ isobars and pions as a function of A. In the case of pions a word of warning is necessary since their light mass may potentially invalidate the convolution formula (2). Fortunately the distribution $f^{\pi}(z)$ is very broad and for most of the pions their life-time is longer than the nuclear interaction time.

All the above elements cast into equ.(2) give the nuclear structure function and consequently the EMC ratio R. The presented scheme is quite restrictive and leaves no free parameters. The resulting curves for four elements are shown together with the data in Fig. 2. One sees that the model describes well the data in the whole range of x.

Since there exist many successful models of the EMC effect one may try to look for other experiments where the change of the "nucleon" structure function could manifest itself. The best known example is the Drell-Yan process on nuclear and nucleon targets, where the cross-section is directly proportional to the target structure function. Many analyses have been performed but when compared with the data they are not very conclusive as concerning the discrimination of the models[17].

A promising possibility has been proposed recently [18] to look at the "polarized" EMC effect, although the experiment may be rather difficult. The idea is to measure the ratio of the spin dependent structure functions

$$R^{\uparrow}(x) = \frac{\frac{1}{A} g_1^A(x)}{g_1^P(x)} \tag{3}$$

These functions appear in the cross-section when both the beam and the target are polarized.

Theoretical construction of these structure functions is straightforward, at least in the N-Δ-π model. The polarized EMC effect seems to be even stronger than the standard one, as seen in Fig. 3.

To summarize this short and by no means complete report, the EMC effect has found many successful explanations of which a large part seems to be equivalent. The distinction between various approaches is possible only when performing a complex analysis which makes use of more than one experiments.

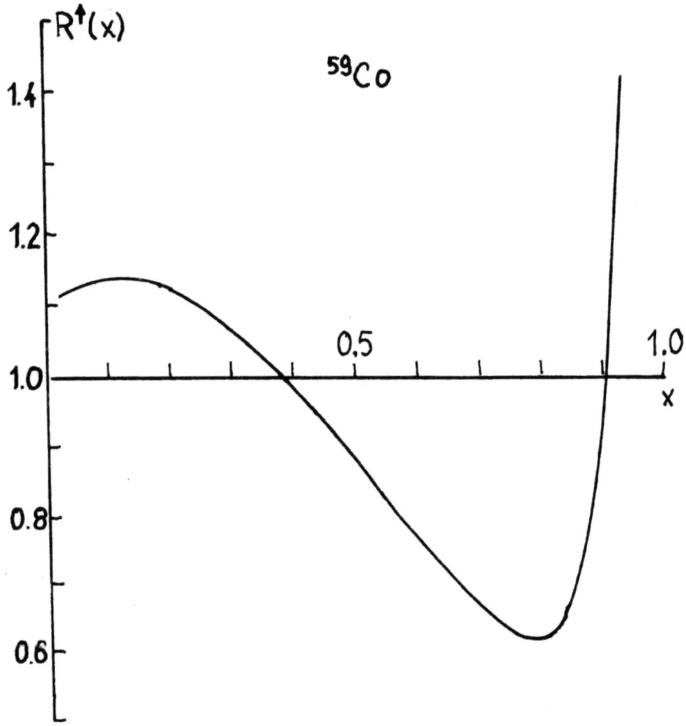

Fig. 3.: The "polarized" EMC effect defined in equ.(3) in the N-Δ-π model [18].

REFERENCES

1. J.J. Aubert et al., Phys. Lett. 123B (1983) 275.
2. A. Bodek et al., Phys. Rev. Lett. 50 (1983) 1431.
3. A. Bodek and J.L. Ritchie, Phys. Rev. D23 (1981) 1070.
4. L.L. Frankfurt and M.I. Strikman, Phys. Rep. 76C (1981) 215.
5. F.E. Close, R.L. Jaffe, R.G. Roberts and G.G. Ross, Phys. Lett. 134B (1984) 449.
6. R.L. Jaffe,Phys. Rev. Lett. 50 (1983) 228. S. Date, Progr. Theor. Phys. 70 (1983) 1683. M. Chemtob and R. Peschanski, Saclay preprint SphT/83/116. G. Cohen-Tannoudji and H. Navelet, Saclay preprint SPhT(83)97; H. Faissner and B. Kim, Phys. Lett. 130B (1983) 322; H. Pirner and J. Vary, Heidelberg preprint UNI-HD-83-02; T. Kanki and O. Miyamira, Osaka preprint OS-GE-84-16; B. Clark, S. Hanna, B. Mulligan and K. Tanaka, Ohio preprint DOE-ER-01545-341 (1984); A.I. Titov, Dubna preprint E2(83)72.
7. W. Furmański and A. Krzywicki, Zeit. f. Phys. 22C (1984) 391.
8. J. Dias de Deus, Munich preprint MPI-PAE/PTh 61/83.
9. M. Staszel, J. Rożynek and G. Wilk, Phys. Rev. D29 (1984) 2638.
10. F.E. Close, R.L. Jaffe, R.G. Roberts and G.G. Ross, Rutherford preprint RAL-84-028; O. Nachtmann and H. Pirner, Heidelberg preprint HD-THEP-84-7.
11. J. Szwed, Phys. Lett. 128B (1983) 245.
12. C.B. Lewellyn-Smith, Phys. Lett. 128B (1983) 107; M. Ericson and A.W. Thomas, Phys. Lett. 128B (1983) 112; E.L. Berger, P. Coaster and R.B. Wiringa, Phys. Rev. 29D (1984) 398.
13. J. Kubar, G. Plaut and J. Szwed, Zeit. f. Phys. 23C (1984) 195; J. Szwed in Proc. of the Moriond Meeting, ed. J. Tran Than Van, Vol. 2, p. 637; in Proc. of the High En. Spin Symp., Marseille 1984; K.J. Heller and J. Szwed, Jagellonian Univ. preprint TPJU 20/84.
14. J.J. Aubert et al., Phys. Lett. 105B (1981) 315; ibid 123B (1983) 123.

15. For references see F. Eisele, Proc. of the Int. Conf. on High En. Phys., Paris 1982.
16. B.L. Friman, V.R. Pandaripande and R.B. Wiringa, Phys. Rev. Lett. 51 (1983) 763.
17. T. Chmaj and K.J. Heller, Jagellonian University preprint TPJU 11/84; Y. Gabellini, J.L. Meunier and G. Plaut, Nice preprint NTH 84/9.
18. L. de Barbaro, K.J. Heller and J. Szwed, Jagellonian University preprint TPJU 24/84.

Acta Physica Austriaca, Suppl. XXVII, 677–683 (1985)

PROPERTIES OF SU(3) SKYRMION+

by

M.A. NOWAK
Institute of Physics
Jagellonian University
Reymonta 4, 30-059 Krakow, Poland

Properties of the $SU(3)_{LEFT} \times SU(3)_{RIGHT}$ Skyrme model are reviewed. Analogies with text-book examples are emphasized.

The old idea that baryons may be topological solitons in a nonlinear sigma model [1] has been recently widely explored in the literature [2,3]. Several static properties of baryons were evaluated for the two-flavour case and the qualitative agreement with the experimental data has been obtained [4]. Other aspects of the model, like consequences for nuclear physics [5], description of resonances [6], or weak current transitions [7] are under study. In this contribution we present the idea of an extension of the model to the real world of three flavours [3,8-10].

We start withthe Skyrme Lagrangian written in terms of left currents $L_\mu = U^{-1}\partial_\mu U$, where $U = U(t,\vec{x})$ is a SU(3) matrix, $F_\pi = 186$ MeV and e is a dimensionless constant:

+Seminar given at the XXIV. Internationale Universitätswochen für Kernphysik,Schladming,Austria,February 20-March 1,1985.

$$\tilde{L} = (F_\pi^2/16)\mathrm{Tr}(L_\mu L^\mu) + (1/32e^2)\mathrm{Tr}[L_\mu, L_\nu]^2 \ . \tag{1}$$

Various properties of this model have been already explained [5]. Let us concentrate on novel aspects of the approach at the SU(3) level. In this case we must add a new term to the Lagrangian, the so-called Wess-Zumino-Witten (WZW) term [3]:

$$L = \tilde{L} - \frac{iN_c}{240\pi^2} \int d\Sigma^{\mu\nu\alpha\beta\gamma} \ \mathrm{Tr}(L_\mu L_\nu L_\alpha L_\beta L_\gamma), \tag{2}$$

where N_c is a number of colours and the integral is over the five dimensional manifold in SU(3) space. Several important features of this term were established by Witten [3]. WZW term removes the redundant, unphysical symmetry of L ('boson number conservation'). The coupling turns out to be quantized and is interpreted as the number of colours. The most remarkable result is the WZW term guarantee that the soliton must be a fermion if and only if the number of colours is odd. The presence of the WZW term is also crucial for the quantization procedure (the classical solutions are the same as in the SU(2) case, because WZW term vanishes for the static Skyrme ansatz (3)). Following the method of collective variables [4] we replace U in the Lagrangian by $\tilde{U} - A(t)U_o(\vec{x})A^{-1}(t)$ where A(t) is an arbitrary, time-dependent SU(3) matrix and

$$U_o(\vec{x}) = \exp(i \sum_{k=1}^{3} \lambda_k \hat{x}_k f(r)), \quad \hat{x} = \vec{x}/r, \quad r = |\vec{x}|, \tag{3}$$

where λ_k are Gell-Mann matrices and $f(0) = \pi$, $f(\infty) = 0$. Second type of embedding of the Skyrme ansatz in the SU(3) group is discussed in [11]. After performing integrations we get the Lagrangian $L = L(A^{-1}\dot{A})$. With Euler-like parametrization of the SU(3) group $A(t) = \exp(iq_m\lambda_m/2)$ we obtain

$$L = -M + \frac{1}{2}g_{mn}\,\dot{q}^m\,\dot{q}^n + \frac{N_c}{2\sqrt{3}}\,\omega_m\,\dot{q}^m\ , \tag{4}$$

where M is the classical mass of a soliton [4]. The last term is nothing but the remnant of the original WZW term. The metric g_{mn} and the 'gauge potential' ω_m are given by

$$g_{mn} = 1/2\ \mathrm{diag}(\alpha^2,\ \alpha^2,\ \alpha^2,\ \beta^2,\ \beta^2,\ \beta^2,\ 0)_{ab}\ e^a_m e^b_n\ , \tag{5}$$

$$\omega_m = 3/2\ \mathrm{Tr}(Y\lambda_a)e^a_m\ , \tag{6}$$

where e^a_m are the vielbeins for the SU(3) group and the explicit form of the parameters α^2, β^2 can be found in [8]. The fact that the metric is not invertible (det $g_{mn} = 0$) indicates that the number of independent coordinates is less than the number we had used. Indeed, since ansatz (3) commutes with the hypercharge operator $Y = \frac{1}{\sqrt{3}}\lambda_8$, the A(t) matrix is determined up to the right multiplication of the $U(1)_Y$ group. Therefore the space of independent variables is rather a seven-dimensional coset space $SU(3)/U(1)_Y$ instead of the original SU(3) manifold. By performing a coordinate rotation $\vec{x}' = \hat{O}\,\vec{x}$, $\hat{O} \in SO(3)$ we identify the remaining symmetries of the model. Since the effect of rotation on the soliton can be compensated by an appropriate transformation in the isospin space

$$U(t,\hat{O}\vec{x}) = A(t)U_o(\hat{O}\vec{x})A^{-1}(t) = A(t)I(\hat{O})U_o(\vec{x})I^{-1}(\hat{O})A^{-1}(t) =$$
$$= \tilde{A}(t)U_o(\vec{x})\tilde{A}^{-1}(t)\ , \tag{7}$$

the Lagrangian remains invariant under $\hat{O}$. It is natural to identify the physical meaning of this $SU(2)_{RIGHT}$ symmetry with the baryonic spin. The remaining $SU(3)_{LEFT}$ symmetry we interpret as a flavour group. Instead of quantizing the Skyrmion on the coset space (see [8]) we present two claryfying analogies which directly lead to the final mass spectrum. The first one is based on the fact that our Lagrangian resembles the quantum mechanical system of a particle moving under the influence of a monopole-type gauge field [3,8,9]. We illustrate the analogy by the following comparison:

MAGNETIC MONOPOLE - SKYRMION ANALOGY

Structure of local Lagrangian on the coset $X = G/H$	
$L = \frac{1}{2} m \dot{x}_i^2 + egA_i\dot{x}_i$	$L = \frac{1}{2} g_{mn} \dot{q}^m\dot{q}^n + \lambda\omega_m\dot{q}^m$
Nontrivial topology	
A_i - singular Dirac potential Homotopy class: $\Pi_1(S_1) = Z$	ω_i - $U(1)_Y$ gauge potential Homotopy class (after dimensional reduction of the WZW term): $\Pi_2(X) = \Pi_1(H) = Z$
Constraint (PS denotes Physical State)	
$\hat{x}_k\hat{J}_k \|PS> = n/2\|PS>$	$\hat{Y}_{RIGHT}\|PS> = N_c/3 \|PS>$, (8)
Quantization condition	
$eg = n/2$ (Dirac condition)	$\lambda = \frac{N_c}{2\sqrt{3}}$

The second analogy [10] consists in an observation that constrained Hamiltonian H($H = p_m\dot{q}^m - L$, $p_m = \frac{\partial L}{\partial \dot{q}^m}$, eighth right generator equals to λ) is identical with the symmetric top Hamiltonian in seven-dimensional configuration space. Because of the constraint originating from the WZW term only selected SU(3) representations (i.e. satisfying (8) are allowed for SU(3) Skyrmions.

SYMMETRIC TOP - SKYRMION ANALOGY [10]

Hamiltonian	
$H = \frac{1}{2I_a}(J_\xi^2+J_\eta^2)+\frac{1}{2I_b}J_\rho^2$	$H = \frac{1}{2\alpha^2}\sum_{i=1}^{3}(2r_i^2)+\frac{1}{2\beta^2}\sum_{i=4}^{7}(2r_i^2)$
	r - generators of $SU(3)_{RIGHT}$
Symmetries	
$SU(2)_{LEFT}$ - Symmetry of space	$SU(3)_{LEFT}$ - Flavour group
	$SU(2)_{RIGHT}$ - Spin group
$U(1)_{RIGHT}$ - Symmetry of the top	$U(1)_{RIGHT}$ - 'right hypercharge Y' group
Eigenfunctions	
$\psi = D^J_{KM}$	$\psi = D^{p,q}_{AB}$
J lables SU(2) irreducible representation	p,q label SU(3) irredicible representation
M - value of the third component of $\hat{J}$	A ≡ (I, I_3, Y) - set of isospin and hypercharge components
K - value of the scalar $(\hat{J}.\hat{n})$, $\hat{n}$ - axis of the top	B ≡ ($J, -J_3, Y_R=N_c/3$) - constrained set of spin components

The lowest allowed representation is an octet with spin 1/2, next decuplet with spin 3/2. Hence the model reproduces known SU(3) representations of baryons giving the mean energy of the baryon multiplets

$$E = M + \frac{1}{2}\,(\alpha^{-2} - \beta^{-2})\; C_2^S + \frac{1}{2}\,\beta^{-2}(C_2^F - N_c^2/12)\;, \qquad (9)$$

where C_2^F, C_2^S are Casimir operators for flavour and spin, respectively. Numerical predictions [8,9,12] yield the static properties of baryons with 30 % accuracy. This qualitative agreement suggests that further improvement of the model is needed.
One possibility is to quantize not only the zero-modes of A(t) but also the remaining ones [13]. Another approach attempts to furnish the Skyrme soliton with a quark-bag core [14]. Encouraging results [14] make plausible further studies in this field.

This work was supported in part by M. Curie Foundation under contract No.F7-071-P.

REFERENCES

1. T.H.R. Skyrme, Proc. Royal Soc. A260 (1961) 127.
2. A.P. Balachandran et al., Phys. Rev. Lett. 49 (1982) 1124, Phys. Rev. D27 (1983) 1153.
3. E. Witten, Nucl. Phys. B223 (1983) 422, 433.
4. G.S. Adkins, C.R. Nappi, E. Witten, Nucl.Phys. B228 (1983) 552.
5. R. Vinh Mau, Lectures at this School.
6. J.D. Breit, C.R. Nappi, Princeton preprint (1984).
7. J. Trampetic, Munich preprint MPI-PAE/PTh32/84 (1984).
8. P.O. Mazur, M.A. Nowak, M. Praszalowicz, Phys. Lett. B147 (1984) 137, P.O. Mazur, Cracow preprint TPJU-5/84 (1984).
9. E. Guadagnini, Nucl. Phys. B236 (1984) 35.

10. S. Jain, S.R. Wadia, Bombay preprint (1984).
11. H. Gomm, F. Lizzi, G. Sparano, Syracuse preprint SU-4222-298 (1984).
12. E. Guadagnini, Pisa preprint (1984), M. Chemtob, Saclay preprint SPhT/84/100 (1984), A. Manohar, Harvard preprint 84/A30 (1984), J. Bijnens, H. Sonoda, M.B. Wise, Caltech preprint 68-1096 (1984), M.S. Sriram, H.S. Mani, R. Ramachandran, Allahabad preprint (1984).
13. Z. Hlousek, Brown preprint HET-531 (1984).
14. For a recent review see M. Rho, lecture at Varenna School 1984, Saclay preprint SPhT/84/123 (1984).

Acta Physica Austriaca, Suppl. XXVII, 685–690 (1985)

NUCLEON-NUCLEON INTERACTION IN THE FRAMEWORK OF LATTICE QCD+

by

M. MEINHART
Institut für Radiumforschung und Kernphysik,
Österreichische Akademie der Wissenschaften,
Boltzmanngasse 3, A-1090 Wien

H. MARKUM
Institut für Kernphysik, Technische Universität Wien,
Schüttelstraße 115, A-1020 Wien

INTRODUCTION

Since QCD is a very complex theory the description of hadrons and their interactions had to rely on QCD-based models like potential, bag and soliton models. The fundamental requirements for any reasonable calculation are local gauge invariance, asymptotic freedom and confinement. In the last few years it was demonstrated that the formulation of QCD on space-time lattice can provide for all these features. Many of the calculations investigated the gluon exchange between a quark and an antiquark in the static approximation [1,2]. We extend these computations in

+Seminar given at the XXIV. Internationale Universitätswochen für Kernphysik, Schladming, Austria, February 20-March 1, 1985.

order to study the interaction energy between two baryons each consisting of three quarks moving in harmonic oscillator orbits on the lattice.

THEORY

Our treatment of nucleonic forces assumes that the time scale of fluctuations of the color fields is much more rapid than the time scale of quark motion. This means that the quarks are static sources of color fields. We construct quark wave functions by distributing the static sources on orbits. The dependence of the potential energy between nucleons versus distance can be obtained from the numerical evaluation of the Polyakov line formalism. The free energy F of a system of N quark fields $\psi_a^+(\vec{r}_i,t)$ with color components a in the presence of a gluonic heat bath $|s\rangle$ with temperature T can be written as [1]

$$\exp(-F/T) = 3^{-N}\sum_s \langle s|\sum_{a_i} \psi_{a_1}(\vec{r}_1,0)\ldots\psi_{a_N}(\vec{r}_N,0)\exp(-H/T)\psi^+_{a_N}(\vec{r}_N,0)\ldots \times \psi^+_{a_1}(\vec{r}_1,0)|s\rangle . \quad (1)$$

The sum over states is restricted to physical states satisfying a generalized Gauss law. The time evolution of the operators ψ^+ creating quarks with three colors

$$\psi^+(\vec{r},t) = \hat{T}\exp\{i\int_0^t dt' \vec{\lambda}\vec{A}_0(\vec{r},t')\}\psi(\vec{r},0) , \quad (2)$$

where $\hat{T}$ denotes the time ordering operator, is obtained by solving the static Dirac equation

$$(\partial_t - i\vec{\lambda}\vec{A}_0(\vec{r},t))\psi(\vec{r},t) = 0 . \quad (3)$$

A_o is the time component of the dynamical gluon field in the fundamental representation of SU_3 and λ are the corresponding Gell-Mann matrices. The free energy of a system of static quarks (1) with the vacuum energy subtracted can be rewritten as a thermodynamic expectation value of Wilson-Polyakov lines

$$\exp(-F/T) = \langle L(\vec{r}_1) \ldots L(\vec{r}_N) \rangle . \tag{4}$$

The Wilson-Polyakov line $L(\vec{r})$ is proportional to the trace of the quark propagator (2)

$$L(\vec{r}) = \frac{1}{3} \operatorname{tr} \hat{T} \exp i \int_o^{1/T} dt \vec{\lambda}\vec{A}_o(\vec{r},t) . \tag{5}$$

Placing the gauge fields on the links of a space-time lattice with N_t points in Euclidean time and N_s points in the spatial directions and denoting the gauge variables by $U_\mu(\vec{r},t)$ the energy (4) can be evaluated from a path integral

$$\exp(-F/T) = \frac{\int \prod_{\vec{r}t\mu} dU_\mu(\vec{r},t) L(\vec{r}_1,t) \ldots L(\vec{r}_N,t) \exp(-\beta S(U))}{\int \prod_{\vec{r}t\mu} dU_\mu(\vec{r},t) \exp(-\beta S(U))} , \tag{6}$$

where S(U) is the Wilson action including only the spinless gluon field and $\beta = 6/g^2$ denotes the coupling. Our choice of $\beta = 5.85$ corresponds via the renormalization group equation to a lattice constant $a \sim 0.23$ fm; according to some uncertainty in Λ_L recent calculations suggest a lattice constant being only half as big. We have chosen a lattice of size $16^3 \times 6$ which should provide reasonable space for two nucleons.

RESULTS

The quarks are assumed to move in harmonic oscillator orbits which are discretized on the lattice and truncated to spheres with radii R equal to one, two and three lattice constants a. For simplicity we take only Gaussian orbits of the form

$$\Psi(\vec{r}) = \frac{1}{K} \exp(-\vec{r}^2/2R^2), \tag{7}$$

where K is provided to normalize the total probability of Ψ on the lattice sphere to unity.

The baryon wave function is the product of three such oscillators. Because the static quarks carry no spin and flavor all baryons are degenerate to a 56-plet which contains the nucleon octet and the delta decuplet. Choosing for the radius R = 3a yields for the baryon on the lattice a center of mass corrected root mean square radius of about 2a. This means a baryonic r.m.s. radius of around 0.46 fm which is a reasonable size for a pure nucleon without a pion cloud.

Treating the baryon-baryon system we start from the Born-Oppenheimer approximation in which the distance d between the centers of mass of the two baryons is fixed and the corresponding energy of the six-quark system is calculated. The total wave function is taken to be a product of six Gauss functions with two centers. Because our baryons have no spin and isospin a decomposition into partial waves cannot be performed.

We investigate the dependence of the nucleon-nucleon forces from the size of the nucleon. The results for spheres with radii R = 1a, 2a, and 3a are presented in the figure. We find an attractive interaction for all examples of spheres. This can be explained by the formation of local color singlets and the exchange of glueballs between the

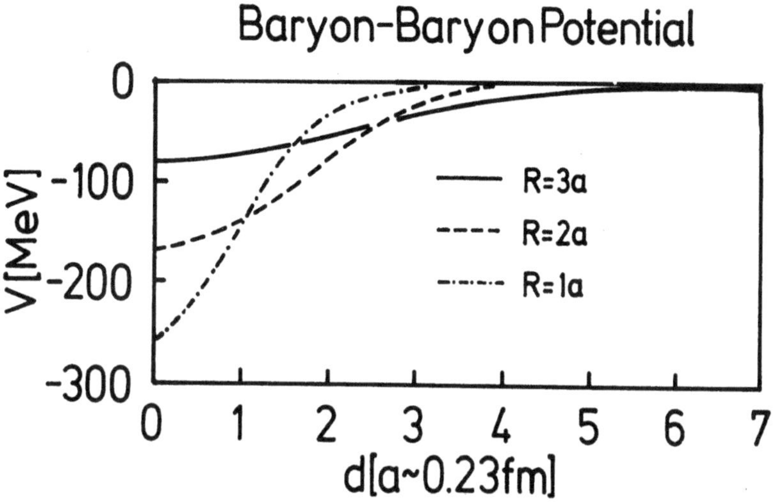

Fig.1: Potential between two baryon spheres of various radii R as a function of distance d in units of the lattice constant a.

two three-quark clusters. The range of the interaction is limited to the extension of the cavities. For separated spheres there appears practically no interaction. This means that color saturation is the dominant effect [3].

The attraction of the potential seems to have its origin in the confining spinless gluon field. Computations in the potential model indicate that the inclusion of spin can change the attractive potential to a repulsive one [4]. Thus it is of great interest to extend the underlying gluon Lagrangian to full QCD Lagrangian at least to a more realistic approximation than in the presented static theory.

REFERENCES

1. L.D. McLerran, B. Svetitsky, Phys. Rev. D24 (1981) 450.
2. C.B. Lang, M. Wiltgen, Phys. Lett. 131B (1983) 153; J.D. Stack, Phys. Rev. D29 (1984) 1213; D. Barkai, K.J.M. Moriarty, C. Rebbi, Phys. Rev. D30 (1984) 1293.
3. H. Markum, M. Meinhart, G. Eder, M. Faber, H. Leeb, Phys. Rev. D (1985) in press.
4. N. Isgur, see these proceedings.

Acta Physica Austriaca, Suppl. XXVII, 691–697 (1985)

THE LOCAL HAMILTONIAN MONTE-CARLO METHOD IN 1+1 DIMENSIONS APPLIED TO FIELD-THEORETIC LATTICE MODELS⁺

by

A. SCHILLER
Sektion Physik
Karl-Marx-Universität
DDR-7010 Leipzig

In this contribution I review some recent results [1,2] concerning the application of the local Hamiltonian Monte Carlo method [3] to field theoretic models of fermions, gauge bosons and scalars on the lattice in 1+1 dimensions. The lattice regularization is at the moment the most attractive possibility to study non-perturbative aspects of field theories. The local Hamiltonian method is very efficient to study low dimensional models with fermions since in the Monte Carlo procedure both bosons and fermions are handled locally. In this approach all properties of a model are obtained calculating expectation values as

$$\langle A\rangle = Tr(Ae^{-\beta H})/Z, \qquad Z = Tr(e^{-\beta H}) \tag{1}$$

where A is an operator, H the lattice Hamiltonian and β the

⁺Seminar given at the XXIV. Internationale Universitätswochen für Kernphysik, Schladming, Austria, February 20-March 1, 1985.

inverse temperature. To evaluate the traces one splits the imaginary time into L small intervals ($\beta = \Delta\tau L$) and inserts complete sets of states

$$Z = \sum_{t_1,\ldots,t_L} \langle t_1|e^{-\Delta\tau H}|t_L\rangle\langle t_L|e^{-\Delta\tau H}|t_{L-1}\rangle \ldots \langle t_2|e^{-\Delta\tau H}|t_1\rangle. \tag{2}$$

Furthermore the Hamiltonian H is decomposed into a finite number of pieces and additional intermediate states are inserted such that the remaining matrix elements become local (see[1]). Typically we neglect contributions of order $O(\Delta\tau^3)$ to the partition sum. A convenient basis for the chosen Susskind fermions and gauge bosons is a state vector given by occupation numbers.

We have studied a Schwinger type model of interacting Higgs scalars ϕ, Dirac fermions ψ and abelian gauge bosons A_μ given by the Lagrangian

$$L = \frac{1}{2}\partial_\mu\phi\partial^\mu\phi - V(\phi) + i\bar{\psi}\not{\partial}\psi + G\bar{\psi}\phi\psi - \frac{1}{4} F_{\mu\nu}F^{\mu\nu} + \frac{1}{2}\frac{\theta}{2\pi} g\, \varepsilon_{\mu\nu} F^{\mu\nu}. \tag{3}$$

Here the real Higgs field ϕ is Yukawa coupled with the coupling G to ψ. The double well Higgs potential

$$V(\phi) = \frac{1}{2}\lambda^2(\phi^2 - \phi_o^2)^2 \tag{4}$$

allows soliton formation. The fermion is gauge coupled with charge g to the gauge potential A_μ. The last term in (3) is added to introduce a background electric field θ.

In the continuum one expects [4] collective excitations

of the original soliton and fermionic degrees of freedom with global fractional charge ($g = 0$). Gauging this global charge ($g \neq 0$) the fractional charged solitons become confined except for a liberation angle $\theta = \pi/2$. In condensed matter physics, such objects are of phenomenological relevance in polyacetylene [5].

The lattice Hamiltonian is derived from (3) by standard methods. We have defined [1] an expectation value on the lattice $\langle S_{1/2} \rangle$ measuring the stability of half-integrally charged solitons. As an example we present in Fig. 1 $\langle S_{1/2} \rangle$ as function of the iteration number at the background field value $\theta = \pi/2$. As starting configuration we choose a trial vacuum without solitons. After about 5000 iterations fractionally charged solitons are created from the vacuum ($\langle S_{1/2} \rangle \neq 0$) and exist as free objects. Details of the Monte Carlo calculation can be found in [1]. We are able to realize on the lattice all the mentioned properties of the continuum model.

Another interesting application of the method is the non-perturbative study of supersymmetric field theoretic models. The problem of constructing supersymmetric models on the lattice has been attracted some attention recently. Since the continuum supersymmetric algebra contains the momenta as continous translation operators it does not exist on the lattice. To avoid, at least partly, this problem a Hamiltonian lattice formulation of supersymmetric models was proposed in [6,7]. Here a subalgebra of the full one is used which involves only the Hamiltonian.

We have studied numerically the 1+1 dimensional $N = 1$ and $N = 2$ Wess-Zumino models on the lattice defined in the continuum by the Lagrangian

$$L = \frac{1}{2}\partial_\mu\phi\partial^\mu\phi + \frac{i}{2}\bar{\psi}_M\not{\partial}\psi_M - \frac{1}{2}(W'(\phi))^2 - \frac{1}{2}W''(\phi)\bar{\psi}_M\psi_M \tag{5}$$

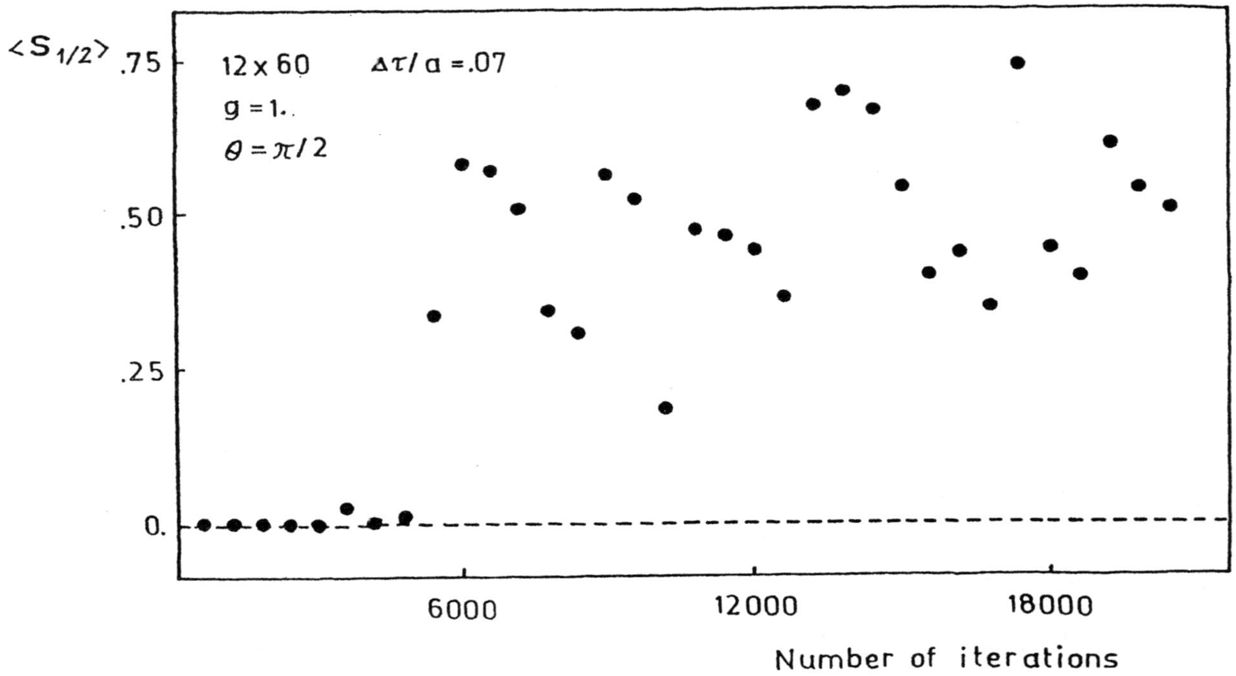

<u>Fig.1:</u> Fractionally charged soliton expectation value $<S_{1/2}>$ as function of the iteration number for the model (3) at $\theta = \pi/2$, $g = 1$, $G = 4.5$ and potential parameters (4) $\lambda = 2$, $\phi_o = 0.8$.

and

$$L = \partial_\mu T^* \partial^\mu T + i\bar{\psi}_D \not{\partial} \psi_D - \frac{\partial W(T^*)}{\partial T^*} \frac{\partial W(T)}{\partial T} - \bar{\psi}_D \frac{1+\gamma^5}{2} \psi_D \frac{\partial^2 W(T)}{\partial T^2} - $$

$$- \bar{\psi}_D \frac{1-\gamma^5}{2} \psi_D \frac{\partial^2 W(T^*)}{\partial T^{*\,2}} \qquad (6)$$

where ϕ is a real scalar, ψ_M a Majorana spinor, $T = \frac{1}{\sqrt{2}}(A+iB)$ a complex scalar field and ψ_D a Dirac spinor.
The lattice Hamiltonians are derived using the mentioned subalgebra and discretizing the supercharges forming that subalgebra. In Fig. 2 we present as an example the measured total energy of model (5) for the derivative of the superpotential

$$W'(\phi) = 2\phi^2 + \alpha \qquad (7)$$

as function of the parameter α. We find support for supersymmetry breaking ($E > 0$) for $\alpha < 0$ in some parameter range, where in the continuum no breaking is expected at the tree level of the effective potential. This result is in qualitative agreement with expectations in an alternative lattice formulation [8].

We summarize our results:
For the $N = 1$ Wess-Zumino lattice model in $d = 1+1$ we find support for Susy breaking for functions $W'(\phi)$ with even numbers of zeros in some parameter range. No breaking of supersymmetry occurs for functions $W'(\phi)$ with odd numbers of zeros. In the case of the $N = 2$ model we do not see Susy breaking within our accuracy both for functions $W'(T)$ with odd or even numbers of zeros. In the case of $N = 1$ we find free supersolitons with half-integral fermionic charge which, however, are not responsible for the observed Susy breaking effect. For model (6) we see indications for

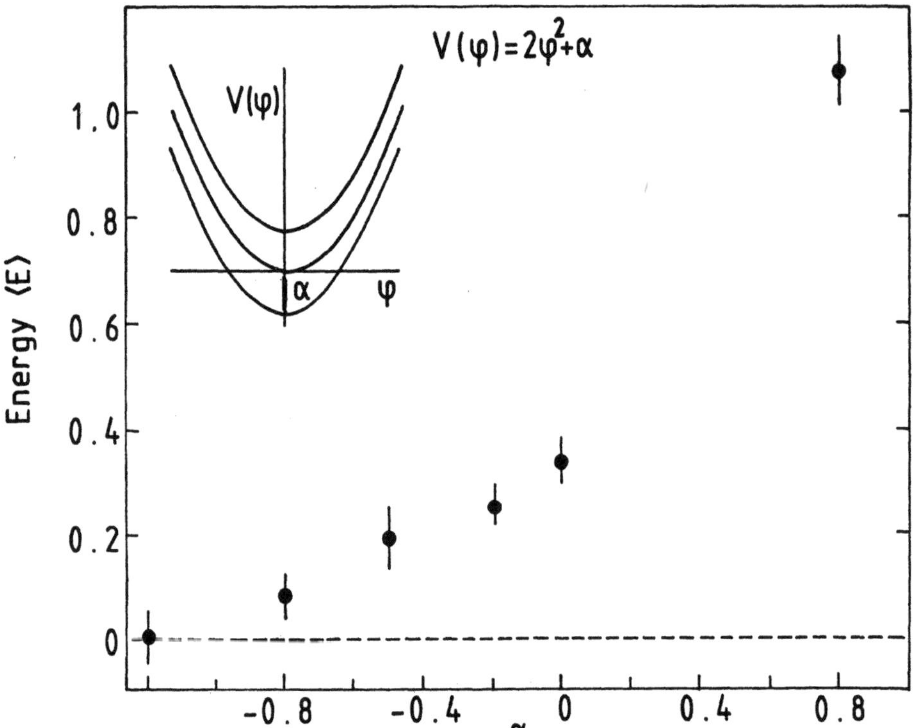

Fig.2: The energy expectation value <E> for model (5) with $V(\phi) = W'(\phi) = 2\phi^2 + \alpha$ as function of the parameter α.

non-trivial topological structures which do not seem to break supersymmetry.

The local Hamiltonian method having the great advantage of local fermion handling fails at higher dimensions since negative weights appear in the Monte Carlo updating which cannot be avoided by special boundary conditions. The Projector Monte Carlo method [9] could be a way to avoid this problem preserving the locality of the fermion updating. More work in studying this method or similar approaches for models including fermions in higher dimensions would be desirable.

The results presented here have been obtained in collaboration with J. Ranft.

REFERENCES

1. J. Ranft and A. Schiller, Phys. Lett. 140B (1984) 239; J. Ranft and A. Schiller, Nucl. Phys. B251 [FS13](1985) 764.
2. J. Ranft and A. Schiller, Phys. Lett. 138B (1984) 166; A. Schiller and J. Ranft, submitted to J. Phys. G.
3. J.E. Hirsch, R.L. Sugar, D.J. Scalapino and R. Blankenbecler, Phys. Rev. B26 (1982) 5033.
4. W.A. Bardeen, S. Elitzur, Y. Frishman and E. Rabinovici, Nucl. Phys. B218 (1983) 445.
5. A. Heeger, Comm. Sol. St. Phys. 10 (1981) 53.
6. V. Rittenberg and S. Yankielowicz, CERN preprint TH 3263 (1982).
7. S. Elitzur, E. Rabinovici and A. Schwimmer, Phys. Lett. 119B (1982) 165.
8. J. Bartels and J.B. Bronzan, Phys. Rev. D28 (1983) 818.
9. R. Blankenbecler and R.L. Sugar, Phys. Rev. D27 (1983) 1304.

Acta Physica Austriaca, Suppl. XXVII, 699–706 (1985)

CHIRAL GAUGE THEORY AND LATTICE REGULARIZATION+

by

H. SCHLERETH
Niels Bohr Institute
University of Copenhagen
Blegdamsvej 17, DK-2100 Copenhagen, Denmark

ABSTRACT

We discuss for a nonabelian gauge theory with only left handed fields in the continuum the lattice regularization. Global gauge invariance is kept and we argue that, if perturbative scaling according to the one loop β-function for a dimensionful and physical quantity is observed, the theory behaves as being effectively gauge invariant. Measurable signatures for an intermediate, approximative restoration of local gauge invariance and a problem of the requirement of reflection positivity are discussed.

Many important theories presently considered are chiral gauge theories. In most cases the Higgs phase of these theories is physically relevant. Examples are the Salam-Weinberg model [1] and left-right symmetric theories [2]. Also supersymmetric theories [3] have in general a chiral

+Seminar given at the XXIV. Internationale Universitätswochen für Kernphysik, Schladming, Austria, February 20-March 1, 1985.

sector. Subconstituent models of quarks, leptons and the weak gauge bosons [4] sometimes employ the idea of "Chiral Confinement [5] . Here it is assumed that a chiral gauge theory can exist in a gauge invariant confining phase.

To reveal the existence or nonexistence of a gauge invariant phase and possible transitions to a Higgs phase a nonperturbative study is necessary. The only technical tool available is the lattice regularization [6]. Because of the Nielsen-Ninomiya theorem [7] there is a basic obstacle in the fermion sector of chiral theories however.

For subconstituent models it would be interesting to look at models with bosonic and fermionic subconstituents with the color and flavour dependence carried exclusively by the bosons leaving for the fermions only spin and hypercolor. In this type of model the spectrum of weak gauge bosons, composite scalars and the sinus squared of the Weinberg angle [1] can be calculated in a "quenched" approximation from the bosonic sector alone in a lattice regularization. Thus an important part of the model becomes calculable without the need to discuss the fermionic sector. In passing I want to remark that in such models there are no flavour anomalies [8] at the subconstituent level and to cancel the anomalies at the composite level one has to make the "spectators" [8] themselves physical composite particles of the theory. Whether such a model can truely be constructed free of contradiction (to make it "observable" one must be able to lower the compositness scale down to Fermi scale) is not investigated in the necessary details [9].

Finally of course one cannot avoid to discuss the chiral sector on the lattice. In the following I describe in short an attempt to do this.

The theorem of ref. 7 says that under certain conditions, which we do not want to violate, there exists to every quantized, conserved and local charge an equal number of left and right handed fields in a lattice regularization. To see the consequences of this consider, as an example, SU(3) as the gauged symmetry of some chiral gauge theory to be put on the lattice (in this pedagogical discussion we do not care for the theory to be anomaly free. This is discussed in more detail below). The global Cartan subgroup generates the charges Q_3 and Q_8. Since then on the lattice to each of these charges there exists an equal number of left and right particles it follows that in order to keep the chiral structure on the lattice the global and a forteriori, also the local gauge symmetry must be broken.

In ref. 10 a lattice regularization of the Salam-Weinberg model which is gauge invariant was done. There the species doublers [7] decouple in the continuum. Then the Higgs doublet allows to write Wilson and kinetic terms gauge invariantly.

In this note it is suggested to keep global gauge invariance by making the species doublers [7] of the Cartan charges decouple in the continuum limit. To achieve this a violation of local gauge invariance is accepted. However this occurs in a specific way and arguments are given suggesting that physical information about the gauge invariant phase (if it exists) can be extracted from this formulation. This approach is necessary in the absence of a suitable scalar field.[+)]

Before proceeding to discuss a concrete model let us remark that we are bound to use Wilson fermions [6]. Susskind fermions [11] are not suitable. This can be seen from the fact that the spin diagonalization transformation [12] which relates Wilson to Susskind fermions $\psi(n) = \gamma_1^x\gamma_2^y\gamma_3^z\gamma_4^t\ \chi(n)$, n = (x,y,z,t) a lattice point, can mix left and right fields.

[+)] Note that the r-term of ref.10, of the symbolic form $\bar{R}\ \phi\ L$ (ϕ fundamental Higgs), does not reproduce the ABJ-anomaly in the continuum. A term of the form $\bar{R}\phi U_\mu L$ has the same disadvantage for the Abott and Farhi model [14].

Now we take the anomaly free group SU(2) (see however below) as gauge group and adopt the following lattice action (herm. conj. to be added)

$$S = \beta \sum_P \mathrm{tr}\, U_P - K \Big[\sum_{x,\mu} \bar{L}(x) U_\mu(x) \gamma_\mu L(x+\mu) + \sum_{x,\mu} \bar{R}(x) \gamma_\mu R(x+\mu)$$

$$- r \sum_{x,\mu} \bar{L}(x) U_\mu(x) R(x+\mu) \Big] - m \sum_x \bar{R}(x) L(x) - k \sum_{x,\mu} \mathrm{tr}\, U_\mu(x) \tag{1}$$

The notation is as usual and will not be explained in detail. L(x) and R(x) clearly denote left and right handed fields. One recognizes the Wilson r-term suppressing species doubling with respect to the chiral charge. The "fine tuning" parameters K and m play the same role as in QCD, except that the physical mass to which they are to be tuned is zero. The last term in (1) is to tune away a gluon mass. There is a further counterterm (of order zero in the lattice spacing) not written in (1).

These possibilities can arise because the third term in (1) is a bare kinetic term for the R(x)-fields which breaks local gauge invariance, if we decide to transform R(x) and L(x) in the same way. This leaves the Wilson r-term gauge invariant. Any other choice giving a gauge invariant kinetic term violates gauge invariance of the Wilson term.

Our choice ensures that the perturbative one loop β-function is insensitive to the violation of local gauge invariance. This function is calculated via the renormalization group equation from the vacuum polarization diagrams up to second order in the bare coupling. In this order the L(x) as well as the R(x) propagator entering the diagrams does not pick up a gluonic dressing. Hence to this order the contributing diagrams form a separately gauge invariant subset of the corresponding diagrams of the gauge invariant theory which is obtained from (1) by the replacement $\bar{R}(x)\gamma_\mu R(x+\mu) \rightarrow$ $\rightarrow \bar{R}(x)U_\mu(x)\gamma_\mu R(x+\mu)$. The difference is the absence of the $\bar{R}$ - R loop contribution in the case of (1) which reflects the left-right asymmetric structure in this order. At finite cut-off also the Wilson term contributes to the diagrams. If its vertices were not gauge invariant the one loop β-function were sensitive to the gauge breaking.

These considerations show that _if_ a mass (e.g. for the "meson" state $\bar{L}(x)\Gamma_\mu L(x)$, $\Gamma_\mu = \gamma_\mu$ or $\gamma_5\gamma_\mu$) calculated from (1) scales in some range of the bare coupling according to the renormalization group with the one loop β-function it provides (preliminary) evidence for the existence of the left handed sector of (1) in a gauge invariant continuum phase.

Looking at _reflection positivity_ [13] it turns out that there is a clash between the chiral structure and the Wilson r-term. In fact it turns out that the sufficient condition for reflection positivity to hold given in refs. 13 requires $r = 0$ in (1). But it is likely that this signals a general trouble. In ref. 10 reflection positivity of the Salam-Weinberg action is not investigated, however the same problem arises there [14].

Since we must keep $r \neq 0$ [6] it is important to note that reflection positivity is violated by an irrelevant term only. Hence the existence of a sound continuum theory is not precluded.

Furthermore the r-term has the property to shift the masses of the chiral doublers to infinity in the continuum limit [6] ($m \sim \frac{r}{a}$, a = lattice spacing). One would expect that if a physical "meson" mass M, say, obeys $M \ll \frac{r}{a}$ already in some calculationally accessible region off the continuum limit then it will depend only weakly on r because the states with the mass $\sim \frac{r}{a}$ should develop a tendency to decouple from the "low" mass state with mass M.

So the relation

$$<0|\bar{L}(x)\Gamma_\mu L(x)\bar{L}(0)\Gamma_\mu L(0)|0>\sim\bar{\chi}(\vec{x})\chi(0).e^{-Mt} \; ; \qquad (2)$$

($x = (\vec{x},t)$ and $\chi(\vec{x})$ the spatial wave function)

valid for large t, can be used to extract the lowest mass state if M has the above properties. Strictly speaking of course this must be proven by constructing for (1)the analogue of the transfer matrix. This is under consideration.

Since the violation of local gauge invariance is transmitted to the left handed sector in (1) only by the r-term (after tuning the mass to zero) a weak r-dependence of M is the precursor of the effective restoration of gauge invariance and of renormalization group scaling also.

The classical long wave length (naive continuum limit) of (1) is given by the Lagrangian

$$L = \frac{1}{g^2} \text{tr } (F_{\mu\nu}(x))^2 + \bar{L}(x)\not{D}\, L(x) + \bar{R}(x)\not{\partial}R(x); \qquad (3)$$

with $D_\mu = :\partial_\mu - A_\mu(x)$; (the notation is obvious). So the right handed fields decouple and leave an interacting left handed system. However (3) is inconsistent because of topological obstructions [15]. A simple way to remedy this is to add a left handed isospin $\frac{3}{2}$ field. Since topological obstructions can only be discussed after the restoration of gauge invariance it is not included in (1). The obstruction of ref. 15 can only become visible close enough to the continuum where topological gauge field configurations occur sufficiently often. A perturbative β-function exists for (3) and if scaling sets in at large enough g^2 (as e.g. in QCD for the string tension) then (1) can be applied as it stands to study the phenomenon of gauge invariance restoration in this region. In general to study how this effect sets in at large enough

g^2 one may take (1) even for SU(N) and consider the limit $N \to \infty$ not even taking care of the usual anomalies.

Quantities to be calculated (e.g. by Monte Carlo [16]) to check for the onset of restoration of gauge invariance are for example the following correlators: $\langle 0|\bar{R}(x)\Gamma_\mu R(x) . \bar{R}(0)\Gamma_\mu R(0)|0\rangle$ which should approximately factorize into two-point functions if R(x) becomes a free neutrino field, $\langle 0|\bar{R}(x)L(x)\bar{R}(0)L(0)|0\rangle$ which should be roughly constant (no exponential decay with time) showing the absence of a left-right bound state and the correlator (2) as discussed above.

Non "color"-singlet correlators which because of Elitzurs theorem [17] are ideal to "measure" gauge invariance are not well suited for Monte Carlo but for analytical calculations. The decoupling of R(x) may be enhanced by writing in (1) an improved r-term of the symbolic form $r\sum \bar{R}UL - r'.\sum \bar{R}UUL$ containing two link connections and determining r and r' such that $O(a^2)$ corrections drop.

Restoration of global supersymmetry on the lattice in a spirit related to our's is studied in ref. 18 . It is interesting that taking SU(N) in (1) one gets for $g^2 \to \infty$, $N \to \infty$ that at the equation of motion level R(x) obeys a free neutrino equation and $\langle 0|\bar{R}(x)L(x) + \bar{L}(x)R(x)|0\rangle = 0$ [14].

ACKNOWLEDGEMENT

For helpful and lively discussions I thank Holger B. Nielsen and for intensive conversations and his interest in a numerical treatment Dr.M. Meinhart. I also thank Dr.A. Schiller for a conversation of ref. 18 and G. Schierholz for pointing out ref. 10 to me in the course of preparation of this work.

REFERENCES

1. A. Salam, Proc. 8th Nobel Symposion, Ed. N. Svartholm, Stockholm (1968); S. Weinberg, Phys. Rev. Lett. 19 (1967) 1264.
2. M.A. Beg, R.V.Budny, R. Mohapatra and A. Sirlin, Phys. Rev. Lett. 38 (1977) 1252.
3. R. Fayet and S. Ferrara, Phys. Rep. 32C (1977) 1; J. Wess and J. Bagger, Supersymmetry and Supergravity, Princeton Uni. Press (1983).
4. For an overview see: W. Buchmüller, Inv. Talk included in these Proceedings.
5. L.F. Abott and F. Farhi, Nucl. Phys. B189 (1981) 547.
6. K.G. Wilson, Phys. Rev. D10 (1974) 2445.
7. H.B. Nielsen and M. Ninomiya, Nucl. Phys. B185 (1981) 20 and ibid 193 (1981) 173.
8. G. 't Hooft, Recent Developments in Gauge Theories, Plenum Press, New York, 1980.
9. I acknowledge conversations with G. Zoupanos about this type of model.
10. P.V.D. Swift, Phys. Lett. B145 (1984) 256.
11. L. Susskind, Phys. Rev. D16 (1977) 3031.
12. J. Kawamoto and K. Shigemoto, Nucl. Phys. B237 (1984) 128.
13. D.C. Brydges, J. Fröhlich and E. Seiler, Ann. of Physics 121 (1979) 227; J. Fröhlich, R. Israel, F. Lieb and B. Simon, Comm. Math. Phys. 62 (1978).
14. In preparation by the present auhtor; note that the r-term is also necessary to get the correct ABJ anomaly in an external field, see: E. Seiler and I.O. Stamatescu, MPI-PAE/PTh 66/81 preprint.
15. E. Witten, Phys. Lett. B117 (1982) 324.
16. M. Meinhart, B. Markum and H. Schlereth, in preparation.
17. S. Elitzur, Phys. Rev. D12 (1975) 3978, J. Fröhlich, G. Morchio and F. Strocchi, Nucl. Phys. B190 (1981) 553. 553.
18. A. Schiller and J. Ranft, Phys. Lett. B138 (1984) 166 and KMU - HEP 84 - 07 preprint.

Acta Physica Austriaca, Suppl. XXVII, 707–715 (1985)

POLARISATION EFFECT OF ρ PRODUCED IN $\bar{p}p$ INTERACTIONS AT 5.7, 12 AND 22.4 GeV/c⁺

(Alma Ata-Dubna-Helsinki-Kosice-Liverpool
Moscow-Prague-Tbilisi collaboration)

by

I. HERYNEK
Institute of Physics
Czechoslovak Academy of Sciences
CS-18040 Praha 8

ABSTRACT

The ρ meson spin alignment is studied in $\bar{p}p$ interactions at 22.4 GeV/c and in the reaction $\bar{p}p \to 2\pi^+ 2\pi^-$ + neutrals at 5.7 GeV/c. An essential ρ^o meson spin alignment is observed. The values of the ρ_{00}^T element of the meson spin density matrix in the transversity frame are equal to 0.56±0.07, 0.53±0.05 and 0.54±0.04 at the above mentioned interactions, respectively. The value ρ_{00}^T = 0.64±0.06 for $\rho^{\pm}$ from the channel $\bar{p}p \to 2\pi^+ 2\pi^- \pi^o$ at 5.7 GeV/c was obtained. An increase of ρ_{00}^T with transverse momentum is obtained.

⁺Seminar given at the XXIV. Internationale Universitätswochen für Kernphysik, Schladming, Austria, February 20- March 1, 1985.

Study of spin dependence of multiparticle production is of considerable interest due to unexpected strong polarisation effect in many experiments [1]. Our first indication of the ρ^o spin alignment at 22.4 GeV/c we observed in a preliminary study [2] and confirmed in ref.3. Our present analysis is based on inclusive $\bar{p}p$ data at 22.4 and 12 GeV/c and on fitted channels $\bar{p}p \rightarrow 2\pi^+ 2\pi^-$ + neutrals of 5.7 exclusive data (45 000, 90 000 and 35 000 events, i.e. 1.2, 2.2 and 3.3 ev/μb).
The 5.7 GeV/c data consist of 574 ev. $\bar{p}p \rightarrow 2\pi^+ 2\pi^-$, 6 435 ev. $\bar{p}p \rightarrow 2\pi^+ 2\pi^- \pi^o$ and 28 540 ev. $\bar{p}p \rightarrow 2\pi^+ 2\pi^- X^o$. The data are from the exposures of 2m HBC at CERN and JINR 2m HBC at Serpukhov by RF separated $\bar{p}$ beams. More details of the experiments could be found elsewhere [5-7].

The effective mass spectrum in the range (.6-.9) GeV/c^2 is described using standard fitting formula

$$\frac{d\sigma}{dM} = \sigma_\rho . BW_\rho(M) . BG(M)/I_\rho + \sigma_{BG} BG(M)/I_{BG} \qquad (1)$$

where σ_i and I_i are corresponding cross-sections and normalization integrals, the relativistic p-wave B.-W. function $BW(M) = \Gamma(M).(M/q)/[(M^2-M_\rho^2)^2 + M_\rho^2.\Gamma^2(M)]$, $\Gamma(M) = \Gamma_\rho.[q/q(M_\rho)]^3.(M_\rho/M)$ and $q = q(M)$ is the decay momentum.
For the combinatorial background the exponential parametrization $BG(M) = q.\exp(-\beta M)$ has been used, where the factor q takes into account the $\pi^+\pi^-$ phase space. The fixed value Γ_ρ= 155 MeV/c^2 was used. (The fit yields M_ρ = 776±3, 767±5 and 764±6 MeV/c^2 for the 5.7, 12 and 22.4 GeV/c data respectively.) For the further analysis also M = 770 MeV/c^2 was fixed leaving only 3 parameters to be fitted (the background slope β, and the cross-sections σ_ρ, σ_{BG}). The effective mass distributions of $\pi^+ \pi^-$ pairs produced in various $\cos\theta_T$ intervals after background subtraction (as illustrated in fig.1) enables the construction of $\cos\theta_T$ distributions (fig.2) of

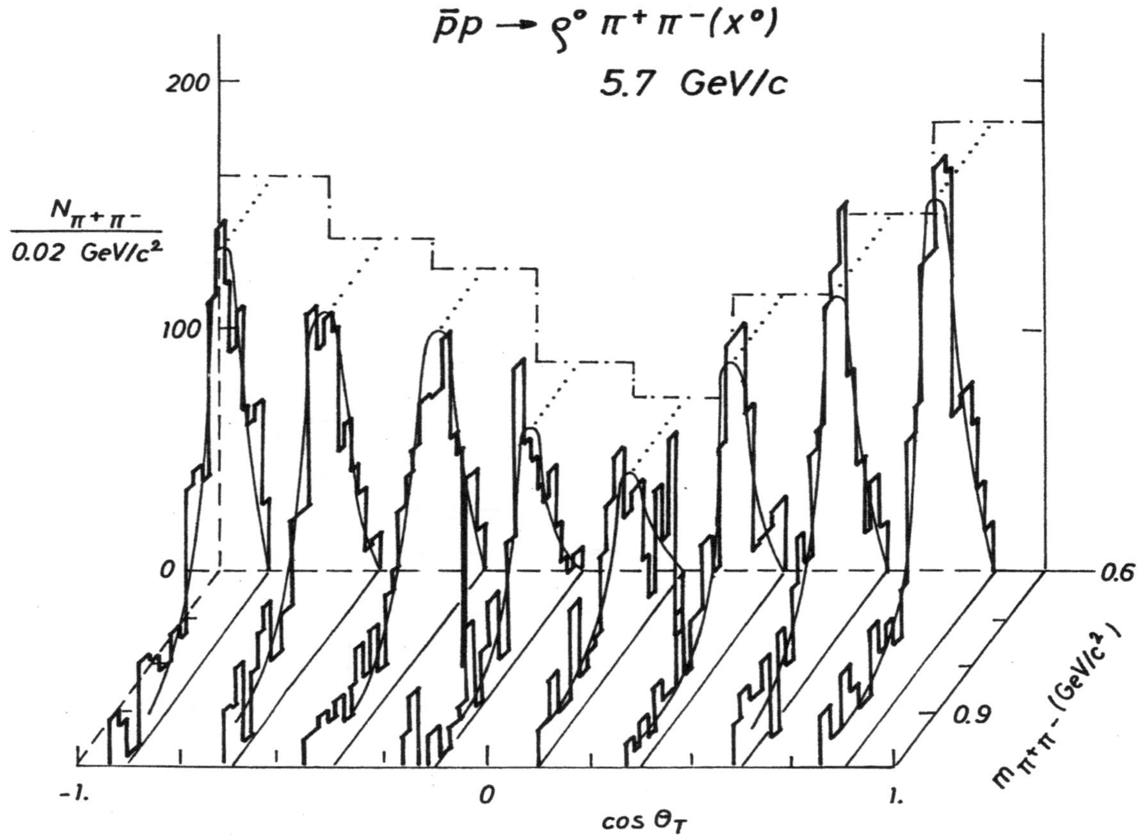

Fig. 1.: Effective mass distributions of $\pi^+\pi^-$ pairs produced in various $\cos\theta_T$-intervals in transversity frame ($\hat{z}||[p_{\bar{p}}\ p_\rho]$) in the reaction $\bar{p}p \rightarrow 2\pi^+2\pi^-$ + neutrals at 5.7 GeV/c after background substraction.

π^+ from ρ^0 decay. The $\cos\theta$ distribution in $\rho\to\pi^+\pi$ decay has the form

$$\frac{d\sigma}{d\cos\theta} = \frac{1}{2}\sigma\ [1 + \frac{1}{2}\ (1-3\rho_{00})(1-3\ \cos\theta)]\ . \qquad (2)$$

The ρ_{00} element corresponds to the probability of the zero ρ meson spin projection on the reference axis ($\rho_{00} = 1/3$ means the isotropy). Detailed study of possible sources of systematical errors is discussed elsewhere [4].

Our results (fig.2 and table 1) point out 5, 4 and 3 standard deviations from the uniform $\cos\theta$ distribution for $\bar{p}p \to \rho^0 + X$ at 5.7, 12 and 22.4 GeV/c respectively with a large probability of zero spin projection on the normal to

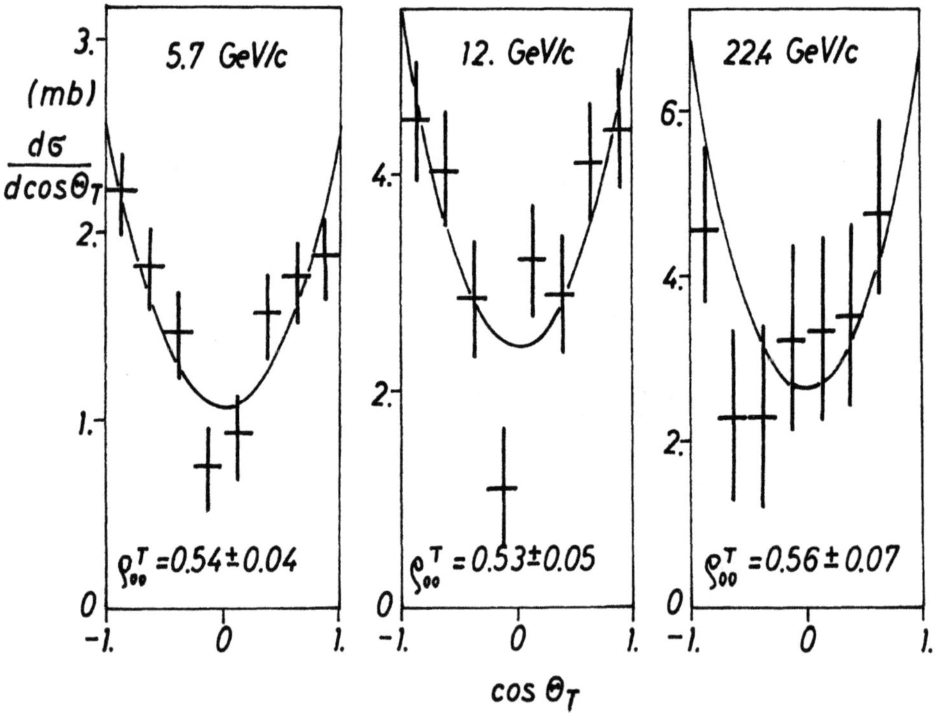

Fig.2.: ρ^0 decay angular distributions in the transversity frame (see fig. 1.).
The curves represent the results of the fits by equ.(2).

the production plane. Similar effect could be seen also for $\rho^{\pm}$ meson 5.7 GeV/c data (Fig. 3a).

It is interesting to note that in $pp \to \rho^{o} + X$ at 24 GeV/c. the $\rho_{oo} \approx 1/3$ corresponding to the isotropy distribution [8]. It makes possible to connect the observed ρ^{o} spin alignment with the annihilation mechanism (but we must pay attention to attribute the difference in $\bar{p}p$ and pp to annihilation channel only [9]). The assumption of negligible spin alignment in nonannihilation channels and roughly equal values of ρ^{T}_{oo} (tab 1.) for 5.7, 12 and 22.4 GeV/c data (with $\approx$100 %, $\approx$70 %, and $\approx$50 % annihilation, respectively) however would indicate an increase of the ρ^{o} alignment in annihilation channels with increasing primary momenta. If such an increase saturates at some primary energy we can expect a decrease of ρ^{T}_{oo} for higher primary momenta of $\bar{p}$ [13].

Significant increase of ρ^{T}_{oo} with increasing $p_{\perp}^{2}$ at 5.7 and 12 GeV/c data is observed (Figs. 4 and 5).

Some models for polarisation mechanism in $\bar{p}p$ interactions could be found elsewhere [10-12].

Table 1.: Fitted values of ρ_{oo} and (χ^{2}/NF) for the same reactions as given in Fig. 2.

p_{lab} GeV/c	Transversity frame	Gottfried-Jackson frame	Helicity frame	Adair frame
5.7	0.54±0.04 $(\frac{8}{6})$	0.21±0.04 $(\frac{3}{6})$	0.25±0.04 $(\frac{7}{6})$	0.24±0.04 $(\frac{10}{6})$
12	0.53±0.05 $(\frac{10}{6})$	0.15±0.05 $(\frac{12}{6})$	0.27±0.05 $(\frac{3}{6})$	0.15±0.05 $(\frac{12}{6})$
22.4	0.56±0.07 $(\frac{9}{6})$	0.19±0.08 $(\frac{5}{6})$	0.29±0.08 $(\frac{7}{6})$	0.33±0.08 $(\frac{4}{6})$

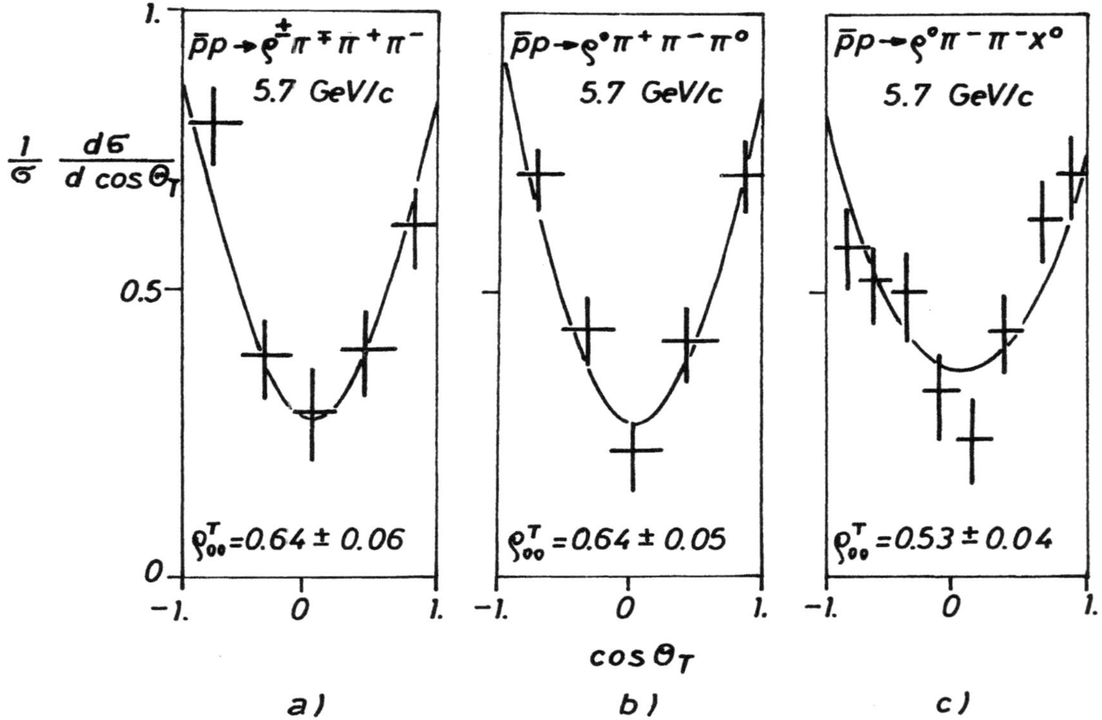

Fig. 3.: ρ decay angular distributions in the transversity frame for the exclusive channels at 5.7 GeV/c.

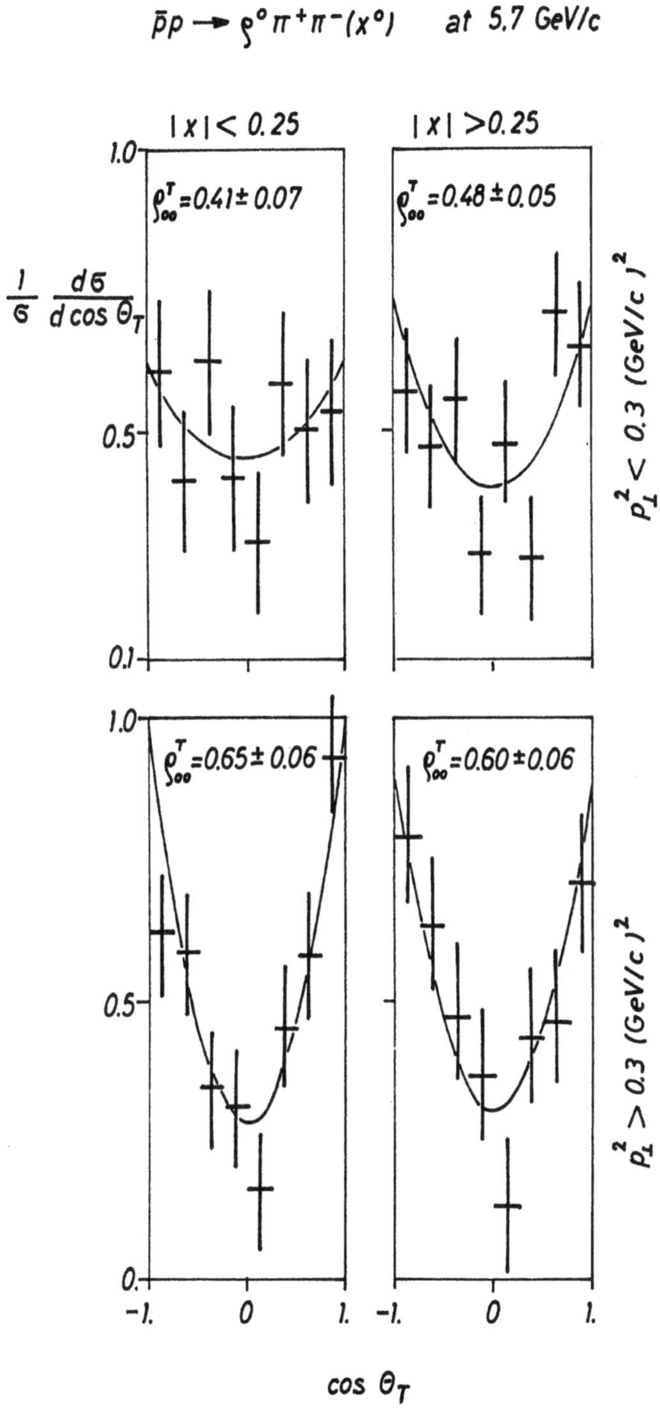

Fig.4.: ρ^0 decay angular distributions in the transversity frame for the data 5.7 GeV/c in different intervals of kinematical variables.

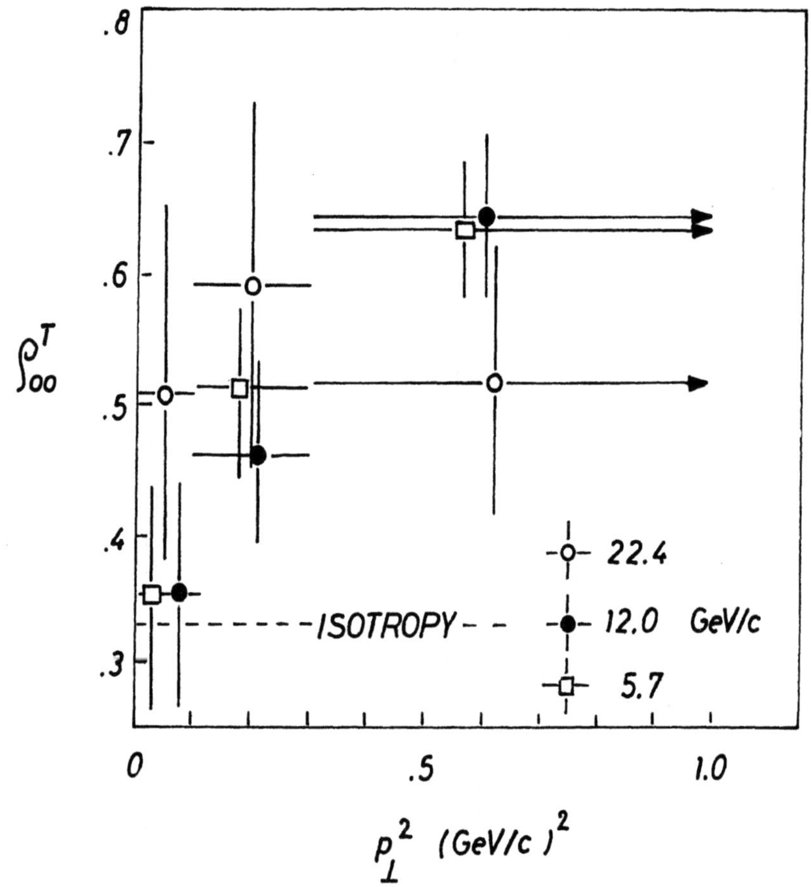

Fig.5.: $p_\perp^2$ dependence of ρ_{oo}^T of ρ^o meson produced in the same reactions as in Fig. 2.

REFERENCES

1. See e.g. AIP Conf. Proc., No. 99 Part. and Fields, Subseries No. 28, New York, 1983: High Energy Spin Physics 1982.
2. D.I. Ermilova et al., Ludmila Collaboration, Nucl. Phys. B137 (1978) 29.
3. B.V. Batyunya et al., Ludmila Collaboration, Czech. J. Phys. B31 (1981) 1353; Proc. of the Conf. on Multiparticle Dynamics, Notre Dame, USA, August 1981, p. 201.
4. B.V. Batyunya et al., Ludmila Collaboration to be published.
5. E.G. Boos et al., Ludmila Collaboration, Nucl. Phys. B121 (1977) 381, B174 (1980) 45.
6. P. Johnson et al., AHLS Collaboration, Nucl. Phys. B173 (1980) 77; B.M. Whyman et al., AHLS Collaboration, Z. Phys. C12 (1982) 203.
7. H.W. Atherton et al., Nuovo Cim. 30A (1975) 505; Nucl. Phys. B103 (1976) 381; B113 (19) 378. J. Cvach, V. Simák, Czech. J. Phys. B30 (1980) 72.
8. V. Blobel et al., Phys. Lett. 48B (1974) 73; more detailed comparison of $\bar{p}p$ and pp data, using the same method of analysis, will be published elsewhere.
9. V.V. Kniazev et al., IHEP 77-106, Serpukhov, 1977; P.V. Chliapnikov et al., Nucl. Phys. B148 (1979) 400; S. Nandi et al., Phys. Rev. D17 (1978) 1336.
10. J. Ftáčnik et al., Czech. J. Phys. B33 (1983) 889.
11. A.V. Efremov, O.V. Teryaev, Preprint JINR P2-82-832, Dubna, 1982.
12. R. Lednický, Czech. J. Phys. B33 (1983) 1177; Preprint JINR E2-84-12 Dubna, 1984.
13. V.V. Babintzev et al., Yad. Fiz. 39 (1984) 1189.

Acta Physica Austriaca, Suppl. XXVII, 717–724 (1985)

THE DECAY OF HEAVY QUARKS[+]

by

H. PIETSCHMANN
Institut für Theoretische Physik
Universität Wien

ABSTRACT

The decay of heavy quarks (b and t) is studied with the help of spectral functions. As much information as possible is directly taken from experiment. Older computations in the free quark approximation are supported. This is important for the extraction of Kobayashi-Maskawa matrix elements from experiments.

While the existence of decay products from a particle containing a top-quark is still investigated [1], life-time and branching ratios from b-decay are surprisingly well-known [2]. These data form an important tool to determine the parameters of the Kobayashi-Maskawa mixing matrix. In order to do so, the life-time of a (physical) particle containing a heavy quark (b or t) has to be related to the life-time of the quark and this, in turn, to the elements of the Kobayashi-Maskawa matrix, $U_{qq'}$. For the first step,

[+]Seminar given at the XXIV. Internationale Universitätswochen für Kernphysik, Schladming, Austria, February 20-March 1, 1985.

we shall assume - in accord with the literature - the spectator model. For heavy quarks, its validity should be quite good.

The second step is usually computed with the help of a free quark model [3]. Following reference 4, we shall try to extract as much information as possible directly from experiment. We restrict the free quark assumption to the emission of the virtual W-boson and describe its transition into final state particles by spectral functions (see Fig. 1). In this way, the decay rate for a heavy quark q

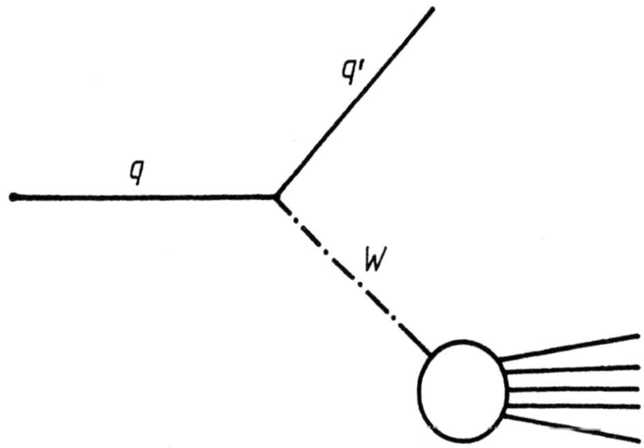

Fig.1.: Decay of a heavy quark q.

into a light quark q' (plus other particles) becomes

$$\Gamma(q,q') = \frac{G^2 |U_{qq'}|^2 m^3}{16\pi^3} \int_0^{(1-a)^2} \frac{d\eta}{\eta}\, w(1,a^2,\eta) \ .$$

$$\cdot\ \{[(1-a^2)^2+(1+a^2)\eta-2\eta^2]\rho_q(\eta) + [(1-a^2)^2-(1+a^2)\eta]\rho_{q'}(\eta)\} \quad (1)$$

where

$$a = m_{q'}/m_q \ . \quad (2)$$

w is the notorious kinematical function (for definition see for example reference 5), ρ and ρ' are the spectral functions for the conserved (ρ) and non-conserved (ρ') part of the charged weak current [5]. In their dimensionless form of equ. (1) they are taken to be

$$\rho_q(\eta) = \rho(\eta m_q^2) = \rho(t) \ . \quad (3)$$

For free quarks a and b the spectral functions can easily be computed to give

$$\rho_{ab}(t) = \frac{|U_{ab}|^2}{(2\pi)^2} \frac{w(t,m_a^2,m_b^2)}{t^2} [2t^2 - t(m_a^2+m_b^2) + (m_a^2-m_b^2)^2] \quad (4)$$

$$\rho'_{ab}(t) = \frac{3|U_{ab}|^2}{(2\pi)^2} \frac{w(t,m_a^2,m_b^2)}{t^2} [t(m_a^2+m_b^2) - (m_a^2-m_b^2)^2] \ . \quad (4')$$

Insertion into equ. (1) is nothing but a more complicated way to derive the standard formula of ref. 3, used in the analysis to extract Kobayashi-Maskawa matrix elements from experiment [6]. For b-quark decay, we obtain

$$\tau_b = \frac{\tau_b^o}{c_{bc}|U_{bc}|^2+c_{bu}|U_{bu}|^2} \quad (5)$$

with

$$\tau_b^o = \frac{192\pi^3}{G^2 m_b^5} = \tau_\mu \left(\frac{m_\mu}{m_b}\right)^5 . \tag{6}$$

Using the mass values of ref. 3, i.e.

$$\begin{aligned} m_b &= 5 \text{ GeV} \\ m_c &= 1.75 \text{ GeV} \\ m_s &= 0.5 \text{ GeV} \\ m_\tau &= 1.8 \text{ GeV} \end{aligned} \tag{7}$$

and all other masses taken to be zero, the coefficients in equ. (5) become $c_{bc} = 2.75$ and $c_{bu} = 7.7$. They are composed of the various contributions collected in table 1.

Table 1.: Contributions to the coefficients of Equ. (5)

	c_{bu}	c_{bc}
$e\bar{\nu}_e$	1	0.41
$\mu\bar{\nu}$	1	0.41
$\tau\bar{\nu}_\tau$	0.39	0.08
$\bar{u}d_c$	3.9	1.6
$\bar{c}s_c$	1.4	0.25

Notice that the last two rows in table 1 have been multiplied by an enhancement factor 1.3 due to QCD-corrections aside from the obvious color factor 3.

The weakness of the free quark approximation (in the low energy region) is clearly seen, when we compare equ.(4') with the narrow resonance approximation (assuming pion saturation of ρ')

$$\rho'(t) = f_\pi^2\, t\, \delta(t - m_\pi^2)\ . \tag{8}$$

Following Ref. 4, we shall thus try to relate the spectral function ρ to experiments assuming ρ' to be given by equ. (8). To do so, we recall

$$R(t) = \frac{\sigma(e^+e^- \to \text{hadrons})}{\sigma(e^+e^- \to \mu^+\mu^-)} = \frac{12\pi^2}{t}\,\rho_{e.m.}(t)\ . \tag{9}$$

The vector part of ρ is related to the isovector part of $\rho_{e.m.}$ by CVC. If we relate isovector and isoscalar parts of $\rho_{e.m.}$ by SU_3 below and SU_4 above charm threshold, and assume spectral function sum rules [7], we obtain

$$\rho(t) = \frac{t}{3\pi^2}\, R(t)\, \frac{1+b(t)}{1+a(t)} \tag{10}$$

where a is 1/3 below and 11/9 above charm threshold whereas b vanishes below and is unity above charm threshold. With the mass values of equ.(7), the coefficients of equ.(5) become

$$c_{bc}\ (\text{equ. } 7) = 2.95$$

$$c_{bu}\ (\text{equ. } 7) = 7.75 \tag{11}$$

in comforting agreement with the free quark model (notice that QCD enhancement is now implicit in the exerimental information apart from corrections which do not factorize according to Fig. 1; they appear to be small). The agreement also lends support to the assumptions leading to equ. (10) which were also used in earlier computations of the Z^o-decay [8].

Using improved mass values

$$\begin{aligned} m_{u,d} &= 0.34 \\ m_s &= 0.51 \\ m_c &= 1.8 \\ m_b &= 5.2 \end{aligned} \qquad (12)$$

and the experimental lepton masses, the coefficients of equ. (5) become

$$\begin{aligned} c_{bc}\ (\text{equ. } 12) &= 3.03 \\ c_{bu}\ (\text{equ. } 12) &= 7.51 , \end{aligned} \qquad (13)$$

But we can go one step further: Some integral information on ρ and ρ' is contained in the hadronic decay rate of the τ-lepton. It can be used in the following way [4]:

$$\Gamma(q \to q' + \text{hadr.}) = |U_{qq'}|^2 \Gamma_\tau \left(\frac{m_q}{m_\tau}\right)^3 \cdot$$

$$\cdot\{\text{B.R.}(\tau \to \text{hadr.}) + \frac{12\pi^2}{m_\tau^2}\,\text{B.R.}(\tau \to e)[\Delta_{\tau,q} + \Delta_{q'}]\} \qquad (14)$$

where $\Delta_{\tau,q}$ and $\Delta_{q'}$ are two contributions stemming from the larger phase space in q-decay ($\Delta_{\tau,q}$) and from the correction due to the non-vanishing mass of q' as compared to $\nu_\tau(\Delta_{q'})$. Their analytic form can be found in Ref. 4.

Using the mass values of equ. (12), the coefficients of equ. (5) become

$$c_{bc}\ (\text{equ. }12,\ \tau) = 3.04$$
$$c_{bu}\ (\text{equ. }12,\ \tau) = 7.52 \tag{15}$$

in wonderful agreement with equ. (13).

The branching ratio into a specific lepton channel is predicted to be about 14 % which is slightly better than the free quark model (15 %) but still a bit high compared to experiments [2].

Let us now proceed to top-decay [9]. The analog of equ. (5) is now

$$\tau_t = \frac{\tau_\mu (m_\mu/m_t)^5}{c_{tb}|U_{tb}|^2+c_{ts}|U_{ts}|^2+c_{td}|U_{td}|^2} . \tag{16}$$

We know already, that U_{ts} and U_{td} can be neglected compared to U_{tb} [10]. Their coefficients c_{ts} and c_{td} are fairly independent to the method as well as the top-mass, varying just between 8.9 and 9.0. For c_{tb}, the different methods change the value also insignificantly (for details see Ref. 9), but it is somewhat dependent on the top-mass. Values are given in table 2.

Table 2: The coefficient c_{tb} as a function of the top-quark mass

m_t (GeV)	35	40	45
c_{tb}	7.5 - 7.6	7.9	8.1

REFERENCES

1. J. Strauss, Contribution to this Winter School.
2. C. Klopfenstein et al., Phys. Lett. 130B (1983) 444;
   E. Fernandez et al., Phys. Rev. Lett. 51 (1983) 1022;
   M.S. Alam et al., Phys. Rev. Lett 51 (1983) 1143;
   N.S. Locker et al., Phys. Rev. Lett. 51 (1983) 1316;
   A. Chen et al., Phys. Rev. Lett. 52 (1984) 1084;
   M. Althoff et al., Z. Phys. C22 (1984) 219.
3. M.K. Gaillard and L. Maiani, Cargèse Lectures Vol. 61, eds. M.Lévy et al., Plenum Press (1980).
4. H. Pietschmann and H. Rupertsberger, Z. Phys. C27 (1985) 73.
5. H. Pietschmann, Weak Interactions - Formulae, Results and Derivations (Springer Wien, New York, 1983).
6. For example K. Kleinknecht and B. Renk, Phys. Lett. 130B (1983) 459.
7. C. Bernard et al., Phys. Rev. D12 (1975) 792.
8. R. Bertlmann and H. Pietschmann, Phys. Rev. D15 (1977) 683.
9. M. Klein, H. Rupertsberger, H. Pietschmann, Phys. Lett. 153B (1985) 341.
10. L. Wolfenstein, Phys. Rev. Lett. 51 (1983) 1945.

Acta Physica Austriaca

Stochastic Methods and Computer Techniques in Quantum Dynamics

Recent Developments in High-Energy Physics

MIX
Papier aus verantwortungsvollen Quellen
Paper from responsible sources
FSC® C105338

If you have any concerns about our products,
you can contact us on
ProductSafety@springernature.com

In case Publisher is established outside the EU,
the EU authorized representative is:
Springer Nature Customer Service Center GmbH
Europaplatz 3, 69115 Heidelberg, Germany

Printed by Libri Plureos GmbH
in Hamburg, Germany